Environmental Management Handbook

for the Hydrocarbon Processing Industries

Gulf Publishing Company • Book Division • Houston, London, Paris, Tokyo

Environmental Management Handbook

for the Hydrocarbon Processing Industries

Compiled by
James D. Wall
Gas and Solid Fuels Editor
Hydrocarbon Processing

Environmental Management Handbook for the Hydrocarbon Processing Industries

Library of Congress Cataloging in Publication Data

Main entry under title:

Environmental management handbook for the hydrocarbon processing industries.

Includes bibliographical references.

1. Petroleum refineries—Environmental aspects—Handbooks, manuals, etc. 2. Petroleum chemicals industry—Environmental aspects—Handbooks, manuals, etc. I. Wall, James D.

TD195.P4E56 661'.81 80-16294

ISBN 0-87201-265-4

Contents

Part I.

Regulation

Use These "Keep-Up" Resources for Environmental Management

Key organizations, literature references and government agencies described here can help you to maintain a constant inflow of useful, timely information

C. H. Vervalin, HYDROCARBON PROCESSING

THE HPI MANAGER needs daily access to timely information covering the spectrum of environmental management. He should know the views and actions of government, legislators, "environmentalists," professional societies, etc. So, here are major literature and other information sources describing the who, what, when, where and how of keeping up with environmental management developments — day-by-day, month-by-month.

First, let's look at resources in the United States — then at those outside the United States.

UNITED STATES

Federal government agencies are the focal point of major environmental laws affecting the HPI. Foremost among them is the Environmental Protection Agency (EPA).

EPA has authority in the areas of air, water, pesticides, solid wastes, radiation hazards and noise. On request, EPA will put you on its list to receive news releases and its newsletter publication, *Environmental News.* It will also send you announcements of upcoming conferences.

There are 10 regional EPA directors. Each regional office has an information staff to help you. For addresses of these officers, write to EPA Headquarters, Office of Public Affairs, Washington, D.C. 20460.

The Council on Environmental Quality (CEQ) supervises the federal government in all actions that might affect the environment. All federal projects are studied by CEQ, and "impact statements" of those plans are available to the public. At the regional offices you can view them. Or you can get on the list to receive the monthly publication, *102 Monitor.* (CEQ address: 722 Jackson Place N.W., Washington D.C. 20006.)

See, also, CEQ's Annual Report. The most current one — *Environmental Quality* — is available from the Superintendent of Documents, U.S. Government Printing Office, Washington, D.C. 20402.

National Technical Information Service (NTIS) of the Department of Commerce publishes the bi-weekly *Environmental Pollution and Control.* This is useful for keeping up with projects that have environmental implications.

The Federal Register is a comprehensive bulletin that reports on major legislation and regulations affecting the environment, and occupational safety and health. Subscription from Superintendent of Documents, U.S. Government Printing Office, Washington, D.C. 20402.

The U.S. Government Organization Manual provides names, functions, addresses and phone numbers of all federal bodies in the United States. The current issue has dozens of listings under "Environment." It is revised each year. Available from Superintendent of Documents (address above).

U.S. Government Research and Development Reports is available from National Technical Information Service, Department of Commerce. Order this from the Springfield, Va., office.

NTIS also offers an information search service. More than 200,000 reports are on file. For a fee, you can get detailed abstracts of all reports relating to the subject on which you inquire. The service is called NTIS Search.

NTIS also publishes a weekly paid-subscription newsletter, *The Environmental Awareness Reading List.* This is an official publication of the U.S. Department of Interior.

The National Industrial Pollution Control Council (NIPCC) of the Department of Commerce advises the government on environmental quality improvement policies and programs that affect industry.

For a list of publications, write to U.S. Department of Commerce, NIPCC, Washington, D.C. 20230. Many of the publications have excellent bibliographies that provide still more information resources.

Subcommittee hearing transcripts of the U.S. Congress are a good source of environmental information. You can get these transcripts from the Superintendent of Documents. For a modest annual fee the office will send you a monthly catalog of publications.

The State Air Pollution Control Board and State Water Quality Board in your own state are good sources of information. Rules and regulations sometimes differ from federal laws, so it is important to know demands in your own area.

State universities are also setting up various environmental activities. Three such organizations are Pennsylvania State University's *Office of Environmental Quality Programs,* Virginia Polytechnic Institute and State University's *Center for Environmental Studies* and the University of Michigan's *Institute for Environmental Quality.*

The professional societies. A number of professional organizations serving the HPI have special environmental management programs. Get on their mailing lists for announcements of seminars, programs, special publications, etc.

American Institute of Chemical Engineers, 345 E. 47th St., New York, N.Y. 10017, has an Environmental Division which, among other things, is the guiding group for the institute's "pollution solution" groups.

American Petroleum Institute has formed a Committee on Environmental Affairs. Produces a report annually on refining industry environmental expenditures. For more information, contact API at 2101 L St. N.W., Washington, D.C.

Chemical Manufacturers Association is also particularly active in environmental management activities. 1825 Connecticut Ave., N.W., Washington, D.C. 20009. Produces an excellent monthly newsletter, *Chemecology.*

Ecology and conservation groups are also a good source of information, whether you agree with all of their policies and views or not. Many of these groups use newsletters to gain public support of their goals. These can be useful information resources.

Two good ones are *National News Report* from the Sierra Club and *CF Letter* from the Conservation Foundation.

The National Wild Life Federation publishes the annual *Conservation Directory,* recognized as one of the most complete listings of groups and officials involved in environmental interests.

The four categories in the directory are U.S. government groups: international, national and interstate organizations and commissions; state and territorial agencies and citizen's groups and a personal name index of people active in the environmental field. The directory even lists other directories of interest. You can purchase this from the National Wildlife Federation, 1412 16th St. N.W., Washington, D.C. 20036.

National Environmental Development Association is an HPI-oriented not-for-profit organization that conducts meetings and conferences on industrial environmental management. It produces an excellent newsletter, *Balance,* covering current events and legislative action. Write to NEDA, National Press Building, Washington, D.C. 20045.

Abstracts, indexes. Finding the journal or article you want can be simplified by using several abstract references and indexes covering environmental interests. Some useful ones are:

1. *Pollution Abstracts,* from Pollution Abstracts, Inc., P.O. Box 2369, La Jolla, Calif. 92037.
2. *U.S. Government Research and Development Reports,* from NTIS, Department of Commerce, Springfield, Va. 22151.

For delving even more deeply into a literature search, be sure to check out the Reference Division of the Library of Congress for the Science and Technology unit. This is open for inter-library loans. Ask about this at your library.

HPI-oriented research. Two organizations are of particular interest to the HPI, both for their research efforts and their information resources. They are Chemical Industry Institute of Toxicology, P.O. Box 12137, Research Triangle Park, N.C. 27709 and the American Industrial Health Council, 1075 Central Park Ave., Scarsdale, N.Y. 10583.

U.S. information resources. The following list of groups, information and literature does not include all available sources. But here you will get a comprehensive cross-section of a wide variety of inputs, ranging from scientific research reports to newsletters from "radical" environmentalist groups. From this, you can select the right blend of resource material you need for a balanced view of ongoing developments.

Directories

Conservation Directory
The National Wildlife Federation
1412 16th Street, N.W.
Washington, D.C. 20036

Directory of Conservation Environmental Groups
Greenwich Environmental Service
P.O. Box 95
Pleasantville, N.J. 08232

Directory of Organizations Concerned with Environmental Research
Lake Erie Environmental Studies
State University College
Fredonia, N.Y. 14063

Pollution Control Directory
American Chemical Society
1155 16th St., N.W.
Washington, D.C. 20036

A Resource Guide on Pollution Control. Federal, State and Local Agencies and Organizations that Deal with Environmental Problems.
American Association of University Women
2401 Virginia Ave., N.W.
Washington, D.C. 20037

Yellow Pages, A Guide to Organized Environmental Efforts
Environmental Resources, Inc.
2000 P St., N.W.
Washington, D.C. 20036

Newsletters and other periodicals

Archives of Environmental Health
American Medical Association
535 North Dearborn St.
Chicago, Ill. 60610

BioScience
American Institute of Biological Sciences
3900 Wisconsin Ave. N.W.
Washington, D.C. 20016

Bulletin of Environmental Contamination and Toxicology
175 Fifth Ave.
New York, N.Y. 10010

Catalyst for Environmental Quality
274 Madison Ave.
New York, N.Y. 10016

CF Letter
The Conservation Foundation
1717 Massachusetts Ave., N.W.
Washington, D.C. 20036

Conservation Report
National Wildlife Federation
1412 16th Street, N.W.
Washington, D.C. 20036

Currents
Chemical Manufacturers Association
1825 Connecticut Ave., N.W.
Washington, D.C. 20009

Ecology Today
P.O. Box 180
West Mystic, Conn. 06388

Environment
P.O. Box 755
Bridgeton, Mo. 63044

Environment Action Bulletin
Rodale Press, Inc.
22 E. Minor St.
Emmanus, Pa. 18049

Environment and Behavior
Sage Publications
275 South Beverly Drive
Beverly Hills, Calif. 90212

Environment Reporter
The Bureau of National Affairs, Inc.
1231 25th St., N.W.
Washington, D.C. 20037

Environmental Science and Technology
American Chemical Society
1155 16th St., N.W.
Washington, D.C. 20036

FDA Papers
Superintendent of Documents
Washington, D.C. 20402

Hydrocarbon Processing
(see especially each October issue beginning in 1972, and review annual indexes published in each December issue or available from Reprint Department, P.O. Box 2608, Houston, Tx. 77001.

Journal of the Air Pollution Control Association
4400 Fifth Ave.
Pittsburgh, Pa. 15213

Journal of Water Pollution Control Federation
Water Pollution Control Federation
Prince and Lemon Streets
Lancaster, Pa. 17604

Research, conservation and environmental groups

This list includes both information groups and action groups. Some of the organizations seek a membership fee as a prerequisite of receiving their newsletters and other publications. Write to the individual organization of interest for information.

Air Pollution Control Association
4400 Fifth Ave.
Pittsburgh, Pa. 15213

American Association for Conservation Information
c/o Russell McKee
Department of Natural Resources
Lansing, Mich. 48926

American Conservation Association, Inc.
30 Rockefeller Plaza
New York, N.Y. 10020

American Institute of Biological Sciences
3900 Wisconsin Avenue, N.W.
Washington, D.C. 20016

Citizens for Clean Air
40 West 57th St.
New York, N.Y. 10019

Committee for Environmental Information
438 N. Skinker Blvd.
St. Louis, Mo. 63130

Committee of Two Million
Room 1032
760 Market Street
San Francisco, Calif. 94102

Conservation Foundation, The
1717 Massachusetts Ave., N.Y.
Washington, D.C. 20036

Ecology Action
3029 Benvenue
Berkeley, Calif. 94709

Ecology Center Communications Council
P.O. Box 21072
Washington, D.C. 20009

Environmental Action, Inc.
Room 200
2000 P St., N.W.
Washington, D.C. 20036

Environmental Defense Fund, Inc.
162 Old Town Road
East Setauket, N.Y. 11733

Environmental Research Institute
Box 156
Moose, Wyo. 83012

Environmental Resources, Inc.
Room 200
2000 P St., N.W.
Washington, D.C. 20036

Friends of the Earth
30 E. 42nd St.
New York, N.Y. 10017

Institute of Environmental Sciences
940 East Northwest Highway
Mount Prospect, Ill. 60056

International Union for Conservation of Nature and Natural Resources
1110 Morges
Switzerland

National Association of Conservation Districts
1025 Vermont Ave., N.W.
Washington, D.C. 20005

National Organization to Insure a Sound-Controlled Environment
Suite 623
4545 Connecticut Ave., N.W.
Washington, D.C. 20008

National Parks and Conservation Association
1701 18th St., N.W.
Washington, D.C. 20009

National Water Resources Association
897 National Press Building
Washington, D.C. 20004

National Watershed Congress
Room 1105
1025 Vermont Ave., N.W.
Washington, D.C. 20005

National Wildlife Foundation
1412 16th St., N.W.
Washington, D.C. 20036

Public Interest Research Group
1025 15th St., N.W.
Washington, D.C. 20005

Rachel Carson Trust for the Living Environment, Inc.
8940 Jones Mill Road
Washington, D.C. 20015

Resources for the Future, Inc.
1755 Massachusetts Ave., N.W.
Washington, D.C. 20036

Sierra Club
1050 Mills Tower
San Francisco, Calif. 94104

Washington Ecology Center
2000 P St., N.W.
Washington, D.C. 20036

Water Pollution Control Federation
3900 Wisconsin Ave., N.W.
Washington, D.C. 20016

EUROPE

A *must* contact for the HPI manager in Europe is Stichting CONCAWE — otherwise known as the Oil Companies' International Study Group for Conservation of Clean Air and Water (Western Europe), 60, Van Hogenhoucklaan, The Hague 2018, The Netherlands. Stichting CONCAWE has member companies representing about 80 percent of the refining capacity in Western Europe. The group's basic function is to examine and to promote the use of means for preventing air, water, soil and noise pollution attributable to oil industry activity and the use of oil products. Its territorial scope is limited to Western Europe.

The Council of CONCAWE, representing all the participating companies, directs the group's activities.

The objectives of the group are:

1. The collection and dissemination to participating companies of scientific, technical and legal information from worldwide sources dealing with causes and effects of air, water, sea and soil pollution attributable to the oil industry and to the use of petroleum products, and with equipment and processes used exclusively for purifying liquid and gaseous effluents and for abatement of noise.

The study of sea pollution is restricted to the fate of the oil from the moment it enters the surface water until its final collection, dispersal or destruction.

2. The promotion of active cooperation between the participating companies to secure the flow of information.

3. The provision of work programs and funds to research institutes to develop new knowledge about pollution by the oil industry and by its products and its abatement.

4. To make available to others, and in particular to the governments, interested administrations, international intergovernmental organizations, research institutes, etc., the results of studies and the considered views of the study group.

Details of the organization's makeup, services and goals appear in its staff report, *What is CONCAWE?* (Report Nr. 1/72, revised edition). You can get a copy from Stichting CONCAWE, address above.

CONCAWE is concerned only with "traditional" oil refining and does not handle problems associated with petrochemicals manufacturing. An attempt is being made by CEFIC (Conseil Européen des Fédérations de l'Industrie Chimique) to fill this gap with an organization similar to CONCAWE. However, this is still in the formulative stage and no Secretariat exists. The address of CEFIC is, Avenue Louise 250 Bte 71, Brussels 1050, Belgium.

No EPA in Europe. There is no equivalent of the U. S. Environmental Protection Agency in Europe. Each country has its own ministry or department, with associated services, to deal with different aspects of the environment. In the past, responsibility was divided between various ministries depending on the subject.

There is now a tendency to create specialized departments/ministries to co-ordinate over the entire field, e. g. the Department of the Environment in the United Kingdom. Further, in some countries such as the Federal German Republic and Switzerland, legal competence in these fields may rest with the individual federal state or even the municipalities (as in The Netherlands), so that advice must be sought from the local authorities.

Across national boundaries. Intergovernmental organizations have become very active in the past years. These are the Organization for Economic Co-operation and Development in Paris, the U. N. Economic Commission for Europe in Geneva, Council of Europe in Strasbourg, NATO Committee on the Challenges of Modern Society and, most important of all, the Commission of the European Communities (Common Market) in Brussels. A brief review of the purpose and activities of some of these organizations is contained in a paper "The International Viewpoint" published by Stichting CONCAWE, address above.

This paper is out of date, but the broad outline remains unchanged. The most important changes are in the European Commission where a Service for Environmental and Consumers Protection has been established (Services Environement et Protection Consommateurs). There are three divisions within this service: Prevention of Pollution and Nuisances, Quality of Life and Consumer Protection.

The first of these will deal with air and water pollution problems from pulp and paper, chemical, petrochemicals, heavy metal, iron and steel. In addition, the Hydrocarbon Directorate of the General-Directorate for Energy is also concerned with problems of pollution associated with the manufacture and supply of hydrocarbons.

A key source. The most comprehensive source of information on air pollution is *"Profile Study of Air Pollution Control Activities in Foreign Countries"; First Year Report"; NAPCA Publication No. SPTD-0601, National Air Pollution Control Association.*

Table 1 lists the code numbers of the organizations of value, at least to the refining industry. The code numbers refer to the 1970 edition of the above report.

Other contacts. A number of organizations not listed in APTD-0601 may also be useful sources of information and advice: *France* Union des Chambres Syndicales de l'Industrie du Pétrole, 16, Avenue Kléber, Paris 16, *Belgium* "Ecochem", Fédération des Industries Chimiques de Belgique, 49, Square Marie Louise, Brussels, *Norway* (Norsk Petroleums Institutt, Bygdy Allée 8, Oslo 2).

Table 2 lists the European government ministries and departments dealing with environmental matters.

TABLE 1 — Organizations listed in NAPCA Publication No. APTD-0601

Country	Code	Organization
Belgium	B.008	
	B.014	
	B.023	
	B.032	—Belgian Petroleum Federation
	B.034	
Denmark	DK.001	
	DK.011	—Oliebranchens Faellesrepraesentations Petroleum Industry Association.
France	F.006	
	F.052	
	F.086	
West Germany	D.016	
	D.021	
	D.035	
	D.036	
	D.040	—Mineraloel Wirtschafts Verband (MWV)
	D.402	
	(D.038)	DGMK
United Kingdom	GB.005	
	GB.039	
	GB.055	—Institute of Petroleum
	GB.101	
	GB.170	
	GB.171	
Italy	I.002	
	I.011	
	I.041	
	I.050	—Unione Petrolifera
Netherlands	NL.002	
	NL.003	
	NL.009	
	NL.017	
	NL.019	
Norway	N.001	
	N.013	
	(N.014)	—transformed into Norsk Petroleums Institutt
Sweden	S.002	
	S.011	
	S.018	—Svenska Petroleum Institutet
Switzerland	CH.021	
	CH.022	—Swiss Petroleum Association.

TABLE 2—European government ministries/departments dealing with environmental matters

Country	Ministry/Department
Austria	Ministry for Health and Environmental Protection Vienna
Belgium	Ministère des Affaires Economiques, Brussels Ministère de la Santé Publique, Brussels Ministère de l'Emploi et de Travail, Brussels
Denmark	Ministeriet for Forureningsbekaempelse Milsosturelsen (Ministry for Environmental Protection) St. Kongensgade 47I Copenhagen
Finland	Finnish Environmental Protection Committee Aleksanterinkatu 31 00170 Helsinki 17 Council for the Air Conservation and Noise Abatement, Työterveyslaitos, Haartmaninkatu 1 Helsinki 29
France	Service de l'Environnement Industriel Ministère de la Protection de la Nature et de l'Environnement 13 rue de Bourgogne Paris 7ème
Italy	Ministro per l'Ambiente (Minister without portfolio for the Environment) Via del Tritone, 142, 00100 Rome
Norway	Roykskaderädet (Smoke Damage Committee) Oslo—Dep. Oslo 1
U.K.	Department of the Environment London

JAPAN

CONCAWE and other key organizations mentioned do not deal with HPI environmental problems in Japan. The main agency there, which can supply consultation and full information on Japanese environmental law, is Environment Agency of Japan, 100 No. 3-1-1, Kasumigaseki, Chiyoda-ku, Tokyo, Japan. Other major groups are Central Headquarters for Pollution Control and the Public Nuisance Prevention Corp. The latter is a quasi-official organization, but CHPC is under direct supervision of the Prime Minister.

Other HPI-related contacts include Chemical Society of Japan, 1-5 Kanda Surugadai, Chiyoda-ku, Tokyo 101; Japan Chemical Industry Association, Tokyo Club Building., 3-2-6 Kasumigaseki, Chiyoda-ku, Tokyo 100, and Petroleum Association of Japan, Keidanren Building, 1-9-4, Ohte-machi, Chiyoda-ku, Tokyo 100.

Some insight into Japan's approach to environmental

management problems can be gleaned from observations of Professor Dr. Saburo Yanagisawa, Faculty of Engineering, Keio University, Tokyo, Japan. He spoke at a panel session on national and international standards for the environment, at a meeting of International Organization for Standardization.

Establishment of international standard values for air quality would be convenient, Dr. Yanagisawa said. Such standards would, presumably, represent the best and most advanced procedures. They might lead to improvement in standards in countries where the standards are not yet very strict. International standards also would permit accurate comparisons of air pollution among countries and would provide a sound basis for countermeasures against it.

There would be disadvantages. Because of differences in weather, geographical features, industry, population density, style of living, and the like, the international air quality standards might no be appropriate for particular nations. International standards might be abused; countries with standards below the international level might be blamed for air pollution problems in other countries. And the fixing of a method sometimes tends to obstruct the development of still better methods.

On the whole, however, in Dr. Yanagisawa's opinion, the advantages of international air quality standards outweigh the disadvantages.

World Health Organization. At the United Nations Conference on Human Environment, held in Stockholm in 1972, the World Health Organization was asked to establish the primary protection standards. WHO, which had been working on such standards since the 1963 WHO Symposium, has already developed air quality criteria and guides for urban air pollutants and has proposed a five-year plan for further investigation and establishment of criteria. In 1969 the Joint International Labor Organization/World Health Organization Committee on Occupational Health recommended safe concentration zones, agreed to by all countries, for 24 components of the atmosphere. These efforts may provide a basis for the establishment of international air quality standards.

The International Organization for Standardization should cooperate with WHO, Dr. Yanagisawa suggested, in expressing pollutant concentrations not only chemically but also in terms of the international standard of weights and measures. The two organizations should work together to standardize the continuous monitoring methods necessary to meet the standards.

The establishment of international air quality standards, Dr. Yanagisawa declared, would "enable us to compare and understand the conditions of air pollution in the world by continuous monitoring using calibration standard gas." This would be a major step toward a "healthy and enjoyable environment," which would "satisfy the objective of both ISO and WHO."

WATER POLLUTION

A vast array of organizations and agencies influence water pollution management. But perhaps the group best prepared to offer information is the *International Association on Water Pollution Research.* Membership details are available from Dr. J. H. Denysschen, P.O. Box 395, Pretoria, South Africa. The association is a nonprofit international organization. It was formally constituted in 1965 with the following general objectives:

1. To encourage international communication, cooperative effort, and a maximum exchange of information on water pollution research and water quality management.

2. To sponsor regular international meetings and conferences where reports on important research in water pollution are presented.

3. To provide a scientific medium for the publication of research reports and activities of the association.

4. To shorten the time lag between development of research findings and their application in engineering design.

Individual Membership is open to all persons interested in the field of water pollution research, water management and waste water treatment, and to research workers and employees of Associate Members. Professional organizations may apply for Individual Membership by nominating an individual as their official representative.

AIR POLLUTION

The key groups to contact for air environmental management are those affiliated with the *International Union of Air Pollution Prevention Associations,* Graf-Recke-Strasse 84, 4 Dusseldorf 1, Federal Republic of Germany.

Under some circumstances, it may be best to contact individual delegates of corporate members of IUAPPA or affiliate members. Detailed descriptions of member organizations' activities, goals and services appeared in the November 1970 issue (pp. 747-752) of the *Journal of the Air Pollution Control Association,* 4400 Fifth Ave., Pittsburgh, Pa. 15213.

Of special interest in that same issue is the article "Foreign Profiles in Air Pollution Control Activities: Special Sources of Information," (pp. 753-755).

LAND POLLUTION

Valley of the Drums and the Love Canal have become household terms, along with increasing concern for land pollution. New solid-waste-disposal legislation is being formulated. And the Toxic Substances Control Act in the U.S. calls for ever-tightening controls. However, no known association or society deals only with land-pollution problems. The best resource for the HPI, meanwhile, continues to be the Chemical Manufacturers Association and the American Petroleum Institute, both in Washington, D.C.

What You Need to Know About Impact Statements

The rules are involved—but experts can help with preparations

Byron H. Willis and **W. Michael Henebry,**
Environmental Research & Technology, Concord, Mass.

ENVIRONMENTAL MATTERS are a very serious business element confronting HPI management. Expansion, modernization and new construction projects confront management with environmental impact statements and the surrounding body of regulatory practice. Because these matters are complex, and evolving, we have chosen a question and answer format so that you can get practical information in short order.

WHEN IS AN ENVIRONMENTAL IMPACT STATEMENT REQUIRED?

On Jan. 1, 1970, the National Environmental Policy Act (NEPA) was signed into law (PL 91-190). NEPA requires each federal agency to prepare an Environmental Impact Statement (EIS) in advance of each major "action," recommendation or report on legislation that may significantly effect the quality of the human environment. This law applies to an extremely broad range of public and private activities, including new highway construction, harbor dredging or filling, large-scale spraying of pesticides, river channeling, munitions disposal, airport expansions, new plant construction, the expansion or modification of existing plants and so forth. NEPA and associated regulations require federal agencies to identify and respond to public opinion concerning estimated environmental consequences (impacts) prior to "actions."

The EIS is the heart of a federal administrative process intended to ensure achievement of national environmental goals. Each EIS must assess in detail the potential environmental impact of a proposed action, and *all* federal agencies are required to prepare statements for matters under their jurisdiction. As early in the decision process as possible, and in all cases prior to agency decision, the agency prepares a "Draft EIS" for review by other federal agencies, by state and local environmental agencies and by interested parties.

After comment from agencies and interested parties, the "Final EIS" is prepared, incorporating all comments and objections received on the Draft EIS and indicating how significant issues raised in the course of the commenting process have been resolved. Both the Draft EIS and the Final EIS are filed with the President's Council on Environmental Quality (CEQ) and made available to the public.

The Hydrocarbon Processing Industry (HPI) becomes involved in this federal process principally through application for construction and operating permits sought from federal agencies, and through certain other actions of federal agencies such as leasing of federal lands, licensing, certification or other entitlement. With the increasing body of environmental regulations, permit requirements and performance standards, it is quite likely that any expansion, modernization or grassroots project will involve one or more federal agencies in an action which may require an Environmental Impact Statement (EIS).

If a company is required to supply data on a project because of federal involvement, the information is usually in the form of an environmental report. The most commonly required type of report is called an Environmental Impact Assessment (EIA). The EIA report is the document which the involved federal agency will review as the basis for administrative action. A common type of administrative action is the issuance of a permit. Depending on company-supplied data contained within the EIA, and upon other factors, an agency will either undertake to prepare an Environmental Impact Statement, or issue a "negative declaration" stating that an Environmental Impact Statement is not required.

WHY IS AN EIS REQUIRED?

The National Environmental Policy Act is intended to be an environmental "full disclosure" process. The Environmental Impact Statement is the principal mechanism which the President's Council on Environmental Quality (CEQ) employs for coordinating environmental policy. The EIS, a detailed statement on the effects (impacts) of proposed actions on the environment, is intented to be the information source from which decisions can be made. A project or various aspects of a project, can have positive impacts, negative impacts or no significant impacts at all. An EIS is officially prepared by a concerned federal agency, under the guidance of CEQ. The mandated primary role of CEQ is to supervise the enforcement of NEPA functions and to coordinate a single environmental policy. It is not the role of CEQ to make detailed evaluations of each EIS; however, CEQ does often review all or part of selected ones. This function is performed by interested federal agencies such as the Environmental Protection Agency (EPA), state agencies and the public.

The public and various governmental agencies will review the data contained within the Draft and Final EIS, and thus become aware of the potential impacts of the action which is the subject of the Environmental Impact Statement. It is, therefore, important that your EIA report be complete, so that the EIS stands the best chance of also being complete. It is important to note that the existence of negative, or adverse, impacts does not imply the non-approval of a project. The National Environmental Policy Act (NEPA) does not specifically prohibit any actions, but requires public consideration or assessment of predicted environmental impacts before any actions that would "significantly" affect the human environment are undertaken. This is to include comparison of alternatives. Thus, detailed analyses must be made of all options, *including* the no-action alternative.

HOW DOES EIS RELATE TO APPROVALS?

The fate of a project depends upon granting of permits by various agencies. These agencies base their consideration of the applicant's project upon two categories of information. The first is a set of technical data contained within various forms, applications and so forth filed by the project sponsor (applicant). The second is a review of environmental matters relating to the project. Each cognizant agency will, in reaching its decision, depend upon the data supplied within the EIA or EIS, and often also depend upon a hearing process which involves discussion and commentary upon information contained in the EIS.

HOW SOON SHOULD YOU START?

In the more than six years since the inception of NEPA, there has been an increase in the experience and sophistication of both agencies and permit applicants. In particular, agencies are now more sophisticated in their review of projects. However, the number and complexity of environmental regulations that may constrain the size, location or operations of a proposed project have increased since the inception of NEPA.

The body of law and regulations, and the administration of these, must be taken seriously at the inception of any project because they *will* effect the project. It is much to the applicant's advantage to start the process as early as possible in the life of a project, because the environmental review and permitting process requires several sequential steps: data collection, monitoring, data analysis, impact prediction, reviews and hearings. Execution of these steps extends over many months.

It is also necessary to attempt to anticipate problems (whether they be technical, environmental or scheduling problems) which may arise from interacting agency procedures. Initial project concept planning must include environmental matters; the serious investigation and consideration of environmental matters should commence as early as any other aspect of a project.

HOW CAN PROBLEMS AND DELAYS BE MINIMIZED?

The best way to minimize problems and delays in a project is to make environmental planning an integral part of over-all project planning and to start environmental planning at the very inception of the project.

Project definition and selection of the site for a project have always involved consideration of many technical, logistical, operational and economic factors. The consideration and projection of these factors is an integral part of feasibility planning for a project. The inter-relationship of these factors and the constraints which they may impose on a project must be anticipated and identified.

Likewise, possible environmental constraints must also be anticipated and identified by starting early with an Environmental Reconnaissance (ER) as a first step in the over-all environmental process. An ER may include the screening of potential plant sites, the identification of permits and other regulatory requirements and an assessment of the attitudes displayed by cognizant agencies, thus providing the possibility for early feedback of information into the project design phase so that changes in project concepts can be undertaken while the applicant's options are still economically open.

The next step in the environmental process is to interface with cognizant agencies to identify baseline data requirements which they will impose, requirements for considerations necessary in the preparation of the environmental impact statement, etc. Certain data requirements may be satisified only by collection periods that extend over many months. In particular, baseline studies involving air quality, water quality, terrestrial ecology and aquatic ecology are in this category. The collection of such data should be started at the earliest possible time. An assessment of community attitudes and concerns may be required for an Environmental Impact Statement, or may be taken as a prudent step in the total planning of a project. Such assessment of attitudes and concerns should be undertaken at the earliest possible time so that information about a project can be released that is responsive to community concerns, minimizing the possibility of opinions and fears arising out of misinformation. Once all data concernng a project are in, the preparation of the Environmental Impact Assessment report can be completed.

The final step is the preparation of the Environmental Impact Assessment (EIA) report, which is the applicant's submission of information required to prepare an Environmental Impact Statement. The entire process of gaining construction and operating permits cannot effectively go forward until the applicant's EIA report is prepared.

HOW DO YOU PREPARE AN ENVIRONMENTAL IMPACT ASSESSMENT?

An EIA is the applicant's report, to the concerned federal agency, which describes the project and its potential impacts upon the physical, social and economic facets of the environment. There are a variety of guidelines[1] available to describe the necessary contents of an environmental impact assessment and these guidelines make clear that an EIA must be far more than simply a collection of data. Essentially, a prediction of impacts is required. The subjects which must be treated are:

- A detailed description of a proposed action (arising

from the applicant's project) including information and technical data needed to permit a careful assessment of environmental impact

- Discussion of the probable impact upon the environment, including any impact on ecological systems and any direct or indirect consequences that may result from the action (possible mitigating measures must also be discussed)
- Any adverse environmental effects that cannot be avoided
- Alternatives to the proposed action that might avoid some or all of the adverse environmental effect, including cost analyses and environmental impacts of these alternatives
- An assessment of the cumulative, long-term effects of the proposed action including its relationship to the short-term use of the environment vs. the long-term productivity of the environment
- Any irreversible or irretrievable commitment of resources that might result from the action or which would curtail beneficial use of the environment.

To carry out the necessary work the applicant must define specific work packages and then bring to bear the necessary technical expertise. This is not a simple problem because the range of expertise generally ranges from process engineering to sociology and includes such diverse areas as land use planning, hydrology, aquatic and terrestrial ecology and meteorology.

HOW TO ORGANIZE FOR IMPACT ASSESSMENT

It is critical that top management assign a senior manager to oversee environmental work necessary to a project. This manager must understand how serious is the need to make environmental planning an integral part of total project work. Environmental planning is critical for a number of reasons, and he should understand the need to assure coordination of environmental work with other aspects of project planning and design; environmental matters should be his prime concern. Further, it is necessary to manage and coordinate the internal and external resources a company must bring to bear on environmental matters. In particular, this manager must have enough "clout" to assure timely access to required data concerning the project. These data will include engineering design information, process data and information concerning eventual operations of the project. Finally, identification of this senior manager is necessary to assure that the applicant will have an appropriate level person available for interfacing with agency personnel.

HIRE AN ENVIRONMENTAL CONSULTING FIRM?

To carry out the many tasks required for environmental assessment and to successfully negotiate in the permitting process, an applicant must decide whether to do the job internally or whether to hire an environmental consultant. Considerations of credibility and objectivity are very important. Whether justified or not, intervenors tend to jump upon corporate prepared information as non-objective and self-serving. If you get into hearings, or into court, the defensibility of your environmental data is of prime importance. They will be defensible if they are complete, of high quality, responsive to adverse impacts, responsive to alternatives and if they can be attested to by recognized expert witnesses.

Also, the matter of availability of appropriate and necessary in-house expertise gives rise to a set of questions: a) Are all of the necessary environmental disciplines available in-house? (this can cover a very wide range); b) Are there areas of special analysis required? (such as a social survey); c) Are there special monitoring requirements? (air, water, etc.).

Generally, the environmental impact process is an infrequent or one-time experience for the applicant. However, prior experience and familiarity with regulations and agency requirements is necessary to the speedy conclusion of the process. The group undertaking environmental aspects of a project must be aware of all relevant environmental regulations; they must be experienced in doing Environmental Impact Assessment studies; they must have the ability to interpret agency requirements to define the appropriate level of effort for EIA studies; and they must understand the implications of study efficiencies, scheduling and agency relationships.

An applicant's ability to devote proper personnel to the project for the necessary time interval to carry forward environmental studies is frequently a serious problem. The size of a project and its complexity determine the range of skills, the time duration and the agencies involved. If a project is simple, chances that the applicant can make the necessary in-house skills available are pretty good. If the project is complex, it is unlikely that these resources would be available in their entirety or could be devoted to the project for the time needed.

The applicant may also need to satisfy corporate mandates to utilize available corporate resources. This raises a variety of questions concerning: desires to use in-house laboratory facilities; whether the right kind of person is really available to undertake the required tasks; the question about how maximum use can be made of in-house expertise, experience and facilities. In general, it is not a simple task to untangle whether or not use of these in-house resources or facilities can help or hinder with regard to costs, time and performance of a project.

Depending upon the applicant's resources and relevant corporate mandates, the entire environmental undertaking can be performed in-house, it can be performed with the assistance of a consultant or a consultant may be retained to manage the entire process, making best use of the applicant's personnel and other resources.

HOW TO SELECT ENVIRONMENTAL CONSULTANT

Similar to other contracting involved in a project, the HPI firm applying for permits needs to define the scope of service to be contracted for. Secondly, the firm should generate a list of candidates, considering their prior experience and exposure to projects such as his, references from other firms in the industry and references or comments from agencies.

The next step is to cull this list to a manageable number of possibilities by interviews, by examination of the consultant's literature, by scrutiny of his qualifications and experience, by review of his recommendations and by looking at special considerations (such as regional presence, special skills or particular experience). The next step is to invite a limited number of proposals based upon the scope of services required. Next, these proposals should be evaluated by examining each proposer's skills, management abilities, prior relevant experience, proven ability to perform and familiarity with regulations and agencies. It is advisable to meet the proposed project manager and some of the senior personnel to be supplied by the consultant. Further, it may be desirable to inspect his facilities, laboratories and equipment.

The time at which you select a consultant can vary, depending in part upon the type and scope of services desired. Site evaluation and project feasibility studies occur very early in the life of a project. Baseline monitoring comes later. An Environmental Impact Assessment comes still later. These various tasks may be contracted for at different times.

HOW DO YOU DEAL WITH THE AGENCIES?

Interfacing with the agencies having cognizance of the project is an evolving and continuing process that should be started as early as possible. An applicant must also recognize that the relationship with agencies will be on different levels at different stages of development.

Basic interaction with an agency revolves around permits. Once the required construction permits and operating permits are identified (and thus the cognizant agencies are identified), the applicant must develop an efficient strategy for obtaining the needed permits.

At early, conceptual stages (when feasibility is still a consideration), the applicant may not have a full definition of the project and may be concerned, constrained or apprehensive about how much to reveal to agencies (or to the public) about plans. However, some information must be revealed to agencies in order to get useful information in return. For example, at this stage you will want to know what permits will be required and you will want to know what attitudes agency personnel will have towards projects similar to yours. You will also want to know about environmental problems or constraints of particular concern to each agency. Some of this information may be readily obtained without having to be too explicit, but the more explicit you can be, the better and more useful will be the responses or information from the agencies. Experience shows that agencies are willing to be helpful and are generally respectful of the proprietary nature of your corporate project plans.

Likewise, the more willing you are to *talk* to agency personnel about potential plans or options, the more responsive and trustful they will be. This is not to say that they do not have their own viewpoints about control technology, etc. They have their mandates and objectives which may represent views counter to yours.

The objective of discussions at the early stages should be to determine if the proposed project has reasonable potential for complying with environmental regulations, and whether the project has reasonable chances for obtaining *all* necessary permits without incurring undue difficulties or delays (for example, public controversy, expensive control measures, lengthy permitting procedures, etc.). Information obtained at an early stage can be helpful in over-all project planning and scheduling; for example, helpful in preparing design schedules, plans for ordering equipment and preparation of construction schedules.

Contact with agencies at an early stage of project development is usually one portion of an Environmental Reconnaissance (ER). The other major portion of an Environmental Reconnaissance is concerned with a rapid survey or screening of potential sites in order to identify site-related environmental problems that would also bear on potential difficulties in obtaining permits. The information needed to complete an Environmental Reconnaissance is most often available from public sources, such as environmental regulatory agencies, and can be obtained without *any* reference to a particular project.

Dealing with agencies at later stages of the project will require an even greater degree of interaction and will involve even more detail in describing the project. For example, the next distinct step is the active search for (and evaluation of) potential sites, providing that any such site alternatives are available or feasible. Discussions at this stage will likely require interactions with agencies not previously contacted. Further, determining compatibility of a project with a specific site will require having more detailed information about the project and revealing more about the project to the agencies.

Selection of a site next leads to detailed studies necessary for impact assessment. At this stage the applicant should be prepared to discuss the project with agencies openly and in detail. These discussions commence with definition of the scope of the environmental monitoring programs and EIA studies that will be required by concerned agencies.

One often-voiced concern is that such discussions can lead to exorbitant demands by agencies for monitoring and impact studies. Another concern is that revealing areas of potential environmental problems may adversely influence agency attitudes towards the project. These concerns must be referenced to the fact that the EIA report is to be a *full disclosure document*. In the absence of adequate information about a project, agencies may very well have a tendency to ask for more information rather than less.

It is recommended that the applicant fully scope the extent of potential problems and define a study program adequate to examine these problems. Further, it is recommended that developed work programs be taken to the agencies for comment and approval rather than simply asking for agency personnel to specify a work program. If the applicant has been open, and has established credibility with the agencies by acknowledging potential problems, it is highly likely that mutual agreement will be reached upon a work program that is not excessive.

It is dangerous not to acknowledge or reveal potential environmental problems early-on in discussions with the agencies. Such problems usually come to light eventually and it is a far better strategy to identify them before an agency or intervenor does. Early identification underscores the fact that you're aware of the problems and are trying to do something about them. In addition, legal entanglements over Environmental Impact Assessments are almost always the result of *not* having described the potential problems and impacts of the project, rather than the result of revealing adverse impacts.

By the time you begin the EIA study program, your project will most likely be known to the public and will begin to be watched closely by interested individuals and groups. Thus, throughout the conduct of the Environmental Impact Assessment studies it is advisable to maintain frequent and close communication with agencies so that no "surprises" are presented to agencies upon submission of the EIA report which they must review.

WHICH AGENCIES WILL BE INVOLVED?

The type of project and where it is located will both influence what kinds of permits will be required (air pollution, water discharge, solid waste disposal, dredge and fill, etc.) and what levels of government will be involved. Some projects will involve local, state and federal agencies.[2] Because there is no standard, universal answer to the question of which agencies will be involved on a project, it is necessary to determine which permits are required through a careful review of the applicability of local, state and federal regulations. Many states have their own environmental policy acts.

In many cases it is possible to list rather quickly the more obvious permits likely to be required. However, it may not be easy to determine which permits will *actually* be required. Likewise, some required permits may not be obvious immediately. Because of these uncertainties it is recommended that the applicant make a list of potentially required permits and then verify this list through discussions with potentially involved agencies as part of an Environmental Reconnaissance. Early identification of all required. permits is important for two reasons. First, one of the permitting agencies will eventually become the "lead" agency having responsibility for the environmental review. Second, failure to identify a required permit may lead to delays later on in the life of a project. Such delays may arise because of:

- Failure to generate appropriate design information
- Failure to study relevant environmental concerns
- Requirements to modify project plans, or design, especially at a time when options have already been limited
- Failure to identify and analyze alternatives.

WHAT IS A 'LEAD' AGENCY?

When premits are required from more than one agency of the federal government, a single agency is designated as the "lead" agency and takes on responsibility for preparation of the Environmental Impact Statement (if required), and for coordinating environmental reviews and comments by all other involved federal agencies. Often the choice of a lead agency is obvious, due to the particular significance of some environmental aspect of the project, or its siting.

If two agencies are both equally involved, or have equal interests, they will generally reach a mutual agreement as to which will become the lead agency. If they cannot agree, then the President's Council on Environmental Quality (CEQ) makes the final determination of the lead agency.

In addition to this "classical definition," the term "lead agency" is also used in other situations involving nonfederal agencies that must issue permits. In such cases of overlapping jurisdiction, one agency at a local or state level will assume responsibility for coordinating environmental reviews, primarily on the basis of established procedures within a state, and on the basis of which environmental reviews are required.

While in most cases the designation of a lead agency is not a matter open to choice, in those situations where a decision must be made, the applicant may request a particular agency to take the lead. Such a preference can be based upon consideration of differences in procedures, or "time-lines" or availability of agency staff, etc. However, there is no guarantee that the applicant's request will be honored.

WHAT IS ROLE OF NON-PERMITTING AGENCIES?

At all levels of government there are agencies that do not directly issue permits but which may be involved by virtue of their special areas of concern or knowledge that are invoked by a project. For example, a state geologist may be asked to review aspects of the project that will affect geology or ground water, even though he does not issue any permit or any formal approvals.

The Environmental Impact Assessment report (EIA) will be forwarded to such agencies at all levels of government for review and comment, to assure that all potential environmental concerns have been completely and adequately addressed in the report. In some cases, explicit approvals (but not permits) may be required; for example, by a letter of clearance or a statement that regulations have been complied with. Thus, review procedures are designed to assure full disclosure of all relevant matters of environmental concern.

HOW DO YOU DEAL WITH THE PUBLIC?

It is evident that a large degree of latitude exists for interaction with the public at various stages of a project. Recognizing that the EIA is a full disclosure document, and that it will ultimately be made public, and further recognizing that a project which impacts a community will be of public interest, the EIA must be written to be effective as a document read by the public.

Perhaps the greatest concern of corporations is how to interact with the public in the early stages of project development prior to release of the EIA. There are many strategies which may be employed, ranging from a very elaborate and highly visible public relations program to a very low profile. The degree of such interactions and the amount of information given about the project is largely a matter of corporate choice.

Moreover, if desired, techniques such as surveys can be employed to sample public attiudes and thus provide some basis for determining how best to interact with the public. Regardless of which strategy a corporation chooses to employ, we recommend that a few general points be observed:

- Prepare in advance the type and amount of information you choose to release to the public.
- Make sure the information released is consistent with actions by the corporation that are observable by the public.
- Make sure that information given by various corporate sources is consistent between the sources.
- Be prepared to discuss your project in a manner that is responsive and which will not be viewed as evasive.
- Don't oversell the project, or avoid discussion of po-

TABLE 1—Areas for consultation with agencies

Air quality and air pollution control	Flood plains and watersheds	Transportation and water quality
Weather modification	Mineral land reclamation	Congestion in urban areas, housing and building displacement
Environmental aspects of electric energy generation and transmission	Parks, forests and outdoor recreational areas	Environmental effects with special impact on low income neighborhoods
Natural gas energy development, generation and transmission Toxic materials	Soil and plant life, sedimentation, erosion and hydrologic conditions	Rodent control
Pestcides	Noise control and abatement	Urban planning
Herbicides	Chemical contamination of food products	Water quality and water pollution control
Transportation and handling of hazardous materials	Food additives and sanitation	Marine pollution
Coastal areas: wetlands, estuaries, water fowl refuges and reaches	Microbiological contamination	River and canal regulation in stream channelization
Historic and archeological sites	Radiation and radiological health	Wildlife
	Sanitation and waste systems	
	Shellfish sanitation	
	Transportation and air quality	

tential problems, but do be prepared to give positive responses about what you're doing about public concerns.

HOW LONG DOES THE PROCESS TAKE?

Roughly speaking, there are four stages in the permitting process of which preparation of the Environmental Impact Assessment is only one. There is a preliminary stage during which there is still some flexibility about a project. During a second stage the Environmental Impact Assessment is made, and a report prepared. In the third stage the lead agency, and other agencies, perform an environmental review and make a decision as to whether a "negative declaration" will suffice or whether preparation of an Environmental Impact Statement (EIS) is required. In the fourth stage administrative action is taken on permit applications which relate to the project.

The specific involvement by an applicant in any one stage of the project is highly dependent upon many factors including project type, size, complexity and location. Details in these categories will determine the permits that are required and thus the agencies that are involved. Correspondingly, different technical and procedural requirements (or options) may apply, which implies different amounts of time to conduct monitoring programs, impact assessments and the length of agency review.

Time spent on preliminary environmental studies, for example, Environmental Reconnaissance (site selection, identification of permits, etc.), will depend upon corporate attitudes and the pace of project planning and development activities. However, reconnaissance activities may reasonably be carried out in a period of two months or so. Environmental inputs into the site selection process can also be carried out in reasonably short periods and usually this is not a critical time factor unless there is a corporate decision to obtain some environmental baseline data for input to the decision process during this time.

It is often the case that environmental concerns are not on the most critical path in project planning and development until a corporate decision is reached to apply for permits. Consequently, this and subsequent agency environmental review activities leading to action on the permit applications become highly visible to management, and the time required to accomplish this permitting activity becomes of concern.

It has happened, occasionally, that project planning has developed target dates for construction without full awareness of the time requirements for this stage of the environmental process. The consequence has been either that too little time is allocated to perform Environmental Impact Assessment studies, or schedules for construction, engineering design and equipment procurement have had to be delayed.

Actual time requirements to carry out EIA studies to the point of submitting an EIA report for agency review depends upon the status and availability of project design data, and upon the availability of, and requirements for, environmental baseline data. An applicant should anticipate that the time to prepare an EIA will range from a minimum of about 6 months (when there is minimum requirement for obtaining *new* environmental monitoring and baseline information) to about 18 months when agency requirements exist for obtaining new data (often involving 12-month measurement periods).

It should be noted that because of a general increase in the availability of environmental baseline data, either from prior environmental impact studies or from state or federal data collection and monitoring programs, there is a clear trend toward shorter time requirements to complete major project environmental impact assessments. Time requirements for agency environmental review processes are highly dependent upon the agencies involved and, in particular, upon the procedural requirements of the "lead" agency. Agency review can take extremely short periods of time if a negative declaration is issued (typically on the order of two to four months), assuming a federal agency is involved and that NEPA applies.

If an Environmental Impact Statement (EIS) is required then the schedule includes time for preparation of the Draft EIS, review of the Draft EIS by agencies, review by CEQ, review by the public, public hearings and preparation of a Final EIS. In the EIS process there are also built-in mandatory waiting periods as specified by CEQ guidelines. NEPA requires each federal agency to "consult with and obtain the comments of any other agency (including state and local, as well as federal) which has jurisdiction by law or special expertise with respect to any environmental impact involved" in the course of preparation of an impact statement.

In its guidelines for preparing statements, CEQ lists a variety of agencies which must be consulted on this basis in the areas shown in Table 1. The guidelines also call for Draft Environmental Impact Statements to be made available for public comment. Many individual agency procedures allow for direct solicitation of such comments from interested parties and private organizations. If there are no significant public controversies and no legal issues or entanglements, this EIS process will take about 12 months from submission of the applicant's EIA report.

If there is no federal involvement in the permitting process for a project, then environmental study requirements (if any) will be dictated by state (or local) regulations. Time schedules are therefore highly dependent upon such regulations. For example some states now require environmental impact studies of substantial scope even in the absence of federal involvement.

In summary, an applicant firm can expect preparation of the EIA report, and review by agencies, to require many months. Just how many months depends upon the project. The time period is also affected by the quality and adequacy of the EIA report. It is clearly desirable to minimize the necessity to correct deficiencies in the EIA report after it is prepared, or to make design modifications or revisions in order to comply with environmental constraints. All such activity adds to the time for the environmental review, and the review will be on the "critical path" of project development.

By making environmental considerations an integral part of project feasibility studies (that is, when environmental problems are not on the critical path), and by continuing to pay close attention to environmental factors and concerns at all stages of project planning and development, potentially serious environmental problems can be avoided. The chances for encountering serious delays in the environmental review by agencies can thus be minimized.

WHAT DOES IT COST?

In considering the cost of environmental studies, the total impact of environmental considerations on project cost must be recognized. Obviously, there are the direct costs for performing EIA studies. Less direct, but perhaps more significant, are project costs associated with the choice or necessity of environmental pollution control measures. Other indirect costs can arise from delays in project completion. Possible project delays and control measures involve far larger amounts of money than do environmental studies.

Doing things right means that decisions are made prior to major commitments in the project. Environmental studies, especially at the early project planning stages, can permit project management to anticipate serious problems and to avoid them through changes in design or site location before major commitments of effort or resources are made. Not taking a hard look early in a proposed project means that risks are higher later on, and that environmental control measures will be more costly if changes have to be made to deal with environmental problems which surface at later stages of a project. There is also the risk that non-recoverable expenditures will have been made.

Project costs will be significantly impacted by time delays. Such time delays may result from public opposition arising out of certain environmental matters, requests by reviewing agencies for additional data or the need to correct deficiencies in an EIA report (for example, by giving further consideration to project alternatives). In the extreme, environmental problems may be encountered which are of such significance as to require rejection of a chosen site, modification of design or alteration and addition of controls.

Not all such needs for information can be anticipated in advance, but it can generally be stated that with more effort devoted to preparation of an EIA of high quality with adequate coverage of all potential environmental issues, and with such efforts applied at early stages of the project, the risks of encountering difficulties with the agency review process can be greatly reduced. Consequently, the amount to be spent directly on preparation of an EIA can have *significant influence* on total project costs. Corporate policies will be a governing factor here.

A strategy for seeking permits must be determined early-on: For example, spending minimum dollars on an EIA; or minimizing risks to get permits; minimizing time to get permits, or minimizing the commitment to environmental control devices. The decision as to which choice is preferable cannot be suggested except on a case-specific basis. For example, the strategy of minimum risk to get permits may be more costly and more time-consuming than other strategies, since it may require more environmental studies than necessary, or may mean more control expenditures than might actually have been required. Recognition that such tradeoffs exist is important, and key to a cost-effective environmental planning process.

As to the direct question of the cost of an Environmental Impact Assessment, a simple answer doesn't really tell the story because the effort will be dictated by the time and scope of baseline data collection and impact assessment studies required to meet agency demands for any particular project. Correspondingly, agency demands will be dictated by the size, complexity and potential for impact for a given project. Thus the range of potential costs varies too widely to be stated, except on a case-specific basis.

WHAT ARE THE CHANCES OF SUCCESS?

Agencies cannot, and will not, give any assurance that permits will be granted prior to a full review as required by established procedures. Agencies generally will provide help in specifying baseline data collection and impact assessment requirements, for example, to assure that an adequate EIA is submitted for their review. However, even if all standards and other environmental regulations are met, there is still no assurance that the project will be granted all necessary permits. If there is sufficient public opposition, legal actions can cause significant delays, possibly to the point of effectively killing the project. Or, such matters can bring about conditions to be imposed on the granting of permits that make the project economically infeasible.

Clearly, in seeking permits, there are many risks and no guarantees. However, some of the risks can be reduced, primarily through serious attention to environmental concerns starting early in the project planning stage and continuing throughout project development.

LITERATURE CITED

[1] Council on Environmental Quality, *Guidelines for Preparation of Environmental Impact Statements*, Federal Register, 1AUG73, Vol. 38, #147, Part II, 20550-20562.

[2] Council on Environmental Quality, *Environmental Impact Statements: An Analysis of Six Years Experience by Seventy Federal Agencies*, CEQ, March 1976.

Before the Environmental Assessment

This handy checklist can give you valuable pre-environmental-impact-assessment guidelines and head off trouble before it begins. Use it to save money and time

C. W. Moores, Badger America, Inc.,
Cambridge, Mass.

DELAYS AND even abandonment of hydrocarbon processing construction projects can be avoided when potential environmental problems are identified early. One way to implement an early identification program is to use the checklist here, well before preparation of the environmental impact statement.

Use an environmentalist. Development of new construction projects progresses at different rates, dependent on myriad factors. At key points, management decisions must be made regarding the addition of members of various disciplines to the project team.

Usually, at the site selection phase of the project, it is wise to use an experienced environmentalist. Occasionally the project environmental responsibilities are assumed by a well-meaning, cost-conscious, environmentally inexperienced member of the project team. But too often, due to project momentum, environmental guidance is deferred until "later." Despite the reasons, neglect of critical environmental factors during the early stages of a project results in loss of credibility, project delays, prolonged negotiations and unnecessary frustrations. All of these translate into increased cost.

Prepare a checklist. An environmental checklist can be used by an experienced environmentalist to minimize the possibility of oversights. Or, the checklist may be used by an inexperienced engineer as an information-gathering device to expand the capability of an experienced environmental supervisor. A checklist is applicable to many phases of projects. Examples:

- Select sites
- Plan meetings with regulatory agencies
- Obtain a declaration on environmental impact assessments
- Determine the degree of environmental control required for construction and process emissions
- Prepare construction permits
- Prepare for public hearings.

The examples cited below—based on actual projects—suggest exigencies that can be minimized through the use of a checklist.

The environment is only one of the many factors that must be considered during the selection of a new project site. Given two or three alternate locations, all site factors can be evaluated in a semi-objective fashion resulting in the selection of a prime site. But to overlook environmentally sensitive aspects makes all other positive aspects of the site worthless. A good checklist will help reveal areas of extreme environmental sensitivity before real commitments are made.

Early start. Have you ever had your site preparation subcontractors start operations in the field only to be notified you haven't received all the necessary permits? Conduct preliminary discussions with various regulatory agencies as early as possible. These meetings will establish which agencies are "lead" agencies and will help to predict whether environmental permit procedures will become schedule determinants. Information from an environmental checklist will make these preliminary meetings more valuable. You will also begin to establish credibility that is invaluable in dealing with regulatory agencies.

Have you ever paid a premium to expedite long-delivery equipment only to find that your projected nine-month environmental permit schedule has been changed into an 18-month Environmental Impact Assessment? An environmental checklist is invaluable in discussions regarding Environmental Impact Assessments (EIA). The checklist will not serve as an EIA. However, it will give an indication as to whether an EIA is required and will suggest what schedule changes may be needed.

Spot costly items. Have you ever been told that the receiving stream passing through your property cannot assimilate your process waste water, no matter what kind of treatment you provide? Although the U.S. government has set effluent guidelines for various types of "new sources," state and local regulatory groups or even local environmental action groups can require more stringent controls. A checklist should help identify sites which require unusually large expenditures for pollution controls.

Have you ever been asked by a regulatory agency how you intended to control turbid rainwater runoff during construction on a site covered with silty clay? An environmental checklist is helpful in planning the construction-related environmental problems. These must be considered as a part of construction permits.

Have you ever been asked in a public hearing if the

3,000 cars and trucks transporting the construction workers will cause traffic jams? Oversights of this type can cause anxiety among local residents, resulting in opposition and a general loss of credibility. Again, the environmental checklist will force critical preliminary thinking.

Keep checklist relevant. If the "Have you evers" above give you moist palms, consider using an environmental checklist. While generally reliable, the checklist below may require modification to fit specific types of projects. It must be continually up-dated to reflect changes in legislation and control technology.

Project Environmental Checklist

GENERAL SITE DATA

1. Where is the project site? ______
2. Is alternate site available? Yes ____ No ____
3. Real estate status—proposed, optioned, owned ______
4. Size of the site: ______ acres
5. Describe previous and present land use: ______
6. Estimate percent cover:

   wooded ___% exposed soil ___%

   shrub ___% exposed bedrock ___%

   wetland ___% open water ___%

   open grassland ___%

   Make sketch showing this distribution on the site. For wooded, shrub and wetland areas, specify on the sketch vegetation density and height in the following manner: Density—scattered, normal, dense; Height—tall, average, short.
7. Provide freehand sketch showing topographic description including slope and direction of runoff.
8. Provide USGS topographic map. If not available sketch major area topographical features.
9. Is the site wetland? Yes ____ No ____
10. Is the site adjacent to a wetland?

    Yes ____ No ____
11. Is the site subject to flooding or ponding?

    Yes ____ No ____
12. If yes, give 100-year flood history: ______

    ______
13. Soil permeability: ______ in./hr.
14. Soil erodibility: (check one)

    slight ____ moderate ____ severe ____
15. Depth to bedrock: ______ ft. Depth to water table ______ (state range)
16. General soil description: ______
17. Any on-site or adjacent ecologically unique areas?

    Yes ____ No ____

    If yes, specify: ______
18. Is the site on or adjacent to a wildlife: Breeding area ____ habitat ____

    migration route ____ feeding area ____

    If yes, specify: ______
19. Specify any known rare or endangered species on the site or in the area ______

    ______
20. Do drainage paths or receiving bodies support aquatic life? Yes ____ No ____

    If yes, specify: ______
21. Is the site a potential wildlife sanctuary?

    Yes ____ No ____
22. Has a previous ecological study been made in the area? Yes ____ No ____

    If yes, are results applicable to this site?

    Yes ____ No ____

    If yes, are data available? Yes ____ No ____

    Where ______
23. How is the area zoned? ______
24. When was the last change in zoning:

    ______(estimated)
25. Describe the surrounding properties, land uses and give property values, in general terms (use map or sketch). ______
26. What is the estimated population density of the area: ______ per sq. mile
27. Describe the nearest population centers, giving names, direction and distances from the site (use map): ______
28. Give distance and direction of nearest private residences and their property values (use map or sketch). ______
29. Describe briefly the existing pollution sources in the region (use map or sketch): ______

    ______
30. Describe any previous public and/or private pollution complaints: ______

    ______
31. To what extent is the area a tourist site: ______

    ______
32. Describe any educational, cultural, historical or recreational sites nearby (use map or sketch): ______

    ______

ENVIRONMENTAL CHECKLIST

33. Identify any active environmental groups in the area. Include names, addresses and telephone numbers: ______

34. Describe briefly general communities' attitudes toward new industry: ______

35. Describe economic well-being of nearby population centers: ______

36. Describe employment level of the region: ______

37. Compare this level with local, state and national norms: ______

38. Describe the tax base of the region: ______

39. What is the prevailing assessment rate: ______

40. What is the anticipated growth for the region: ______

41. Describe any recent building activities in the area: ______

42. Describe anticipated land use changes: ______

43. For the following, estimate the ability of municipal facilities and services to absorb the influx of personnel (workers and their families) during site construction and plant operation:

	Construction Phase		Operation Phase	
Number of personnel	______		______	
	(avg/max)		(avg/max)	
Housing	Yes	No	Yes	No
Schools	Yes	No	Yes	No
Medical facilities	Yes	No	Yes	No
Police and fire protection	Yes	No	Yes	No
Roads	Yes	No	Yes	No
Wastewater treatment	Yes	No	Yes	No
Water supply	Yes	No	Yes	No
Solid waste disposal	Yes	No	Yes	No
Recreational facilities	Yes	No	Yes	No

44. What is the estimated schedule for the project:
    - Site selection ______
    - Permit preparation ______
    - Engineering ______
    - Procurement ______
    - Construction start ______
    - Construction finish ______
    - Startup ______

CLIMATOLOGY

45. Nearest weather station location: ______

46. Distance and direction from the site: ______ mi.

47. Temperature:
    - Average ______ min. ______ max.
    - Extreme ______ min. ______ max.

48. Relative humidity:
    - Average ______
    - Extremes ______

49. Precipitation:
    - Total—yearly average ______
    - Snow—yearly average ______

50. Rainfall intensity data:
    - 15 min. max. ______ in. frequency ______
    - 1 hr. max ______ in. ______
    - 24 hr. max. ______ in. ______

51. Average number of days per year of:
    - 2.0 in. rain ______
    - 1.0 in. snow ______
    - thunder ______
    - fog ______
    - clear ______
    - p. cloudy ______
    - cloudy ______

52. On a separate sheet, provide wind rose.

53. Give strength and frequency of inversion occurrence ______

54. Hurricanes, tornadoes, cyclones:
    - Frequency: occurrences per year ______
    - Wind velocity Avg. ______
    - Max. ______

WATER POLLUTION

55. Identify and describe the drainage paths from the site, giving major body of water into which they flow (sketch or map): ______

56. Are these drainage paths visible to the general public: Yes ______ No ______

57. If aqueous discharge is to surface waters, circle the appropriate description:

drainage ditch	small stream
river	man-made waterway
landlocked lake	lake with outlet
tidal estuary	ocean or gulf

58. Is site at or near delicate marine ecosystem?
    Yes ______ No ______
    If yes, specify: ______

59. Will the project involve effluent discharge?
    Yes ______ No ______

60. Receiving body.
    - Low flow: ______ gal./day
    - Average flow: ______ gal./day
    - Maximum flow: ______ gal./day

61. Give direction and strength of currents that would interact with effluent: ______

62. If stratification is present, specify cause, depth, extent and variation: ______

ENVIRONMENTAL CHECKLIST

63. What is the approximate dilution provided by re-
____
ceiving body at minimum flow: ______________

64. Ambient water quality standards for receiving body:
dissolved oxygen:
________ (max.) ________ (existing average)
dissolved solids:
________ (max.) ________ (existing average)
bacteria*: ________(max.)________(existing average)
*specify: total coliform or fecal coliform
pH: ________ (max.) ________ (existing average)
turbidity:
________ (max.) ________ (existing average)
temperature:
________ (max.) ________ (seasonal range)

65. Specify classification of receiving body:
contact recreation non-contact recreation
fish and wildlife habitat domestic raw water supply
other ______________________________

66. Specify state and/or local regulations concerning:
taste - odor ______________________________
color ______________________________
oil ______________________________
toxic substances ______________________________
heavy metals ______________________________
other ______________________________

67. Specify any source standards that are applicable:
____

68. Give any additional federal, state and/or local water pollution standards: ______________________________
____

69. Is there an on-site or nearby aquifier recharge area? Yes ____ No ____

70. Is there an on-site or nearby ground water discharge area? Yes ____ No ____

71. Specify the quality of the ground water: ________

72. Specify the existing or potential project interaction with surrounding water table: ______________
____

73. Specify area ground water use: ______________
____

74. Give distance to the nearest ground water drinking supply (use map): ______________________________
____

75. List sources (quantity and quality) of water pollution associated with project: ______________________________
____
____

AIR POLLUTION

76. Give state ambient air quality standards, specifying maximum permissible concentrations and existing concentrations for the following. List sampling time for existing concentrations:
particulate matter:
________(max.) ________(existing) ________(time)
CO:
________(max.) ________(existing) ________(time)
Hydrocarbon:
________(max.) ________(existing) ________(time)
NO_2:
________(max.) ________(existing) ________(time)
SO_2
________(max.) ________(existing) ________(time)
photochemical oxidants:
________(max.) ________(existing) ________(time)
other:
________(max.) ________(existing) ________(time)

77. Will traffic and/or parking facilities be subject to regulations as an air pollution source during construction and/or operation? Yes ____ No ____

78. Is the project subject to stationary source emission standards? Yes ____ No ____
If yes, include standards.

79. Is there an odor control regulation?
Yes ____ No ____

80. List any additional state and/or local air pollution regulations: ______________________________
____

81. Describe and give location of any nearby sensitive receptors of air pollution: ______________________________
____

82. Describe any on-site existing background odor: ____
____

83. Will odors from the project be contained within site property? Yes ____ No ____

84. Give location and emission type of existing nearby air pollution sources: ______________________________
____

58. List quantity and quality of air pollution associated with the proposed project. Identify sources. ____
____

SOLID WASTES

86. Is the site located on an area suitable for sanitary landfill? Yes ____ No ____

87. Is a landfill operated near the site? Yes____No____
If so, who operates landfill site and how far away is it? ______________________________

88. State type, classification and limitations of landfill:
____

89. Is an acceptable contract disposal service available for solid wastes: Yes ____ No ____

90. List quantity and quality of solid wastes associated with construction and operation of the proposed project. Identify sources. ______________________________
____

ENVIRONMENTAL CHECKLIST

NOISE

91. List any regulations regarding community noises: ______

92. Describe any existing background noise: ______

93. Describe type, location and duration of noise sources currently affecting site: ______

94. Locate (use map or sketch) critical noise receptors near site: ______

95. Have noise complaints occurred regarding existing noise? Yes ____ No ____

96. List anticipated noise sources to be associated with proposed project: ______

TRANSPORTATION

97. Identify access routes into the site (use map): ____

98. Describe existing traffic patterns, particularly noting points of congestion (use map): ______

99. Will transportation net support movement of construction equipment, personnel and materials?
    Yes ____ No ____
    If no, explain: ______

100. Will transportation net support operation of the facility? Yes ____ No ____
    If no, explain: ______

101. Will chemical de-icing of roadways be necessary?
    Yes ____ No ____

102. Is dust a problem on roadways?
    Yes____ No ____

103. How will the following be delivered to or removed from the site? Estimate frequency or volume of movements.
    Construction materials ______
    Feedstocks ______
    Misc. chemicals ______
    Maintenance equipment ______
    Products ______
    Waste materials ______
    Spoil materials ______
    Refuse ______
    Sanitary wastes ______

104 Will the project involve marine transport?
    Yes ____ No ____
    If yes, is oil/chemical spill cleanup equipment available locally? Yes ____ No ____
    If yes, specify the name, address and their capacity: ______

105. Will the project involve construction of docking facilities? Yes ____ No ____

106. If marine transport and/or docking facilities are planned, will they interfere with existing navigation practices? Yes ____ No ____

CONSTRUCTION

107. Identify nearby sources of construction material: ______

108. Describe available construction work force: ______

109. Specify extent local construction labor will satisfy demand: ______

110. Describe normal method of sanitary sewage disposal during construction: ______

111. Is open burning permitted? Yes ____ No ____

112. During construction, will the facility use any existing sanitary landfill sites? Yes ____ No ____

113. Will procedures and/or equipment be used which will produce excessive noise, ie. blasting, piledriving, compressors, riveters, large mobile equipment. If so, list types, numbers, frequency and duration of operation. ______

114. Describe method of spoil disposal: ______

115. Describe method of refuse disposal: ______

AESTHETICS

116. Will development of the site create visual/aesthetic disruption? Yes ____ No ____
    If yes, describe best aesthetic configuration of facilities on the site. Attach diagram if necessary. ______

117. List sizes of visually prominent equipment used during operation and construction: ______

118. List and describe sources of light or glare such as general plant lighting and flares: ______

119. Will the completed project be visible from site boundaries? Yes ____ No ____

REGULATORY AGENCIES

120. List names, addresses, telephone numbers and key personnel of lead regulatory agencies:
    Water Pollution: ______
    Air Pollution: ______
    Solid Waste: ______
    Noise: ______
    Aesthetics: ______

Refinery Cyanides: A Regulatory Dilemma

Control objectives are thwarted by inadequate understanding, restrictive sampling requirements and unrefined test methods. But the problems are real and the needs must be met

R. G. Kunz and **J. P. Casey,** Air Products and Chemicals, Inc., Allentown, Pa., and
J. E. Huff, Armak Co., Chicago

REFINERY CYANIDE CONTROL is a regulatory dilemma for both agencies and refiners. Inadequate understanding of refinery cyanide chemistry and inherent limits in control mechanics confuse the problem. Proposed alteratives provide inadequate solutions and acceptable solutions lack adequate technological answers. The ultimate answer is yet to be found. Meanwhile the problem is being "handled."

The petroleum industry uses an average of 18 gpm water per gallon of crude processed.[1] Eighty to ninety percent of it is for indirect cooling, thus uncontaminated except for exchanger and other equipment leaks.[1,2] Resultant petroleum wastewaters are usually high-volume, low-concentration streams containing dissolved and suspended solids, metals, organics identified as chemical and biochemical oxygen demand, phenolics, oil, and inorganics such as ammonia, sulfides and, in certain cases, cyanides.[3]

Nearly all modern refineries release cyanides into receiving streams. However, the Environmental Protection Agency (EPA) has proposed[4] a discharge restriction of key pollutants, among them cyanides.

CYANIDE SPECIES

Cyanide compounds are characterized by a cyano (-C≡N) group. There are three categories of compounds:

- Free—Hydrocyanic acid (HCN) plus the cyanide ion, CN^-. The equilibrium between HCN ($pK_a = 9.21$)[5] and CN^- is a function of pH (Fig. 1).

- Simple—Compounds of the form $A(CN)_x$, where A is an ammonium group, alkali, alkaline earth, or heavy metal. These ionize or hydrolyze in aqueous solutions at low concentrations, and are mostly present as HCN below pH 8.

- Complex—Compounds of the form $A\,[M^{n+}(CN)_x]^{(x-n)-}$, where A is an ammonium group, an alkali, alkaline

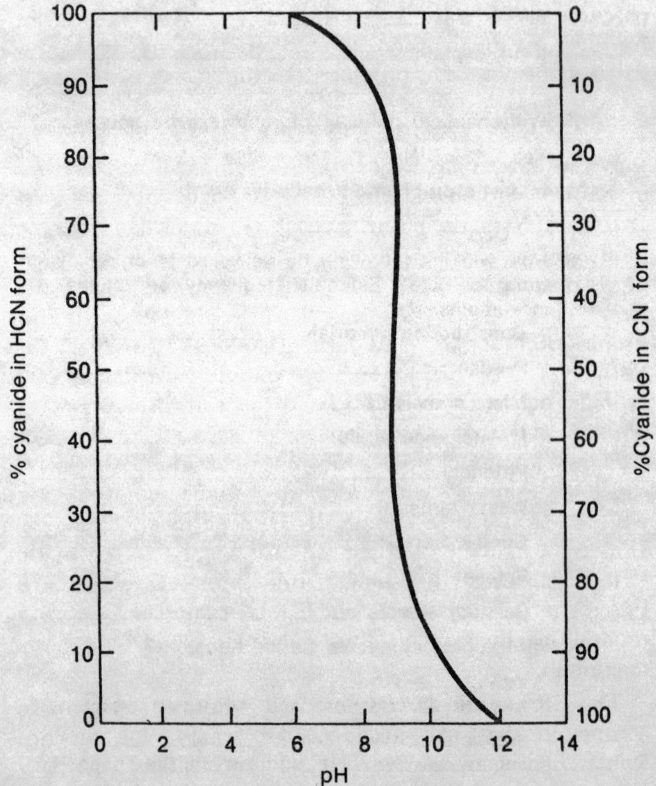

Fig. 1—Effect of pH on HCN dissociation at 25°C.

earth, or heavy metal; M is the central atom of the complex ion. Complex cyanide stability varies from nearly completely ionized compounds ($K_2Cd(CN)_4$ and $K_2Zn(CN)_4$), through an intermediate group ($K_2Cu(CN)_3$ and $K_2Ni(CN)_4$) to compounds where almost none of the cyanide ionizes ($K_4Fe(CN)_6$, $K_3Fe(CN)_6$ and $K_3Co(CN)_6$).

Total cyanide is the sum of all cyanides present.

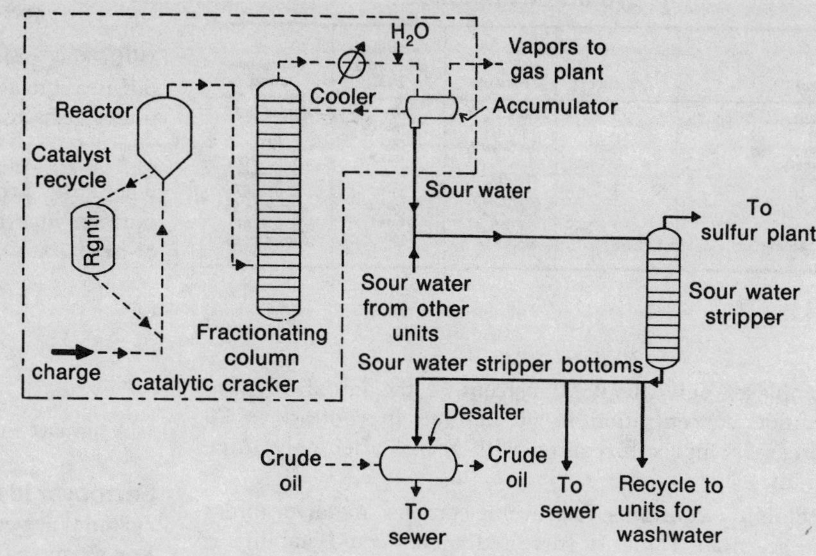

Fig. 2—Cyanide generation and disposal in refinery.

CYANIDE TOXICITY

Cyanide compounds vary in toxicity to aquatic life. Free cyanide is toxic.[6] Hydrocyanic acid shows acute toxicity at 0.25 mg/l to minnows (24 hr. LC_{50}). Chronic toxicity evidenced by failure of minnows to survive 28-day exposure is approximately one-third that level (0.081 to 0.106 mg/l).[8] Adverse growth effects occur between 0.035 and 0.062 mg/l, egg fertility is reduced at 0.044 to 0.073 mg/l, and egg production is decreased at 0.020 mg/l. The maximum acceptable toxicant concentration (MATC), (highest no-effect level), for minnows is estimated at 0.012 to 0.020 mg/l. Spawning data[9] on the more sensitive brook trout show a cyanide level of 0.006 to 0.011 mg/l to be tolerable, indicating a general ratio of 96 hr. LC_{50}: MATC of about 10:1 for trout as well as minnows.

The toxicity of simple and complex cyanides increases with the degree of cyanide ionization, or the instability of complexes as evidenced for $Cd(CN)_4^{-2}$ and $Ni(CN)_4^{-2}$. Toxicity evaluation of the more stable cyanides may be obscured by effects of heavy metals themselves (Ag in $Ag(CN)_2^-$). Furthermore, cyanide toxicity may be aggravated by heavy metal presence as with zinc or cadmium.[10]

CYANIDE REGULATIONS

While most state water quality standards are general in nature (prohibit the presence of toxic substances, or 1/10 to 1/20 of the 96-hr. LC_{50m}), 13 states have specific total cyanide water quality standards. These specific standards range from 0.005 mg/l to 0.2 mg/l. Only one state, New York, currently specifies different standards for simple (0.1 mg/l) and iron cyanides (0.4 mg/l as $Fe(CN)_6$).[11] Several states have established effluent total cyanide standards in addition to the water quality standards. Delaware[12] has adopted an effluent standard of 0.05 mg/l total cyanide. Missouri[4] considers 0.05 mg/l the lower limit to which cyanide can be removed in a wastewater stream economically with current technology. Illinois[14] originally adopted a 0.025 mg/l total cyanide effluent standard; however, a rule change in early 1978 raised the standard to 0.1 mg/l total cyanide based on a monthly average and 0.2 mg/l cyanide based on a daily maximum.

The EPA originally proposed to regulate "any cyanide compound that will produce free cyanide ion or molecular hydrogen cyanide (HCN)" in effluents at a level not to exceed 0.1 mg/1 (measured as CN^-).[4] For locations where the receiving water flow is less than 10 times the effluent flow, the specified maximum effluent concentration was to be 0.01 mg/1.

More recently, several environmental groups sued EPA for failure to implement regulation. An agreement to settle the suit proposes restrictions on 65 toxic pollutants, including "cyanides." Although no standards have been promulgated, the proposed reduction in water quality criteria from 0.005 to 0.001 mg/1,[15] based on total cyanide, indicates increasingly stringent effluent standards.

THE PROBLEM

Part of the organic nitrogen from the crude oil enters in the feed to the catalytic cracker (Fig. 2), where a large amount of this nitrogen is liberated as ammonia and a small amount as cyanide.

All reactor gas goes overhead in the distillation column, including cyanide and ammonia where water is injected into the overhead line for corrosion control. This water is collected in accumulators and is pumped to a stripping column along with other "sour water."

Significant concentrations of total cyanide also are found in the wastewaters from coker units. In most refineries all the cyanide bearing streams are fed to a sour water system.[16]

Many refineries routinely monitor effluent cyanide concentrations of which the total may vary from 0 percent to 100 percent daily. Three refineries reported a long-term average of 55, 61 and 71 percent CN^- in the effluent.[17]

Complex cyanides. The simple cyanide ion (CN^-) readily forms numerous complex anions with many metals such as the copper, nickel, iron and zinc likely to be found in petroleum wastewater. Simple qualitative experiments reported previously[18] indicate that cyanide and ferrous ions begin to form a detectable amount of $[Fe(CN)_6]^{-4}$ immediately on contact even at room temperature. However, rate of complex formation with iron is slower than that of nickel, copper or zinc; iron

TABLE 1—Stability constants of several metal cyanide complexes

Metal	Valence	Formula	Stability constant at 25°C	Literature cited
Cobalt	3	$Co(CN)_6^{-3}$	10^{64}	quoted in 84
Iron	3	$Fe(CN)_6^{-3}$	10^{52}	25
Iron	2	$Fe(CN)_6^{-4}$	10^{47}	25
Mercury	2	$Hg(CN)_4^{-2}$	$10^{41.1}$	5, p. 5-47
Nickel	2	$Ni(CN)_4^{-2}$	$10^{31.3}$	5, p. 5-47
Copper	1	$Cu(CN)_4^{-3}$	$10^{30.3}$	5, p. 5-47
Cadmium	2	$Cd(CN)_4^{-2}$	$10^{16.9}$	quoted in 84
Zinc	2	$Zn(CN)_4^{-2}$	$10^{16.7}$	5, p. 5-47
Manganese	3	$Mn(CN)_6^{-3}$	$10^{9.7}$	quoted in 84

complexes only about 50 percent of the initial 10 ppm cyanide concentration in 60 minutes in contrast to 90 percent complex formation with nickel after only three minutes.[19]

Stability constants for several complex metal cyanides are given in Table 1. Increased metal-ligand stability is indicated by a larger stability constant. Aside from the very stable hexacyanocobaltate complex, the most stable among these listed are the iron complexes, ferricyanide $[Fe(CN)_6]^{-3}$ and ferrocyanide $[Fe(CN)_6]^{-4}$, whose iron component can originate either in the feed or from the carbon steel processing equipment.

The following series of events is believed to lead to the formation of $[Fe(CN)_6]^{-4}$ as a corrosion product.[20-22]

- Hydrogen sulfide reacts with steel to form an iron sulfide scale and atomic hydrogen; in the presence of sulfide, the formation of molecular hydrogen is retarded.

$$H_2S + Fe^0 \rightarrow FeS + 2H^0 \quad (1)$$

- Any hydrogen cyanide present then reacts with the protective ferrous sulfide film to regenerate H_2S.

$$FeS + 2\,HCN \rightarrow Fe^{++} + 2\,CN^- + H_2S \quad (2)$$

With the protective coating removed, atomic hydrogen diffuses into the metal and may result in blister formation under certain conditions.[23]

- Ferrocyanide forms under dilute alkaline conditions where the product of the concentrations of ferrous and cyanide ions is a maximum. An experimental optimum of pH 9 has been noted.[19]

$$Fe(OH)_2 \rightleftarrows Fe^{++} + OH^- \quad pK_{sp} = 15.1 \quad (3)$$

$$HCN \rightleftarrows H^+ + CN^- \quad pK_a = 9.21 \quad (4)$$

$$Fe^{++} + 6\,CN^- \rightarrow [Fe(CN)_6]^{-4} \quad (5)$$

A similar mechanism is proposed in a recent paper.[24]

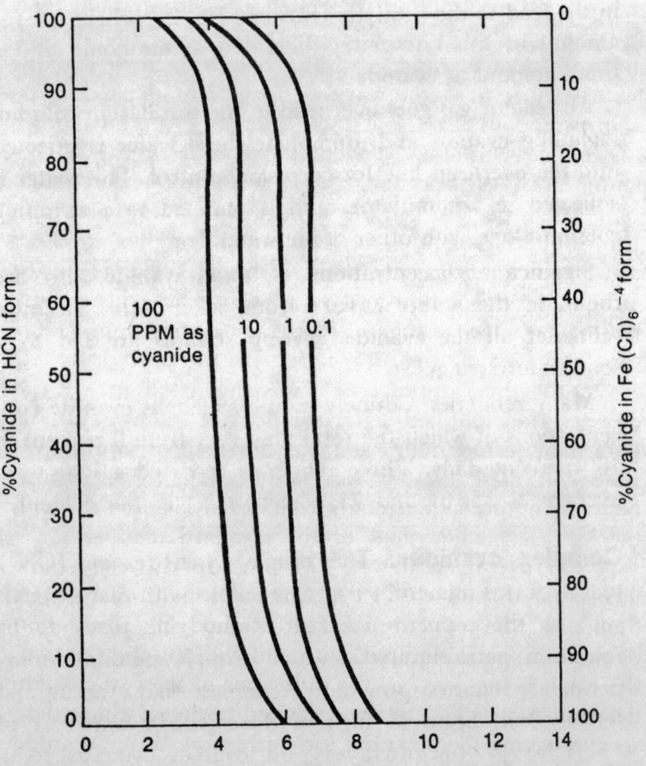

Fig. 3—Effect of pH on ferrocyanide dissociation.

Ferrocyanide complex stability. Once formed, iron cyanides are extremely stable to pH and chemical changes. For example, the dissociation of ferrocyanide can be estimated from the following material balance and equilibrium equations (with all concentrations in moles/liter):

Total cyanide species =

$$[T] = [HCN] + [CN^-] + 6\,[Fe(CN)_6^{-4}] \quad (6)$$

Total iron species =

$$[t] = [Fe^{++}] + [Fe(CN)_6^{-4}] \quad (7)$$

The initial condition: $[T] = 6\,[t]$ (8)

Equilibrium conditions:[25,5]

$$[Fe(CN)_6^{-4}] / [Fe^{++}]\,[CN^-]^6 = 10^{47} \quad (9)$$

$$[H^+][CN^-]/[HCN] = 6.2 \times 10^{-10} \quad (10)$$

Distribution of cyanide species is depicted as a function of pH (Fig. 3).

Given sufficient time, the initial ferrocyanide concentration (mg/1 expressed as cyanide) will equilibrate between residual ferrocyanide, volatile HCN, and less than 0.0025 mg/1 of cyanide ion. Based on these calculations, at equilibrium there is virtually no dissociation for ferrocyanide at the 100 mg/1 cyanide level above a pH of about 6, but the degree of dissociation increases with decreasing initial ferrocyanide concentration.

However, upon exposure to sunlight, ferrocyanide dissociates more extensively than indicated (Fig. 3). These were calculated using a stability constant,[25] based on long-term equilibration of test solutions in the dark. Ferrocyanide breakdown has been noted under the action of sunlight in a natural body of water[26,27]; however, the breakdown rate may be a function of U.V. absorption properties. The same effect can be produced in the laboratory by ultraviolet light.[28] Clear ferrocyanide solutions turn pale yellow on standing.[29] The actual breakdown mechanism in all these cases may involve the ferroaquapentacyanide ion as an intermediate.[29]

$$Fe(CN)_6^{-4} + H_2O \underset{\text{dark}}{\overset{\text{light}}{\rightleftarrows}} Fe(CN)_5H_2O^{-3} + CN^- \quad (11)$$

Cyanide-induced corrosion. Severe cyanide-promoted corrosion can occur in compressor interstage coolers, debutanizer overhead condenser shells, fractionator trays and absorber bottoms.[30] Specifically, one refiner uses ferrocyanide in water drawoffs to monitor corrosion and hydrogen activity.[31] Another found deposits of ferric ferrocyanide $Fe_4\,[Fe(CN)_6]_3$ (Prussian Blue) in distillation columns of a gas-concentration unit.[32]

In laboratory tests, ferrocyanide and other corrosion products were identified in the stripper bottoms when potassium cyanide was present in the synthetic sour water feed.[24] The presence of complex cyanide was suggested by a 1972 API study on refinery sour water strippers[33] which found that the average cyanide removal was only 37 percent by stripping—much less than would be predicted if all the cyanide were simple. Commercial experience with severe corrosion in sour water strippers containing cyanide is summarized in API Publication 944.[34, 35]

Corrosion inhibited with polysulfide. To mitigate cyanide corrosion, sodium and ammonium polysulfide have been used as corrosion inhibitors in some catalytic cracking systems.[22] The polysulfide combines with cyanide, forming thiocyanate.[36]

$$CN^- + S_x^= \rightarrow SCN^- + S_{x-1}^= \qquad (11)$$

While some refineries have found that polysulfide injection in the fluid catalytic cracking (FCC) unit successfully converted over 90 percent of the cyanide ions, severe fouling and plugging resulted in the sour water strippers.[37] Removal of 90 percent of the cyanide via polysulfide injection did not result in a corresponding reduction of cyanide in the effluent. A limited amount of data[16] may correlate with the lack of success of polysulfide experienced by some refineries. Linear regression analysis[17] of these data showed no correlation ($R = 0.000$) between the inlet and outlet cyanide concentrations from the sour water stripper. The lack of correlation is evident (Fig. 4). This result suggests that cyanide removal via stripping may result in an irreducible minimum concentration, independent of the inlet cyanide level. If this non-strippable cyanide is due to corrosion product cyanide complexes, this would suggest that polysulfide injection or other in-plant control processes (i.e., catalytic oxidation of the cyanide gases in the FCC unit overhead) would have to be extremely efficient (> 95 percent) before any significant reduction in effluent cyanide were realized.

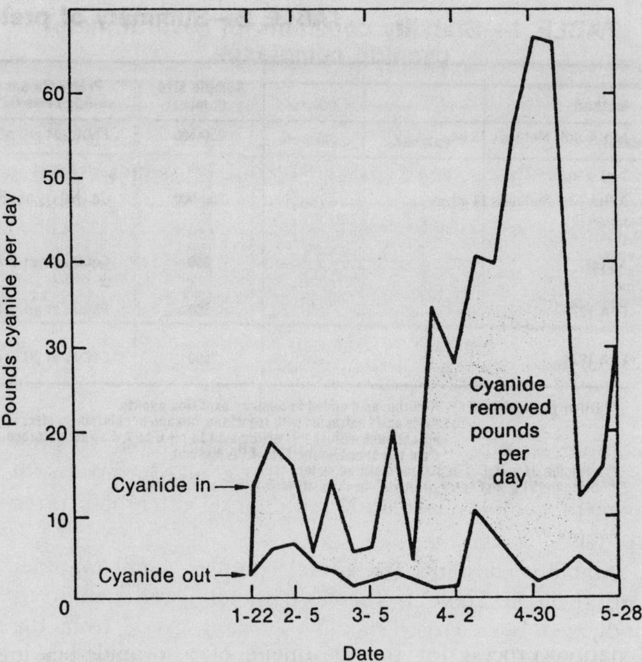

Source: Prather and Berkemeyer, 1975.

Fig. 4—Sour water stripper cyanide removal.

ANALYTICAL PROBLEMS

Since the EPA's proposed regulations in 1973 to limit compounds producing cyanide ion or molecular hydrogen cyanide in plant effluents,[4] there has been been much activity to identify the preferred analytical procedure which will isolate only "simple" cyanide. Analytically, "simple" cyanide includes free, simple and weakly complexed cyanides, as defined above.

Unfortunately, the methods[38–40] specified by EPA[41] and subsequent modifications[42–44] fail to differentiate between "simple" and strongly complexed cyanides. Available methods which do make the distinction are subject to interferences by materials found in refinery waste streams. Consequently, regulations with the stated intent to limit only CN^- and HCN in the effluent in effect ban all cyanide species when such non-specific analytical methods are prescribed. The problem is, therefore, how to identify and eliminate complex cyanides which would be measured as free cyanide.

Among the numerous ways to analyze for cyanide,[45] those which have received the most attention in recent years are variations of the Serfass procedure[38, 39] and techniques based on the Williams method.[40, 42, 43] The Standard Methods 14th Edition[44] procedure appears to be a cross between the two, with supplementary introduction of ion electrode techniques.

To eliminate interferences in the ultimate cyanide titrimetric, ion electrode or colorimetric determination, these procedures call for sample pretreatment to remove sulfides, fatty acids and oxidizing agents before refluxing with sulfuric acid and various metal salts depending on the method (Table 2). The salts catalyze the breakdown of complex cyanides; the preferred salt is $MgCl_2$ because $HgCl_2$ is toxic and Cu_2Cl_2 is subject to air oxidation to cupric ion, causing a negative interference.[46] In the distillation step, both simple and complex cyanides are converted into volatile hydrogen cyanide to be stripped from solution and scrubbed by aqueous sodium hydroxide, which is then analyzed for cyanide ion. Reflux time and type of apparatus vary with the procedure.

One of these approaches[38] was evaluated in a series of tests using distilled water solutions of sodium cyanide, potassium ferrocyanide, and varying reagent additions.[18] Based on these tests, it was not possible to distinguish between simple and strongly complexed cyanide through mechanical modifications of the standard method.

A variation on the ASTM procedure[40, 42] and now incorporated into the EPA Method[43] and the APHA Standard Methods[44] is ASTM Method B. Distillation of two samples is required, with and without chlorination, the result being given by the difference of the two cyanide determinations. This test was devised to detect all cyanide species having the potential for conversion to free cyanide simply by dilution or pH adjustment. Its principal use is said to determine the performance of an alkaline chlori-

TABLE 2—Summary of pretreatment and reflux procedures

Method	Sample size, ml	Pretreatment for sulfide removal* †††	Salts added to distillation flask	Reflux procedure
APHA Std. Methods 13 ed.	250-500	$PbCO_3$ at pH >11	$HgCl_2$ 20 ml** $MgCl_2.6H_2O$ 10 ml***	1 hr. reflux after boiling commences. Stop heat but continue airflow. Cool for 15 min. Analyze. Repeat reflux as necessary.
APHA Std. Methods 14 ed.	250-500	Cd $(NO_3)_2$ at pH >12	Cu_2Cl_2 10 ml† or $MgCl_2.6H_2O$ 20 ml*** (option when SCN present)	Same as above††
ASTM	500	$CdCO_3$ at pH >11	Cu_2Cl_2 10 ml†	2 hr. reflux after boiling. Cool for 15 min. Analyze. Repeat reflux as necessary.††
EPA 1971	500	$PbCO_3$ at pH >11	$HgCl_2$** 10 ml $MgCl_2.6H_2O$*** 40 ml	1 hr. reflux after boiling. Stop heat but continue airflow. Cool for 15 min. No repeat reflux called for.
EPA 1974	500	$CdCO_3$ at pH >12	Cu_2Cl_2 10 ml†	Same as above††

* Other pretreatment (a) Ascorbic acid added to destroy oxidizing agents.
(b) Fatty acids extracted with isooctane, hexane or chloroform after pH adjustment with (1+9) acetic acid to pH 6 to 7. Extraction procedure not specified for 1971 EPA Method.

** Dissolve 34 g $HgCl_2$ in 500 ml distilled water.

*** Dissolve 51 g $MgCl_2.6H_2O$ in 100 ml distilled water (510 g in 1 liter).

† Dissolve 20 g Cu_2Cl_2 by washing with 2 250-ml portions of (1+49) H_2SO_4 then twice with water, add 250 ml water, and conc. HCl until salt dissolves. Dilute to 1 liter. Store in tightly stoppered bottle with pure copper wire or rod extending from the bottom to the mouth of the bottle.

†† For "Cyanides Amenable to Chlorination," perform test twice on samples before then after chlorination.

††† For highly alkaline samples a suggested modification is CO_2 removal by $CaCl_2$ addition to precipitate $CaCO_3$ followed by filtration. Then cadmium acetate is added to precipitate CdS.

nation process for the treatment of a cyanide-bearing waste.[42] This procedure actually measures all free and simple cyanides completely, some complexes partially, and the iron or cobalt complexes not at all. Destruction by chlorination of cyanide complexed by certain metals such as nickel is strongly time-dependent,[42,44] making the desired analytical result a function of complex type in the sample even at constant reaction time. For samples consisting primarily of complexed cyanide, the method suffers in accuracy because the easily chlorinated species are determined by difference often between two imprecise large numbers. Under these conditions, it is not unusual to obtain a negative number for "amenable cyanides."

Thiocyanate interference. In determining cyanide concentrations well below 1 mg/1, interferences not previously noted have come to light. One of these is thiocyanate (SCN^-) which is present in many refinery streams. Thiocyanate is detected as cyanide in the colorimetric method specified for analysis of solutions containing such low cyanide concentrations. It was previously felt that this interference is eliminated in the distillation step.[47] Several laboratories have indicated a positive interference of as much as 100 percent at high SCN^- to CN^- ratios.[18] Recent quantitative work[46] showed a 1:20 equivalent interference as a 10 mg/1 CN^- readout was obtained for 200 mg/1 SCN^- taken through the distillation procedure.

Several explanations which have been advanced for this phenomenon include HSCN ($pK_a = 0.85$) codistillation from strong acid, entrainment of SCN^- at high reflux rate and decomposition of SCN^- into cyanide and sulfide in the presence of oxidizable Cu_2Cl_2. A thiocyanate interference with SCN^- at concentrations greater than 0.5 mg/1 is reported[44] when the distillation is carried out with cuprous chloride. Substitution of magnesium chloride is recommended. An additional sulfide-induced interference may be H_2S/HCN volatilization followed by sulfide oxidation to $S_x^=$ and SCN^- formation in the caustic scrubber medium. Sample pretreatment, preferably with cadmium acetate to induce CdS precipitation, will remove sulfide prior to acidification and distillation.

Results of testing steel industry wastes plus prepared solutions were reported in 1975.[48,49] These wastes contain relatively low cyanide concentrations in the presence of high thiocyanate. The 1971 EPA Method[39] gave the best recoveries of total cyanide without distinction between the free and complex forms. The ASTM procedures[40,42] gave high results for total cyanide because of thiocyanate interference and negative values for cyanides amenable to chlorination. Reference was made to similar negative results on steel plant effluents, refinery wastes and synthetic samples.

Similar results showing the EPA Method[38] to be somewhat more accurate for total cyanide than the ASTM for refinery effluents were reported by Shell Oil in hearings before the Illinois Pollution Control Board.[50] A significant thiocyanate interference in both methods was found by Shell Oil when measuring distilled water solutions containing sodium hydroxide and sodium thiocyanate.

Recent experimental modifications[46] on coke oven wastes eliminated a positive thiocyanate-induced sulfide interference by precipitating CdS in the scrubber solution before CN^- measurement.

Modified Roberts/Jackson Method. A method which overcomes all of these problems, by Roberts and Jackson,[51,52] determines small amounts of cyanide ion in the presence of ferrocyanide by distillation of a sample with zinc acetate under reduced pressure. Ferrocyanide is not decomposed in the process.

The procedure was adapted to the equipment of the 1971 EPA method[39] and is now known as the Wood River Modification[50] of the Roberts and Jackson Method or simply the Modified Roberts/Jackson Method. It is currently being tested before publication as ASTM Method F.

The procedure as originally published and unchanged in the modification calls for neutralization of the sample to the phenolphthalein end point by HCl, addition of one extra milliequivalent of the acid plus 10 cc. of 10 percent zinc acetate solution and refluxing for one hour. The pH in the stillpot varied among several different samples from 5.4 to 7. The final pH, where measured, was within this range. Pyridene-barbituric acid is used instead of pyridine-pyrazolone for color development.

Over a two-year period, the modified Roberts/Jackson

Method was used to analyze the final effluent from Shell's Wood River (Ill.) refinery. Consistent values were obtained at the $\mu g/l$ level, showing approximately ⅔ of the total cyanide as determined by the EPA 1971 method[34] to be simple cyanide. The precision of a single sample was about 4 percent ($1 \mu g/l$ standard deviation from eight determinations whose average was $26 \mu g/l$).

Consistent values were also obtained for known blends or spiked samples, for samples in an EPA round-robin evaluation of methods, and for samples in a cooperative study of several refineries by the Illinois Petroleum Council (IPC). The Shell data from the IPC study show a ratio of simple to total cyanide similar to the ratio from their own refinery.

Nickel and copper complexes were reported to be decomposed when using the zinc acetate reagent.[50] This agrees with another study which found that several complexes including those of nickel and zinc were completely recovered as simple cyanide.[49] The Illinois refineries found that a Modified Roberts/Jackson Method was essentially free of interference from thiocyanates and more accurate and reproducible than the methods evaluated for total cyanides.[50]

In another study,[53] the Roberts and Jackson Method among others showed no interferences from thiocyanate or ferricyanide in prepared solutions. All described methods indicate that the refinery effluent tested contained primarily strongly complexed cyanide. This refinery effluent showed a low toxicity to fish. When KCN equal to the total cyanide determined was added, the fish died rapidly.

A recent round-robin study of several test procedures by the ASTM Cyanide Task Group found the Modified Roberts/Jackson Method to be the best of the methods tested. However, the method is not without its shortcomings: the pH changes during distillation and ultraviolet light causes breakdown of ferrocyanide. The pH problem has been solved by using a buffer system. Care must be taken in handling samples in the presence of laboratory fluorescent lights or other U.V. sources.[54]

The Roberts and Jackson method apparently comes closest to the original EPA guideline to identify and restrict only free cyanide in refinery effluent streams. Along with measuring all of the simple cyanides which completely ionize to CN^- at dilute concentrations, this method also recovers the labile complex cyanides.

CYANIDE REMOVAL

Most existing cyanide treatment methods are effective in removing simple cyanides. However, refinery wastes contain both simple and complex cyanides with the percent complex highly variable with respect to time.[55] Thus, any proposed treatment scheme must be capable of removing more than just simple cyanides from refinery wastewaters.

Common treatment methods for cyanide include

- Chemical oxidation
- Biotreatment
- Adsorption and catalytic oxidation
- Ion exchange
- Precipitation.

Chemical oxidation. Oxidants include chlorine, hydrogen peroxide and ozone. Alkaline chlorination[56-59] is very common both in the electroplating industry and in steel industry coke plants, where the staged oxidation of cyanide is practiced:

$$CN^- + HOCl \rightarrow CNCl + OH^- \quad (12)$$

$$CNCl + 2\,OH^- \rightarrow CNO^- + Cl^- + H_2O \quad (13)$$

$$2\,CNO^- + 3\,OCl^- + H_2O \rightarrow 2\,CO_2 + N_2 + 3\,Cl^- + 2\,OH^- \quad (14)$$

For complex cyanides, elevated temperatures, longer holding times and more oxidant are required.[60-65] The reaction of chlorine at ordinary temperatures with ferrocyanide stops at ferricyanide in which only the oxidation state of the iron has been changed, while leaving the accompanying cyanide ligands intact.[56,64] Although noncatalytic ferrocyanide destruction by chlorination at elevated temperature and long residence time in a batch system[60-62] has been claimed, both this and a similar batch chlorination of ferrocyanide in the presence of a soluble mercury catalyst are not practical for refinery effluents.

Simple cyanide can be destroyed by hydrogen peroxide.[65] As with chlorine, the reaction with simple cyanide proceeds in stages but the reaction rate is apparently slower and a copper salt is sometimes used as a catalyst.[66] Stoichiometric hydroperoxide oxidation of hydrocyanic acid to cyanate is maximally efficient at pH 10.4, the midpoint of acid pK_a's (9.21 for NCH, 11.65 for H_2O_2)[5]:

$$HCN + {}^-OOH \rightarrow CNO^- + H_2O \quad (15)$$

Serial hydrolysis to ammonia and CO_2 is best carried out at lower pH:

$$CNO^- + 2H_2O \rightarrow HCO_3^- + NH_3 \quad (16)$$

One proprietary variant[67] involves formaldehyde use and intermediate cyanohydrin formation in treating simple and weakly complexed cyanide. Hydrogen peroxide oxidizes ferrocyanide to ferricyanide in acid solution;[68] in basic solution ferricyanide is reduced to ferrocyanide.[64]

Ozone treatment[69-72] may become preferred over chlorination because oxidation to cyanate is not accompanied by side reactions generating toxic chlorinated hydrocarbons.[73] Competitive reactions do occur, however, generating oxidized organic intermediates. Simple cyanide and copper cyanide complexes are easily oxidized by ozone, but a residual of 4 mg/l of cyanide complexed with iron resisted all attempts at further oxidation.[74] The ferrocyanide is easily oxidized to ferricyanide, but no significant further oxidation by ozone is observed within a reasonable time.[75] Ozone oxidation in the presence of ultraviolet radiation has been demonstrated in laboratory investigations to decompose ferrocyanide.[75-77] Whether ultraviolet light in combination with oxidizing agents weaker than ozone can destroy ferrocyanide sufficiently fast to be of practical value remains to be seen.

One recent study[63] investigated ozone plus U.V. treatment of stripped sour water for cyanide removal. The U.V. light was found to increase the rate of oxidation of cyanide in this study. However, the presence of high concentrations of organics in the waste stream limited the practicability of ozone, with or without U.V. radiation in waste streams containing high organic levels. The competitive oxidation rates of the organics resulted in excessive chemical consumption (3,500 lb. O_3/lb. cyanide destroyed) in order to destroy the cyanide to acceptable levels.[63] Competitive oxidation can be expected with the other oxidants as well. Thus, chemical oxidation has limited potential in refinery wastewaters unless most organics are first removed.

Biotreatment. Biological wastewater treatment facilities remove cyanide by air stripping, chemical condensations, bio-adsorption and bio-oxidation.[79] Low concentrations of simple cyanides can be oxidized by acclimated organisms[80] to carbon dioxide and ammonia in activated sludge but become toxic as the cyanide concentration is increased.[81,82] Ferrocyanide, in contrast, exhibits different behavior.[83] Up to 500 mg/l of ferrocyanide from photographic processing solutions was observed to pass through indoor bench activated sludge units without adverse effects on the biomass but also with little or no ferrocyanide breakdown. In tests conducted outdoors, where decomposition to free cyanide in sunlight was possible, 10 to 40 mg/l of ferrocyanide was passed through a 20,000-gpd pilot plant on a daily basis for over 2½ years with no apparent problems. Sewage containing a mixture of simple and complex cyanides at the μg/l level showed a decrease in both simple and complex cyanides through several treatment plants with a higher degree of removal reported for the simple cyanides.[84]

Adsorption and catalytic oxidation. Activated carbon has been investigated both as an adsorbent and as a catalyst for the detoxification of cyanide wastes. Passing potassium cyanide-bearing wastes over coal or coke resulted in 98 percent removal.[85] High efficiency is believed to result from catalytic oxidation as well as adsorption. Furthermore, removal of nickel and iron-cyanide complexes was attributable to only adsorption and not to any oxidation.

Activated carbon utilized as a catalyst for cyanide oxidation is patented.[86] The process requires an alkaline pH with a mixture of air and the cyanide-bearing waste pumped in a pulsating manner through a bed of activated carbon. The type of cyanide effectively removed is not identified, but is probably only simple cyanides.

Calgon developed a similar method,[87] except that cupric ions are added to the wastewater along with oxygen prior to passing the cyanide-bearing waste through the granular activated carbon columns. In addition to improving the catalytic oxidation of the cyanide, "the presence of cupric ions results in the formation of copper cyanides, which have a greater adsorption capacity than copper or cyanide alone"[87] increasing carbon capacity from 2 to 3 mg CN^-/gm carbon to 25 mg CN^-/gm carbon without any dissolved oxygen present. With oxygen maintained in the system, the active sites were continuously regenerated by cyanide oxidation. Tests with 100 mg/l cyanide in the feed resulted in cyanide effluent levels of less than 0.05 mg/l and low effluent copper levels as long as the adsorptive capacity of the copper on the carbon was not exceeded.[87] Iron cyanides can also be oxidized, but at a slower rate than the copper cyanides.

In a noncatalytic adsorption study[19] activated carbon did not adsorb any appreciable amount of simple cyanide but was effective in removing nickelo- and ferrocyanides from synthetic solutions. Carbon removed all of 100 mg/l of CN^- complexed with nickel at high dosages of 30 g/liter or greater; 95 percent of cyanide in the ferrocyanide form was adsorbed in continuous column tests. However, it can be inferred from Calgon work[87] that the ferrocyanide complex is adsorbed to a lesser extent under more realistic conditions.

Similar high removal efficiency was obtained[88] for the nickel cyanide complex at a loading of but 1 to 10 mg [Ni]/gm carbon. Ferrocyanide was adsorbed at a modest 2 to 7 mg $[Fe(CN)_6]^{-4}$ mg/gm carbon corresponding to a range of 23 to 99 percent removal in batch isotherm tests using pure potassium ferrocyanide solutions. In an actual petrochemical wastewater test, no ferrocyanide was adsorbed. One refinery has investigated adsorption with catalytic oxidation and concluded[37] that pH control and equalization are required and that 51,000 pounds per day of carbon would be exhausted in the removal of 6.3 pounds per day of cyanide.

Another adsorption/oxidation variant has been investigated in a laboratory study.[89] Here powdered activated carbon (PAC) and cupric chloride were added directly to an activated sludge aeration basin. The reported advantages of this approach were very low cost for cyanide removal, improved performance of the activated sludge unit and the extreme flexibility of the process. The PAC/$CuCl_2$ process was reported best suited for refineries requiring less than a 60 percent reduction in effluent cyanide levels, and especially refineries with low organic loadings entering the aeration basin because of competitive adsorption between the organics and cyanide.

Ion exchange. Metal-cyanide complexes are successfully removed from plating wastes using ion exchange.[90,91] However, elution of cyanide complexes from the strong base anion resins employed is difficult at best or impossible, resulting in a continual loss of capacity through repeated cycles.[90-94] This problem is apparently overcome by use of weak base anion resins.[95] Laboratory experiments show removal of ferrocyanide from synthetic solutions and an industrial wastewater to well below 1 mg/l as cyanide.[96,97]

A three-bed commercial ion exchange is installed for removing complexed cyanide in the metal finishing industry using the sequence: strong acid solution, weak base anion, strong base anion.[92] As a rule, however, anion exchange resins are also susceptible to severe fouling by insoluble oil and dissolved organic compounds,[98] especially (aromatic) carboxylic acids.[99] Ion exchange for the removal of cyanide has limited application in the refining industry because of the organics present in the wastewaters. Fouling and rapid deterioration of the resin preclude the use of ion exchange unless extensive pretreatment of the organics is practiced.[100] Furthermore, even a successful ion exchange system will merely concentrate the mass of complexed cyanide in a smaller volume regeneration stream whose disposal must still be addressed.

Precipitation. A final alternative treatment method is precipitation which is limited to concentrated cyanide streams. This limitation is due to the solubility of the metal cyanides formed during the precipitation reactions which is approximately 4 mg/l as CN^{-}[100] Laboratory precipitation of the iron complex by adding other metal ions such as zinc and copper has reduced iron complex concentration from 750 ppm to about 1 to 2 ppm in the best cases.[101] Therefore, precipitation alone will not lower the cyanide content below the EPA guideline of 0.1 mg/l and deliberate addition of precipitating agents is not recommended[42] because of the toxic sludge that is produced.

PROSPECTS

The establishment of cost-effective effluent and water quality standards for cyanides is a difficult task. There is little question that the free cyanide form is extremely toxic, especially compared to compounds such as ferro- and ferricyanides. One can make an excellent case for standards limited to simple cyanides. Based upon the research of the last few years, accurate analysis of the simple cyanides is now readily achieved.

However, there are still several issues that must be resolved before regulatory agencies can accept this approach. Sample preservation for simple cyanide analysis has not been adequately addressed. Can all simple cyanides be adequately preserved for 24 hours or more? Many regulatory agencies require a minimum of 24 hours between sample collection and analysis of routine samples. Industrial discharges can easily begin analyzing in less than one hour from sampling, but this is just not possible for regulatory agencies.

Another concern with a standard limited to simple cyanides is the fate of the complex cyanides in receiving streams. What fraction of the iron cyanides undergo photochemical decomposition producing free cyanide, and what effect do variables such as clarity, sunlight intensity, flow rate and other pollutants have on the rate of photodecomposition?

While there has been speculation that streams receiving large amounts of the iron-cyanide complexes will not generate enough free cyanide to be lethal to fish,[6,53] actual stream data are severely lacking. The regulatory agencies need to be assured that the rate of free cyanide formation is indeed slow enough under existing stream conditions such that lethal levels of free cyanide are not attained.

There are indeed areas where additional research is needed. In the meantime, maintaining stringent standards on total cyanide is not totally desirable. While treatment of simple cyanide is well established in the electroplating and steel industries, treatment of complex cyanide is more difficult and not established in any industry. The shortcomings in regulating simple cyanides (preservation and fate) do not justify the adoption of uniform effluent standards for total cyanide based upon the known toxicity of free cyanides. There is a dilemma facing both regulatory agencies and refineries.

Continued research on sample preservation for simple cyanides and the fate of complex cyanides in receiving streams may provide additional insight into this dilemma. When such additional data are obtained, then and only then can a rational basis for cyanide standards be taken.

ACKNOWLEDGMENT

Presented at the Third Annual Conference on Treatment and Disposal of Industrial Wastewaters and Residues, Houston, April 18-20, 1978.

LITERATURE CITED

1 Nemerow, N. L., "Liquid waste of industry—theories, practices, and treatment, p. 436 Addison-Wesley, Reading, Mass, 1971.

2 "Manual on disposal of refinery wastes—volume on liquid wastes," 1 ed., pp. 1-3, American Petroleum Institute, Washington, D.C., 1969.

3 Beychok, M. R., "Aqueous wastes from petroleum and petrochemical plants," pp. 3-24, Wiley, London, 1967.

4 "Environmental protection agency water programs proposed toxic pollutant effluent standards," *Federal Register*, 40 CFR 129; 38 FR 35388, par. 129.05-129.05C, Dec. 27, 1973.

5 Dean, John A., editor, *Lange's Handbook of Chemistry*, 11 ed., McGraw Hill, New York, 1973.

6 Doudoroff, P., "Toxicity to fish of cyanides and related compounds—A review," EPA-600/3-76-038, US EPA Envir. Research Lab., Duluth, Minn., PB-253 528, p. 133, NTIS, Springfield, Va., April 1976.

7 Doudoroff, P., "Some experiments on the toxicity of complex cyanides to fish," *Sew. Ind. Wastes*, Vol 28, No. 8, pp. 1020-1040, August 1956.

8 Lind, D. T., L. L. Smith, Jr., and S. J. Broderius, "Chronic effects of hydrogen cyanide on the fathead minnow," *J. Water Poll. Cont. Fed.*, Vol. 49, pp. 262-268, 1977.

9 Koenst, W. M., L. L. Smith, Jr., and S. J. Broderius, "Effect of chronic exposure of brook trout to sublethal concentrations of hydrogen cyanide," *Env. Sci. Technol.*, Vol. 11, No. 9, pp. 883-887, 1977.

10 Doudoroff, P., G. Leduc, and C. R. Schneider, "Acute toxicity to fish of solutions containing complex metal cyanides, in relation to concentrations of molecular hydrocyanic acid," *Trans. Amer. Fisheries Soc.*, Vol. 95, pp. 6-22, 1966.

11 New York State classifications and standards, official compilation of codes, rules and regulations," Title 6, Parts 700-702, Amended Feb. 21, 1974; Sept. 20, 1974.

12 Delaware Water and Air Resources Commission, Delaware Water Pollution Control Regulations, March 15, 1974.

13 Missouri Code of State Regulations, Title 10, Department of Natural Resources, Division 20, Clean Water Commission, Chapter 7, Water Quality, Appendix I, June 29, 1974, amended April 11, 1975.

14 Illinois Pollution Control Board, Water Pollution Regulations of Illinois, Rule 408, Jan. 6, 1972.

15 U.S. EPA *Quality Criteria for Water*, pp. 65-69, July 1976 and toxic pollutant water quality criteria proposed draft documents.

16 Prather, B. V. and R. Berkemeyer, "Cyanide sources in petroleum refineries," 30th Annual Purdue Industrial Waste Conference, West Lafayette, Ind. May 6-8, 1975.

17 Huff, L. L. and J. E. Huff, "Analysis of the benefits and costs of alternative cyanide standards in Illinois," Illinois Institute for Environmental Quality Document No. 75-24, November 1975.

18 Kunz, R. G., R. R. Lessard, and P. K. Starnes, " 'Free' cyanide—dilemma for refiners," *Proc. API, Refining, 40th Midyear Meeting*, pp. 69-82, American Petroleum Institute, Washington, D.C., 1975.

19 Reed, A. K., *et al*, "An investigation of techniques for removal of cyanide from electroplating wastes, pp. 40-48, 12010 EIE 11/71, U.S. Environmental Protection Agency, Washington, D.C., November 1971.

20 Geerlins, J. G. and J. C. Jongebreur, "Corrosion in oil refinery equipment," First International Congress on Metallic Corrosion, London, April 10-15, 1961, pp. 573-584, Butterworths, London, 1962.

21 Gutzeit, J., "Corrosion of steel by sulfides and cyanides in refinery condensate water," *Materials Protection*, Vol. 7, No. 12, pp. 17-23, December 1968.

22 French, E. C., "Water-soluble inhibitors can control severe corrosion rates in FCC units," *Oil Gas J.*, Vol. 74, No. 15, pp. 62-64, April 12, 1976.

23 1972 NPRA Question and Answer Session on Refining Technology, Petroleum Publishing Co., Tulsa, Okla., 1973.

24 Berry, W. E., E. L. White, and W. K. Boyd, "A study of variables that affect the corrosion of sour water strippers," Preprint No. 45-76, API Refining Department 41st Midyear Meeting, Los Angeles, Calif., May 13, 1976.

25 Broderius, S. J., "Determination of molecular hydrocyanic acid in water and studies of the chemistry and toxicity to fish of metal-cyanide complexes," Ph.D. Thesis, Oregon State University, Corvallis, Ore., University Microfilms 73-21, 229, Ann Arbor, Mich., 1973.

26 Burdick, G. E. and M. Lipschuetz, "Toxicity of ferro and ferricyanide solutions to fish," *Trans. Amer. Fisheries Soc.*, Vol. 8, p. 192, 1942.

27 McKee, J. E. and H. W. Wolf, editors, *Water Quality Criteria*, 2 ed., pp. 245, 246, 267, California State Water Quality Control Board Publication 3-A, Sacramento, Calif., 1963.

28 Kruse, J. M., and L. E. Thibault, "Determination of free cyanide in ferro- and ferricyanides," *Anal.Chem.*, Vol. 45, No. 13, pp. 2260-2261 1973.

29 American Cyanamid Co. *The Chemistry of the Ferrocyanides*, p. 33, Beacon Press, Inc., New York, 1953.

30 Congram, G. E., "Refiners squelch corrosion in cat gas recovery plants," *Oil Gas J.*, Vol. 73, No. 19, pp. 104-105, May 12, 1975.

31 1972 NPRA Question and Answer Session on Refining Technology, pp. 120-121, Petroleum Publishing Co., Tulsa, Okla., 1973.

32 Anon., "NPRA Q & A-3, Refiners discuss FCCU hardware," *Oil Gas J.*, Vol. 74, No. 15, pp. 83-92, April 12, 1976.

33 "Report WBWC 3064, 1972 sour water stripping evaluation," pp. 6, 20-28, American Petroleum Institute Publication 927, Washington, D.C., June 1973.

34 "1972 survey of materials experience and corrosion problems in sour water strippers," American Petroleum Institute Publication 944, Washington, D.C., November 1974.

35 Foroulis, Z. A., and N. C. Carter, "Refiners analyze corrosion in sour water stripper units," *Oil Gas J.*, Vol. 73, No. 31, pp. 82-83, Aug. 4, 1975.

36 McCoy, J. W., "Chemical analysis of industrial water," pp. 125-134, Chemical Publishing Co., New York, 1969.

37 Bernickas, J. V., Testimony presented in Regulatory Hearing R-74-15 and 16, before the Illinois Pollution Control Board, 1975.

38 "Standard methods for the examination of water and wastewater," 13 ed., pp. 397-403, American Public Health Assn., Washington D.C., 1971.

39 EPA Water Quality Office, "Methods for chemical analysis of water and wastes, 1971," Cincinnati, Ohio, 1971.

40 *Annual Book of ASTM Standards*, Part 23 Water, Standard D 2036-72 American Society for Testing and Materials, Philadelphia, Pa., 1972.

41 "Environmental Protection Agency Regulations on Test Procedures for the Analysis of Pollutants," *Federal Register*, 40 CFR 136; 38 FR 28758, par. 136.1-136.2 and Table I, Oct. 16, 1973.

[42] Annual Book of ASTM Standards, Part 31 Water, Standard D2036-74, Appendix, American Society for Testing and Materials, Philadelphia, Pa., 1974.
[43] EPA National Environmental Research Center, "Methods for chemical analysis of water and wastes," EPA-625-16-74-003, pp. 40-50, U.S. EPA Office of Technology Transfer, Washington, D.C., 1974.
[44] "Standard methods for the examination of water and wastewater," 14 ed., pp. 361-387, American Public Health Association, Washington, D.C. 1976.
[45] Crossey, M., "Bibliography—methods for the determination of cyanide and thiocyanate 1954-1975," Unpublished report NUS Corporation Cyrus Wm. Rice Division, Pittsburgh, Pa., July 1976.
[46] Barton, P. J., C. A. Hammer and D. C. Kennedy, "Analysis of cyanides in coke plant wastewater effluents," *J. Water Poll. Cont. Fed.*, Vol. 50, pp. 234-239, 1978.
[47] Ludzak, F. J., W. A. Moore and C. C. Ruchhoft, 'Determination of cyanides in water and waste samples," *Anal. Chem.*, Vol. 26, No. 11, pp. 1784-1792, Nov. 1954.
[48] Gonter, C. E., 'Cyanide and thiocyanate in steel industry effluents," Paper No. 89, Division of Environmental Chemistry, 169th American Chemical Society National Meeting, Philadelphia, Pa., April 6-11, 1975.
[49] Gonter, C. E., "Cyanide and thiocyanate." Paper No. 21, Division of Environmental Chemistry, First Chemical Congress of the North American Continent, Mexico City, Mexico, Dec. 2, 1975.
[50] Sanner, G. L., Testimony presented in Regulatory Hearing R74-15 and 16, before the Illinois Pollution Control Board, 1975.
[51] Roberts, R. F. and B. Jackson, "The determination of small amounts of cyanide in the presence of ferrocyanide by distillation under reduced pressure," *Analyst*, Vol. 96, pp. 209-212, March 1971.
[52] Anon., "Determination of simple cyanide," *Effluent and Water Treatment Journal*, Vol. 16, No. 6, pp. 309-311, June 1976.
[53] Anon., "Gas dialysis speeds cyanide analysis," *Env. Sci. Technol.*, Vol. 11, No. 1, pp. 30-31, 1977.
[54] Fitzgibbons, W. O., Private Communication, Feb. 23, 1977.
[55] Union Oil-Monthly Cyanide Variance reports submitted to the Illinois EPA, Springfield, Ill., Dec. 10, 1974, and Feb. 23, 1975.
[56] Chamberlin, N. S. and J. B. Snyder, Jr., "Technology of treating plating wastes," Proc. 10th Ind. Waste Conf. Purdue Univ., May 9-11, 1955, published in *Engineering Bulletin Purdue Univ.*, Vol. XL, No. 1, pp. 277-305, March 1956.
[57] Cheremisinoff, P. N., and Y. H. Habib, "Cyanide—an assessment of alternatives for water pollution control," *Water and Sewage Works*, Vol. 120, Ref. No., R95-R107, 1973.
[58] Laws, B. C., "Control of cyanides in plating shop effluents," *Plating*, Vol. 59, No. 5, pp. 394-400, 1972.
[59] "Alkaline chlorination of cyanide waste liquors," Technical Service Applications Bulletin No. 102, Allied Chemical Industrial Chemicals Division, Morristown, N.J.
[60] U.S. Patent 3,101,320.
[61] U.S. Patent 2,981,682.
[62] German Patent 1,215,607 (*Chem. Abs.*, Vol. 65, 3552d, 1966).
[63] Huff, J. E., "Cyanide destruction using ozone with ultraviolet light," Mobil Oil Bimonthly Variance Report on Cyanide to the Illinois EPA, Springfield, Ill., August 1975.
[64] Remy, H., "Treatise on inorganic chemistry," edited by J. Kleinberg, translated by J. S. Anderson, Vol. II, p. 286, Elsevier, N.Y., 1956.
[65] Masson, O., "The action of hydrogen peroxide on potassium cyanide," *J. Chem. Soc. (London)*, Vol. 19, pp. 1449-1473, 1907.
[66] U.S. Patent 3, 617,567.
[67] Anon., "New process detoxifies cyanide wastes," *Env. Sci. Technol.*, Vol. 5, No. 6, pp. 496-497, 1971.
[68] Schumb, W. C., C. N. Satterfield, and R. L. Wentworth, "Hydrogen peroxide," A.C.S.. Monograph Series, p. 404, Reinhold Publishing Co., New York, 1952.
[69] Tyler, R. G., *et. al., Sewage and Ind. Wastes, Vol.* 23, p. 1150, 1951.
[70] Walker, C.A. and W. Zabban, "Disposal of plating room wastes, VI. treatment of plating room waste solutions with ozone," *Plating*, Vol. 40, No. 6, pp. 777-780, 1953.
[71] Bober, T. W. and T. J. Dagon, "Ozonation of photographic processing wastes," *J. Water Poll. Cont. Fed.*, Vol. 47, No. 8, pp. 2144-2129, 1975.
[72] Selm, R. P., "Ozone oxidation of aqueous cyanide waste solutions in stirred batch reactors and packed towers," in *Ozone Chemistry and Technology*, pp. 66-77, American Chemical Society Adv. in Chem., Series No. 21, Washington, D.C., 1959.
[73] Ingols, R. S., "Chlorination of water—Potable possibility: Wastewater, no.!", *Water and Sewage Works*, Vol. 122, No. 2, pp. 82-83, February 1975.
[74] Hardisty, D. M. and H. M. Rosen, "Industrial wastewater ozonation," 32nd annual Purdue Industrial Waste Conference, West Lafayette, Ind., May 10-12, 1977.
[75] Garrison, R. L., C. E. Mauk, and H. W. Prengle, Jr., "Cyanide disposal by ozone oxidation," AD-775 152, NTIS, Department of Commerce, Springfield, Va., February 1974.
[76] Prengle, H. W., *et al.* "Ozone/UV process effective wastewater treatment," *Hydrocarbon Processing*, Vol. 54, No. 10, pp. 82-87, October 1975.
[77] Mauk, C. E. and H. W. Prengle, Jr., "Ozone with ultraviolet light provides improved chemical oxidation of refractory organics," *Pollution Engineering*, Vol. 8, No. 1, pp. 42-43, January 1976.
[78] Nebel, C. and P. C. Unangst, "Ozone for pollution control," *Int. Pollut. Control Mag.*, Vol. 1, pp. 28-31 & 126-127, 1972.
[79] Raeff, S. F., *et al.* "Fate of Cyanide and related compounds in aerobic microbial systems I. Chemical reastion with substrate and physical removal II. Microbial degradation," *Wat. Res.*, Vol. 11, pp. 477-492, 1977.
[80] Technical Data Sheet 1377A, "Phenobac mutant bacterial hydrocarbon degrader," Polybac Corporation, New York, 1977.
[81] Ludzack, F. J. and K. B. Schaffer, "Activated sludge treatment of cyanide, cyanate and thiocyanate," *J. Water Poll. Cont. Fed.*, Vol. 34, pp. 320-341, 1962.
[82] Murphy, R. S., "The activated-sludge assimilation of the cyanide ion," Ph.D. Thesis, Pennsylvania State University, University Microfilms No. 64-7733, Ann Arbor, Mich . 1963.
[83] Dagon, T. J., "Biological treatment of photo processing effluents," *J. Water Poll. Contr. Fed.*, Vol. 45, No. 10, pp. 2123-2135, 1973.
[84] Lue-Hing, C., *et al.* "Report on cyanide studies—An evaluation of analytical methodology, effluent toxicity and compliance of publicly owned treatment works with existing effluent discharge criteria," Metropolitan Sanitary District of Greater Chicago Report No. 75-10, Chicago, Ill., June 1975.
[85] Bucksteeg, W., "Decontamination of cyanide wastes by methods of catalytic oxidation and adsorption. 21st Annual Purdue Industrial Waste Conference, Purdue University, Lafayette, Ind., May 1966
[86] Kuhn, R. G., "Process for detoxification of cyanide-containing aqueous solutions," U. S. Patent 3,586,623, June 22, 1971.
[87] Bernardin, F. E., "Cyanide detoxification using adsorption and catalytic oxidation on granular activated carbon," *J. Water Pollut. Contr. Fed.*, Vol. 45, No. 2, pp. 221-231, February 1973.
[88] Kunz, R. G. and J. F. Giannelli, "Activated carbon adsorption of cyanide complexes and thiocyanate ion from petrochemical wastewaters," *Carbon;* Vol. 14, pp. 157-161, 1976.
[89] Huff, J. E. and J. Bigger, "Cyanide removal from refinery wastewater using powdered activated carbon," Illinois Institute for Environmental Quality Document No. 77/08, June 1977.
[90] Keating, R. J., R. Dvorin and J. V. Calise, "Applications of ion exchange to plating plant problems," 9th Purdue Industrial Waste Conference, West Lafayette, Ind., May 11, 1954.
[91] Paulson, D. F. and A. B. Mindler, "Use of ion exchange in the water-treatment field," *Chem. Eng. Progr. Symp. Ser. No. 14*, Vol. 50, pp. 93-96, 1954.
[92] Ciancia, J., "New waste treatment technology in the metal finishing industry," *Plating*, Vol. 60, No. 10, pp. 1037-1042, October 1973.
[93] Burstall, F. H., *et al.* "Ion exchange process for recovery of gold from cyanide solutions," *Ind. Eng. Chem.*, Vol. 45, No. 8, pp. 1648-1658, August 1953.
[94] Talmadge, J. A., "Ion exchange treatment of electroplating wastes," *Ind. Eng. Chem. Process Design and Develop* , Vol. 6, No. 4, pp. 419-423, October 1967.
[95] Sussman, S., F. C. Nachod and W. Wood, "Metal recovery by anion exchange," *Ind. Eng. Chem.*, Vol. 37, No. 7 pp. 613-624, July 1945.
[96] Avery, N. L. and W. Fries, "Selective removal of cyanide from industrial waste effluents with ion-exchange resins," *Ind. End. Chem. Prod. Res. Dev.*, Vol. 14, No. 2, pp. 102-104, 1975.
[97] U.S. Patent 3,788,983.
[98] Tilsley, G. M., "Clean-up of fouled ion exchange resin beds," *Effluent and Water Treatment Journal*, Vol. 15, No. 11, pp. 560-563, 1975.
[99] Abrams, I. M., "Removal of organics from water by synthetic resinous adsorbents," *AIChE Symp. Ser. No. 97*, Vol. 65 pp. 106-112, 1969.
[100] Irvin, J. W. and H. D. Tomlinson. Testimony presented in Regulatory Hearing R74-15 and 16, before the Illinois Pollution Control Board, 1975.
[101] Henrikson, T. N. and L. G. Daignault, "Treatment of complex cyanide compounds for reuse or disposal," pp. 76-92, EPA-R2-73-269, U.S. Environmental Protection Agency, Washington, D.C., June 1973.

Part II.

Water Control

Zero Blowdown—Is It Feasible?

Cooling tower operating costs can be reduced when the blowdown is reduced to zero.

J. M. Brooke, Sweeny, Texas

FOR EVERY MILLION BTU'S of heat dissipated in a cooling tower, an average of 35 gallons of circulating cooling water is wasted to blowdown. This wasted water represents money—money to obtain the water, money to pretreat it, money for scale and fouling prevention, money for corrosion protection, money for pumping it from place to place, and money to treat it before it can be released as waste.

Obviously, cooling tower operation economy can be achieved by decreasing the blowdown. The ultimate economy comes when the blowdown is reduced to zero.

Zero blowdown operation of a cooling system means operating with no voluntary blowdown and with the involuntary blowdown (drift and leaks) at as low a value as possible.

Soluble salts, such as chlorides, are allowed to concentrate in the circulating water without restraint. Salts with limited solubility, such as calcium compounds and silica, are kept at nonscaling levels in the circulating water by treating a portion of that circulating water through a lime-soda softener to reduce the calcium and magnesium hardness and silica. The treated water is then returned to the circuit.

Inhibitors. Economics dictate that the inhibitors used in the circulating water be of such a nature that they are not removed in the softener. This precludes the use of such materials as phosphates, zinc, HEDP and AMP.

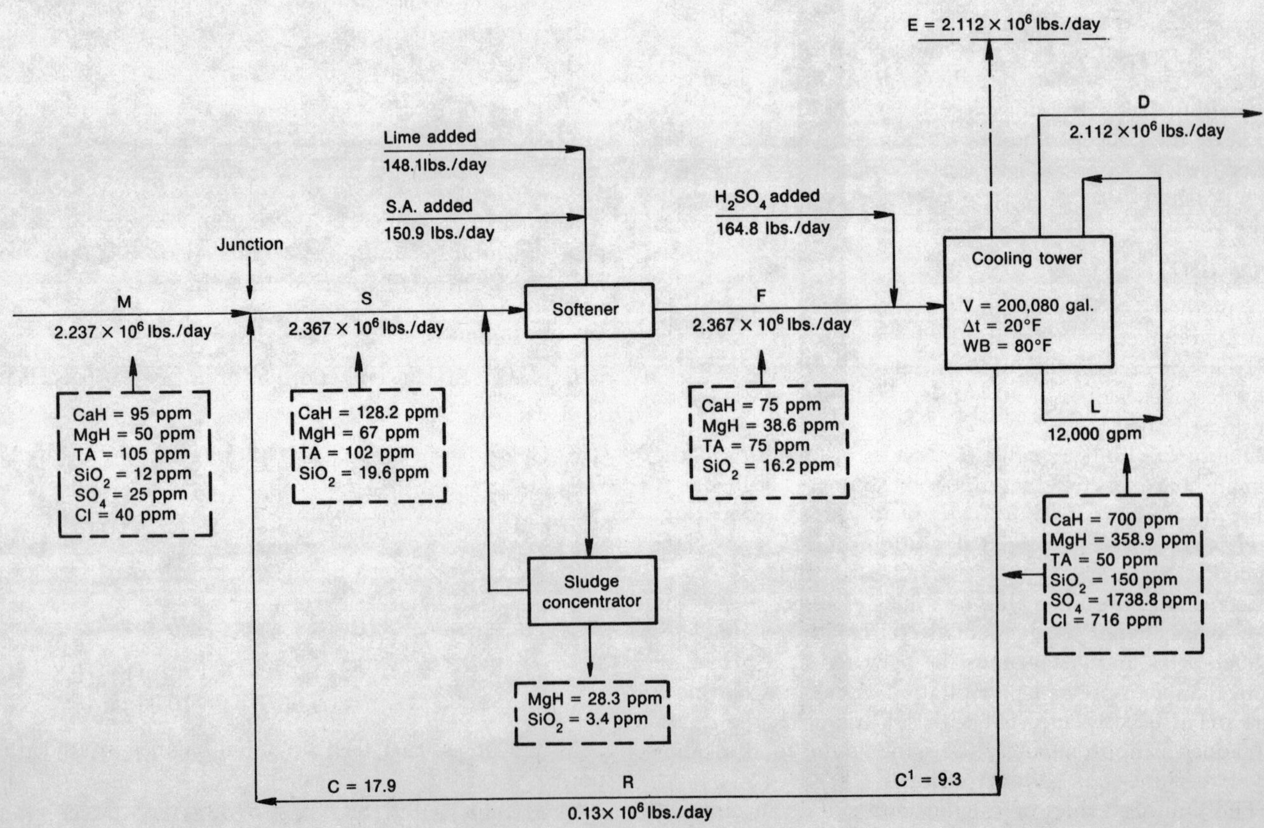

Fig. 1—Material balance diagram of example system.

Fortunately, the best known corrosion inhibitor, chromate, can be used.

There are four modes of combining a cooling tower with a lime-soda softener. It is assumed that softener sludges are concentrated and the liquid is returned to the system; i.e., the blowdown loss from softener operation is negligible.

Mode 1 is applicable to makeup waters having low calcium hardness (less than 100 ppm) and low silica content (less than 25 ppm). The softener is located in the recycle stream.

Mode 2 is the usual configuration used. The softener is located in the tower feed stream handling both makeup and recycle water (See Fig. 1).

Mode 3 has been used, but there is not enough data available to determine its advantages, if any. Softeners are located in both the recycle stream and in the feed stream.

Mode 4 probably has no advantages. One softener handles only recycle water and another handles only makeup water.

There are two operating ratios involved in zero discharge operation calculations; Cycles of concentration, C, is the volume per unit time of makeup water, M, divided by the volume per unit time of the drift plus involuntary blowdown, D. It is also the quotient of the concentration of the chlorides in the circulating water Cl_R divided by the concentration of the chlorides in the makeup water Cl_M.

$$C = M/D = (Cl)_R/(Cl)_M$$

Drift, the circulating water and the recycle stream are assumed to have the same composition.

Cycles of operation, C^1, is the volume per unit time of makeup water divided by the volume per unit time of drift plus recycle, R. It is also the quotient of the concentration of the chlorides or other salt (calcium hardness, magnesium hardness, silica) in the circulating water divided by the concentration of the same salt in the cooling tower feed, F.

$$C^1 = F/(D+R) = (Cl)_R/(Cl)_F = (CaH)_R/(CaH)_F$$

Since $F = M + R$, $C^1 = (M+R)/(D+R)$ and $R = (M - C^1D)/(C^1 - 1)$.

A method for determining the numerical values of M and D are given in Appendix 1.

Cycles of concentration in excess of 100 are theoretically possible. Values of 40-50 are excellent and 20-25 are considered good.

One of the limiting values for cycles of operation is the calcium hardness of the lime-soda softener effluent. A value of 35 ppm CaH is theoretically possible, 50 ppm CaH is considered excellent operation and 75 ppm CaH is good.

The other limiting value of cycles of operation is the allowable calcium hardness of the circulating water.

Such a cooling system must be operated at a pH of no more than 7 to prevent precipitation of calcium hardness. The pH is usually adjusted with sulfuric acid; the quantity added is approximately the same as the total alkalinity of the lime-soda softener effluent.

The limiting values of calcium and sulfate in circulating cooling water is traditionally given as $[Ca^{++}][SO_4^{=}] = 500{,}000$. This is unquestionably conservative because it prevents the deposition of calcium sulfate in pure boiling water. The solubility of calcium sulfate increases with decreasing temperature and with increasing sodium chloride content.

Using the data of Denman,[1] it is believed that under zero blowdown conditions, an upper limit of calcium hardness of 1,000 ppm ($Ca^{++} = 400$ ppm) and a solubility product of $[Ca^{++}][SO_4^{=}] = 1$ million is acceptable.

Klein[2] reports that if the CaH is maintained below 700 ppm, the SO_4 can go as high as 50,000 ppm before scale will form in waters with a high total dissolved solids content.

Example. Zero blowdown operation can best be illustrated by an example. The following assumptions are made:

1. The cooling tower has a 12,000-gpm circulation. It operates with a 20° F range at 80° F wet bulb, normally at six cycles of concentration. The calculations in Appendix 1 give: Capacity = 200,080 gallons; evaporation = 176 gpm; drift = 10.4 gpm; makeup = 176 + 10.4 = 186.4 gpm.

2. Makeup is Mississippi River water with the analysis:

Calcium hardness	95 ppm as $CaCO_3$
Magnesium hardness	50 ppm as $CaCO_3$
Total alkalinity	105 ppm as $CaCO_3$
Silica	12 ppm
Sulfate	25 ppm
Chloride	40 ppm

3. Limiting concentrations in circulating water; calcium hardness 700 ppm; $[Ca^{++}][SO_4^{=}] = 1$ million; silica 150 ppm.

4. The pH of the circulating water is adjusted with sulfuric acid; the control is $(TA)_R = 50$ ppm.

5. The lime-soda softener will operate in Mode 2 and will produce an effluent with 75 ppm each calcium hardness and total alkalinity. The removal of 100 ppm magnesium hardness will remove 12 ppm silica.

Calculations

1. Cycles of concentration: $C = M/D = 186.4/10.4 = 17.9$

2. Chlorides in circulating water: $Cl_R = C\,Cl_M = 17.9\,(40) = 716$ ppm

3. Cycles of operation: $C^1 = (CaH)_R/CaH)_F = 700/75 = 9.3$

4. Recycle rate:

$$R = (M - C^1D)/(C^1 - 1)$$
$$= [186.4 - 9.3\,(10.4)]/(9.3 - 1)$$
$$= 89.68/8.3 = 10.8 \text{ gpm}$$

5. SiO_2 in softener feed; material balance around junction:

$$M(SiO_2)_M + R(SiO_2)_R = S(SiO_2)_S$$
$$[186.4\,(12)] + [10.8\,(150)] = (186.4 + 10.8)\,(SiO_2)_S$$
$$(SiO_2)_S = 19.6 \text{ ppm}$$

6. Allowable SiO_2 in feed: $(SiO_2)_F = (SiO_2)_R/C^1 = 150/9.3 = 16.2$ ppm

7. Silica to be removed: $19.6 - 16.2 = 3.4$ ppm

8. Magnesium hardness that must be precipitated in softener (see Item 5 under assumptions and Item 7 above) $(3.4/12) \times 100 = 28.3$ ppm

9. Calcium hardness in softener feed; material balance around junction: $M(\text{CaH})_M + R(\text{CaH})_R = S(\text{CaH})_S$
$186.4\ (95) + 10.8\ (700) = (186.4 + 10.8)\ (\text{CaH})_S$
$(\text{CaH})_S = 128.1$

10. Total alkalinity in softener feed; material balance around junction:

$$M(TA)_M + R(TA)_R = S(TA)_S$$
$$186.4\ (105) + 10.8\ (50) = (186.4 + 10.8)\ (TA)_S$$
$$(TA)_S = 102 \text{ ppm}$$

11. Magnesium hardness in drift, circulating water and recycle; material balance around entire system:

$$M(\text{MgH})_M = D(\text{MgH})_D + S(\text{MgH removed, see Calculation 8})$$
$$186.4(50) = 10.4(\text{MgH})_D + (186.4 + 10.8)(28.3)$$
$$(\text{MgH})_D = 359.5 = (\text{MgH})_R$$

12. Magnesium hardness in feed:
$(\text{MgH})_F = (\text{MgH})_R/C^1 = 359.5/9.3 = 38.6$

13. Magnesium hardness in softener feed; material balance around junction:

$$M(\text{MgH})_M + R(\text{MgH})_R = S(\text{MgH})_S$$
$$186.4(50) + 10.8\ (359.5) = (186.4) + 10.8)\ (\text{MgH})_S$$
$$(\text{MgH})_S = 67 \text{ ppm}$$

Check:
$(\text{MgH})_S - (\text{MgH})_F =$ MgH to be removed
$67 - 38.6 = 28.4$ same as Calculation 8

14. Sulfuric acid needed:
$(TA)_F - (TA)_R/C^1 = \text{ppm } H_2SO_4$
$75 - 50/9.3 = 69.62$ ppm $H_2SO_4 \simeq 68.2$ ppm SO_4
$69.62\ (2.367) = 165$ lbs./day

15. Sulfates in drift, circulating water and recycle; material balance around entire system:
$M(SO_4)M + F(SO_4 \text{ added}) = D(SO_4)_D$
$186.4(25) + (186.4 + 10.8)(68.2) = 10.4(SO_4)_D$
$(SO_4)_D = 1{,}741.3 \text{ ppm} = (SO_4)_R$

16. Calcium sulfate solubility product in circulating water:

$$0.4(\text{CaH})_R(SO_4^=)_R = 0.4(700)\ 1{,}741.3 = 487{,}564$$

If hydrochloric acid is substituted for sulfuric acid for pH control, the need for a lime-soda softener may be eliminated. If we take a somewhat conservative maximum calcium sulfate product in the circulating water to be $(Ca^{++})(SO_4^=) = 750{,}000$, the permissible cycles of concentration will be

$$C = \sqrt{\frac{750{,}000}{0.4(\text{CaH})_M \times (SO_4)_M}}$$

If this value is equal to or greater than the measured cycles of concentration (Calculation 1), this method may be applicable.

In this case, the silica content of the circulating water becomes the control. But, without the softener, we can treat the circulating water with a low molecular weight polyacrylate polymer (Δ) and carry the concentration of silica at 250 ppm. In this case, the permissible cycles of concentration will be $C = 250/(SiO_2)_M$. If this is also equal to or greater than the measured cycles of concentration, the method is applicable.

In our problem:

$$C = \sqrt{\frac{750{,}000}{0.4(95)25}} = 28.1 \text{ and } 250/12 = 20.8$$

Both values are greater than the measured value of 17.9. So, this cooling tower can operate at zero blowdown simply by adjusting the pH to 6.5-7 with hydrochloric acid and adding a polymer to control silica deposition. Fig. 1 is a materials balance diagram of the example system.

APPENDIX 1—Cooling tower calculations

1. The evaporation rate is calculated from the formula

$$100E = L\Delta t\ (0.00036WB + 0.00014\Delta t + 0.0416) \quad \ldots\ldots\ldots (3)$$
$$E = 12{,}000(20)\ [(0.0000036)\ (80) + (0.0000014)\ (20) + 0.000416] = 176 \text{ gpm}$$

2. Blow the tower down heavily until the cycles of concentration are about one-half normal. Block in the blowdown and determine the chromate concentration. Add a known amount of chromate, wait an hour and again determine the chromate concentration.

Example: Initial chromate concentration = 2 ppm. Add 200 pounds of a treatment containing 10 percent chromate, equivalent to adding 20 pounds of chromate.
Final chromate concentration = 14 ppm.
Calculate the capacity of the tower

$$V = \frac{\text{lbs. chromate added } (10^6)}{\text{chromate increase}} = \frac{20 \times 10^6}{(14-2)} = 1{,}666{,}667 \text{ pounds or } 200{,}080 \text{ gallons}$$

3. With the blowdown blocked in, determine the chromate concentration at measured intervals until the cycles of concentration return to normal. Resume normal operation.

Example: Chromate concentration after 24 hours = 13 ppm. Chromate concentration after 48 hours = 12 ppm.

Calculate the drift loss:
$$D = 2.309\ (\log I/F^1)\ V/T$$
$$D_1 = 2.309\ (\log 14/13)\ (200{,}080/1{,}440) = 10.3 \text{ gpm}$$
$$D_2 = 2.309\ (\log 14/12)\ (200{,}080/2{,}880) = 10.7 \text{ gpm}$$
$$D_{avg.} = 10.5 \text{ gpm}$$

4. $M = E + D = 176 + 10.5 = 186.5$ gpm.

APPENDIX 2—Lime-soda softener calculations

Space does not allow a comprehensive discussion of the lime-soda softening of water. It is a process designed to reduce the hardness of water by reactions with lime and soda ash, the hardness being calcium and magnesium salts designated as CaH and MgH. If the alkalinity is greater than the hardness, lime will also react with the excess alkalinity. An excess of both chemicals is required to drive the reactions to equilibrium. The reactions involved are: (an underlined compound designates a precipitate or removal of that compound)

1. $Ca(HCO_3)_2 + Ca(OH)_2 \rightarrow \underline{2CaCO_3} + 2H_2O$

2. $CaSO_4 + Na_2CO_3 \rightarrow \underline{CaCO_3} + Na_2SO_4$

3. $Mg(HCO_3)_2 + 2Ca(OH)_2 \rightarrow \underline{2CaCO_3} + \underline{Mg(OH)_2} + 2H_2O$

4. $MgSO_4 + Ca(OH)_2 + Na_2CO_3 \rightarrow \underline{CaCO_3} + \underline{Mg(OH)_2} + Na_2SO_4$

5. $NaHCO_3 + Ca(OH)_2 \rightarrow \underline{CaCO_3} + NaOH + H_2O$

Calcium and magnesium bicarbonates are called temporary hardness (because they can be reduced simply by heating) and are designated as CaH(T) and MgH(T). Equation 1 is the reaction of temporary calcium hardness with lime. Equation 3 is the reaction of temporary magnesium hardness with lime.

Other calcium and magnesium salts are called permanent hardness and are represented by CaH(P) and MgH(P). Equation 2 is the reaction of permanent calcium hardness with soda ash. Equation 4 is the reaction of permanent magnesium hardness with lime and soda ash. Both of these permanent hardnesses are shown as the sulfate. They can also be chlorides, nitrates or other soluble salt.

The equilibrium hardness of the softener effluent is 34 ppm calcium hardness and 32 ppm magnesium hardness. The precipitated magnesium hydroxide will adsorb silica from solution, thus decreasing its concentration. The removal of 100 ppm magnesium hardness will remove about 12 ppm silica. If there is not enough magnesium hardness removable from the makeup water to reduce the silica by the amount needed, additional magnesium compounds such as magnesium oxide can be added to the softener.

There are three possible combinations of hardness and alkalinity for the softener feed water.

1. The total alkalinity is equal to or greater than the total hardness. Thus, it contains only temporary calcium and magnesium hardness. The difference between the total alkalinity and the total hardness is excess alkalinity, $TA(x)$. The lime requirement (as $CaCO_3$) for softening Type 1 water is equal to the calcium hardness reduction plus twice the magnesium hardness reduction plus the excess alkalinity plus 45 ppm excess. No soda ash is needed.

2. The total alkalinity is less than the total hardness but greater than the calcium hardness. It contains temporary calcium hardness plus both temporary and permanent magnesium hardness. The lime requirement (as $CaCO_3$) for softening Type 2 water is equal to the calcium hardness reduction plus twice the temporary magnesium hardness reduction plus the permanent magnesium hardness reduction plus 45 ppm excess. The soda ash requirement (as $CaCO_3$) is equal to the permanent magnesium hardness reduction plus 5 ppm excess.

3. The total alkalinity is less than the calcium hardness. It contains temporary and permanent calcium hardness plus permanent magnesium hardness. The lime requirement (as $CaCO_3$) for softening Type 3 water is equal to the temporary calcium hardness reduction plus the magnesium hardness reduction plus 45 ppm excess. The soda ash requirement (as $CaCO_3$) is equal to the permanent calcium hardness reduction plus the magnesium hardness reduction plus 5 ppm excess.

To recapitulate:

Type 1 water; $TH \leq TA$

Contains	Lime requirement	S.A. requirement
CaH(T) = CaH	CaH reduction	None
MgH(T) = MgH	2MgH reduction	None
$TA(x) = TA - TH$	$TA(x)$	None
	45 ppm excess	

Type 2 water; $TH > TA >$ CaH

Contains	Lime requirement	S.A. requirement
CaH(T) = CaH	CaH reduction	None
MgH(T) = TA − CaH	2MgH(T) reduction	None
MgH(P) = $TH - TA$	MgH(P) reduction	MgH(P) reduction
	45 ppm excess	5 ppm excess

Type 3 water; $TA <$ CaH

Contains	Lime requirement	S.A. requirement
CaH(T) = TA	CaH(T) reduction	None
CaH(P) = CaH − TA	None	CaH(P) reduction
MgH(P) = MgH	MgH reduction	MgH reduction
	45 ppm excess	5 ppm excess

For the problem under study, the lime and soda ash requirements for the Type 3 water are:

Hardness	Concentration in Softner feed	Softener effluent	Quantity removed	Lime requirement $CaCO_3$	S.A. requirement $CaCO_3$
CaH(T)	102	75	27	27	
CaH(P)	26.2	0	26.2		26.2
MgH(P)	67	38.6	28.4	28.4	28.4
Excess				45	5
Total				100.4	59.6

Conversion factors:

ppm $CaCO_3$ (0.823) = ppm 90% hydrated lime
ppm $CaCO_3$ (0.623) = ppm 90% burnt lime
ppm $CaCO_3$ (1.070) = ppm 99% soda ash

Lime requirement: 100.4 (0.623) 2.367 = 148.1 lbs./day
Soda ash requirement: 59.6 (1.07) 2.367 = 150.9 lbs./day

NOMENCLATURE

C	Cycles of concentration = M/D
C^1	Cycles of operation = $F/(D+R)$
D	Drift or involuntary blowdown, gpm
E	Evaporation, gpm
F	Cooling tower feed, gpm = $S = M + R$
F^1	Final concentration, ppm
I	Initial concentration, ppm
L	Cooling tower circulation, gpm
M	Makeup water, gpm = $E + D$
R	Recycle, gpm = $(M - C^1D)/(C^1 - 1)$
S	Softener feed, gpm = $M + R$
T	Time, min.
Δt	Cooling range of tower, °F
V	Volume of cooling system, gal.
WB	Wet bulb temperature, °F
$(X)_y$	ppm component X in stream y; (C_2) = ppm chloride in makeup
CaH	Calcium hardness, ppm $CaCO_3$
CaH(P)	Permanent calcium hardness, ppm $CaCO_3$
CaH(T)	Temporary calcium hardness, ppm $CaCO_3$
Cl	Chlorides, ppm
MgH	Magnesium hardness, ppm $CaCO_3$
MgH(P)	Permanent magnesium hardness, ppm $CaCO_3$
MgH(T)	Temporary magnesium hardness, ppm $CaCO_3$
SiO_2	Silica, ppm
SO_4	Sulfate, ppm
TA	Total alkalinity, ppm $CaCO_3$
$TA(x)$	Excess total alkalinity, ppm $CaCO_3$
TDS	Total dissolved solids, ppm
TH	Total hardness, ppm $CaCO_3$

LITERATURE CITED

[1] Denman, W. L., "Maximum Re-use of Cooling Water," *I & E Chem* 53 (10) 817-22, (1961).

[2] Klein, E. F., "Calcium Sulfate Solubility in Dynamic Cooling Tower Systems," presented NACE annual meeting, March 22-6, 1976, Houston.

[3] Brooke, J. M., Cooling Water, P. R. Puckorius Associates, Cleveland, Ohio, 1977.

[4] Norman, L. R., R. E. Ashman, and D. F. Bollinger, "High Silica Control in a Recirculating Water System," *Western Energy Supply & Transmission Associates Report*, Dec. 15, 1978.

Use Side Stream Softening to Reduce Pollution

Design procedures are available for minimizing cooling tower blowdown

D. T. Reed, E. F. Klen and **D. A. Johnson,**
Nalco Chemical Co., Oak Brook, Ill.

SIDE STREAM softening can be used to greatly reduce cooling tower blowdown, reduce pollution and save water. By-pass lime softening appears to be a realistic technique for use both with new and existing cooling tower systems while synergistic chromate formations can be expected to adequately protect cooling water systems in high dissolved solids applications if properly designed and controlled.

A method is available for designing lime softening systems for use in zero blowdown applications and for predicting the steady state chemistry conditions. However, each situation should be tested to determine the extent of the various effects (softener interferences, corrosion, etc.) on a potential design.

Systematic analysis of the zero blowndown (ZBD) design consists of first applying the simple mass balance relationships in a logical manner, followed by static (limited) and dynamic (pilot) testing to approve the conceptual design before deciding on the equipment, the specific mode of operation of the system, and the chemicals required.

Variables that have substantial impact on design costs and operating chemical usage have been identified and quantified where possible. Accurate calculation of the sidestream softening required for a given system gives maximum safe concentration and softenability of dissolved species *at the steady state concentrations* of non-softenable ions. This steady state is determined by the make-up water content, the system hydraulics and the chemical treatment program.

A number of conceptual design[2,3] have been proposed based on using ion removal systems as the sidestream process. Reverse osmosis, electrodialysis, freeze crystallization and distillation techniques are being investigated although installation, operating, and maintenance costs can be prohibitive for large volume water users.[4] In addition, operational reliability of some methods is yet to be proven in cooling tower applications.

Experience with a zero blowdown is limited. Fowlkes reported the first operating experiences[5,6] and corrosion studies concerning a cooling system employing make-up

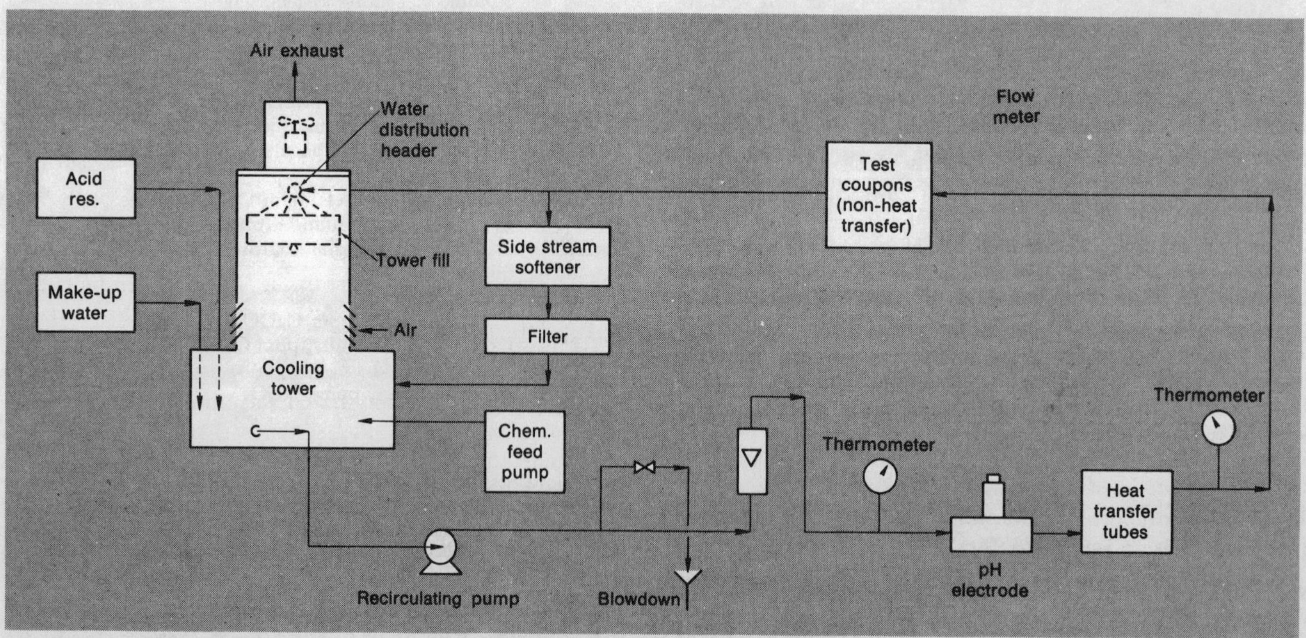

Fig. 1—Pilot cooling tower suitable for zero blowdown operation.

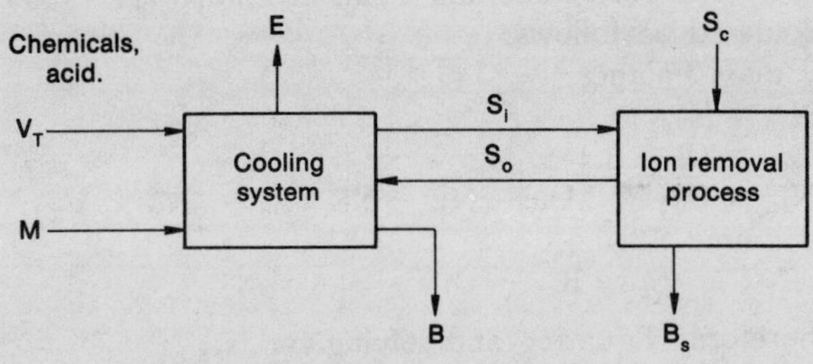

Fig. 2—Conceptual representation of a zero blowdown operation.

and sidestream lime-soda softening. Although this system has the potential to be a closed loop, sufficient fire service water was apparently withdrawn so that the increase in concentrations achieved, over the conventional operation of the system, was only slight. While this system was not entirely closed loop, there was a reduction in the chromate-based corrosion inhibitor consumed and elimination of blowdown of chromate-containing water to a receiving stream.

Perry and Matson[7] reported a study of the application of sidestream softening to a pilot cooling tower using sidestream lime softening to limit the hardness levels and carbon dioxide for pH control. The reasons for using CO_2 were:

- Favorable economics since flue gas was readily available
- Control of pH with Co_2 preserves alkalinity, allowing efficient calcium removal without the addition of soda ash in high-sulfate water
- Use of CO_2 for pH control eliminates need for sulfuric acid, reducing sulfate buildup in the system.

They concluded that the process is feasible for systems with readily available CO_2 sources.

Crits and Glover [8,9] discussed some considerations and variables concerned with minimal discharge from cooling towers. They assumed a hypothetical 500 megawatt electric power plant and theoretically studied the effect of reduced blowdown with the corresponding increase in cycles of concentration. Various types of possible sidestream treatments were discussed in terms of the species they can remove. The papers presented an economic analysis of the various sidestream treatments. Their results indicate that, of the possible solids-removal processes, warm lime softening was the most generally acceptable in that it offered removal of all the potential scaling species (except SO_4^{-2}) with minimum cost.

These investigators demonstrated a limited feasibility of the ZBD concept. In comparison, the controlled dynamic operation of a laboratory pilot ZBD system at various TDS levels and conditions of operating variables permit extention of that ZBD concept into new areas. Previously unreported conditions that are of immediate interest to design engineers for ZBD systems include:

- A method for accurately sizing a sidestream softener
- Projecting softener operations at ultra-high cycles of tower concentrations
- Projecting corrosion problems at ultra-high TDS values.

TEST APPARATUS

The test apparatus was a Pilot Cooling Tower (PCT) similar to that described by Reed and Nass.[10] Fig. 1 shows the basic elements and flows of this system. Installation of a bypass lime softener allowed for closed loop operation. The PCT is ideally suited for ZBD investigations, since it simulates the dynamic operations of a cooling tower system.

This dynamically concentrating system gradually concentrated ions in the same manner as an operating system. Control of pH was maintained within the desired range using a continuous automatic pH controller, similar to those used in field applications. The make-up waters used for PCT zero blowdown operation were natural or synthetic, to simulate the range of actual conditions. Bulk water temperature was kept constant by the evaporation of a portion of the recirculating water as it passed over the tower.

In all cases, except for the tower size, the PCT was operated in a way to duplicate all critical conditions of a full scale field system.

System operation. A thorough understanding of how various conditions such as evaporation, make-up, system blowdown and softening interrelate to effect the steady state system chemistry in a ZBD system is needed. A schematic representation of such a system is shown in Fig. 2, where:

E = Evaporation rate, determined by heat load
M = Make-up rate, defined by evaporation rate and system losses
B = Blowdown, drift and incidental losses, partially determined by system characteristics, but many may be augmented by the waste treatment facilities and system characteristics
S_i = Sidestream flow rate = S_o with minimal softener blowdown
S_c = Rate of addition of softening chemicals.

The evaporation rate (E), is generally fixed for a given situation, as it is determined by the heat load. If the total losses of liquid from the system (B) are small compared to (E), the make-up stream (M) will be approximately equal to (E), the evaporation rate. The *minimum* rate of the total blowdown (B) is fixed by the drift, softener blowdown and incidental losses by the system. It may, however, be desirable, depending on the quality of the make-up stream, to raise (B) above this value.

The size of the sidestream is a major independent variable in the design of a ZBD system. It is this variable, coupled with the make-up rate, the make-up composition, *and the degree to which a given species can be removed in the softener* that determines the steady state level of a species in the system. If these species can contribute to scale or deposition, their steady state levels become highly important with regard to reliability of operation. Therefore, the three chemical variables that determine the minimum rate of the sidestream relative to the make-up rate are:

- The composition of the make-up water
- The level to which potential scalants can be softened

- The maximum level of the scalant that can be tolerated in the cooling system without problems.

The following example will illustrate these points:

A softenable species (A), which is contained in the make-up water in a concentration (A_o), can be softened under the projected operating conditions to the level A_s. A PCT feasibility study has shown that this species can be tolerated in a projected recirculating water at the concentration A_{max} without scaling. To calculate the minimum softener stream (S_o) necessary to limit species A to safe levels, the following mass balance calculation is made:

A_o = Concentration of A in make-up water
A_s = Level to which A can be softened in the th sidestream softner
A_{max} = Maximum safe level of A in the recirculating water
A_{ss} = Concentration of non-softenable ions at steady state.

At steady state:

$$A_{in} = A_{out}$$
$$A_{in} = (M \times A_o) + (S_o \times A_s)$$
$$A_{out} = (S_i \times A_{max}) + (B \times A_{max}) + (B_s \times A_s)$$

However, assuming:

$$S_o \cong S_i \text{ and MB}$$
$$A_{in} = (M \times A_o) + (S_o \times A_s)$$
$$A_{out} = S_o \times A_{max}.$$

Equating these expressions and solving for S_o yields:

$$S_o = M \times \frac{A_o}{A_{max} - A_s}$$

This calculation should be performed for all softenable species in the make-up water. Comparing these calculated S_o values will determine the size of sidestream required to prevent scaling of the most likely scaling species. In effect, the highest S_o value must be chosen as the system S_o to prevent scaling from occurring. The steady state concentrations (A_{SS}) of non-limiting softenable ions can be determined by a similar exercise that yields the relationship:

$$A_{ss} = \frac{A_o \times M}{S_o} + A_s$$

When the steady state concentration of each non-limiting softenable species has been determined, the appropriate dosage of softening chemicals can be determined by standard softening formulas.

Non-softenable species concentrations, e.g., Cl^-, SO_4^{-2}, Na^+, can be worked out by similar methods. While these concentrations do not directly enter into the analysis of system design and operation, they (particularly sulfate) can have major effects both on softening reactions and on the maximum tolerable cooling system concentration. Especially influenced by these ions are the allowable calcium carbonate maximum in the system, and the reduction of calcium and alkalinity by lime-soda softening. Therefore, it is desirable to know the concentrations of non-softenable species at which the system will equilibrate. If these levels are excessive in terms of the inhibition of softening reactions, it may be economical to increase the blowdown.

The non-softenable ion concentrations (N) may be calculated as follows:

By mass balance:

$$N_{in} = N_{out}$$
$$N_{out} = (B + B_s) \times N_{ss}$$
$$N_{in} = (N_o \times M) + (N_{so} - NS_i)\, S_o + (N_T \times V_T).$$

Assuming: $S_i \cong S_o$
$S_c \cong B_s$

Therefore: Equating and solving for N_{ss}

$$N_{ss} = \frac{(N_o \times M) + (N_{so} - N_{si})\, (S_o + (N_T \times V_T)}{B + B_s}$$

Where:

N_{in} = The concentration of non-oftenable ions entering the system
N_{out} = The concentration of non-softenable ions leaving the system
N_{ss} = The concentration of non-softenable ions in the system at steady state
N_o = The concentration of non-softenable ions in the make-up
N_{So} = The concentration of non-softenable ions in the sidestream from the softener to the tower
N_{Si} = The concentration of non-softenable ions in the sidestream from the tower to the softener
N_T = The concentration of non-softenable ions in the chemical treatment
M = Make-up rate
B = Water blowdown rate from tower (including drift and miscellaneous losses)
B_s = Water blowdown rate from clarifier
V_T = Rate of chemical treatment to the tower
S_o = Size of the sidestream.

Using this method (which is similar to the simpler method used for softenable ions), one may calculate the steady state concentrations of all non-softenable ions in the system. These relationships demonstrate the importance of selecting the optimum operating levels of various chemical species in the recirculating cooling tower system. Clearly, the operation of a ZBD system at maximum tolerable scalant levels may not only allow for reduced equipment sizing, but can also reduce operating costs by substantial chemical savings.

Softening efficiency. Efficiency of the softening process will have an impact on both design (capital) and operating costs of the ZBD system. To date, because of an absence of reliable technology in this emerging ZBD type softening technique, many design engineers have applied traditional softening rules to a non-traditional area. Many previously unexpected factors (e.g., high TDS, mechanical operations, various applied cooling water treatments) can interfere with softening efficiencies, so it is necessary to understand these factors if the best conceptual equipment design and system operating conditions are to be achieved. It is difficult to delineate the impact of each of the factors that can interfere with ZBD softening, but the Pilot Cooling Tower provides the dynamic environment for easy understanding of the total impact. As a result of the PCT work completed to date, however, the following general trends and observations can be reported:

- A consistent trend in the loss of softener efficiency was

noted as TDS was increased in the recirculating water. Levels of up to 100,000 ppm TDS were studied and, in some instances, the loss of efficiency was sufficient to render impractical some system designs evaluated.

- Applied scale inhibitor compounds were removed to varying degrees through the softening process, depending on the type of compound and the mode of operation of the system. For example:

Scale Inhibitor	Percent Removed
a	10—50
b	50—75
c	40—60
d	50—100
e	50—100

- Anti-scale compounds had a variable impact on the softening capacity of the system, depending on the type and amount of compound added on the mode of system operation.

- Zinc is generally fully removed through the softener, while chromate removal is minimal.

The degree of impact of these trends was generally not predicted accurately with static (jar) testing. In fact, many important factors were not observable in static testing at all.

Corrosion and corrosion inhibition.

Comprehensive corrosion tests determined the effect high TDS waters, particularly those with substantial chloride and sulfate concentrations, have on corrosion rates of various metals of construction. Test specimens were fabricated into heat transfer and non-heat transfer surfaces, simulating conditions in plant operations.

The recirculating water chemistry was designed to provide an ultra-high TDS level depicting conditions expected by ultimate ZBD systems currently being designed. Analysis of a typical water is shown below:

Recirculating Water Composition

Ca^{+2} (as $CaCo_3$)	500 ppm
Mg^{+2} (as $CaCO_3$)	500 ppm
"M" (as $CaCO_3$)	50
Cl^- (as NaCl)	20,000 ppm
SO_4^{-2} (as Na_2SO_4)	50,000 ppm
pH	6.0
Applied inhibitor	None

This water is very corrosive environment. For example, mild steel corroded at a rate of 82—285 mpy, depending on the presence of heat flux and the bulk water temperature of the system. Similar Admiralty and copper specimens, which ordinarily experience less than one mpy of corrosion in "normal" systems, corroded at 9—71 mpy, while stainless steel (type 316) was the least corroded, showing rates generally less than one mpy.

This evaluation was repeated, employing a proprietary synergistic CrO_4 corrosion inhibitor. The work showed that corrosion control could be achieved on a variety of metals with the application of a proven corrosion inhibitor. For example, mild steel corrosion rates were reduced to less than 2.5 mpy on both heat transfer and non-heat transfer surfaces. Copper and Admiralty corrosion was also substantially reduced to 0.2—2.0 mpy, while stainless steel specimens remained at rates less than one mpy. The low corrosion rates on mild steel and copper alloy specimens should be of special interest to existing installations that have a considerable amount of mild steel and copper related plumbing and process equipment and are considering ZBD modifications. Corrosion data can be valuable in selecting construction materials and in applying and controlling an appropriate inhibitor program. In addition, non-metallic construction materials should be tested to evaluate their performance, e.g., concrete in high sulfate waters.

LITERATURE CITED

1 *Federal Register*, Tuesday, Oct. 8, 1974, Volume 39, #196.
2 Stickney, W. W. and Andrews, W. R., "Brine Concentrators Providing Zero Liquid Discharge." *35th International Water Conference*, Pittsburgh, Pa., Oct. 29-31, 1974.
3 Boies, D. B. and Levin, J. E., "Treat Cooling Tower Blowdown," *POWER*, August 1974.
4 Liptak, B. G., *Environmental Engineer's Handbook, Volume 1, Water Pollution*, 1974, pp. 1243-1285.
5 Fowlkes, C. C., "Softening of Cooling Tower Blowdown for Reuse," *USAEC Report, #KP 4023*.
6 Fowlkes, C. C., "Corrosion Rates to be Expected at Zero Blowdown of Recirculating Water." *Materials Performance*, October 1974, p. 26.
7 Perry, M. E. and Matson, J. V., "Complete Reuse of Cooling Tower Blowdown."
8 Crits, G. J. and Glover, G., "Cooling Blowdown in Cooling Towers." *Water and Waste Engineering*, April 1975, pp. 45-52.
9 Crits, G. J. and Glover, G., "Zero Blowdown from Cooling Towers, Problems and Some Answers." *34th International Water Conference*, Pittsburgh, Pa., Oct. 30, 1973.
10 Reed, D. T. and Nass, R., "Small Scale, Short Term Methods of Evaluating Cooling Water Treatments—Art They Worthwhile?" *36th International Water Conference*, Pittsburgh, Pa., Nov. 4-6, 1975.

Pick the Right Water Reuse System

How well is the plant run? Use these plots to study both existing and new refineries

S. Finelt, Fluor E & C, Inc., Houston, and
J. R. Crump, University of Houston, Houston

HPI OPERATORS faced with increasing fresh water cost may improve environmental protection and save money by directing their water management policies toward recycle of treated water. Such a policy can

- Reduce water cost
- Reduce effluent discharge up to 90 percent
- Establish facilities for wastewater treatment requirements that lead even to "zero discharge."

To accomplish these advantages criteria are needed with which to establish a water reuse policy and to calculate various water rates and cost. An analysis of alternative arrangements for wastewater management shows that properly treated waste effluent can be recycled as makeup to a cooling tower system in well operated plants. However, all cooling water needs will not be met by recycle; some fresh water is required as additional makeup. Furthermore, fresh water cost and refinery size have a high influence on the economics of water reuse.

Rising cost of fresh water plus some stringent effluent discharge standards mean additional water reuse must be practiced. The ultimate in water reuse is recycle of treated wastewater which requires an understanding of the variation in wastewater production rates and quality as affected in various alternatives. Several studies[13,14,15,16,17] have been made to determine the quantity of wastewater generated by petroleum refiners. One concept[17] involves analyzing individual processing units of a refinery to determine their contribution to the total wastewater generated. This approach can be used to represent a typical, well operated refinery in terms of wastewater generation. However, all plants are not well operated as shown by the variation and ratio of total wastewater generated to total fresh water intake for typical refineries (Table 1).

As could be expected, the poorly operated refinery has incentive to recycle wastewater since the savings in fresh water cost is greatest. On the other hand, a water management policy to improve maintenance and housekeeping in such plants is called for as a first step to reduce fresh water intake. But for even well run plants there is sufficient wastewater available for potential reuse.

If all wastewater is treated to a satisfactory level and recovered for reuse as a feed to both cooling and steam generation services, only poorly operated refineries have enough to supply both needs. Average and well operated refineries are not able to eliminate the need for fresh water makeup because atmospheric losses must be compensated for.

The quality criteria to consider in dealing with treated wastewater are those characteristics associated with use as cooling medium or steam generation. A refinery incorporating the "best practical control technology currently available" may expect to treat its raw wastewater to produce a treated effluent with a composition as shown in Table 2.[18] Table 3 compares the treated wastewater for a complex new refinery with generally accepted water criteria for cooling and steam generation applications.[19]

Cooling water quality can be generally specified independent of processing application since the levels of temperatures involved do not appreciatively influence the applicability of the water for the service. On the other hand, required boiler feedwater (BFW) quality is a func-

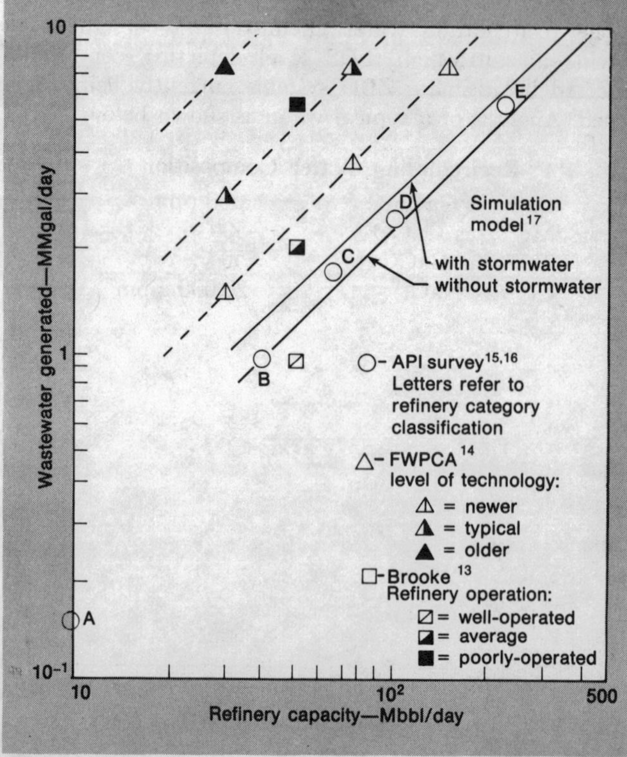

Fig. 1—Effect of refinery size and type on quantity of wastewater production.

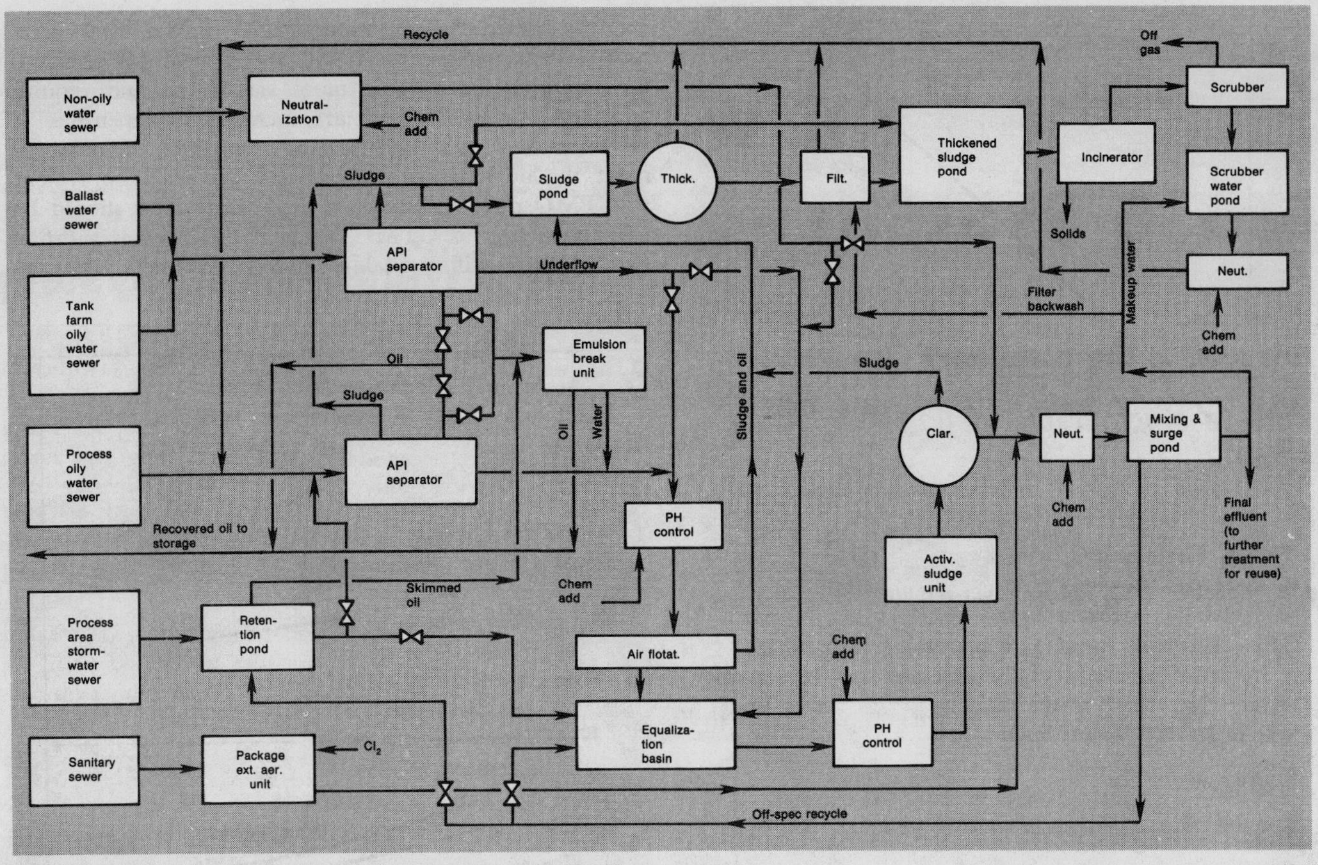

Fig. 2—Complete refinery wastewater treatment plant.

tion of generating pressure, becoming more critical with higher pressure (Table 3). Normally, quality is achieved by external fresh water treatment and maintained by internal treatment at points of application.[20] Use of treated wastewater in place of fresh water requires treatment to the highest purity level (last column, Table 3). The added cost to reach this quality is a function of the difference in component analysis between column 6 and any of the corresponding columns of Table 3. As a first approximation, the cost may be equal for either cooling tower makeup (CTMU) or BFW if the treated waste-

TABLE 1—Effect of operating efficiency on water needs[13]

Mode of Operation	Ratio of Total Wastewater Generated to Total Fresh Water Intake	Ratio of Water Required For Cooling Services to Total Wastewater Generated	Ratio of Water Required For Steam Generation to Total Wastewater Generated
Well-operated	0.347	1.968	0.753
Average	0.506	1.095	0.476
Poorly operated	0.737	0.452	0.304

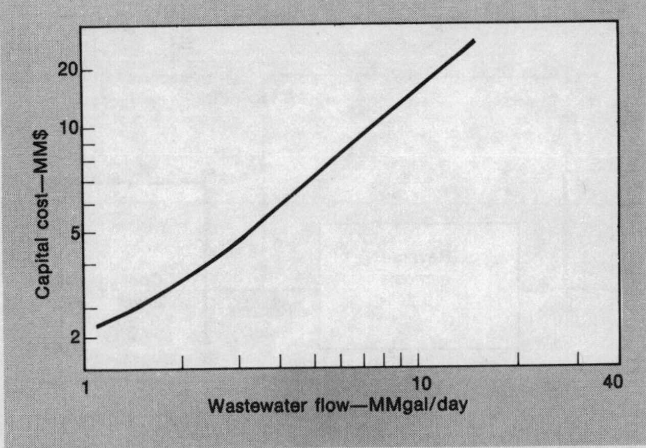

Fig. 3—Effect of treatment plant capacity on cost of a wastewater treatment plant.

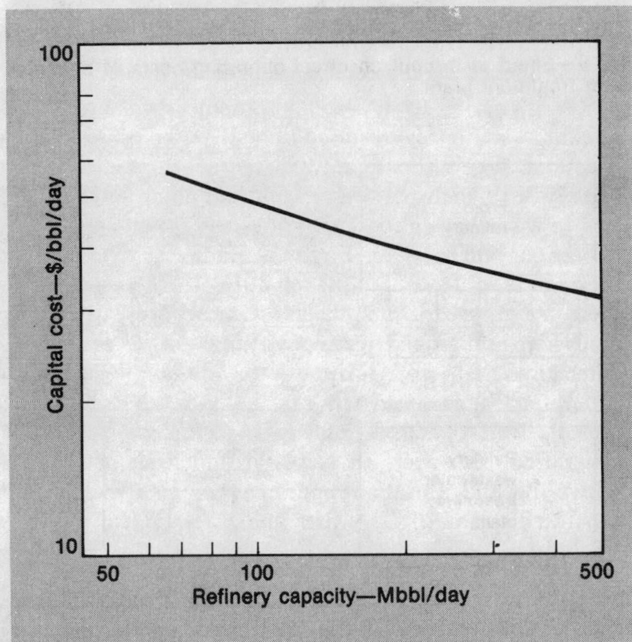

Fig. 4—Effect of refinery capacity on cost of a wastewater treatment plant.

water is limited to application as BFW for the lowest pressure of steam generation, 0-150 psig. Above this pressure, the requirements become more stringent and require additional facilities which are increasingly expensive. If total dissolved solids are used as a yardstick of measurement, the low-pressure steam level is just about equivalent to CTMU quality; hence, treated wastewater should be considered first as CTMU. Only if this need is satisfied and excess is still available, should reclaimed water be

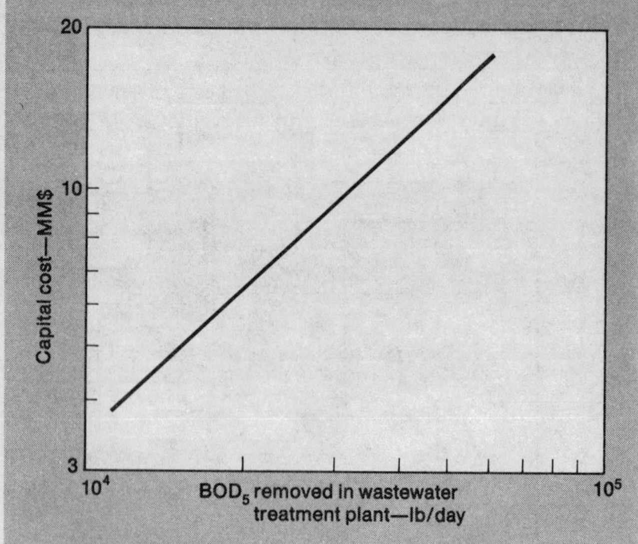

Fig. 5—Effect of BOD_5 removal load requirement on the cost of the wastewater treatment plant.

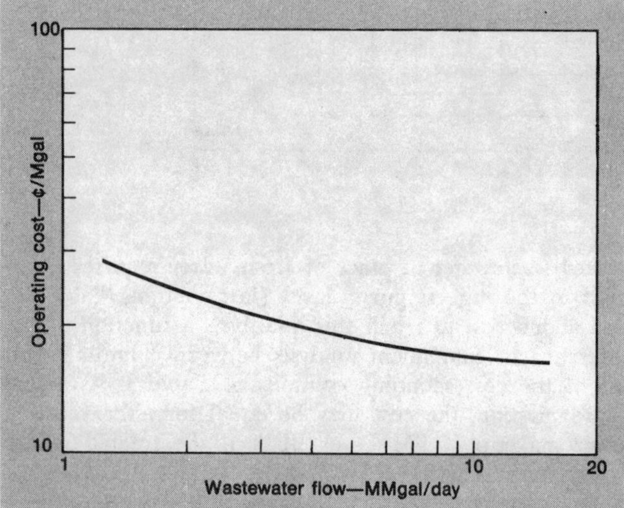

Fig. 6—Effect of thruput on direct operating costs of a wastewater treatment plant.

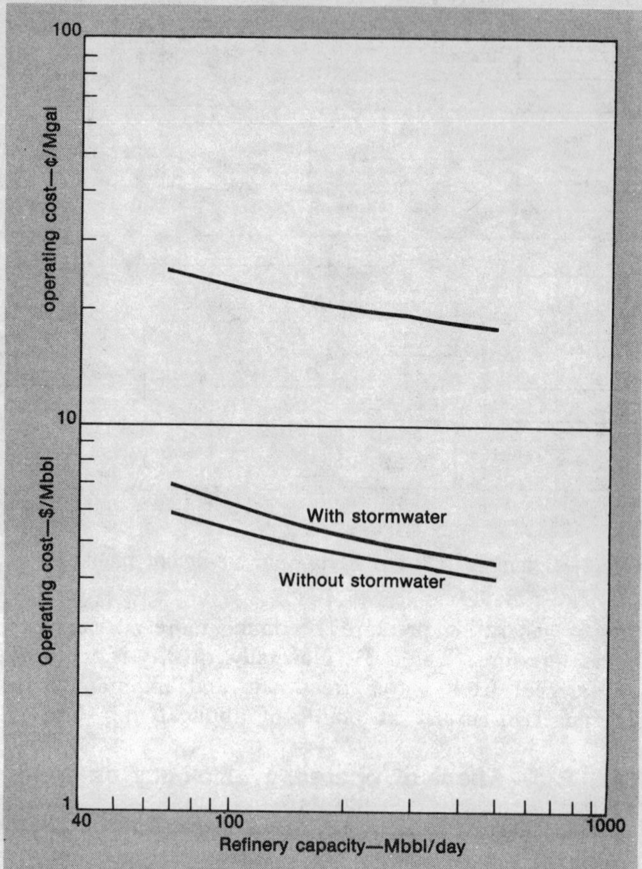

Fig. 7—Effect of refinery capacity on the direct operating costs of a wastewater treatment plant.

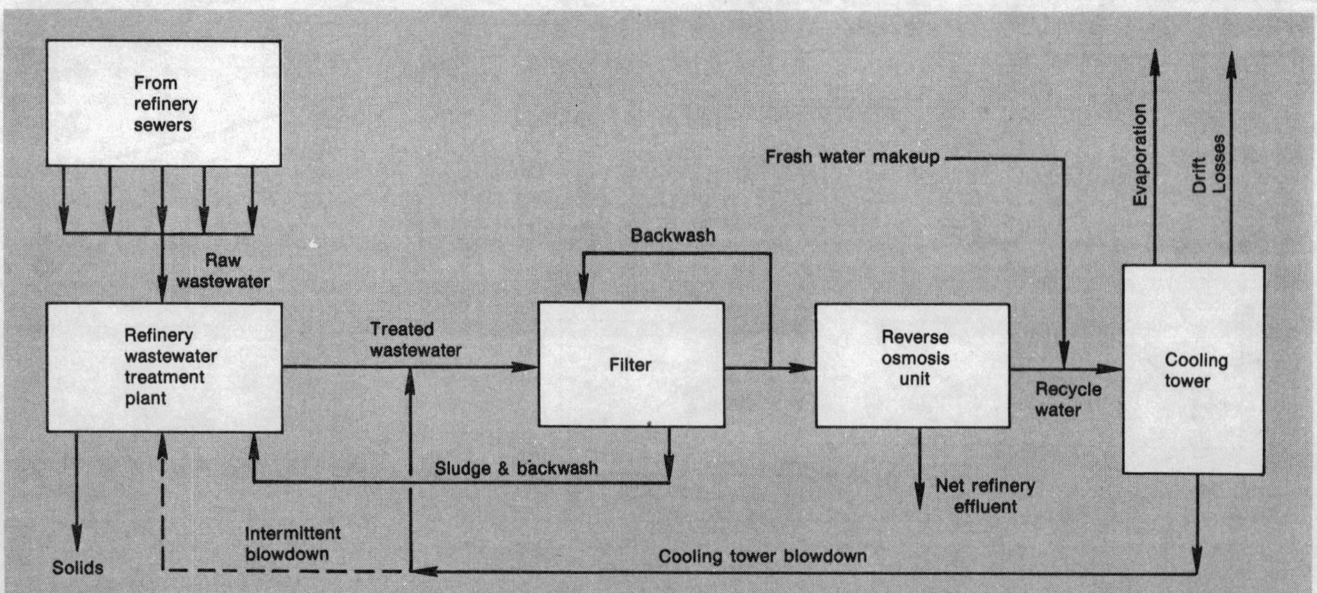

Fig. 8—Use of reverse osmosis can allow total water reuse.

used as BFW for low-pressure steam generation application.

TREATMENT FACILITIES

A refinery in the United States needs to provide facilities (Fig. 2) which are the "best practical control technology currently available" to make reusable water.[17] The components of the facility and the process flows of the several streams are shown in Fig. 2.

Six segregated sewers handle six distinct raw wastewater streams. Primary treatment for oil removal is by API

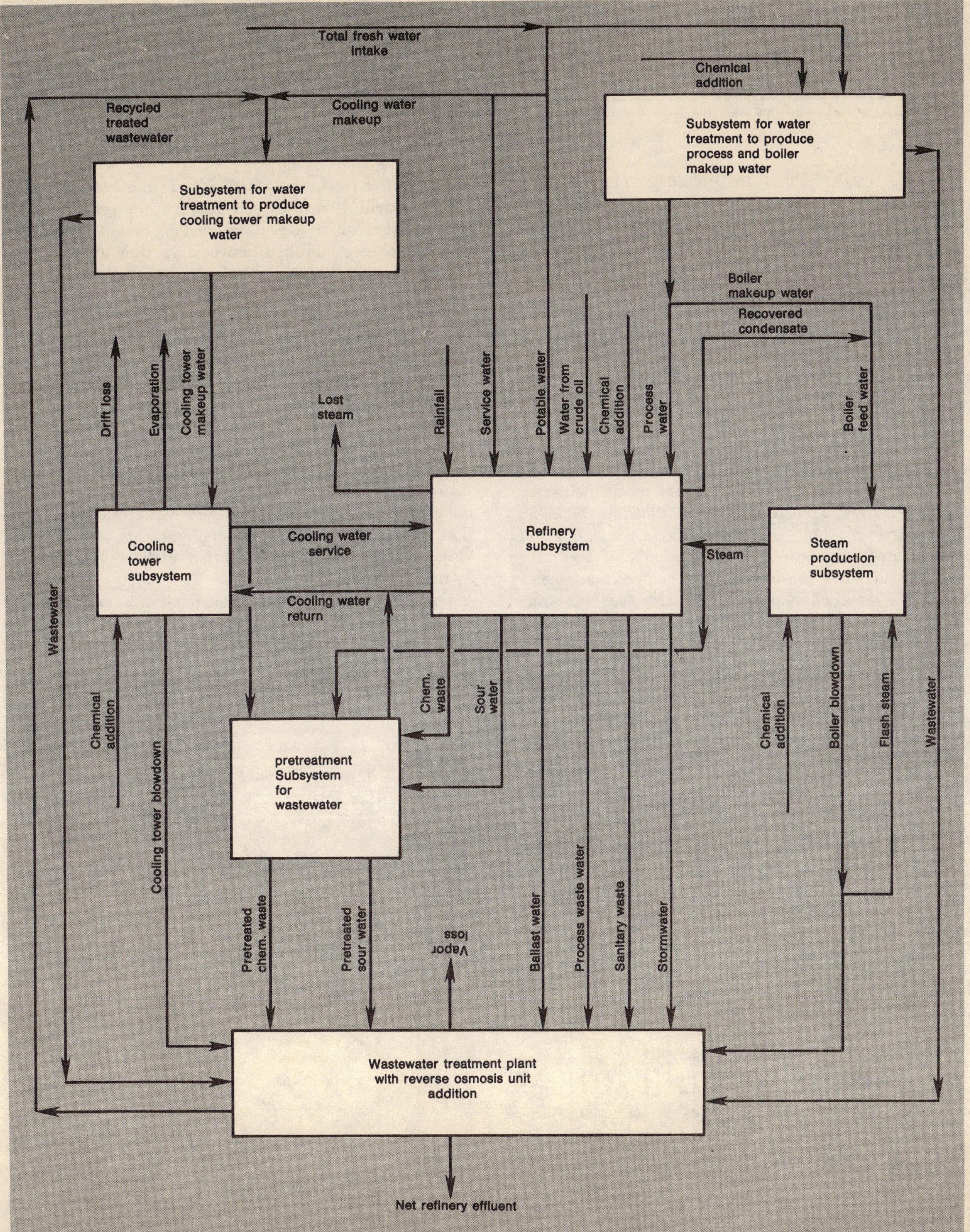

Fig. 9—A refinery wastewater treatment plan for total water reuse.

TABLE 2—Composition of Treated Refinery Wastewater[18]

			COMPOSITION OF WASTEWATER — mg/1																		
			NEW REFINERIES										EXISTING REFINERIES									
Refinery class	Refining capacity Mbpsd	Total Wastewater produced, MMgpd	NH_3 as N	Total organic carbon	Biological oxygen demand	Chemical oxygen demand	Chromium total	Oil & grease	Phenol	Sulfides	Total suspended solids	NH_3 as N	Total organic carbon	Biological oxygen demand	Chemical oxygen demand	Chromium total	Oil & grease	Phenol	Sulfides	Total suspended solids
A	10.0	0.15	13	32	20	122	0.3	13	0.1	0.2	26	34	37	26	146	0.7	20	0.4	1.3	53
B	38.4	0.93	16	40	25	152	0.4	17	0.2	0.3	33	43	70	50	274	0.8	25	0.8	1.7	66
C	102.0	2.59	19	55	35	211	0.5	20	0.2	0.3	39	51	83	59	325	1.0	30	1.0	2.0	78
D	65.5	1.78	22	71	44	269	0.6	22	0.2	0.4	44	82	104	77	426	1.1	33	1.1	2.2	88
E	222.0	5.61	36	128	80	488	0.9	36	0.4	0.6	71	132	175	125	687	1.8	53	1.8	3.6	142

[1] Refinery complexity — A = simplest; E = most complex.

separators, two of which are provided: one to handle process wastewater and another to handle similar streams from storing, shipping and receiving. Although one API separator might serve both purposes, usual refinery arrangements and practical API separator physical sizes lead economically to two separate and distinct units. An emulsion breaking unit treats recovered oil emulsions to separate and recover oil which together with oil from the API separators is returned to the refinery. Sludge for the API separators is processed in the sludge disposal system together with other sludges, to produce inert solids for land disposal.

An air flotation unit is used to reduce the oil content of the primary-treated wastewater as an intermediate treatment.[15] Further treatment occurs in the secondary treatment system, an activated sludge unit. Net biological sludge from this unit enters the sludge disposal system while treated wastewater flows to a mixing and surge pond, the effluent from which has a composition meeting the criteria for discharge (Table 1). However, this water requires additional treatment to upgrade quality for reuse as CTMU.

This water treatment plant (Fig. 2) has been simulated to the extent that the feed streams are related to the refinery from which they have been generated. A relationship has been established between the flow and composition of each stream and the size and configuration of the refinery.

Fig. 3 presents the capital cost (first quarter 1973) of the waste treatment plant of Fig. 2 as a function of the quantity of wastewater handled. These are installed costs representing the selling price of the entire plant and reflect the total cost required to build the plant and have it ready to operate. Cost includes spare equipment but not land.

Cost data in terms of \$/bpod as a function of refinery thruput are shown in Fig. 4. Figs. 3 and 4 can be combined to determine the quantity of wastewater as a function of the selected refinery configuration and thruput. Cost data are also shown in terms of the BOD_5 removed by the plant (Fig. 5).

Operating costs of the treatment plant as related to the several system variables are shown in Fig. 6 as a function of the quantity of wastewater flow. The direct operating

TABLE 3—Guidelines for Water Use[19]

	TOLERANCES—mg/1					
	Makeup to Recirculating Cooling Water	Boiler Feed Water @ Steam Pressure—psig				Treated Wastewater Class D New Refinery
Characteristic		0-150	150-250	250-400	>400	
Copper (Cu)	....	0.5	0.05	0.05	0.05	....
Silica (SiO_2)	50	40	20	5	1	....
Aluminum (Al)	0.1	5	0.1	0.1	0.1	....
Iron (Fe)	0.5	1	0.3	0.2	0.1	....
Manganese (Mn)	0.5	0.3	0.1	0.05	0.01	....
Calcium (Co) or Magnesium (Mg)	50	As Recd	0	0	0	....
Bicarbonate (HCO_3)	24	50	30	5	0	....
Sulfate (SO_4)	200	....	....	....	....	....
Chloride (Cl)	500	....	....	....	....	....
Dissolved Solids	500	700	500	100-150	50	....
Hardness ($CaCO_3$)	130	20	10	0	0	....
Alkalinity ($CaCO_3$)	20	140	100	70	40	....
pH (Min.)	per TDS	8.0	8.4	9.0	9.6	....
BOD	....	....	....	....	....	44
Suspended Solids (Turbidity)	100	20	10	5	1	44
Organics (TOD)	1	....	....	....	....	71
Oxygen Consumed (COD)	75	15	10	4	3	269
Dissolved Oxygen	....	1.4	0.14	0	0	....
Hydrogen Sulfide (H_2S)[a]	....	5	3	0	0	0.4
Aluminum Oxide (Al_2O_3)	....	5	0.5	0.05	0.01	....
Carbonate (CO_3)	....	200	100	40	20	....
Hydroxide (OH)	....	50	40	30	15	....
Color	....	80	40	5	2	....
Phenols	....	....	....	....	....	0.2
Sulfate/Carbonate (Na_2SO_4/Na_2CO_3)	....	1:1	2:1	8:1	13:1	....
NH_3 (NH_4)	....	0.1	0.1	0.1	0.1	22
Cr	....	....	....	....	....	0.6
Oil & Grease	....	....	....	....	....	22

[a] Except when odor in live stream would be objectionable.

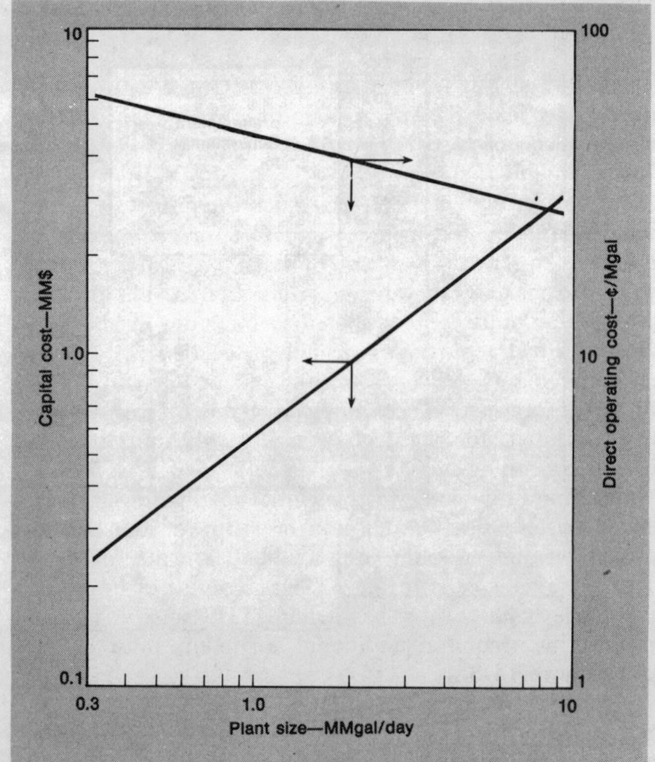

Fig. 10—Costs associated with a reverse osmosis plant.

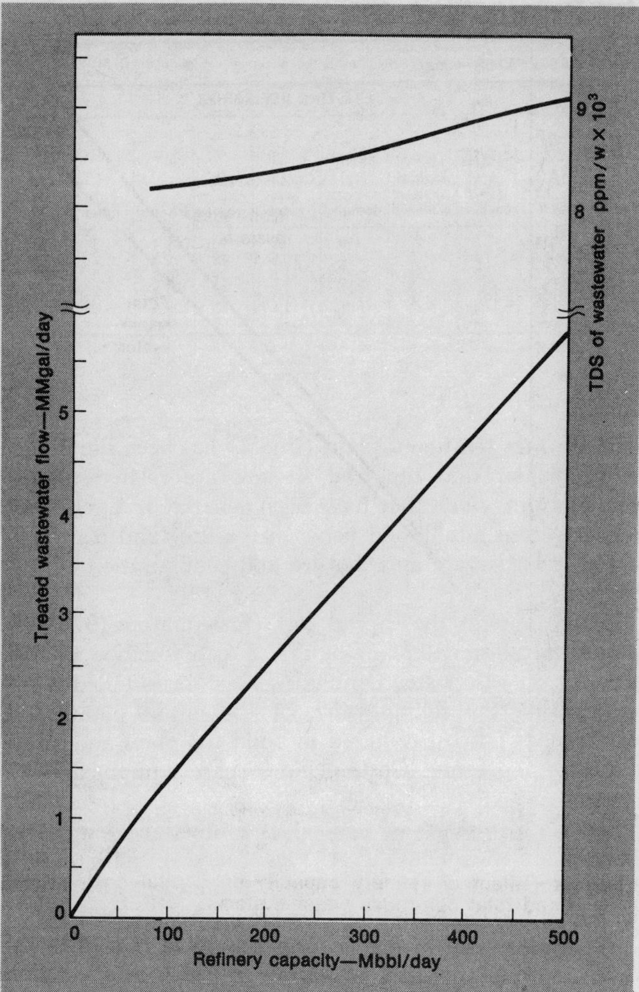

Fig. 11—Effect of refinery capacity on the quantity and total solids of treated water used for recycle.

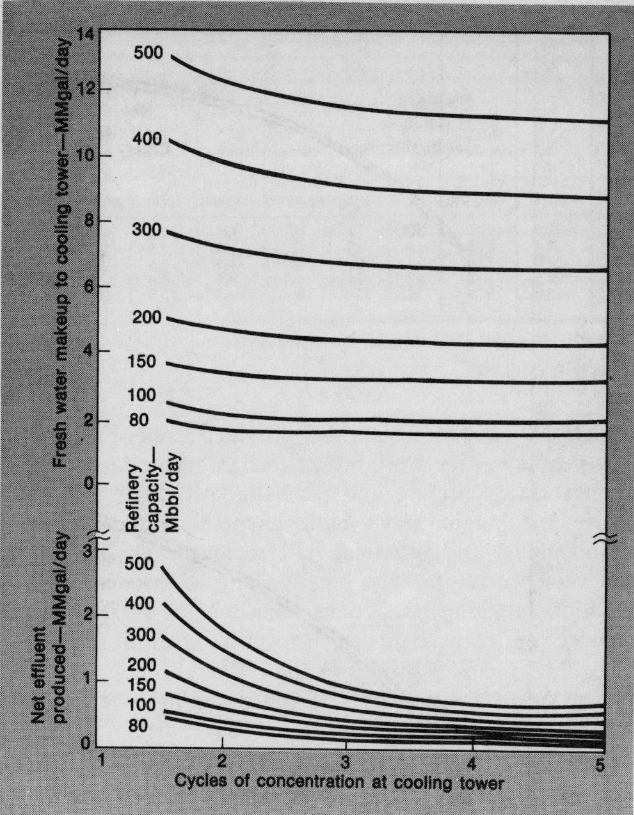

Fig. 12—Effect of concentration cycles on the amount of cooling tower makeup and net refinery effluent.

costs (¢/M gal) are defined as costs associated with utility, labor and chemical but not including maintenance. Cost data in terms of refinery capacity (Fig. 7) include stormwater handling.

Normal treated wastewater effluent requires an additional treatment step to make it usable for cooling tower makeup. This additional cost must be justified by water savings to make total water reuse acceptable. The quality criteria that must be considered is total dissolved solids (TDS), a characteristic not usually associated with waste treatment plants. TDS is basically independent of waste treatment processes since it is the value of raw wastewater adjusted for slight changes in concentrations due to some processing steps.

Makeup of raw wastewater includes cooling tower blowdown, waste process water and/or stripped sour water, stormwater and, in many instances, ballast water. Several of these streams concentrate TDS as a result of evaporation while ballast water has a high TDS since it is seawater. Thus, required additional treatment must essentially remove TDS.

Among the more important water treatment processes used in reducing TDS are demineralization and reverse osmosis (RO), the choice of which depends upon several factors. Demineralization, although well established for reducing TDS[23] in fresh water is not as well recognized for treating wastewater.[24] On the other hand, RO is seldom applied in fresh water treatment except where fresh water is high in TDS,[25] a capability that makes it attractive for use in treating wastewater. Level of present day technology development justifies its use in this application.[26,27,28] Hence, an RO unit capable of reducing TDS of treated wastewater to acceptable levels for reuse as CTMU should be considesed (Fig. 8). The RO unit is placed in series with the waste treatment plant. A filter is located upstream of the RO unit to act as a polishing unit in removing suspended solids from the treated wastewater. Recycle water is fed to the cooling tower, along with fresh water which may or may not be required as additional CTMU. The need for fresh water is dependent upon the quantity of raw wastewater produced by the refinery as compared to the magnitude of the refinery heat load rejected to the cooling tower.

Total water reuse requires a slight change in waste treatment processing: cooling tower blowdown (CTBD) is not treated in the activated sludge unit for BOD and COD removal but is treated at the cooling tower itself. Successful use of cooling towers as an oxidation medium has been described[29,30] but not as yet used widely. However, it may be considered a viable operation for the recycle system since organics present in cooling water are moved in the cooling tower with only occasional intermittent blowdown to the activated sludge unit required for control of organics buildup. The combination of CTBD and treated wastewater from the waste treatment system is passed to the RO unit for dissolved solids reduction.

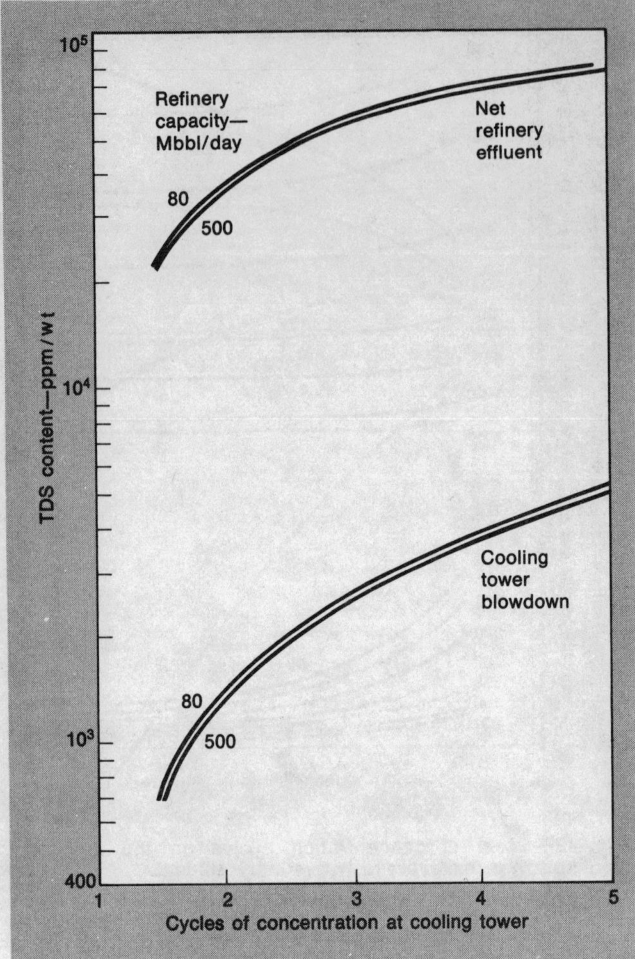

Fig. 13—Effect of concentration cycles on the solids content of refinery effluent and cooling tower blowdown.

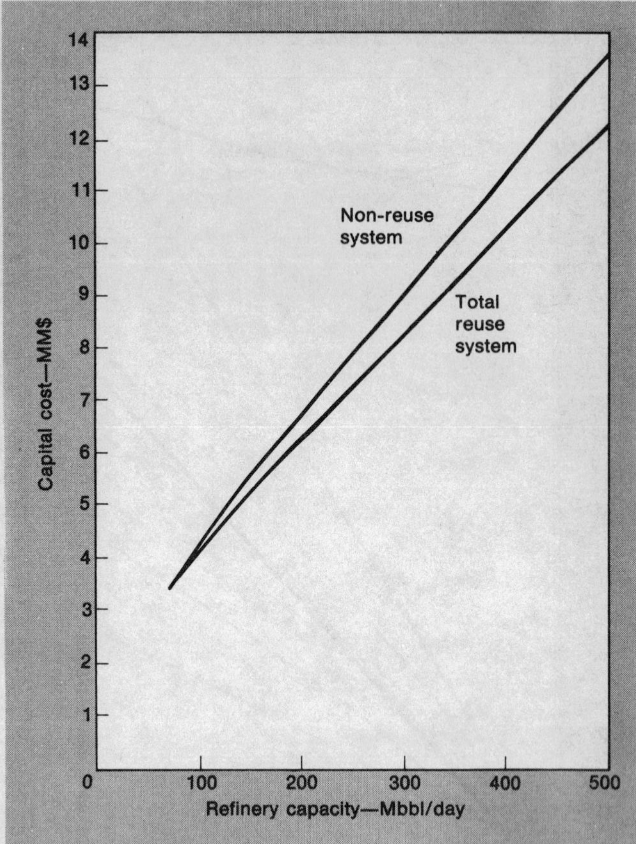

Fig. 14—Effect of refinery capacity on capital costs of total reuse and total non-reuse water systems.

Fig. 9 shows the arrangement for a refinery over-all water distribution and recycle system. It is divided into three basic subsystems: cooling tower, refinery and steam production. Water as needed is fed to each subsystem from fresh water treatment facilities. Fresh water treatment varies from none for service water to the level required for boiler makeup use. A wastewater pretreatment system is included as part of the refinery system to handle sour waters and phenolic caustics to produce recovered products and partially treated wastewaters which form a part of the raw wastewater fed to the waste treatment system.

Additional costs for waste treatment include costs associated with the RO unit and its auxiliaries (Fig. 10). These costs include those of a polishing filter and all necessary auxiliaries required to fit the unit into the system. (Fig. 8)[27,28] other than land cost.

Results of the detailed analysis[17] of the systems are shown in Figs. 8 and 9. Simulation of the system results in data plotted in Figs. 11, 12 and 13. Fig. 11 presents the quantity of treated wastewater, without CTBD, as a function of refinery size. Since CTBD does not enter the basic water treatment plant (it is handled only in the RO unit), it is not considered a portion of the raw wastewater to be treated by the waste treatment plant. Also shown in Fig. 11 is the corresponding TDS of the treated wastewater as a function of refinery size. Comparing the available treated wastewater with the CTMU required reveals that in this refinery the recycle water will be insufficient to supply total CTMU needs.

CTBD is a function of the cycles of water concentration which in turn are a function of the ratio of the TDS in the CTBD to the CTMU. Since the TDS in the CTBD is usually held at a fixed value, dependent in part upon factors other than the system TDS, it follows that the cycles of concentration attainable at the cooling tower is a direct function of the TDS of the CTMU. With a fixed composition of fresh water and with an RO unit operating to produce a constant TDS of the effluent, it is possible to portray the net wastewater produced and the fresh water makeup required as a function of the cycles of concentration achieved at the cooling tower (Fig. 12) which is based upon refinery operations and includes a 40 percent recovery of condensate and a TDS of 500 ppmw in the fresh water.

The TDS of the CTBD and the net effluent produced by the total water reuse system is shown in Fig. 13. The influence of refinery size is relatively insignificant, in terms of TDS, for the two streams.

Comparing Figs. 12 and 13, the quantity of net effluent produced decreases as the cycles of concentration increases, however, the TDS of the effluent increases. In terms of an RO unit, this is the comparison of increased capital costs contrasted with increasing operating costs to remove a fixed quantity of TDS.

Approximately 90 percent reduction in final plant

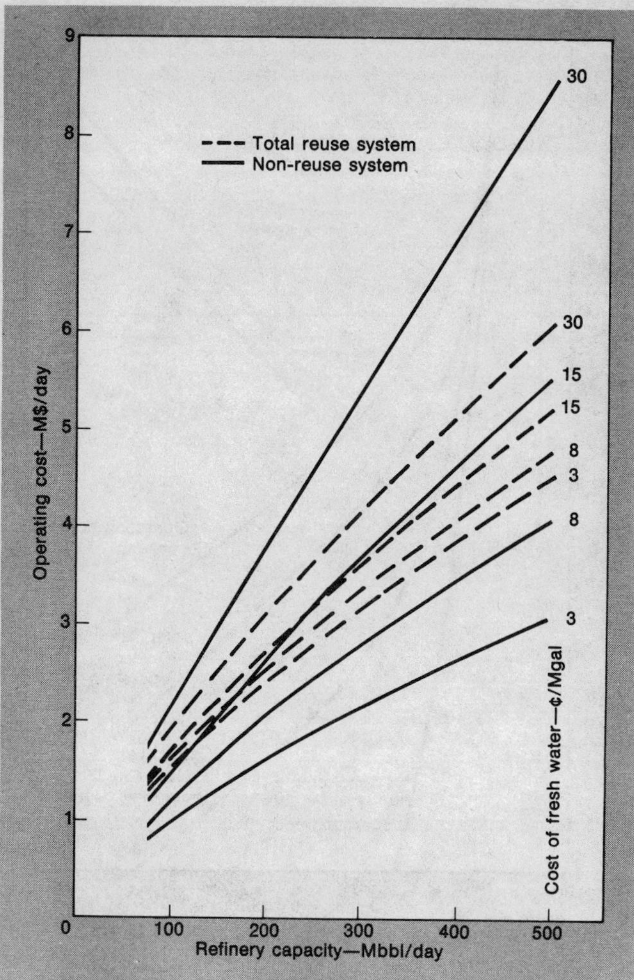

Fig. 15—Effect of refinery capacity on the operating cost of a total reuse and total non-reuse water system.

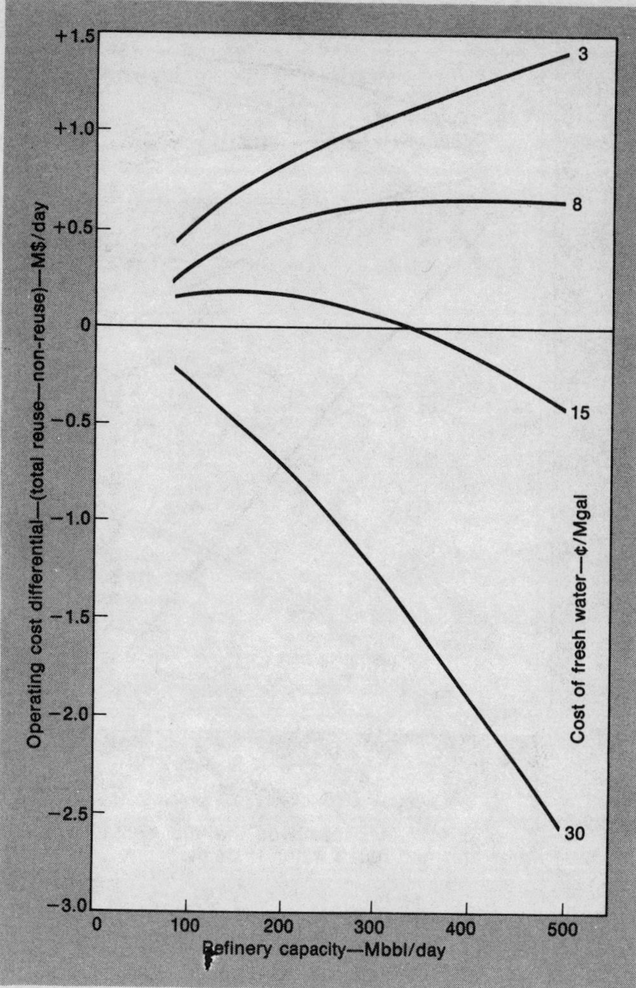

Fig. 16—Effect of refinery capacity on the cost differential for total reuse and non-reuse water systems.

effluent discharge is achieved by the total water reuse program: 2 to 2.5 gal./bbl. compared to 20 to 25 gal./bbl. effluent (Figs. 12 and 13). This compares with an EPA recommendation that refinery effluents be limited to 7 gal./bbl.[31] EPA based the recommendation on use of present technology and equipment.

Total water reuse provides the means of achieving "zero discharge" or at least reducing the net discharge to simply a solids product. By recycling the RO effluent to the incinerator all remaining liquid can be converted to vapor, leaving behind only an inert solids product for disposal.[30]

ECONOMICS

New refineries may be evaluated for total water reuse which consists of comparing the cost of the non-total reuse operation with the total reuse operation where·

Total Cost of System = Cost of Fresh Water + Capital Cost of Treatment Facilities for Fresh Water + Operating Cost of Fresh Water Treatment Facilities + Capital Cost of Waste Treatment Facilities + Operating Cost of Waste Treatment Facilities.

Reuse reduces the need for fresh water but increases the cost of treatment required to raise the quality level to the treated effluent to the equivalent of fresh water. Although not considered in the cost equation, an imposed cost of discharge could be considered. This term is a cost imposed by regulating authorities or subsequent municipal treatment facilities and is related to the quantity and quality of the refinery effluent discharge.

Capital costs of system facilities for both reuse and non-reuse arrangements are affected by refinery capacity (Fig. 14). Costs include all capital items associated with both fresh water and raw effluent treatment. Capital costs of the reuse system is less than for an equivalent non-reuse system for all refineries above 80,000 bpod capacity largely because of a reduction in the activated sludge unit capacity since most of the CTBD bypasses this unit in the water reuse system. These relationships do not apply to refineries below 80,000 bpod since the arrangement of downstream processing units are significantly different compared with the arrangement used in this study.

Operating costs for the two systems are shown as a function of refinery capacity (Fig. 15). Cost of fresh water is plotted for 3, 8, 15 and 30 ¢/Mgal. to include the effects of this variable upon cost. The cost of fresh water has a greater influence on the operating cost of one system as compared to the other, thus, at 3 ¢/Mgal. the operating cost of the non-reuse system is lower than a corresponding reuse system for all refinery capacities up to 500,000 bpod. However, at 30 ¢/Mgal. fresh water, the reverse is true (Fig. 16). The point of equal operating

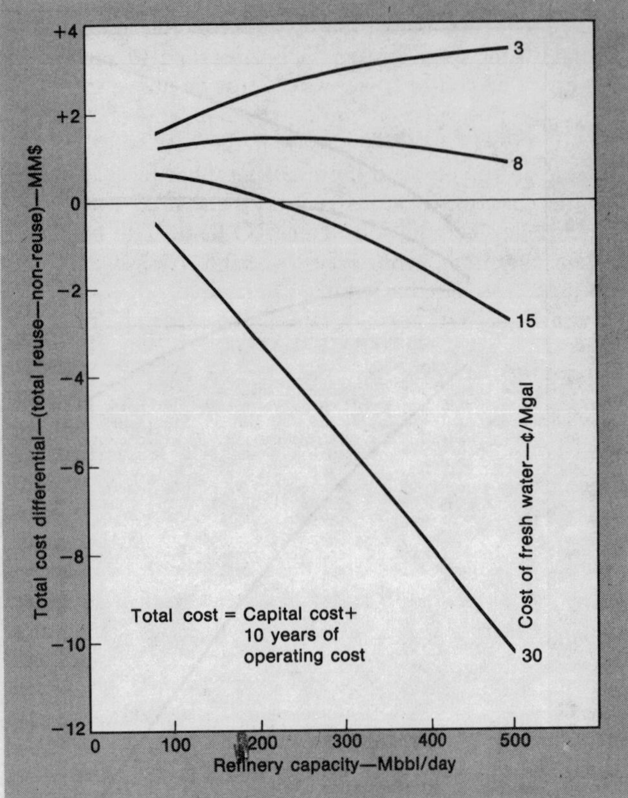

Fig. 17—Effect of refinery capacity on the total cost differential for total reuse and non-reuse water systems.

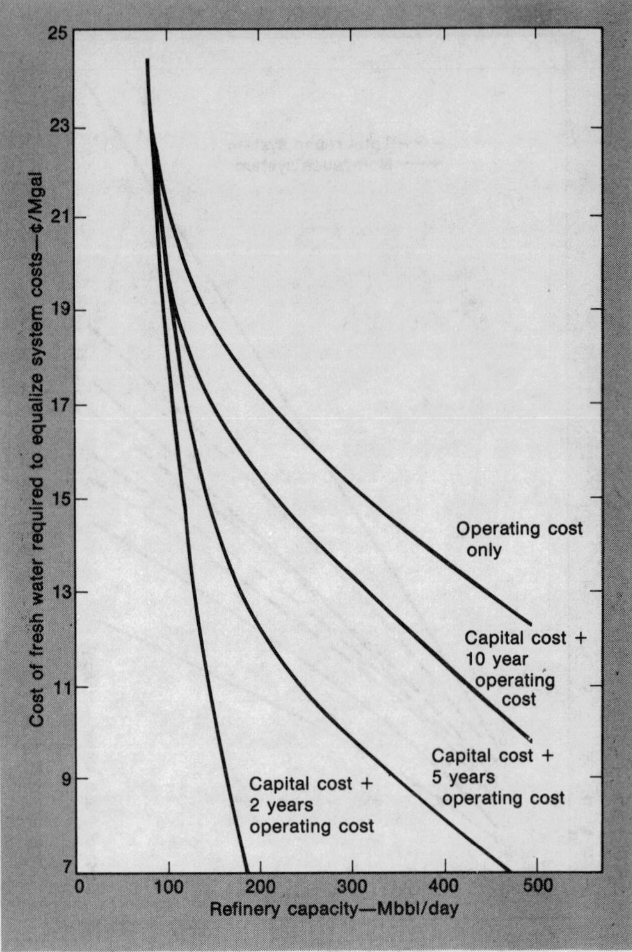

Fig. 18—Effect of refinery capacity on the minimum cost of fresh water required to equalize costs between total reuse and total non-reuse water systems.

costs, i.e., a difference of zero, is crossed by the 15 ¢/Mgal. fresh water cost at about 330,000 bpod. Both 3 and 8 ¢/Mgal. are above zero for all capacities, whereas 30 ¢/Mgal. is below zero over the entire range of capacities considered.

Increased operating costs of reuse are offset by lower capital costs (Fig. 17). Favorable capital cost differential of the reuse systems tend to lower the curves at all fresh water costs; thus, the 15 ¢/Mgal. curve now crosses the zero line at a refinery capacity of about 210,000 bpod. Hence, when total costs are considered, a smaller refinery will find it economical to institute total water reuse policies.

The most significant cost is that of fresh water. It determines the minimum size of refinery above which reuse is indicated to be an attractive policy and below which it would not prove economically justified. The minimum fresh water cost at which a given sized refinery would find it economically attractive to institute a policy of total reuse is illustrated in Fig. 18. Since both operating and capital costs are being considered, parameters of operating periods are shown as 2, 5 and 10 years operating time. In effect this represents the period of useful equipment life or depreciation of all the equipment involved. Fig. 18 represents boundary lines distinguishing the equalization of costs. This establishes the minimum cost of fresh water required to yield a continuous profit for the reuse system as contrasted to the non-reuse system. As an example, considering 10 years of operation, a 500,000 bpod refinery would show an equivalency in total costs when employing the reuse system if fresh water costs are about 10 ¢/Mgal. or above, whereas a 150,000 bpod refinery would require the minimum cost of fresh water to be about 17 ¢/Mgal. before achieving the same state.

As the operating time considered is decreased, the minimum cost of fresh water required decreases, due to the fact that the capital costs are less for the reuse system as contrasted to the non-reuse system. Fig. 18 can be used to determine the minimum cost of fresh water required in order for a refinery to institute an economically attractive total water reuse policy.

The economic incentive to institute total water reuse policies decreases with decreasing refinery capacity. Or stating it another way, the smaller refiner would need to be faced with excessively high fresh water costs before it would pay him to consider large-scale total water reuse. Thus, smaller refiners would find it difficult to reduce effluent discharges just from an economic analysis alone. This may be one of the reasons which explains a recent EPA finding that smaller refineries discharge effluents in a greater proportion to their size than do the larger refineries.[31]

Existing refinery. An existing refinery with waste treatment facilities equivalent to those shown in Fig. 2 needs only to install an RO unit, pumps, filter and other piping modifications to approach total water reuse. Considering only the cost involved for such an RO unit, Fig. 19 presents the payout curves for the cost of the add-on system. The payout period before taxes is shown as a function of the refinery capacity for the applicable fresh water cost parameter. Only a 30¢/Mgal. fresh water

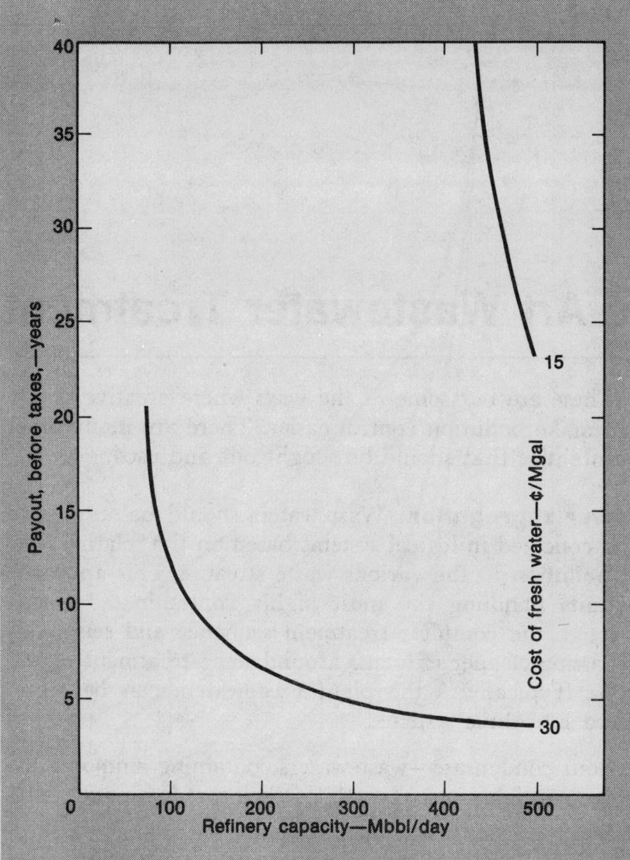

Fig. 19—Effect of refinery capacity on the payout of water reuse systems for an existing refinery.

cost (of the four previous cost costs considered) will show a payout for all refinery capacities. This shows that operating costs are greater for the reuse system than for the non-reuse system at the lower fresh water costs. The 15¢/Mgal. fresh water cost does offer a payout for the 500,000 bpod refinery although it is about 22.5 years.

The fresh water cost that can offer any actual payout time, although it may be excessively high, can be estimated from Fig. 18 using the operating cost curve. This line represents the value of fresh water required at each refinery capacity to equalize operating costs between the two systems. Fresh water cost values above the curve indicate a payout, values below offer no possible payout. Thus for a 400,000 bpod refinery, the minimum cost of fresh water required for a payout is about 14 ¢/Mgal., which represents nearly an infinite payout, whereas, from Fig. 19, at 30 ¢/Mgal. the payout is reduced to about four years. The operating cost curve of Fig. 19 can be used as a first indication of the economic potential for a total water reuse system to be installed in an existing refinery. The cost of fresh water must be above the values shown by this curve in order to have any chance of a payout. Fig. 19 shows a payout if the cost is 30 ¢/Mgal. Note again the problem the smaller refiner faces in justifying the conversion of his refinery to a total water reuse water policy by addition of an RO unit. The economics dictate that only with excessive high fresh water costs can it be considered a feasible alternative.

LITERATURE CITED

[1] Maguire, W. F., "Reuse Sour Water Stripper Bottoms," *Hydrocarbon Processing, 54*, No. 9, 151 (September 1975).

[2] Harpel, W. L., and James, E. W., "Wastewater Use Saves On Cooling Tower Makeup," *Oil & Gas Journal*, 73, No. 35, 118 (Sept. 1, 1975).

[3] Porter, J. W., Blake, S. H., and Milligan, R. T., "Complete Industrial Waste Water Reuse Goal of Refining Study," *Oil & Gas Journal*, 71, No. 39, 70 (Oct. 1, 1973).

[4] Anon., "Grass Roots Pollution Abatement," *Oil & Gas Journal*, 70, No. 2, 60 (Jan. 10, 1972).

[5] Racine, W. J., "Environment Protection Gets First Priority at Cherry Point," *Oil & Gas Journal*, 69, No. 46, 162 (Nov. 15, 1971).

[6] Schmidt, C. J., Kugelman, I., and Clements, E. V., "Municipal Wastewater Reuse in the U.S.," *Journal Water Poll. Control Fed.*, 47, No. 9, 2229 (September 1975).

[7] Harpel, W. L., and James, E. W., "Aspects of Waste Water Reuse as Cooling Tower Makeup," Paper presented at Proceedings of 34th International Water Conference, Oct. 30-Nov. 1, 1973, Pittsburgh, Pa.

[8] Racine, W. J., "Plant Designed to Protect the Environment," *Hydrocarbon Processing*, 51, No. 3, 115 (March 1972).

[9] Sawyer, G. A., "New Trends in Wastewater Treatment and Recycle," *Chem. Eng.*, 79, No. 15, 120 (July 24, 1972).

[10] Aalund. L. R., "Big Heartland Refinery Ready for '70's," *Oil & Gas Journal*, 71, No. 16, 45 (April 23, 1973).

[11] Wegren, A. A., and Burton, F. L., "Refinery Wastewater Control," *Journ. Water Poll. Control Fed.*, 44, No. 1, 117 (January 1972).

[12] Carnes, B. A., Eller, J. M., and Martin, J. C., "Reuse of Refinery & Petrochemical Wastewaters," *Ind. Water Eng.*, 9, No. 4, 25 (June/July 1972).

[13] Brooke, M., "Utilize Utility Dollars," *Combustion*, 41, No. 9, 22 (September 1970).

[14] "The Cost of Clean Water," U.S. Dept. of the Interior, Vol. III, Industrial Waste Profile No. 5, Petroleum Refining, FWPCA Publication No. I.W.P.-5, Washington, D.C. (November 1967).

[15] "1967 Domestic Refinery Effluent Profile," American Petroleum Institute, Committee on Air and Water Conservation (September 1968).

[16] "Petroleum Industry Raw Waste Load Survey," American Petroleum Institute, Committee on Environmental Affairs (December 1972).

[17] Finelt, S., "Petroleum Refining: Environmental, Energy and Economical Impacts," Ph.D. Dissertation, University of Houston, May 1975.

[18] Anon., "EPA Issues Waste-Treatment Guidelines," *Oil & Gas Journal*, 70, No. 46, 30 (Nov. 30, 1972).

[19] Petrasek, A., and Esmond, S., "Industrial Uses of Municipal Wastewater," *Ind. Water Eng.*, 10, No. 4, 26 (July/August 1973).

[20] Barker, P. A., "Wastes Treatment for Steam Generating Systems," *Ind. Water Eng.*, 12, No. 2, 5 (March/April 1975).

[21] API, *Manual on Disposal of Refinery Wastes, Volume on Liquid Wastes*, 1st Edition, American Petroleum Institute (1969).

[22] Beychok, M. R., *Aqueous Wastes from Petroleum and Petrochemical Plants*, 1st Edition, London, England: John Wiley & Sons (1967).

[23] Goldstein, P., and Griffin, R. W., "Water Requirements for Modern Boilers," *Ind. Water Eng.*, 9, No. 6, 20 (October/November 1972).

[24] Wolff, J. J., "Deionization of Waters Containing Organic Matter." Paper presented at 30th Annual Meeting, Int. Water Conf., Oct. 30, 1969, Pittsburgh, Pa.

[25] Cecil, L. K., "Complete Waste Reuse—Industry's Opportunity," National Conference of Complete Waste Reuse," Washington, D.C. April 23, 1973.

[26] Anon., "Reverse-Osmosis System Wins for Cummins Engine Co.," *Power* 119, No. 11, 60 (November 1975).

[27] Newkirk, R. W., and Schroeder, P. J., "RO Can Help with Wastes," *Hyd. Proc.* 51, No. 10, 103 (October 1972).

[28] Cruver, J. E., and Nusbaum, I., "Application of Reverse Osmosis to Wastewater Treatment," *Jour. Water Poll. Control Fed.*, 46, No. 2, 301 (February 1974).

[29] Anon., "Sun Oil Cuts Water Demand, Ups Quality of Effluent," *Petro/Chem Eng.*, 40, No. 3, 22 (March 1968).

[30] Crits, G. J., "Zero Blowdown from Cooling Tower—Problems & Some Answers," Paper presented at Procedings of 34th Int. Water Conf., Oct. 30-Nov. 1, 1973, Pittsburgh, Pa.

[31] "Economic Analysis of Proposed Effluent Guidelines—Petroleum Refining Industry," EPA, Office of Planning & Evaluation, Washington, D.C., June 1973.

State-of-the-Art Wastewater Treatment

Put water pollution control systems into focus for best results in meeting new laws and regulations

Milton R. Beychok, Fluor Corp., Los Angeles

THE BEST SYSTEM for treating a plant's wastewater must be tailor-made for the plant. System design begins with process plant design and includes logical arrangements for separating, collecting and treating the various wastewaters according to individual stream needs. Strategic use of in-plant pretreatment is required for the most highly contaminated wastes and the final treatment trains must be designed to remove oil and supended solids, reduce biological oxygen demand (BOD), and eliminate dissolved organics.

PLANT DESIGN

The chemical engineer who designs a new process plant should devote as much attention to minimizing potential pollution sources as to optimizing the basic process units[1, 2]. Initial plant design can pay for itself in reduced costs for waste treatment facilities through selection of equipment and designs that avoid pollution problems[3]:

• **Air-coolers** minimize the use of cooling water and reduce the amount of wastewater effluent.

• **Reboilers** avoid production of sour condensate when used in place of steam stripping.

• **Surface condensers** cut production of contaminated wastewater when used in place of barometric condensers.

• **Vacuum pumps** eliminate use of water required by barometric and surface condensers.

• **Hydrodesulfurization** produces no spent caustic and similar wastes normally associated with chemical treatment of products.

• **Water reuse** cuts down on fresh water requirements and wastewater production. For example, clean condensate may be recycled to recover concentrated waste components. Sour condensate, after stripping out hydrogen sulfide and ammonia, can be used to make low-pressure dilution steam for olefin cracking furnaces or for washwater in crude desalting. Reuse in desalting has the added advantage of further purifying the water by removing phenols.

These are but some of the ways where creative design can make pollution control easier. There are many other possibilities that should be sought out and used.

Sewer segregation. Wastewaters should be segregated and collected in logical systems based on the relative level of pollution in the various waste streams. This approach permits handling the most highly contaminated wastes through the complete treatment sequence and selectively bypassing cleaner effluents around some treatment operations. Typically[4, 5] the plant wastewaters can be segregated into three systems:

• Sour condensate—wastewaters containing ammonia hydrogen sulfide, and phenols as their predominant pollutants.

• Oily water—wastewaters containing free or dissolved hydrocarbons.

• Clean water—non-oily stormwater from non-process areas and water such as boiler blowdown, which contains only inorganic salts.

In-plant treatment. Maximum use of in-plant pretreatment should be made to remove pollutants from highly contaminated wastes before they are routed to the main wastewater treatment facilities. For example, sour condensates can be steam stripped rather easily (Fig. 1) to remove about 99 percent of the hydrogen sulfide and about 96 percent of the ammonia content. In many cases steam stripping may also remove as much as 20-40 percent of any phenols present.

The usual stripper design parameters are within these

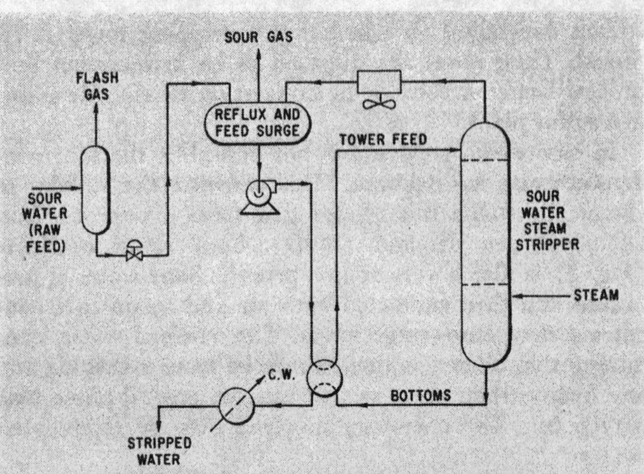

Fig. 1—Typical sour water steam stripper.

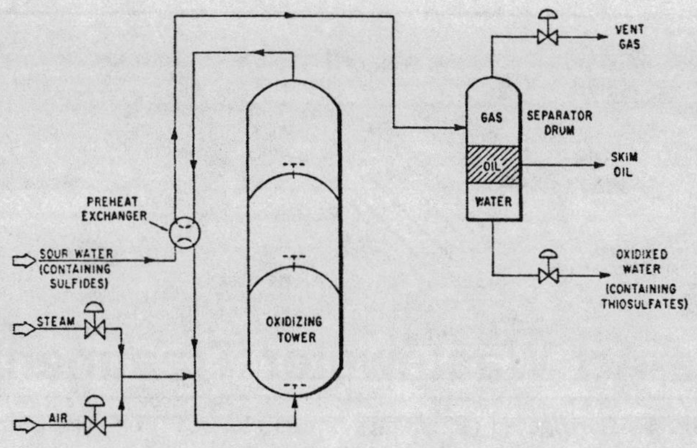

Fig. 2—Sour water oxidizer.

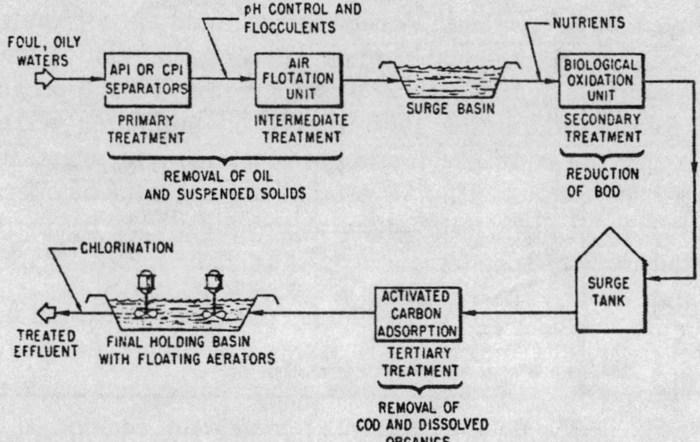

Fig. 3—Complete system for treating a refinery wastewater.

ranges:

Tower bottom pressure	21 psia
Tower bottom temperature	230°F
Number of trays	8-10
Stripping steam rate* (lbs/gal. of tower feed)	0.9-1.1
H_2S removal	98-99+%
NH_3 removal	95-97%
Phenols removal	20-40%

*Excluding requirements to raise the feed temperature to 230° F

A detailed tray-to-tray design procedure for steam strippers has been proposed. [4]

Acid gases may be removed in a relatively wet or dry stream depending on whether the stripping tower is refluxed. These gases are disposed of by incineration in a process heater firebox or by conversion to saleable sulfur in a sulfur plant.

In case steam stripping is not desirable, the sour condensates may be oxidized. This converts the sulfides to chemically stable thiosulfates that have a very low biological oxygen demand (BOD). Sour water oxidizing (Fig. 2) is also a very simple process. Sour water is preheated and then contacted with air and steam in a concurrent flow, three-stage tower. The oxidized water, containing thiosulfates, is then separated from excess air and any hydrocarbon gases as well as from any oil phase that may occur. The chemistry involved may be represented as:

$$2S^{--} + 2O_2 + H_2O \rightarrow S_2O_3^{--} + 2OH^- \quad (1)$$

or:

$$2SH^- + 2O_2 \rightarrow S_2O_3^{--} + H_2O \quad (2)$$

The oxidation process may be applied to ammoniacal sulfides, NH_4SH and $(NH_4)_2S$, as well as to sulfidic spent caustics, NaSH and Na_2S. A detailed design [4] is usually based on the following design criteria:

Temperature, bottom	165-225°F
Pressure, bottom	75-100 psia
Inlet air, actual/theoretical	1.4-2.1
Superficial air velocity, ft/sec bottom	0.08-0.10
Sulfide oxidation rate, (lbs/hr) ft^3 of tower	0.35

More sophisticated, propietary processes are also available for the treatment of sour condensates, including

- WWT process (Chevron Research) for the two-stage distillation of sour condensates.
- AMMONEX process (Howe-Baker) for the extraction of ammonia from sour gases stripped out of sour condensates.
- SULFOX process (UOP) for the catalytic oxidation of sour condensates.

OVER-ALL SYSTEM

There are many options to consider for the sequence or combination of treatment steps in an over-all system. However, the main functions of a good system can be classified as follows:

Removal of oil—The primary removal of oil is usually provided by an API separator or a corrugated plate interceptor (CIP). Effluent from these *primary separators* will still contain about 50-100 ppm of oil. [4]

Removal of suspended solids—Removal of suspended solids and further removal of oil is usually accomplished by chemical sedimentation or air flotation. Effluent from these *intermediate separators* contains about 5-20 ppm of oil and 25-60 ppm of suspended solids. [4, 5]

Reduction of BOD—Some type of biochemical oxidation will be required as a *secondary treatment* to reduce the biological oxygen demand (BOD) of the treated wastewater. A typical wastewater from a complex petroleum refinery may have a 5-day BOD of 150-350 ppm which can be reduced by 75-95 percent in a well designed bio-treater.

Removal of dissolved organics—Some effluents from petrochemical plants may contain dissolved organics which are not amenable to biological degradation. These may require *tertiary treatment* to remove the dissolved organics. Activated carbon adsorption can remove as much as 95 percent of the dissolved organics from typical industrial wastewaters. [6, 7]

A complete system (Fig. 3) also includes a final holding basin downstream of the treatment train. This basin permits the dilution or equalization of any "off-specification" effluent caused by a brief upset and the impounding and re-processing off-specification effluent in the event of a prolonged mal-operation. In some cases small floating aerators may be provided in the basin to provide a positive oxygen (D.O.) content in the treated effluent. A final chlorination step also may be required if fecal coliform or other pathogenic materials are expected (as from a sanitary waste system). Not every refinery or petrochemical plant requires the complete system. However, in view of the ever-growing body of environmental laws and regulations, most industrial plants will soon be expected to provide a system that extends at least through some form of secondary treatment.

System elements. Some of the available options for accomplishing the functional steps in the over-all system are briefly summarized with a typical schematic of each functional step.

Oil-water separators. Oil removal in an API or CPI separator is accomplished by taking advantage of the specific gravity difference between oil and water, and by providing sufficient low-velocity residence time for the physical separation to take place. The CPI separator, developed and licensed by the Shell Oil Co., employs corrugated parallel plates to assist the oil-water separation by providing surfaces upon which the oil globules coalesce and are channeled upward. (Fig. 4).

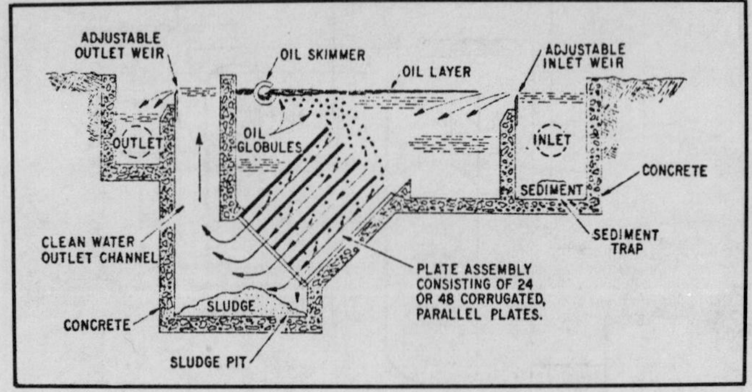

Fig. 4—Corrugated (CPI) Plate Interceptor (CPI) oil separator.

Air flotation. An air flotation unit uses finely dispersed air bubbles which rise to the surface, carrying with them oil which has adhered to the bubbles. The surface of the unit is swept by a rotating skimmer arm which removes the oil-water froth. The clarified water underflows the oil-water froth and exits from the unit. A portion of the clarified effluent is saturated with compressed air and then recycled back to the unit through a flash valve which releases the air in the form of finely dispersed bubbles. Any heavy sediment, that may be present, settles to the bottom and is removed as a wet, oily sludge by a rotating scraper arm.

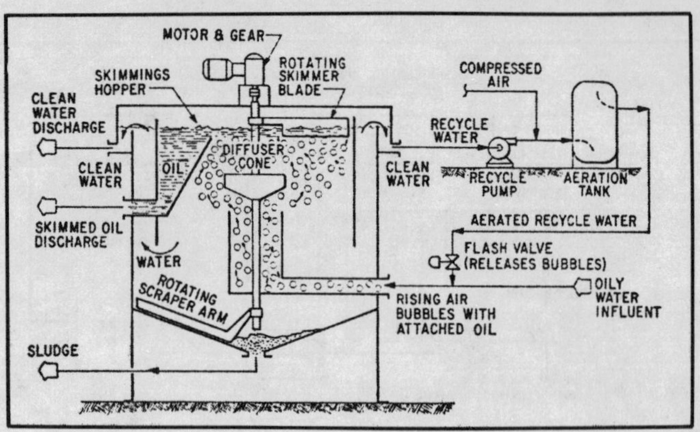

Fig. 5—Air flotation unit with recycle water aeration.

Air flotation units vary widely from cylindrical tanks (Fig. 5) to rectangular concrete pits. However in most installations chemical flocculents (lime or proprietary polyelectrolytes) are added to the air flotation influent to assist in the separation. Typical design criteria are:

Flotation tank retention	15-20 minutes of total flow (influent plus recycle)
Flotation tank rise rate	3 gal/min. per ft^2 of liquid surface, based on total flow
Recycle rate	50% of influent rate (33% of total flow)
Flocculents	About 25 ppm, based on total flow
Air rate	0.25-0.50 SCF/100 gals. of total flow

Biological oxidation. Processes involving the use of bacteria or other microbes for the stabilization (oxidation) of wastes are variously called biological, biochemical or bacterial oxidation processes. They all use oxygen and microbial action in a reaction that may be written:

Organic waste + microbes + oxygen source →
new microbes + end products + excess energy (3)

In an aerobic system where the oxygen source is air or free oxygen, the end products are primarily CO_2 and H_2O. If the waste contains nitrogen, phosphorus and sulfur, then the end products will include NO_3^{--}, PO_4^{---} and SO_4^{--}.

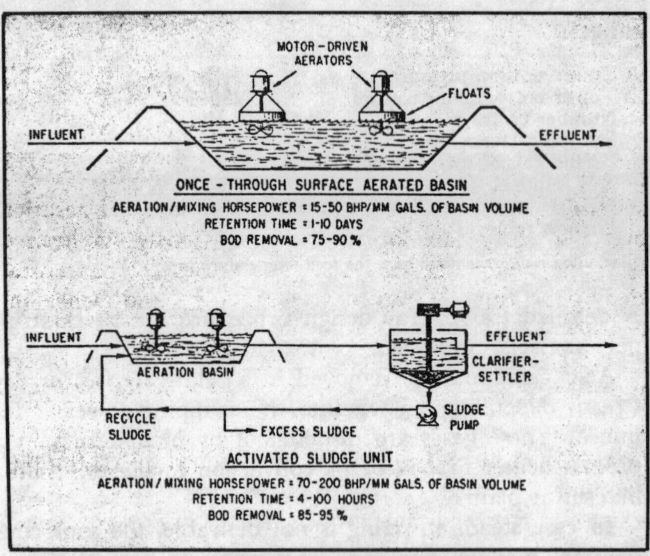

Fig. 6—Biological oxidation systems.

The various oxydation systems may be broadly categorized according to reaction rate:

- Low—Natural oxidation lagoons (with no mechanical addition of oxygen).
- Intermediate—Once-through basins with mechanical addition of oxygen by surface-mounted aerators.[8] No recycle of sludge is employed.
- High—Activated sludge processes involving contacting and mixing aqueous wastes with microbial sludge in a

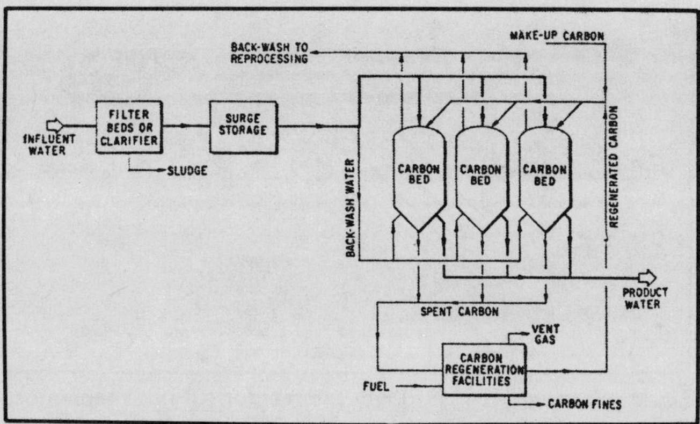

Fig. 7—Wastewater treatment by activated carbon adsorption.

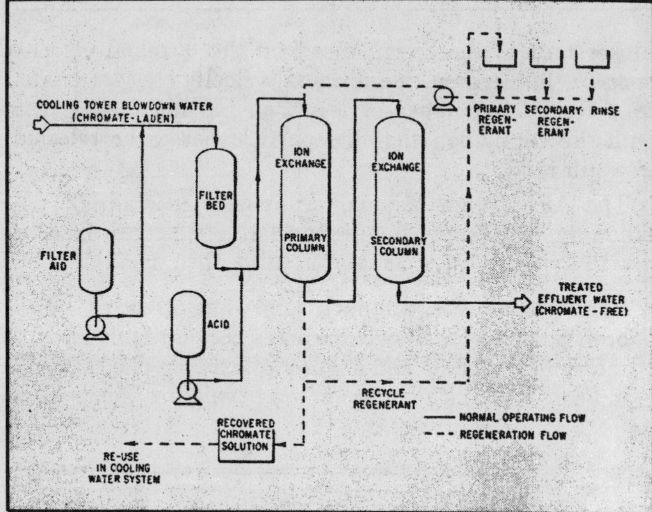

Fig. 8—Chromate recovery by ion exchange.

system supplied with a high level of mechanical aeration and with a means for recycling previously synthesized microbial sludge. Sludge recycle increases the concentration of reactant microbes (Equation 3) and hence increases the reaction effectiveness.

Typical surface aerated basins and activated sludge units (Fig. 6) can reduce a wastewater BOD by 75-95 percent.

Activated carbon adsorption. Removal of pollutants from wastewaters by adsorption onto a fixed bed of granular activated carbon is a relatively new process which has been the subject of intensive development over the past few years. Of the numerous processes under evaluation for the advanced treatment of wastewaters, activated carbon adsorption has emerged as one of the most practical and most promising in terms of performance. Well over a dozen plants, varying in throughput from 50,000 gallons per day to 10 million gallons per day, are currently in operation or in some stage of design and construction. Many of these are pilot or demonstration units, designed to treat effluents from the conventional secondary treatment of municipal sanitation wastes. However, some are in service on industrial wastes[7]. The Atlantic-Richfield refinery at Watson, Calif., has reported completion and operation of a 4.2 million gallon per day unit.[9]

A typical carbon adsorption unit (Fig. 7) is a semi-batch, cyclic process which consists of three main components:

- Pre-filtration or clarification
- Carbon bed adsorbers
- Carbon regeneration facilities.

Some of the major design parameters for the adsorbers are:

Flow rate, gpm/ft^2	5-10
Flow rate, ft/min.	0.67-1.34
Bed depth, ft.	10 minimum
Contact time, minutes	15-50
Carbon capacity, lbs COD/lb. C	0.2-1.2

Chromate removal. Most industrial cooling water systems require dosage of the water with chemicals to prevent scaling and fouling of heat exchangers and to inhibit corrosion. The most effective corrosion inhibitors maintain a chromate concentration of about 10-40 ppm (as CrO_4) in the system. Therefore, the system blowdown water contains chromates which are undesirable from the viewpoint of possible toxicity to aquatic life and unfavorable economics. Annual chromate losses in a system maintaining 20 ppm of CrO_4 might be:

Circulation (gpm)	Concentration Cycles	Blowdown (gpm)	Chromate Losses (lbs/year)
10,000	5	50	4,380
	8	30	2,630
20,000	5	100	8,760
	8	60	5,260

This chromate loss and contamination can be avoided by ion exchange whereby chromates are then recovered in a concentrated solution (2-5 percent CrO_4) when the ion exchange resin is regenerated. The recovered chromate solution is suitable for reuse in the cooling water system. Fig. 8 shows a schematic flow diagram for the ion exchange process. Richardson et al have published an excellent discussion of design and operational parameters.[10] A full-scale system, treating 60 gpm of blowdown from a 24,000 gpm cooling tower, is in successful operation within a large petrochemical plant.[11]

ACKNOWLEDGMENT

Adapted from a paper presented at the "First Technical Environmental Conference" sponsored by the Instituto de Ingenieros Quimicos de Puerto Rico, July 1971, San Juan, Puerto Rico.

LITERATURE CITED

1 Payne, R. A., "Process Design for Pollution Control," *AIChE Workshop, Volume 2, Industrial Process Design for Water Pollution Control*, Houston, Texas, April 1969.
2 Rabb, A., "How to Reduce Wastewater Effluents from Petroleum and Chemical Process Plants Through Initial Design," *ibid.*
3 Thomson, S. J., "Techniques for Reducing Refinery Waste," ASME Petroleum Division, Denver, Colorado, September 1970.
4 Beychok, M. R., *"Aqueous Wastes From Petroleum and Petrochemical Plants,"* John Wiley & Sons Ltd., New York and London, 1967.
5 Beychok, M. R., "Trends in Treating Petroleum Refinery Wastes," *AIChE Workshop, Volume 2, Industrial Process Design for Water Pollution Control*, Houston, Texas, April 1969.
6 "Adsorption as a Treatment for Refinery Effluents," Report for the CDRW Sub-committee on Chemical Wastes, American Petroleum Institute, 1969.
7 Cooper, J. C. and Hager, D. G., "Water Reclamation with Granular Activated Carbon," Water Reuse, *Chem. Eng. Prog. Symposium Series* No. 78, Vol. 63, 1967, A.I.Ch.E.
8 Beychok, M. R., "Performance of Surface-Aerated Basins," Water-1970, *Chem. Eng. Prog.* Symposium Series No. 107, Vol. 67, 1971, A.I.Ch.E.
9 Mehta, P. L., "The Application of Activated Carbon in Tertiary Treatment of a Refinery Effluent," API 36th Midyear Meeting, Division of Refining, San Francisco, May 1971.
10 Richardson, E. W., Stobbe, E. D., and Bernstein, S., "Ion Exchange Traps Chromates for Reuse," *Environ. Sci. & Tech.*, November 1968.
11 Anon., "Louisiana Plant Recovers Chromate," *Oil & Gas Journal*, September 28, 1970.

Data Improves Separator Design

Collected operating data correlate well with performance to give better technique for designing and evaluating separators

S. J. Thomson, Fluor Engineers and Constructors, Inc., Los Angeles

DESIGN OF GRAVITY-TYPE oil water separators is improved by use of an empirical relationship based on operating data. This empirical procedure better describes the performance in a gravity-type oil water separator regardless of its shape. Based on surface area per unit flow as a parameter, the method best predicts the performance of existing separators and gives the most reliable estimate of new separators.

Generalized approaches previously acceptable are no longer adequate because of more stringent regulations which demand covered gravity-type oil water separators in addition to further waste water treatment downstream. Furthermore, quantifying performance by detailed examination of oil droplet size in operating gravity-type oil water separators is impractical if not impossible.

Correlations to account for the departure from laminar flow conditions are shown in the API procedures but relate only to "satisfactory oil removals".

Attempts to correlate effluent oil water content with retention time, Reynolds number, water depth, separator shape, flow velocity and surface area per unit flow have been made. Based on operating data provided by various oil companies, the results gave a meaningful correlation only for effluent oil content as it relates to surface area per unit flow.

PAST PRACTICE

The classical picture of what is usually assumed to happen in a gravity-type oil water separator exists only if flow is in the laminar region (Fig. 1). Flow in a separator 20 feet wide and 6 feet deep is limited to a velocity of less than 0.08 feet per minute at the upper limit of laminar flow,[1] corresponding to a Reynolds number of 2,000. This limitation is necessary to correctly apply Stokes Law to the design based on the terminal velocity concept. In addition, the terminal velocity concept relates to a Reynolds number of less than 0.5 for the particle with the stipulation that the particle has to be released in a still fluid.[2]

The API Design Manual[3] assumes a generalized condition in explaining the performance expected from a gravity-type oil water separator using their method:

"This globule size, although somewhat arbitrary, has been adopted for design purposes because both laboratory experiments and a study of existing plant data indicate that satisfactory oil removals are achieved

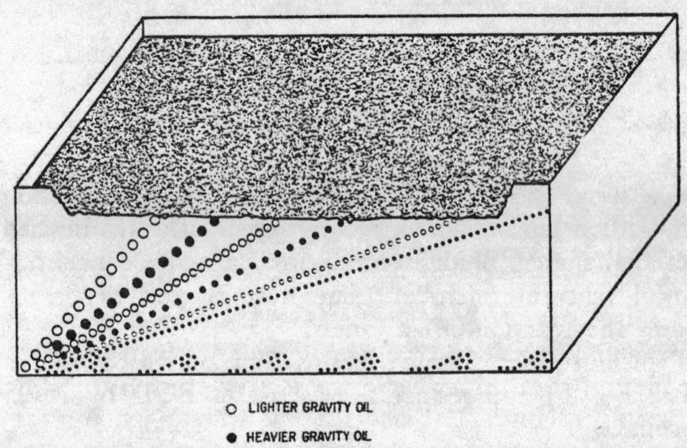

Fig. 1—Classical concept of gravity-type oil water separator.

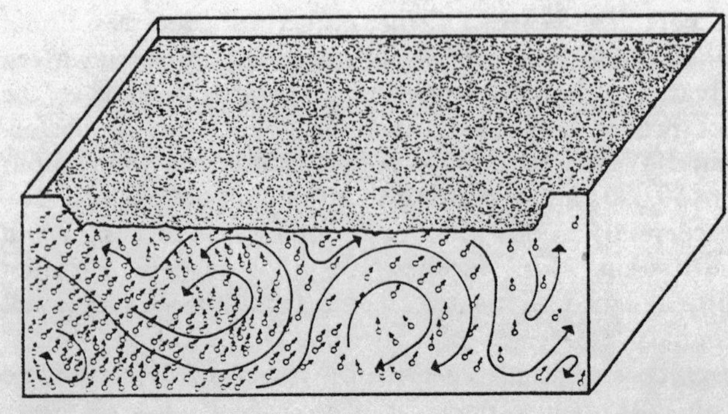

Fig. 2—Actual operation of a gravity-type oil water separator.

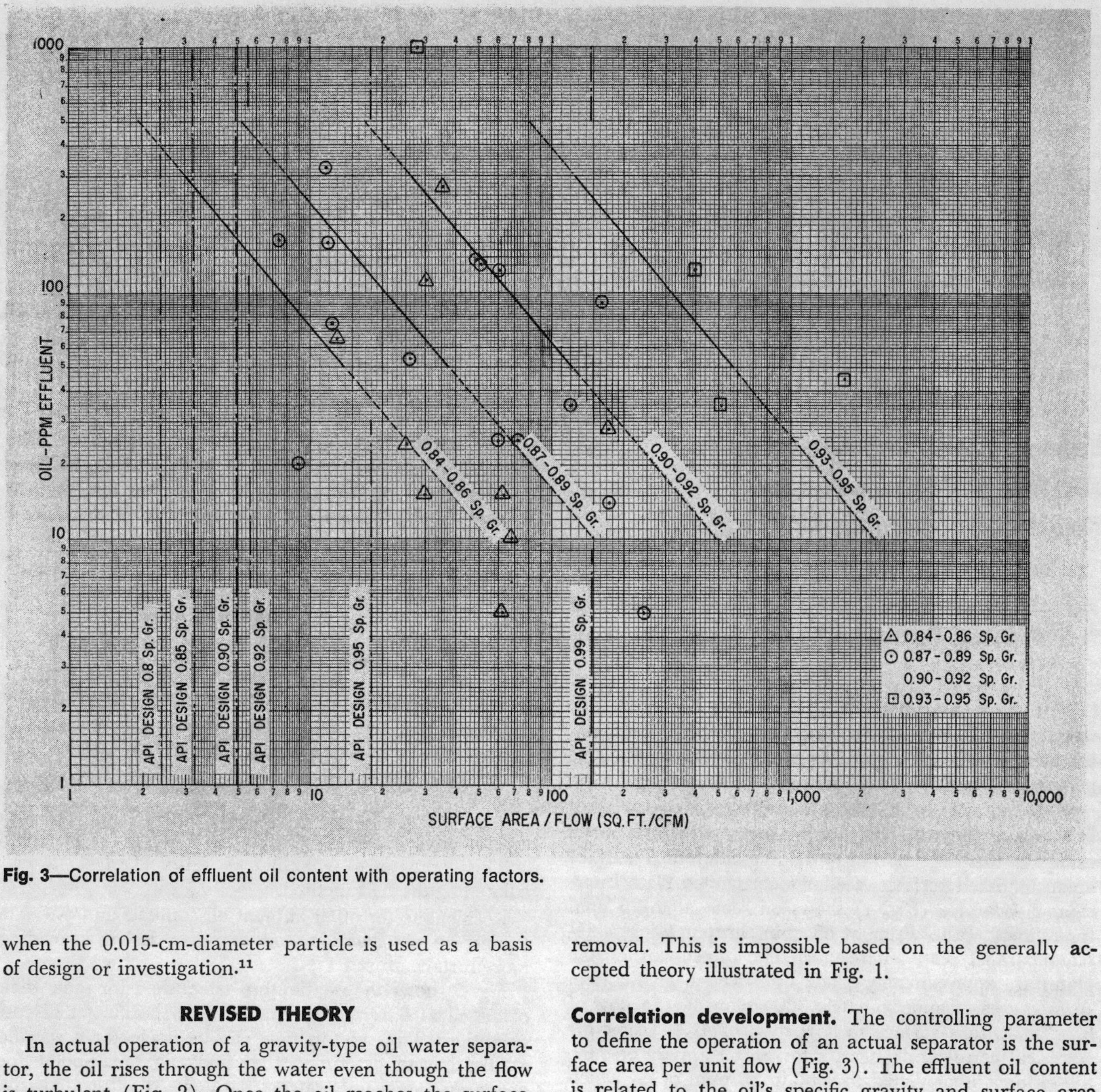

Fig. 3—Correlation of effluent oil content with operating factors.

when the 0.015-cm-diameter particle is used as a basis of design or investigation.[11]

REVISED THEORY

In actual operation of a gravity-type oil water separator, the oil rises through the water even though the flow is turbulent (Fig. 2). Once the oil reaches the surface, it is trapped there as a part of the oil film, unless there is sufficient turbulence to re-entrain the oil. Two separate operating plants have found that by leaving an oil film on the surface they have been able to improve their oil removal. This is impossible based on the generally accepted theory illustrated in Fig. 1.

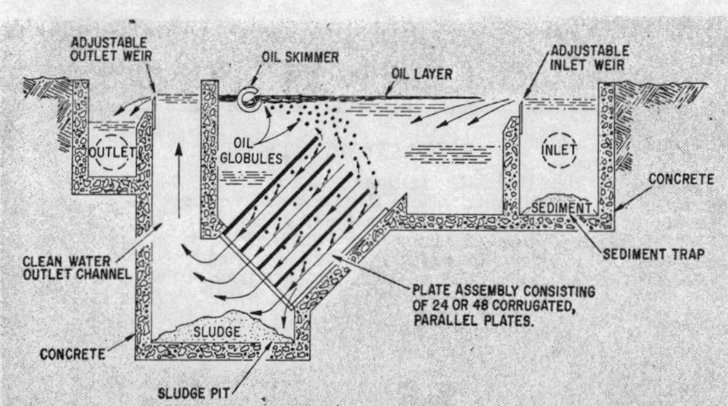

Fig. 4—Corrugated plate interceptor.

Correlation development. The controlling parameter to define the operation of an actual separator is the surface area per unit flow (Fig. 3). The effluent oil content is related to the oil's specific gravity and surface area per unit flow.

Operating data with an inlet oil content of 1,000 ppm or greater was plotted first to establish the oil gravity lines (Fig. 3). Some of the data was based on units with three steps of gravity-type oil removal in series. This data was used to establish the slope of the gravity line after which data were correlated for oil contents of less than 1,000 ppm on an incremental surface basis. The points established on incremental surface bases were then plotted (Fig. 3) to further verify the position of the gravity line.

Next the design of gravity-type oil water separators in accordance with the API Design Manual on disposal of refinery wastes was reviewed. This design does not predict an effluent quality but only provides a separator size for a specified gravity and flow rate. (Several of the separators from which data were received were in accordance with API design, but all were running at a much less than design thruput.) Maximum surface areas, cal-

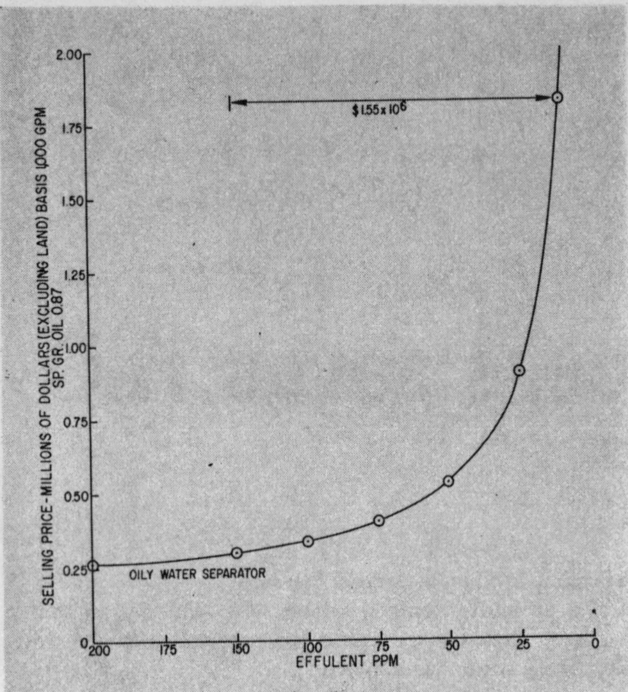

Fig. 5—Study comparison of covered oily water separator with air flotation.

culated by the API procedure, are given (Fig. 3) to show the oil removal expected from a separator calculated in strict accordance with the API procedure. Indications are that separators designed in accordance with the API procedures are only good for rough oil removal.

Since the surface area per unit flow concept seemed to correlate well with operating data, a check was made using another method for providing additional surface area. Published performance for a corrugated Plate Interceptor[4] separator (Fig. 4) shows oil removal to 25 ppm for refinery waste water at 65 gpm thruput using a 24-plate parallel plate interceptor. The dimensions of the plates are approximately 5 feet by 3 feet to give a total surface area of approximately 360 square feet, a flow of 8.7 cfm, with a surface to flow ratio of 41.4. Using the empirical correlation (Fig. 3) for an oil gravity of 0.87, the effluent is predicted to contain 30 ppm of oil which compares favorably with the claim of 25 ppm.

DESIGN PROCEDURE

Design using this correlation adequately describes the performance expected from a gravity-type oil water separator regardless of separator shape. Assuming design conditions for waste water and oil similar to those shown in the example used in the API Design Manual, the design can be accomplished as follows:

Assumptions.

Wastewater:	Maximum flow	800 cfm
	Temperature	105° F
	Specific gravity of water	0.992
	Absolute viscosity	$\mu = 0.0065$ poise
	Oil gravity	0.92 @ flowing temperature
Effluent quality desired:		150 ppm oil

Calculations.

Loading: Read (Fig.3) surface area per unit flow for 150 ppm in effluent, 0.92 oil specific gravity.

$$A_S/Q = 60$$

(Note: For inlet oil contents less than 1,000 ppm. use the difference in surface. Example: For 250 ppm inlet $A_S/Q = 60 - 38 = 22$. Surface area $22 \times 800 = 17{,}600$ square feet.)

Surface area: For a flow of 800 cfm, 800×60 or 48,000 square feet

Dimensions: There are three limits suggested in establishing the dimensions:

- Flow rate should not exceed 2 fpm.
- Length-to-width ratio should be at least 5 to avoid dead spots in the separator.
- Minimum depth set at 4 feet-0 inches.

Assume ratio 5 for length to width to give a channel approximately 100 feet wide by 480 feet long.

Set depth at suggested minimum of 4 feet-0 inches and check velocity:

Cross sectional area 400 feet^2
Velocity $800/400 = 2.0$ fpm

Divide into two 50-feet channels to give length-to-width ratio greater than 5.

This compares to two channels (20 x 112 feet) as calculated by the API method.

Recalculate for other effluent oil contents and evaluate against an air flotation unit or other designs providing more surface area.

Fig. 5 shows the relationship established for separation of an oil at 0.87 specific gravity at a specific set of cost conditions. Plots of this type can be made for a specific application to determine the economic break-point for any gravity-type oil water separator. As illustrated, the additional area required to reduce the oil content of the effluent water from 150 to 10 ppm using the gravity-type separator is $\$1.55 \times 10^6$, which compares with $270,000 for an air flotation unit for the same purpose.

ACKNOWLEDGEMENT

Adapted from a paper presented to the 28th Annual Purdue Industrial Waste Conference, West Lafayette, Ind., May 5, 1973. Based on data gathered by the Environmental Committee of Western Gas and Oil Refiners Association.

LITERATURE CITED

1 Standards of the Hydraulic Institute.
2 Camp, T. R. "Sedimentation and the Design of Settling Tanks", American Society of Civil Engineering Transactions, Paper No. 2285, April 1945.
3 API, "Manual on Disposal of Refinery Wastes—Volume on Liquid Wastes", Chapter 5, Oil-Water Separator Process Design, 1969.
4 Brochure, Heil Process Equipment Corp., Cleveland, Ohio.

Select Aerators Carefully

Vendor ratings of mechanical aerators are frequently misleading because of poor performance testing. Know what you are getting before you buy. Insist on standard test procedures and evaluate the result with knowledge

J. F. Ferrel, Fluor Corp., Houston, and
D. L. Ford, Engineering-Science, Inc., Austin, Texas

AERATORS installed for water treatment frequently fail to do what is expected of them. More units are needed than predicted by design calculations because field performance falls short of vendor claims due to poor vendor evaluation tests. Typically, one group of aerators claimed to have an oxygen uptake capacity of 3 pounds per horsepower hour (lbs./hp-hr.) proved to have a capacity of only 2.3-2.5 lbs./hp-hr. when properly tested. Such inconsistencies suggest a need for standard tests for rating aerators and point to a need for buyer understanding of test conditions.

Standard, scientifically sound tests can avoid these problems. Unfortunately, vendors' testing facilities often are inadequate because of wide variations in size and shape which prevent uniform ratings. Consequently, calculated capacities are suspect and seldom reproducible, and vary with unit sizes and between vendors.

Aerator ratings. Plate, updraft and downdraft areators are normally quoted on the basis of oxygen uptake per horsepower hour calculated from equations based on runs using clean, oxygen scavenged water. Complete mixing is assumed[1] and a transfer coefficient is used that is independent of time.

Brush-type aerators are rated with values which may be converted from oxygen absorption per unit length (lbs./ft.) using the unit length and driver horsepower. Motor shaft horsepower should be used, as it is the basis for over-all power consumption. In the case of combination units, the horsepower of the auxiliary blower must be added to that of the aeration motor horsepower.

Calculations for oxygen uptake are valid only when mixing is complete. Since many vendor test facilities do not meet this criteria, care should be taken in evaluating performance claims. Standard test parameters and uniform specification data (Fig. 1) should be required. Furthermore, all capacities should be based on operation at one temperature, such as 68°F, for ease of comparison.

Test facilities. The general physical criteria for test facilities for mechanical aerators are:

- Basin volume—All water contained in the test tank should be within the turbulent region of the aerator(s) and conversely, the surface area must be large enough to minimize backmixing from the tank walls.

- Interfacial area—The reaction area involved per unit time should be large. Normally this is accomplished by having a large circulation rate and is evaluated by calculation. Physical measurement of the absolute rate is almost impossible but estimates may be made using standard formulae and data from simultaneous samples of the water to and from the aerator.

- Bubble migration—The return rate of air bubbles from the interfacial area to the suction of the aerator must be slow. Assuming complete mixing, this is a function of the diameter of turbulence and the depth of the liquid in the basin. Inadequate diameters of discharge, or too shallow a basin can lead to reduced pumping efficiency caused by air bubbles short-circuiting to the suction.

Adequate test facilities can be assured when test parameters fall within the following limits:

- Power input per basin volume—0.10 to 0.18 hp/1,000 gal. This range covers both complete mixing and aeration requirements unless the suspended solids load is great enough to require more than 0.08 to 0.10 hp/1,000 gal. The exact horsepower level should be stated and compared to actual design values.

- Basin surface area—100 to 200 square feet per horsepower. The circle of mixing cited by the manufacturers average 200 sq. ft./hp.

- Basin depth—10 feet, minimum except for units less than 20 hp when the basin depth can be less. This depth prevents bottom scouring and does not require the use of core extensions. The basin walls should be either vertical or have a slope of 1 to 1 since field construction will probably be of this type because of real estate values.

These parameter ranges are consistent with those found in installed activated sludge or aerobic lagoons; therefore, the deviations between field installations and test results should be reduced to local loading factors:

α—Ratio of oxygen transfer in waste water to transfer in clean water

β—Ratio of oxygen saturation in waste water to saturation in clean water

γ—Biological respiration rate for in situ testing

Other variables such as fixed, floating or anchor post mounting are inherent in machine type. Mechanical details are inherent in vendors' design. Since these effect maintenance, they should be considered in bid evaluation, including plate and impeller diameter; number of blades; propeller pitch; submergence of plate, impeller or brush; RPM; deflectors or vanes; bearings; seals; and lubricating system.

MECHANICAL AERATORS

FOR: ____________

UNIT: ____________ ITEM NO.: ________ P.O. NO. ________

SERVICE: ____________ TYPE DRIVE: ____________

VENDOR: ________ SIZE ________ TYPE ________ NO. REQ'D. ____

EXPECTED OPERATING CONDITIONS:

LIQUID TO BE AERATED ________ TEMP. ____ °F α ______ β ______

OXYGEN REQ'D: ACTUAL ____ Lbs/Hr CLEAN WATER ________ Lbs/Hr

BASIN SIZE: ______ L ______ W ______ D ______ GAL

______ or DIAM

PWR INPUT ________ HP/1000 GPM TURBULENT AREA ____ Ft^2/HP OR ____ Ft Diam/HP

AERATOR TEST DATA: (BY VENDOR)

TEST BASIN DATA ____ L ______ W ______ D ______ GAL

POWER INPUT: ____ V x ____ A x ____ PF x ____ M EFF x ____ D EFF = ____ HP

NO. UNITS ____ HP/UNIT ________ BASIN WATER TEMP. ________ °F

TIME, 0 - 90% SATURATION ________ MIN K LA ________ Hr^{-1}

$K\ LA_{20}$ ________ Hr^{-1} O_2 UPTAKE ________ Lb/Hr EFF ________ Lb/HP/Hr

AT ________ HP/1000GAL AND ________ Ft^2/HP TURBULENT AREA

CONSTRUCTION:

	FLOATING TYPE		BRUSH TYPE		PLATE TYPE
IMPELLER Ø	________	BRUSH Ø	________	PLATE Ø	________
PITCH	________	BRUSH L	________	NO. BLADES	________
FLOAT Ø	________	ATTACH	________	PLATFORM SIZE	________
FLOAT MAT'L	________	SUBMERGENCE	________	PLATFORM MAT'L	________
SUBMERGENCE	________	RPM	________	SUBMERGENCE	________
RPM	________			RPM	________

DRIVER: MOTOR - TYPE ________ HP:ACT ________ CONN ________
BEARINGS ________ SEALS ________ LUBE ________
SHAFT: Ø ________ MAT'L ________ BUSHINGS ________

TYPE DRIVE: DIRECT ______ BELT ______ GEAR ______ MISC ______

MISCELLANEOUS: A WITNESSED PERFORMANCE TEST WILL/WILL NOT BE REQUIRED ON AT LEAST ONE UNIT BEFORE SHIPMENT.

Fig. 1—Specification sheet for evaluating mechanical aerators.

NOMENCLATURE

L	Length, feet
W	Width, feet
D	Depth, feet
V	Volts
A	Amperes
PF	Power factor
M Eff	Motor efficiency, percent
D Eff	Driver efficiency, percent
K LA	Over-all transfer coefficient
$K\ LA_{20}$	Over-all transfer coefficient at 20° C.
ATTACH	Type attachment
SUBMERGENCE	Depth of brush below water surface, inches
ϕ	Diameter
α	Ratio of oxygen transfer in waste water to transfer in clean water
β	Ratio of oxygen saturation in waste water to saturation in clean water
γ	Biological respiration rate for in situ testing

LITERATURE CITED

[1] Water Pollution Control, Eckenfelder and Ford, The Pemberton Press, 1970.

Pontoon System Automated for Slop Recovery

Successful operation saves manpower, improves oil quality

R. G. Gantz and **L. W. Cresswell,** Continental Oil Co., Ponca City, Okla., and **J. F. Gauen,** Continental Oil Co., Billings, Mont.

VAPOR EMISSIONS from API oil-water separators may be reduced while automatically recovering oil from the separator using an improved pontoon floating cover system. Actual operation of such a unit at the Billings, Mont., refinery of Continental Oil Co. has demonstrated the success of the system including operator savings and improved slop oil quality. Further it has provided data from which to recommend improvements for future installations.

Environmental regulations in many areas require control of hydrocarbon emission losses from refinery oil-water separators. Typically these laws apply to separators recovering more than 200 gallons per day oil with a Reid vapor pressure exceeding 0.5-1.5 psi and require some positive cover over the separator. Prompted by the Montana requirement, Continental installed floating pontoon-type covers on a two-channel API separator at its Billings refinery in late 1972. In addition to satisfying environmental requirements, the covers have the unique feature of automatic pump-out for oil accumulating underneath the pontoons. Performance after almost two years of operation has been very encouraging with a significant reduction noted in operator requirements. Recovered slop oil contains very little water and is less emulsified. Consequently, it is much easier to upgrade for recycle to the refinery crude unit. Total installed cost of the covers was about $75,000.

EQUIPMENT PERFORMANCE

Automated pump-out. The automatic pump-out system has performed even better than expected. Pumps have operated automatically as designed and with manual control needed only to clean out the sumps. Never have the emergency override switches provided in the system been needed. Although it was feared that heavy oils dumped periodically might cause problems with the level switch sensitivity, no evidence of this has been seen. In addition, no sludge deposits have built up inside the switch stilling well to cause sticking of the submerged element.

Some problems have been experienced with the pumps, including plugging with weeds, rags or similar objects. Although the sump pumps are oversized for normal operations, several times during turnaround large oil dumps have escaped to the separator afterbays. Oil can enter the separator at a faster rate than the pumps can handle. This can be compensated for partially by modifying the dual float mechanism to initiate earlier pump-out.

Oil recovery in the covered forebays has performed as expected. When the forebays are clean, oil removal is nearly complete such that less than 200 gallons of oil per day is recovered in the afterbays. However, when a significant sludge layer builds up in the forebays, the amount of oil entering the afterbays can exceed this amount. A regular cleanout schedule is necessary to ensure optimum performance.

The recovered oil contains very little water and is much easier to process due to a marked decrease in the amount of oxidation sludge collected. Although confirming data have not been taken, the recovered oil density is believed to be lower due to retention of additional light ends previously lost by vaporization. Recovered slop oil returned to the crude unit contains less BS&W than it did previously.

Sludge removal. Complete separator cleanout is accomplished by supporting the floating covers by chains attached to two pipes laid across the forebays on the concrete side walls. Flow is blocked to the forebay to be cleaned and the liquid contents are pumped out. Finally residual liquids are stripped from the bay using vacuum hose. In addition to this semiannual cleaning, a partial onstream cleaning is achieved twice annually using hoses inserted into the forebays to pump out some of the accumulated sludge.

Conservation seals. Several problems have developed with the conservation seals. They jam against the sides of the forebays during cover removal and are beginning

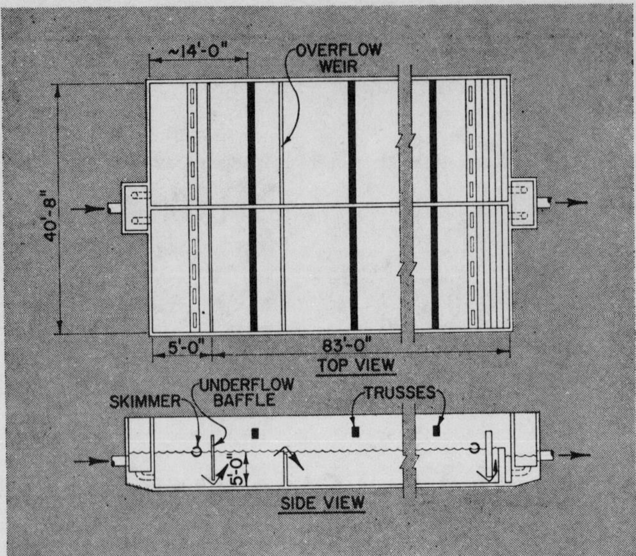

Fig. 1—API separator before modification

to tear loose from the covers. The seals do not fit perfectly against the separator walls, particularly in the basin corners. Deterioration from hydrocarbon exposure is also evident.

Miscellaneous. Paddles installed under the pontoons for sludge breakup in the skimming section are not necessary. The covers apparently inhibit formation of sludge by retaining light ends and reducing oxidation.

Hydraulic imbalance between parallel separator channels has been observed periodically. When a large difference in oil layer depth occurs between forebays, nearly all of the waste water flows through the forebay with the shallower oil layer, thus reducing oil removal efficiency.

No corrosion occurred in a year's operation even though the pH ranged from 2 to 14 for brief periods of time.

The covers are accepted well by supervisors who report that with automated pump-out only about one-third as much time is required to operate and maintain the separator as previously. Total operating requirement is about three man-hours per day.

SYSTEM CONSTRUCTION

The automatic pump-out, floating pontoon system is adapted to a conventional API separator (Fig. 1) which was constructed with two parallel channels each 88 by 20 feet with 5 feet normal liquid depth. The walls of each bay are braced by large concrete trusses spaced about every 14 feet along the length of the basin. These trusses are elevated about 1 foot above the liquid level and posed a real obstacle to covering the entire separator with floating pontoons. The separator was considerably oversized for existing conditions giving hope that only a portion need be covered. The normal flow of 135-150 gpm (not including utility blowdowns which are bypassed around the separator) has an average temperature of 95° F and contains about 100 bpd of 28° API gravity recoverable oil.

Size studies of the separator, as built, indicated dramatic oversize. In fact, two forebays 17 feet × 12.5 feet are large enough for adequate oil separation according to the API separator sizing procedure.[1] Consequently, the basin could be modified for two covered forebays without removing any of the concrete braces.

Early prediction of losses at various rates suggested that the system would be adequate to meet requirements. Predictions made with a new method[2] for estimating API separator effluent quality indicate the forebays should recover all but 85 ppm oil at the normal flow rate of 135-150 gpm and 180 ppm at 300 gpm.

DESIGN DETAILS

With the separator onstream, forebays were constructed at the entrance of each existing bay by the addition of

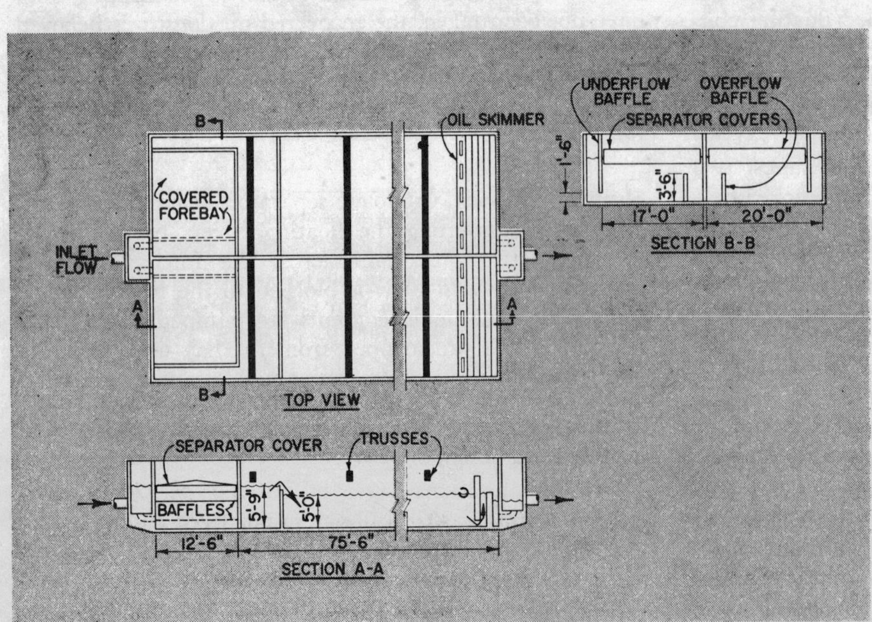

Fig. 2—API separator cover installation

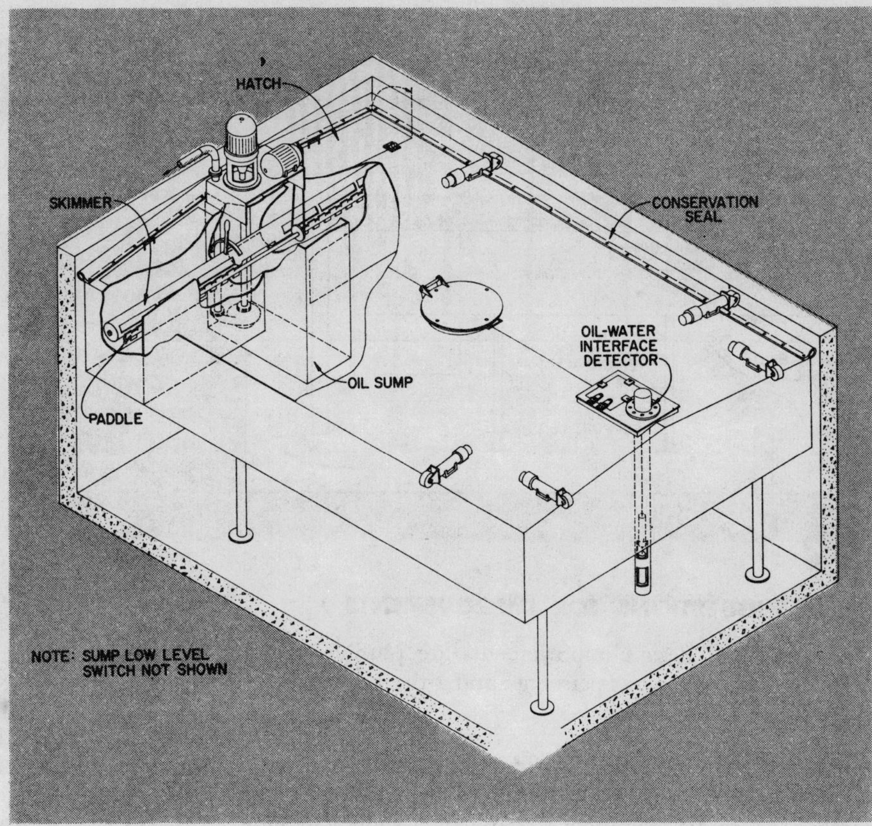

Fig. 3—Pontoon details

two new concrete walls. The first new wall is 12 feet, 6 inches from the existing separator inlet and runs parallel to the entrance wall (Fig. 2). Flow enters as in the original arrangement and turns 90° to pass through the forebays and under a wall 17 feet from the entrance. Overflow weirs installed in the afterbays downstream of the underflow walls set the liquid level in the separator. However during periods of heavy rainfall when outlet sewers are overloaded the liquid level rises approximately 2 feet. A 3-foot, 6-inch-high concrete baffle added at each forebay entrance helps distribute incoming flow.

Pontoon. Each pontoon-type floating cover is 16 feet, 6 inches long, 12 feet wide and 1 foot, 9 inches deep. The covers are constructed of ¼-inch carbon steel with a 3-inch-diameter rubber conservation seal provided around the periphery (Fig. 3). Proper alignment within the forebays is ensured by use of two spring-loaded roller guides mounted on each side of the cover. Electrical supply and pump discharge lines are flexible and sufficiently long to allow necessary vertical movement. Stationary support legs attached to the covers limit downward travel and maintain a 4-foot clearance above the basin floor whenever the separator is drained. These legs are unnecessary if the support chains previously mentioned are always used.

Each pontoon is equipped with a rotatable skimming pipe at the exit end of the cover from which oil flows from both ends into a sump located at the middle of the pipe (Fig. 3). Oil is pumped to the treating facilities located nearby. A slow-moving, electrically driven paddle is available for use when needed at a position in front of the skim pipe to move oil to the skimmer and break up any large chunks of sludge which may form.

The cover is hollow to provide buoyancy except for a portion at the exit end where hatches are provided to observe condition of skimmer, paddles, etc. Unfortunately, this equipment adds extra weight and compounds the buoyancy problem which is partially alleviated by use of a counter-weight added to the exit end of the cover.

Automated recovery. Automation depends on a reliable oil-water interface detector to actuate the slop oil pumps. Since the oil specific gravity averages only 0.875 at 95° F, a dual-ball type mechanism can detect the buoyancy difference between oil and water. The detector is located in a stilling well at the front end of the pontoon (Figs. 3 and 4). When the oil layer under the pontoon reaches 10½ inches, an electric switch is tripped and pump-out begins. The mechanism automatically shuts off the pump when the oil layer is within 1 inch of the pontoon bottom. The skimmer pipe is operated with the inlet lip submerged ½ inch below the liquid level to ensure that the pump sump is always flooded. In case the pump starts when the skimmer lip has been inadvertently rotated above the separator liquid level, a ball float override switch located inside each sump cuts off the electrical supply. This prevents prolonged pumping with the sump empty. A master switch allows the system to be operated manually for pump-out of any heavy oils or emulsions which gradually accumulate under the cover.

At an oil recovery rate of 50 bpd per forebay, the system operates on a 15-hour cycle. Each pump is sized for 35 gpm and requires less than one hour for pump-out. This cycle assumes no significant quantities of water in the recovered slop oil.

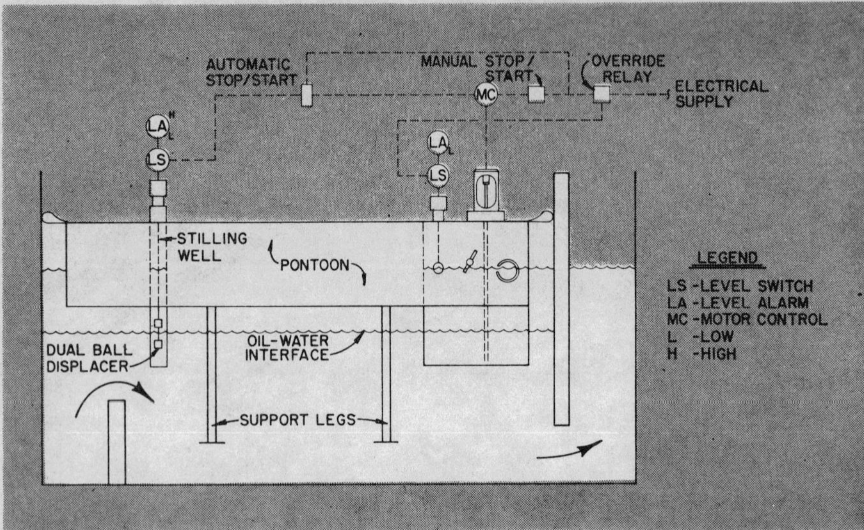

Fig. 4—Automatic oil pump-out system

SUGGESTIONS FOR IMPROVEMENTS

A number of design improvements are possible based on comments from plant personnel and a thorough review of operations. These include:

- Elimination of the sludge paddles which appear to be unnecessary. Sludge is probably due to formation of heavy material from air oxidation. Cover design should include all mountings and fittings to install paddles and motor at a later date, if required, since sludge generation may vary from plant to plant.

- Conservation seals should be installed with adjustable fittings or designed for frequent replacement. A material more resistant to aromatics should be used.

- A counterbalance creates problems and, therefore, should be avoided in future designs. One alternative is to place the sump and pump close to the center of the cover while leaving the skimmer at the downstream end. Another alternative is to attach an additional flotation chamber at the downstream end in place of the counterweight. A combination of both ideas probably gives the most stable cover configuration.

- Flow to each forebay should be set upstream of the separator to ensure independent flow control in a manner unaffected by the amount of oil accumulated in each forebay.

- Sludge removal could be improved by adding several additional access hatches to each cover.

- Skimmer plugging should be avoided by use of wire mesh trash screen wrapped around each skimmer pipe. Pumps should also be designed to handle large pieces of foreign material.

- As discussed earlier, during turnaround or occasional large spill, oil may enter the forebays at a faster rate than the cover pumps can handle. A level indicator should be provided to show the oil-water interface position thus warning operators of impending oil release to the afterbays.

- A nonskid surface should be installed on all covers for personnel protection.

LITERATURE CITED

[1] *API Manual on Disposal of Refinery Wastes,* Chapter 5, 1969, American Petroleum Institute, New York.
[2] Thomson, S. J., "Data Improves Separator Design," *Hydrocarbon Processing,* October 1973, pages 81-83.

How to Design Activated Sludge Units

Data collected from HPI plants help make this design easy

S. J. Thomson, Fluor Engineers & Constructors, Inc., Los Angeles

DESIGN OF activated sludge units for treatment of biodegradeable materials requires considerable flexibility because values of some operating parameters can not be precisely fixed. Some of these parameters are

- Concentration of unknown materials
- Flow of effluent water
- Contaminant concentration
- Biological reactions.

These difficulties can be overcome and units adequately designed based of operating data collected from units in service in the hydrocarbon processing industry (HPI). Furthermore, similar design relationships can be fixed for design of units for other industries by similar collection and evaluation of appropriate data.

ACTIVATED SLUDGE SYSTEMS

An activated sludge treatment system is used on effluent waters to remove biodegradeable materials using microorganisms. Consumption of biodegradeable materials in these aerated biological treating systems varies with

- Temperature
- Population of microorganisms
- Nutrient availability
- Oxygen supply
- Type and concentration of organics (biological oxygen demand or BOD) in feed.

Temperature. Practical designs of biological treating units are limited to a range of 35° F to 100° F, although biological action will occur beyond this range. A temperature range of 70° F to 95° F is preferred for efficient operation.

Population of microorganisms. BOD removal efficiency improves as the total microorganism population in the system is increased. This population, which is dependent on microorganism concentration and aeration basin volume, is controlled by settling characteristics of microorganism population as sludge is formed. Experience indicates that a concentration of approximately 2,500 mg/l (or #-sludge/#-water) provides the most favorable settling characteristics.

Type. Several configurations of activated sludge systems are available among which the complete-mix type is in wide usage. This system is one in which effluent water feed and recycled microorganisms are distributed in such a manner that the concentration of biodegradeable materials and microorganisms themselves are uniform throughout the basin. This system operates efficiently, particularly in industrial applications where effluent water contains concentrations of organic materials which are toxic in varying degrees to the microorganisms.

DESIGN PROCEDURE

This procedure is most applicable to complete-mix type systems.

Pilot and industrial operating data for activated sludge plants in refining and petrochemical industry are shown in Figs. 1 through 4.

Residence time. Fig. 1 is a graph of BOD removal as a function of food to mixed liquor suspended solids (MLVSS) ratio. The residence time can be determined from this graph by fixing the MLVSS concentration in the aeration basin.

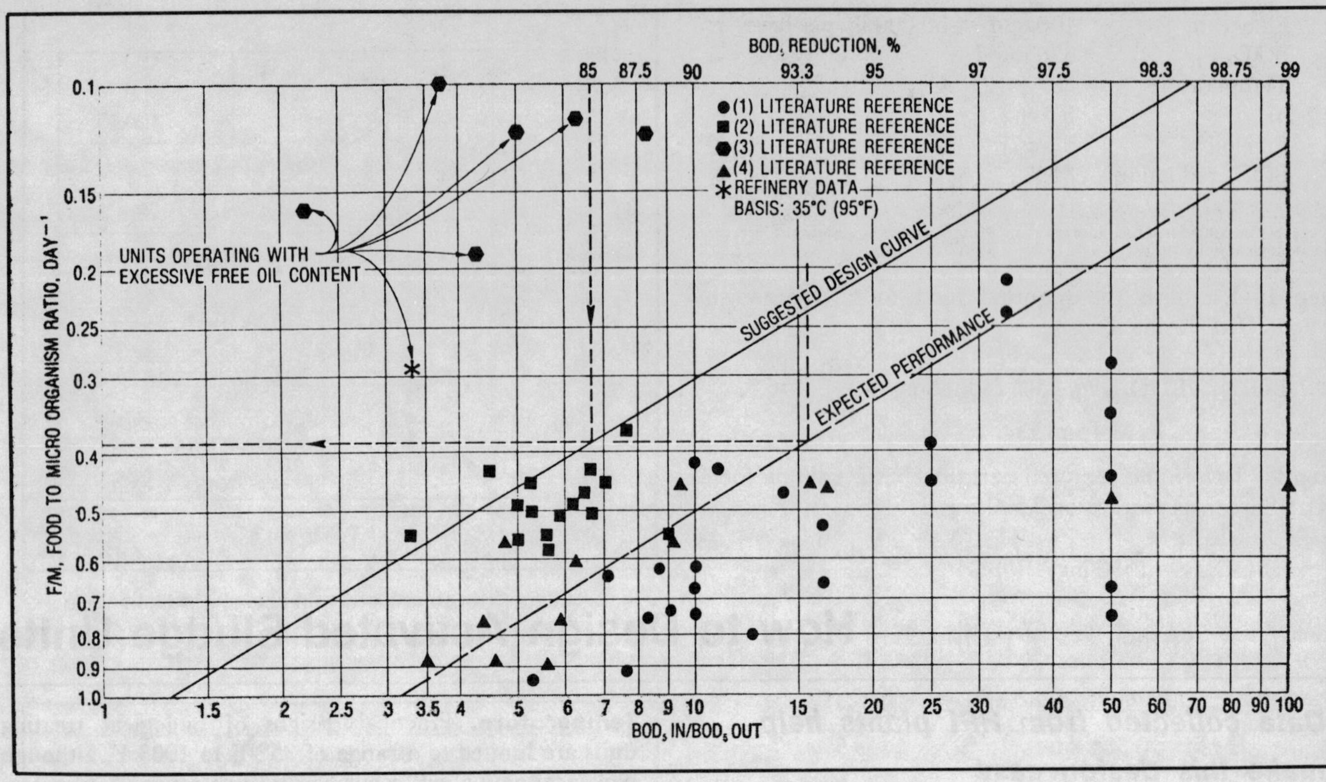

Fig. 1—Relationship between BOD_5 reduction ratio and the food to microorganism ratio.

Sludge production. Fig. 2 is a graph of BOD removal efficiency as a funtcion of pounds of sludge produced per pound of BOD in the influent. A suggested design curve and an expected performance curve are shown. Suggested design curve should be used to determine capacity of sludge handling facilities.

Sludge age. Fig. 3 is a graph of sludge age vs. BOD removal efficiency and provides a method to examine the validity of the design provided by Figs. 1 and 2. Sludge age is calculated by dividing weight of the microorganisms in the aeration basin by weight of sludge wasted each day.

Oxygen requirements. Fig. 4 is a graph of BOD removal vs. pounds of oxygen required per pound of BOD removed. The suggested design curve describes a method to determine necessary horsepower. Expected performance curve predicts BOD removal based on this horsepower.

Nutrient requirements. The theoretical nitrogen requirement is:

$$\text{lb. nitrogen/day} = \frac{\text{BOD lb./day}}{\text{20 lb. BOD/lb. nitrogen}}$$

The theoretical phosphorus requirement is:

$$\text{lb. phosphorus/day} = \frac{\text{BOD lb./day}}{\text{100 lb. BOD/lb. phosphorus}}$$

Equipment for supplying nutrients should be designed to provide 1.4 times theoretical.

Temperature correction. The four design curves (Figs. 1-4) are based on a temperature range of 70° F to 95° F. Operation at temperatures in the 35° F to 70° F range are calculated using the following procedure:

Calculate K_1 from

$$K_1 = \frac{S_o\,(S_o - S_e)}{S_e X_v t}$$

S_o = Influent BOD, mg/l

S_e = Effluent BOD, mg/l

X_v = MLVSS

t = Aeration time, hours

K_1 = Removal rate coefficient at 70° F.

Using K_1 calculate K_2 from

$$K_2 = K_1 \Theta^{(T_2 - T_1)}$$

T_1 = 21° C

T_2 = New temperature

Θ = Temperature coefficient (1.0 @ 21-32° C; 1.04 @ 2° C)

K_2 = Removal rate coefficient at new temperature.

SAMPLE PROBLEM

Interpretation of the graphs is demonstrated by the following sample problem, which is based on the following assumptions:

Flow rate	100 gpm = 50,000 lb. per hr.
BOD_5 in	350 mg/l
Temperature	80° F (27° C)
pH	6.5-8
Free oil*	5 mg/l
BOD_5 reduction	85%

* Free oil in concentrations greater than 5 mg/l causes a significant decrease in biological activity.

Step 1—Establish the required food to microorganism ratio (F/M) using Fig. 1

$$BOD_{5\,in}/BOD_{out} \text{ @ } 85\% \text{ reduction} = \frac{350}{52.5} = 6.7$$

$$F/M \text{ from Fig. } 1 = 0.385$$

Step 2—Determine required aeration basin volume for a MLVSS concentration of 2,500 mg/l

$$BOD_{5\,in} = (350 \text{ mg/l})\left(8.34 \times 10^{-6} \frac{\text{lb./gal.}}{\text{mg/l}}\right)\left(100 \frac{\text{gal.}}{\text{min.}}\right)\left(1{,}440 \frac{\text{min.}}{\text{day}}\right)$$

$$BOD_{5\,in} = 420 \text{ lb./day} = \text{feed} = (F)$$

$$F/M = 0.385 \text{ day}^{-1}$$

$$M = \frac{420 \text{ lb./day}}{0.385 \text{ day}^{-1}} = 1{,}091 \text{ lb.}$$

Calculate basin volume

$$V = \frac{1{,}091 \text{ lb.}}{(2{,}500 \text{ mg/l})\left(8.34 \times 10^{-6} \frac{\text{lb./gal.}}{\text{mg/l}}\right)}$$

$$V = 52{,}326 \text{ gal.}$$

Step 3—Calculate aeration residence time.

$$T_{hrs} = \frac{52{,}326 \text{ gal.}}{100 \text{ gal./min.} \times 60 \text{ min./hr.}} = 8.72 \text{ hr.}$$

Step 4—Establish sludge production (Fig. 2)

Sludge produced = 0.425 lb./lb. BOD_5 @ 85% BOD_5 removal

Calculate lb./day of sludge produced

0.425 lb. sludge/lb. BOD_5 × 420 lb. BOD_5/day × 0.85 (fraction of feed consumed)

Sludge = 152 lb./day MLVSS

Calculate total sludge (use 1.25 factor for inert allowance)

152 × 1.25 = 190 lb./day (about 200 lb./day)

Design sludge handling facilities for 200 lb./day.

Step 5—Check sludge age

$$\text{Sludge age} = \frac{\text{Total lb. MLVSS in basin}}{\text{lb. MLVSS wasted day}}$$

$$\text{Sludge age} = \frac{1{,}091 \text{ lb.}}{152 \text{ lb./day}}$$

Sludge age = 7.2 days

From Fig. 3: Required sludge age 7 days, sludge age of 7.2 days sufficient.

Step 6—Establish oxygen requirement

From Fig. 4: O_2 requirement for 85% removal is 1.14 lb.-O_2 per lb.-BOD_5 removed.

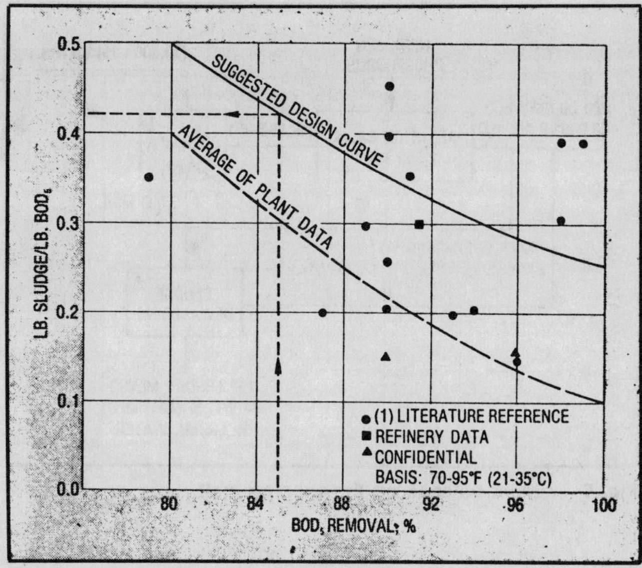

Fig. 2—Effect of BOD_5 reduction on sludge production.

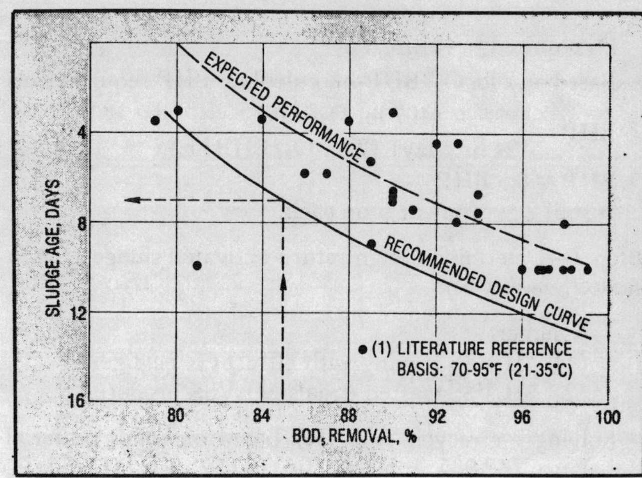

Fig. 3—Effect of BOD_5 reduction on sludge age in the unit.

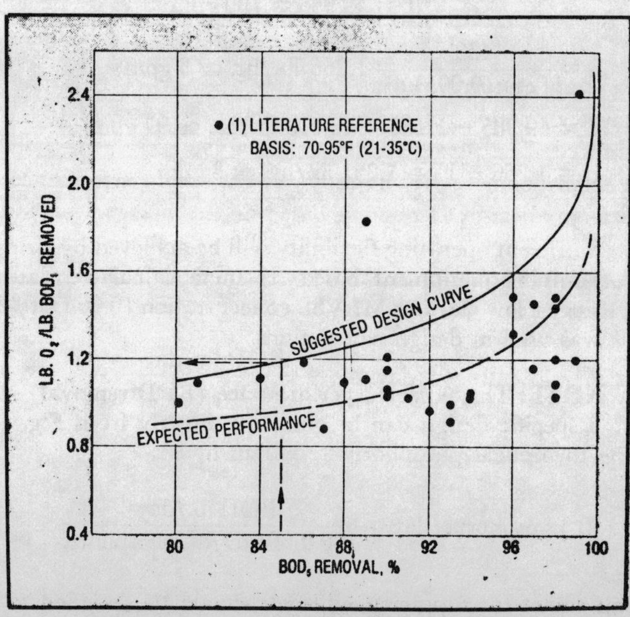

Fig. 4—Effect of BOD_5 removal on oxygen demand.

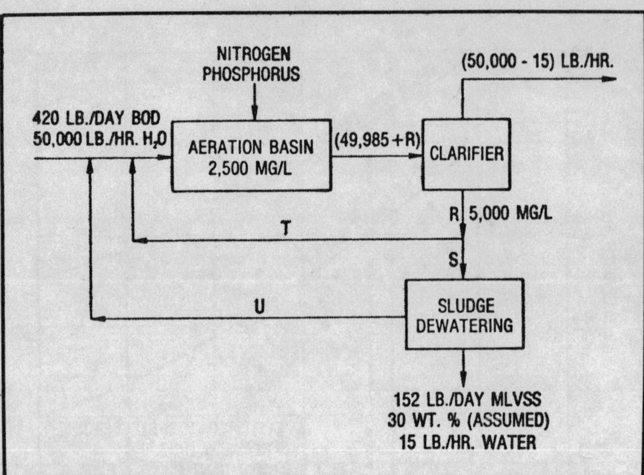

Fig. 5—Weight balance on the example unit.

O_2 required = (420 lb.-BOD/day) × (0.85 fraction removed) × (1.14 lb. O_2/lb. BOD_5 removed)

O_2 required = 407 lb./day

Based on 2 lb. O_2/BHP-hr. calculate BHP requirements

$$BHP = \frac{(407 \text{ lb.-}O_2/\text{day})}{(24 \text{ hr./day})\ (2 \text{ lb.-}O_2/\text{BHP hr.})}$$

BHP = 8.5 BHP

Install 2 aerators at 5 hp each.

Step 7—Determine design return activated sludge (RAS) flow. (See Fig. 5.)

Assume:

(1) MLVSS leaving with effluent is negligible

(2) MLVSS wasted equals MLVSS produced.

Solving for recycle rate (R) based on solids material balance (49,985 + R) lb./hr. (0.0025 lb. MLVSS/lb. water) = R(0,005 lb MLVSS/[16] water)

$$124.9625 + 0.0025\,R = 0.005\,R$$
$$R = 124.9625/0.0025$$
$$R = 49{,}985 = 100 \text{ gpm}$$

$$S = \frac{152}{(24)\ (0.005)} = 1{,}266 \text{ lb./hr.} \cong 3 \text{ gpm}$$

$$T = 49{,}985 - 1{,}266 = 48{,}719 \text{ lb./hr.} \cong 97 \text{ gpm}$$

$$U = 1{,}266 - 15 = 1{,}251 \text{ lb./hr.} \cong 3 \text{ gpm}$$

Sufficient operating flexibility will be achieved by sizing pumping equipment based on these calculated rates, since a low clarifier MLVSS concentration (0.005 wt%) was used in design calculations.

NOTE: The expected performance (BOD removal) on a specific design can be obtained directly from Fig. 1.

Read BOD removal ratio from F/M ratio (0.385) from expected performance curve at 16. $\left(\frac{16-1}{16} = 94\%\right)$

Step 8—Calculate nutrient requirements

Nitrogen = 420 lb. BOD/day × 1/20 lb. nitrogen/lb. BOD × 1.4 (operational flexibility factor) ≅ 29.4 lb./day

Phosphorus = 420 lb. BOD/day × 1/100 lb. phosphorus/lb. BOD × 1.4 (operational flexibility factor) ≅ 5.88 lb./day

Step 9—Temperature correction (for operation at 35° F)

$\Theta = 1$ 70° F-95° F

$\Theta = 1.04$ 35° F

Vary Θ between 70° F and 35° F by straight line interpolation.

Calculate K_1

$$K_1 = \frac{S_o\,(S_o - S_e)}{S_e X_v t}$$

$$= \frac{250 \text{ mg/l } (350 \text{ mg/l} - 53.5 \text{ mg/})}{(52.5 \text{ mg/l})\ (2{,}500 \text{ mg/l})\ (8.72 \text{ hrs.})}$$

$= 0.091$ hrs.$^{-1}$ for 70° F (21° C)

Calculate K_2 for 35° F (1.7° C)

$$K_2 = K_1 \Theta^{(T_2 - T_1)}$$

$$K_2 = (0.091 \text{ hrs.}^{-1})\ (1.04^{-19.4°C})$$

$$K_2 = (0.091)\ (0.46) = 0.042$$

Calculate S_e based on K_2

$$K_2 = \frac{S_o\,(S_o - S_e)}{S_e X_v t}$$

$$K_2 = \frac{350\,(350 - S_e)}{(S_e)\ (2{,}500)\ (8.72)}$$

$$S_e = 97$$

$$BOD_{in}/BOD_{out} = 350/97 = 3.61$$

$$\%\ \text{Removal} = \frac{350 - 97}{350} = 72.3\%$$

BIBLIOGRAPHY

[1] Associated Water and Air Resources Engineers, Inc., "Process design techniques for industrial waste treatment," 1973.

[2] Dickerson, B. W., and Laffey, W. T., "Pilot plant studies of phenolic wastes from petrochemical operations," Industrial Waste Conference, Purdue University, 1959.

[3] Dickenson, R. L., and Giboney, J. T., "Stabilization of refinery waste waters with the activated sludge process: Determination of design parameters," Industrial Waste Conference, Purdue University, 1970.

[4] Gillian, A. S., and Anderegg, F. C., "Biological disposal of refinery wastes," Industrial Waste Conference, Purdue University, 1959.

[5] Kumke, G. W., Conway, R. A., and Creagh, J. R., "Performance of internally clarified activated-sludge process treating combined petrochemical-municipal waste," Industrial Waste Conference, Purdue University, 1968.

[6] Thomson, S. J., Stock, J., and Mehta, P. L., "Investment and operating costs for treatment of oily wastewater," *Petroleum and Petrochemical International*, May 1973.

Activated Carbon Improves Wastewater Treatment

Alternate approaches offer several economical choices

Paschal B. DeJohn and **Alan D. Adams,**
ICI United States, Inc., Wilmington, Del.

ACTIVATED CARBON can be used to successfully treat refinery wastes by either of two methods: contact in a treatment column or direct addition to activated sludge systems. Each method offers substantial performance benefits in treating either total refinery effluent or selected streams from within the refinery complex. All methods present viable and economically feasible alternatives to other methods of treating refinery waste and deserve design attention in selecting pollution control systems.

Carbon used for these two methods is either granular or powdered depending on treatment type. Choice of carbon type is based on physical characteristics with respect to granular size, surface area and pore size. Other factors that affect performance are carbon density and regeneration capabilities.

Granular carbon is made from either lignite or bituminous coal. Lignite carbon has a bulk density of 22 lbs./ft.^{3} compared to a density of 25-26 lbs./ft.^{3}, depending on the mesh size,[1–3] for bituminous granular carbon.

Carbons made from bituminous coal have a large total surface area, a high percentage of which is in small pores (less than 20 Å).[4] Lignite carbons perform better than bituminous coal carbon when treating refinery wastes containing predominately large molecules because they have more surface area in the "transitional" (20-500 Å)[4] pores. On the other hand, bituminous coal carbons, because of the preponderance of small pore surface area perform better than virgin lignite carbons when treating wastes containing predominantly small molecules. However, surface area properties coverge after repeated thermal regenerations such that performance of regenerated carbons is equivalent. Over-all economics usually favor lignite carbon because of its lower density and higher impurity loading capacity.

Powdered carbons contribute several positive benefits in improving operation of activated sludge systems that often result in savings on other chemicals or in sludge handling with little or no capital expenditure. They improve organic removals, aid solids settling and sludge handling and provide protection from toxic or shock loadings. In the face of widely varying influent organic or hydraulic loads, carbon usage can level out the effluent quality.

High density powdered carbons are preferred to minimize carryover from secondary clarifiers and to increase sludge compaction. Such carbons also require less makeup to maintain desired aerator equilibrium levels since carbon is lost only during sludge wasting.

Use of powdered activated carbons does not necessarily increase costs. Savings on defoamers, coagulants, power and labor can often reduce operating expenses.

GRANULAR ACTIVATED CARBON

For granular activated carbon to be an effective treatment mode, it must provide

- Sufficient ability to adsorb impurities to meet effluent standards
- Reasonable (5-10 percent total) regeneration losses
- Minimal pressure drop
- Reasonable operating costs.

TABLE 1

Test	Feed	Effluent	Breakthrough, % removal	Carbon dosage, #/1,000 gal.		Loading, # removed per # carbon	
	COD, ppm			Lignite	Bituminous	Lignite	Bituminous
One	104	31	70	0.93	1.31	0.65	0.46
Two	70	21	70	1.91	2.49	0.21	0.16
	Sulfonate						
Three	1,200	20	98	76	576	0.13	0.017

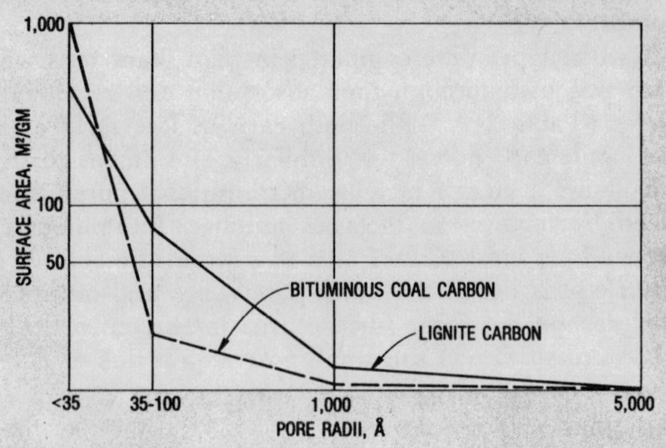

Fig. 1—Surface area distribution for bituminous coal and lignite carbon.

TABLE 2

Installation	Regeneration, number	Iodine number	Molasses number	Ash, %
Pittsburg Activated Carbon Co.	0	942	...	5.3
	10	588	...	11.9
MSA Research Corp.	0	1,090	250	5.7
	3	940	355	9.5
MSA Research Corp.	0	1,090	190	....
	10	630	250	....
Pomona, Calif.	0	1,100	...	....
	7	700	...	....
Pomona, Calif.	0	1,028	...	....
	10	686	...	....
Tahoe, Calif.	0	935	...	5.0
	4	820	...	7.1
Pomona, Calif.	0	1,100	...	....
	4	690	...	....

Impurity adsorption. Comparative pilot studies show that both types of granular carbon can adsorb impurities.

Studies were based on the same feed concentrations to columns containing lignite and bituminous coal carbons and on "breakthrough" at required pollutant percentage removal as the measure of quality.

Because of lignite carbon's lower density, it has to achieve a higher loading of impurities or require a lower dosage to give equivalent performance. This proved true in three tests.

Test one was performed on sand filtered, API separator effluent from an East Coast refinery. Two sets of three 1½-inch ID carbon columns were operated in parallel at a linear flow rate of 2 gpm/ft². Each set of three columns was then operated upflow in series. Empty bed contact time was 18 minutes through each set of three columns. One set contained a 12 x 40 mesh lignite carbon and the other set contained a 12 x 40 mesh bituminous coal carbon.

The lignite carbon required about 29 percent lower carbon dosage and loaded about 41 percent higher than the bituminous coal carbon (Table 1).

Test two was performed on sand filtered, API separator effluent from another East Coast oil refinery. Two sets of four 1½-inch ID carbon columns were operated in parallel at a linear flow rate of 0.5 gpm/ft.². Each set of four columns was then operated downflow in series. Empty bed contact time was 88 minutes through each set of four columns. In this case, both carbons were 8 x 30 mesh.

Lignite carbon required about 23 percent lower carbon

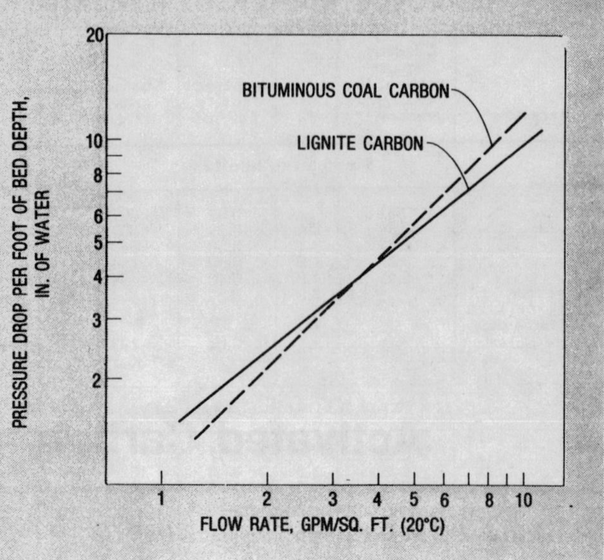

Fig. 2—Pressure drop across bituminous and lignite carbon, 12 x 40 mesh.

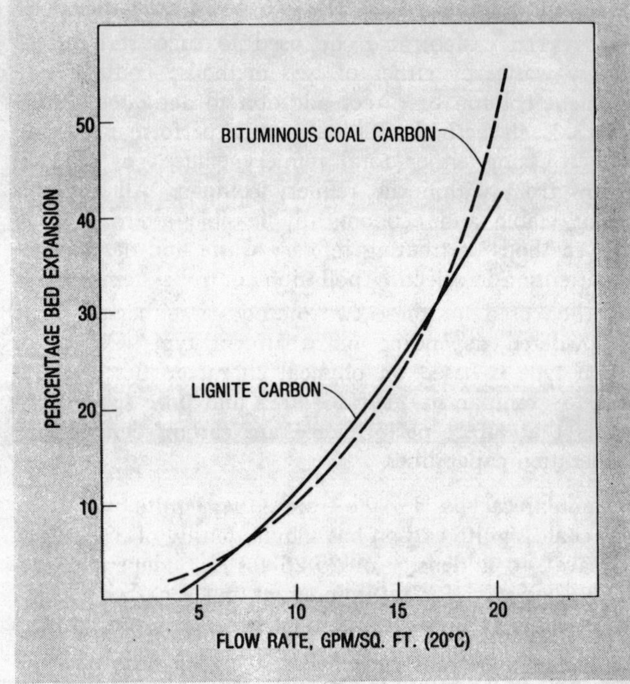

Fig. 3—Backwash bed expansion in bituminous and lignite carbon beds, 8 x 30 mesh.

dosage and loaded about 31 percent higher than did bituminous coal carbon (Table 1).

Test three was performed on a selected sulfonate waste stream from a West Coast refinery. The wastewater was not pretreated prior to carbon treatment. Two sets of four 1¼-inch ID carbon columns were operated in parallel at a linear flow rate of 1 gpm/ft². Each set of four columns was then operated downflow in series. Empty bed contact time was 150 minutes through each set of four columns. One set contained a 12 x 40 mesh lignite carbon, and the other set contained a 12 x 40 mesh bituminous coal carbon.

The lignite carbon required about an 87 percent lower

TABLE 3

Regeneration*	Loading, spent carbon, lb. COD/lb. carbon**	Loss, %	Apparent density, gm/ml	Surface area, m^2/gm	Mol. RE	Ash, %
8 x 35 mesh lignite						
0	0.31	..	0.39	607	111	13.1
1	0.34	2.4	0.39	607	111	15.0
2	0.37	3.1	0.39	553	131	14.2
3	0.37	1.2	0.40	454	119	15.2
4	0.41	4.8	0.40	494	119	15.1
Average loss		2.9				
Change between virgin and 4 times regenerated carbon	....	..	0.01	−123 (−20.3%)	+8	+2.0
12 x 40 mesh bituminous coal carbon						
0	0.23	..	0.47	1149	39	5.0
1	0.29	3.2	0.46	1028	43	8.2
2	0.29	1.6	0.48	952	54	7.9
3	0.30	2.6	0.47	847	69	9.0
4	0.35	3.0	0.47	718	89	9.0
Average loss		2.6				
Change between virgin and 4 times regenerated carbon	....	..	0.0	−431 (−37.5%)	+50	+4.0

* Regeneration in pilot plant rotary tube using virgin carbon Vibrating Feed Apparent Density as a control.

** Feed concentration increased during test period.

TABLE 4

	Virgin carbon	Regenerated carbon	% Change
Iodine number	950—1000	560—680	−27 to −44
Molasses number	230	280	+22
Phenol, % removal	99.9	63	−37
COD, % removal	71	13	−82

TABLE 5

Impurity removal, (lbs./lb. carbon applied)	Virgin carbon	Regenerated carbon	% Δ
TOC	0.17	0.096	−44
COD	0.73	0.35	−52
Phenol	0.04	0.03	−25

carbon dosage and loaded about 665 percent higher than did the bituminous coal carbon (Table 1).

Higher loadings on lignite carbons are achieved because of pore volume and surface area distribution characteristics. Bituminous coal carbon has a higher ratio of small pores (Fig. 1)[5] which is confirmed by its iodine number, a good measure of pores less than 20 Å. Lignite carbons have a higher surface area and pore volume in the transitional pore size range which is measured by the molasses number.

"Higher molasses number," "higher iodine number" or similar quality characteristics must be weighed against the job to be done and size molecule of the impurity to be adsorbed. The amount of total surface of candidate carbon actually **available** in that size is what is important.

Regeneration losses. Granular carbon must be regenerated by the large volume user to be economically competitive with other treatment methods. Thermal regeneration affects carbon performance as reported for bituminous coal carbons[6] and summarized in Table 2.

When a bituminous coal carbon is regenerated, surface area in transitional pores is increased (as shown by increase in molasses number), and total and small pore surface areas are reduced (as shown by decrease in iodine number).

Carbon types were compared in pilot plant tests where each was used through four adsorption and regeneration cycles (Table 3).[6] While both carbons lose surface area, the lignite carbon loses less. Similarly, bituminous coal carbon shows a greater increase in transitional pores as indicated by increase in molasses number. Bituminous coal carbon loses more surface area at a faster rate because lost surface area lies in the small pore range and bituminous coal carbon has more surface area in this range to lose.

Loss of surface in this small pore area is due to:

- Ash that is picked up by carbon during service (Ca, Mg, Fe, etc.)
- Partial elimination of adsorbed organics during regeneration (If temperature is raised high enough to attempt 100 percent removal, burn losses are excessive.)
- Enlarging small pores due to burning.

Plant results at the BP Oil Refinery, Marcus Hook, Pa., confirm these tests. In a 2.2 mgd filtration-carbon adsorption system operating with 8 x 30 mesh bituminous coal carbon results after 18 months service show a significant decrease in iodine number (small pore surface area) and an appreciable increase in molasses number (transitional pore surface area) after successive regenerations (Table 4).[7]

Change in pore size distribution must be considered in designing units when wastewater contains contaminants with predominantly small molecules. Because of loss of small pore surface area during regeneration, column studies to obtain design data should be performed with carbon that has been regenerated 5-6 times. If such carbon cannot be obtained and virgin carbon must be used, lignite carbon is preferred because its properties do not change as much on regeneration. Otherwise, there is a very real risk of significantly undersizing the system, and operating costs can be much higher than anticipated after system installation.

At BP Oil, regenerated carbon iodine numbers of 560-680 show a significant decrease from virgin carbon iodine numbers of 950-1,000, so that adsorption performance of regenerated bituminous coal carbon for phenol and COD removal was sharply reduced. Isotherms comparing regenerated and virgin bituminous coal carbon at current influent concentrations confirm that poorer plant performance should have occurred (Table 5).[7]

Both carbon regeneration rate and carbon dosage had to be increased to maintain purification levels obtained with virgin carbon. This increases operating costs significantly over design.

For treatment of a selected refinery wastewater stream or a total refinery effluent which contained predominantly *small* molecules, virgin bituminous coal carbon should be better than virgin lignite carbon because of its higher surface area in small pore range. However, after a number of regenerations, carbon properties tend to converge and both should perform equally in removing small molecules. On a long service basis, economics may still favor use of lignite carbon. If the wastewater contains predominantly *large* molecules, the lignite carbon should perform best.

Pressure drop. While adsorption and regeneration characteristics of a carbon are important factors in selecting the best carbon, pressure drop and bed expansion properties will affect system operation after carbon is put into service.

Pressure drop curves for 12 x 40 mesh lignite and bituminous carbons in a downflow system show no significant difference (Fig. 2). Similar results are obtained with other mesh sizes and in upflow designs. For regenerated carbons, data indicate that pressure drop problems should not be significantly different for either carbon (Table 6).

TABLE 6

Carbon type	Avg. pressure drop (in. H_2O) after backwashing and draining	
	Lignite carbon	Bituminous coal carbon
Virgin	95	95
Regenerated—1	68	70
Regenerated—2	55	60
Regenerated—3	45	45
Regenerated—4	40	43

Bed expansion on backwashing lignite and bituminous 8 x 30 mesh carbons shows no significant differences in hydraulic characteristics (Fig. 3) since particle density wetted in water is the same for both types of carbon (1.3-1.4 g/cc).

If lignite and bituminous coal carbons are reactivated under optimized conditions, equivalent losses should be obtained. Given reasonably equivalent impurity loadings on both carbons, this optimization includes using either lower temperature and/or less residence time in the furnace for lignite carbon.[8]

POWDERED ACTIVATED CARBON

Petrochemical and refinery wastes can also be treated biologically via the activated sludge process. However, conventional systems often experience effluent quality and operating problems. Oil that is not removed in API separators can pass through aerators and clarifiers and will be measured as TOC or COD. Oil can also entrap biological solids, prevent them from settling and lead to high effluent suspended solids. Surface active agents, such as alkanolamines or petroleum sulfonates often cause foaming in the aerator and on the receiving stream. Shock toxic loads from phenols, heavy metals or cyanide can kill the active biomass. Finally, oily characteristics of petroleum waste sludges make them difficult to dewater and handle.

Addition of powdered activated carbon to activated sludge systems does not involve additional capital expenditure for equipment but does solve many problems.[9–14] Use of powdered carbon may allow a refinery to meet effluent standards and reduce costs. Among possible benefits are:

- Improved organic pollutant removals (BOD, COD and TOC)
- More uniform operation and effluent quality, particularly during periods of widely varying organic and hydraulic loads
- Decreased effluent solids and thicker sludges, resulting in reduced sludge handling costs
- Adsorption of organics, such as detergents, oils and dyes that are refractory to the biological system
- Protection of biological system from toxic waste components
- More effective removal of phosphorous and nitrogen

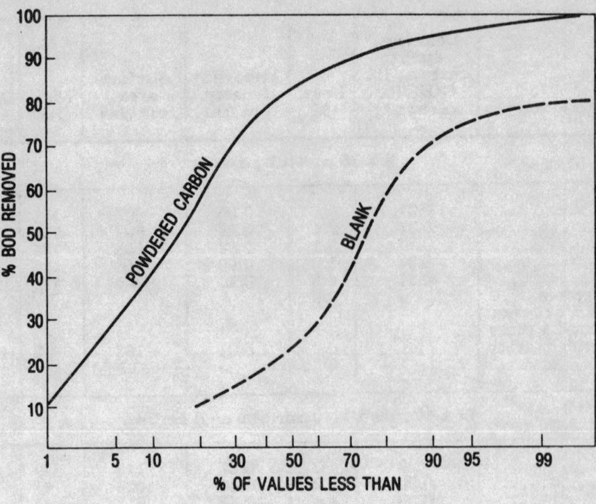

Fig. 4—Effect of powdered carbon on BOD removal.

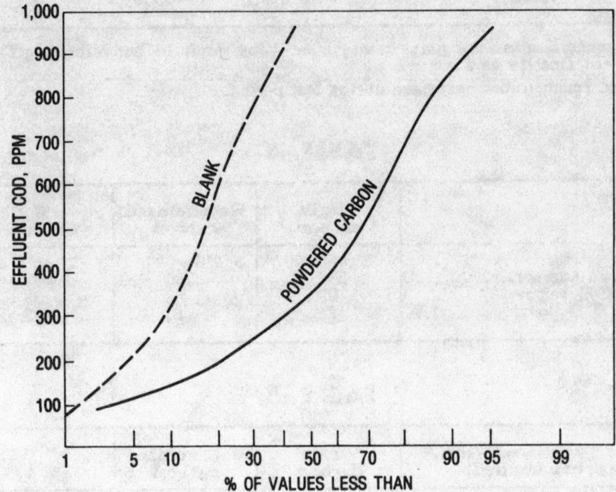

Fig. 5—Effect of powdered carbon on effluent COD.

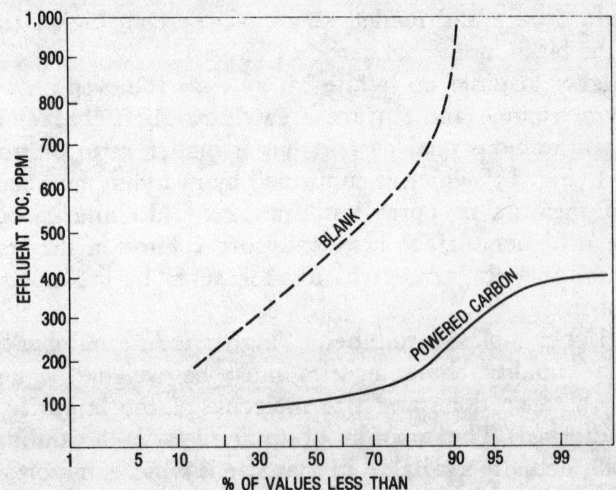

Fig. 6—Effect of powdered carbon on effluent total organic carbon.

- Increased effective plant capacity at little or no additional capital investment
- Savings on operating costs resulting from reduced defoamer, coagulant and power requirements

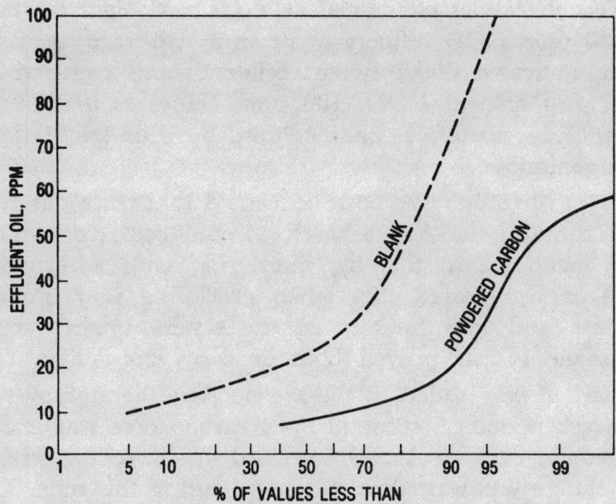

Fig. 7—Effect of powdered carbon on effluent oil.

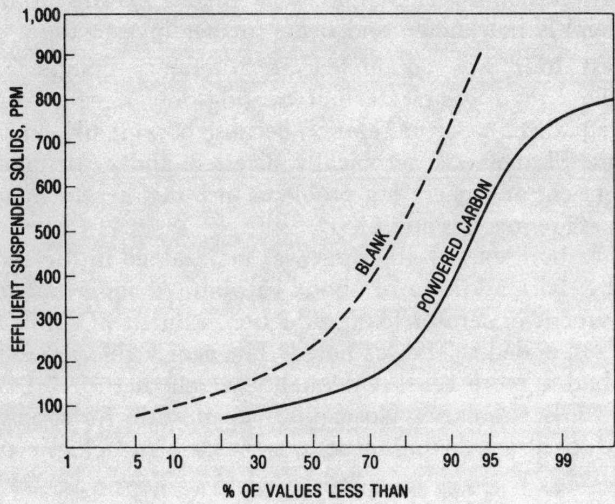

Fig. 8—Effect of powdered carbon on effluent suspended solids.

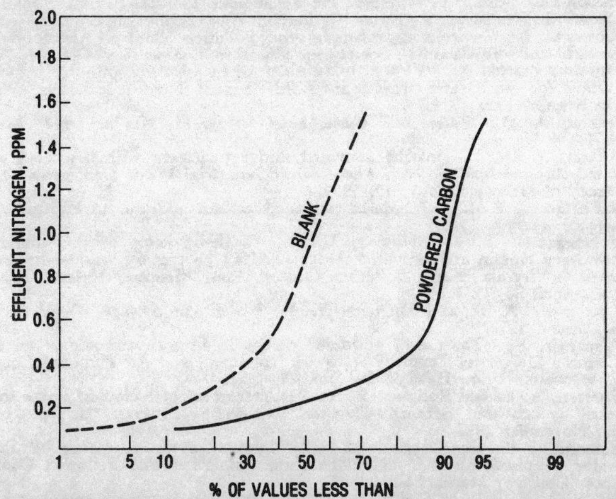

Fig. 9—Effect of powdered carbon on effluent nitrogen.

- Greater treatment flexibility since carbon dosages can be varied to match waste strengths and flow rates.

Powdered carbon improves activated process treatment because of its adsorptive and physical properties. Pollutants and oxygen are both adsorbed, localizing these reactants, thus increasing their concentration and driving the biological reaction further toward completion to improve BOD removals.

Many pollutants, not biologically degraded in conventional systems, may be degraded when contacted longer with the biomass. This contact time is extended from hours to days when the pollutants are adsorbed by carbon and are settled with the sludge. This additional contact lowers effluent COD and TOC values.

Powdered carbon also adsorbs toxic waste components to protect biological system from upset.

Aerator foam is reduced because carbon adsorbs detergents. This improves oxygen transfer and organic removal.

High-density powdered carbons improve solids settling in secondary clarifiers which lowers effluent suspended solids and BODs and yields sludges that are thicker and more easily dewatered. Under high organic load conditions, which normally would lead to sludge bulking, dense carbon acts as a weighting agent, keeping sludge in the system. When dispersed biofloc results due to low organic loads, carbon serves as a seed for floc formation, preventing loss of solids. Higher mixed liquor volatile sludge solids (MLVSS) can be maintained because of improved sludge settling. Both phosphorous and nitrogen removals are enhanced under these conditions.

Finally, carbon is biologically regenerated[15] to some degree such that adsorptive surfaces are continually renewed. This reduces virgin carbon dosage to maintain treatment efficiency.

Powdered carbon is added at any convenient process point to get it into the aerator: to the aerator directly, to sludge return lines, to influent channels or through secondary clarifier. Batch addition either dry or with suitable feeders at any time of the day is satisfactory.

PLANT TESTS

Four refinery and/or petrochemical plants have evaluated the powdered activated carbons in full scale activated sludge systems.

Test one involved treatment of a 2.2 mgd flow with an average BOD of 400 ppm in a 1.2-million-gallon aerator. Mixed liquor solids were maintained at 3,600 ppm (2,880 ppm volatile). Waste activated sludge was digested aerobically, centrifuged and hauled to landfill. Despite a secondary clarifier overflow rate of only 423 gallons/ft.^{2} and use of 22 ppm cationic polymer for secondary solids capture, effluent solids averaged in excess of 100 ppm. Toxic loads caused periodic loss of aerator biosolids. Defoamer costs averaged $200/day for aerator foam control.

A high-density, lignite-based powdered carbon was added to the aerator over a 4½-month period to establish an equilibrium carbon level which reached 1,800-2,000 ppm. This level was maintained with a daily average carbon dose of only 20 ppm at the sludge solids concentration obtained and wasting rates employed.

Average BOD reduction over the entire carbon test period equaled 82% versus 23% during post test control period (Fig. 4). As carbon built up in the system, BOD removals reached 90-95% to achieve required maximum of 30 ppm BOD effluent. Effluent COD was reduced from an average of 1,180 ppm without carbon to 350 ppm with carbon (Fig. 5). Average effluent TOC decreased from 420 ppm to 100 ppm (Fig. 6). Slope of the carbon plots in Figs. 4, 5 and 6 also indicates improved uniformity in effluent quality with carbon present.

This treatment had a dramatic effect on reduction of oil through the system (Fig. 7); effluent concentration was reduced by 75% (average) while narrowing the range.

Both removal of oil by powdered carbon and weighting effect of carbon resulted in lower effluent solids (Fig. 8). Prior to carbon treatment, the plant used polymer at a dosage of 20 ppm, but still experienced poor solids settling. When carbon was added to the system, solids settling improved, and polymer dosage was cut in half. Effective solids settling could not be achieved with use of carbon or polymer alone but combination of the two attained desired results with an operating cost savings. Improved solids settling increased sludge thickening which allowed a 65% reduction in sludge wasting to save centrifuge operation including power and labor.

Addition of powdered carbon eliminated need for aeration defoamer, more than justifying carbon expense. Adsorbing foaming agents eliminated foam problems in receiving stream. Defoamers only suppress foam in the aerator and do not prevent its reappearance in the effluent. Carbon can reduce operating costs by allowing surface aerators to aerate and mix activated sludge rather than expend energy generating foam.

Increased nitrogen removals (Fig. 9) occurred because dense carbon settled nitrifying organisms which normally would float out of the system. Longer solids retention resulted to favor nitrification. Improved solids settling is probably reason for decreased phosphorus levels since phosphorous is precipitated with carbon-biosolids floc and removed in sludge rather than degraded biologically.

Upsets were greatly reduced with carbon in the aerator possibly because carbon adsorbs heavy metals from wastewater.[16–18] Lower zinc levels were noted in this effluent.

Test two was conducted at a plant treating an average 12 mgd flow having a TOC of raw waste from 100-1,000 ppm, averaging about 200 ppm. Major treatment problems included aerator foaming caused by alkanolamines in the wastes; high effluent TOC; oily, difficult to handle sludge, and high effluent solids.

Bench aeration tests were used to determine full scale operating parameters. Mixed liquor solids, including carbon, were maintained at 3,000 ppm. At a 100 ppm carbon level, a 20% reduction in sludge wasting and a 25% decrease in effluent suspended solids were observed. When the carbon level was increased to 500 ppm, effluent became clearer and effluent variability decreased markedly.

These results were confirmed in a full scale test. Unfortunately, uncontrolled sludge wasting prevented building a high aerator carbon level. Still, carbon benefits were apparent. The system was less susceptible to upset with carbon present under conditions that previously would have caused bioactivity loss. One time an excessively low pH caused an upset but the system recovered more rapidly than expected.

Effluent TOCs were maintained below 20 ppm during shock load periods which is well within the standard of 50 ppm. In a post test control phase, effluent quality deterioration was observed as carbon was wasted from the system through sludge.

Test three was conducted at a 2.5 mgd plant treating a 550 ppm COD refinery waste in a two-stage, conventional activated sludge system. Effluent solids averaged 25 ppm and effluent COD, 180 ppm before carbon treatment. Raw waste was characterized by wide fluctuations in organic load.

Plant operators had been concerned that trace oils getting through the API separator would coat the carbon and inactivate it, thereby, interfering with adsorption. Also, it was feared that when processing sour crudes, sulfides and H_2S buildup in the system might occur. Subsequent tests proved both concerns groundless.

Carbon was added to the second stage aerator over a six-week period. A constant daily carbon dose was maintained for each week and increased in succeeding weeks. No sludge was wasted intentionally during this time.

Optimum treatment was found at a relatively high influent dose of 200 ppm. Effluent solids and COD removals increased 40 percent, and BOD removals, already high, increased 10 percent.

Cyanide was removed more efficiently in conjunction with $CuSO_4$ treatment to cut effluent levels from 1 ppm before carbon to 0.05 ppm. The precise nature of this removal is not known and bears further investigation.

Test four was conducted at a refinery designed to handle 2 mgd wastewater but treating only 1 mgd. Waste was equalized prior to aeration because of wide pH fluctuations. Sludge was aerobically digested and centrifuged. Treatment and operating problems included aerator foam, high effluent solids and COD.

A carbon level of 400 ppm was maintained in the aerator by daily addition of about 15 ppm. A significant improvement in aerator foam reduction, effluent BOD, COD and suspended solids was noted. The nearly colorless effluent had a more consistent quality which met 1977 BOD and COD standards. No capital expenditure for sophisticated treatment equipment was needed to achieve this quality.

LITERATURE CITED

1 Bulletin 20-2C, "Filtrasorb 300 and 400 for Wastewater Treatment," Calgon Corp., Pittsburgh, Pa., 1970.

2 Product Data Bulletin, "Nuchar Granular Active Carbon Grade: Nuchar WV-G 12 x 40," West Virginia Pulp and Paper, Covington, Va.

3 Product Data Bulletin, "Nuchar Granular Active Carbon Grade: Nuchar WV-L 8 x 30," West Virginia Pulp and Paper, Covington, Va.

4 Smisek, M. and Cerny, S., *Active Carbon,* American Elsevier Publishing Co., New York, (1970).

5 Internal ICI United Sates Inc. Report, "Competitive Carbon Study."

6 Hutchins, R. A., *Chemical Engineering Progress,* November 1973.

7 McCrodden, B. A., "Treatment of refinery wastewater using filtration and carbon adsorption," presented at the Technology Transfer Seminar, Physical-Chemical Treatment, Activated Carbon in Water Pollution Control, jointly sponsored by Environment Canada, the Pollution Control Association of Ontario and the Canadian Society for Chemical Engineering, Oct. 24, 1975.

8 DeJohn, Paschal B., "Factors to consider when selecting granular activated carbon for wastewater treatment," 29th Annual Purdue Industrial Waste Conference, May 1974.

9 Adams, A. D., *Water and Wastes Engineering,* 11, No. 3, . B-8, March 1974.

10 Adams, A. D., "Improving activated sludge treatment with powdered activated carbon—textiles," 6th Mid-Atlantic Industrial Waste Conference, University of Delaware, Nov. 15, 1973.

11 Robertaccio, F. L., "Powdered activated carbon addition to biological reactors," *idem.*

12 Foertsch, G. B. and Hutton, D. G., "Scale-up tests of the combined powdered carbon and activated sludge (PACT) process for wastewater treatment," Virginia Water Pollution Control Assn. Meeting, Natural Bridge, Va., April 30, 1974.

13 Scaramelli, A. B. and DiGiano, F. A., *Water and Sewage Works,* p. 90, September 1973.

14 Alspaugh, T., "The use of powdered carbon in an activated sludge system," Clemson University Textile Wastewater Treatment and Air Pollution Control Conference, Hilton Head, S.C., Jan. 22, 1975.

15 Perrotti, A. E. and Rodman, C. A., "Enhancement of biological waste treatment by activated carbon," *Chemical Engineering Progress,* 69, No. 11, p. 63, November 1973.

16 Esmond, S. E. and Petrasek, A. C., "Removal of heavy metals by wastewater treatment plants," WWEMA Industrial Water and Pollution Conference, Chicago, March 14-16, 1973.

17 Sigworth, E. A. and Smith, S.B., "Adsorption of inorganic compounds by activated carbon," *Journal AWWA, Water Technology/Quality,* June 1972, p. 306.

18 Linstedt, K. D. et al, "Trace element removals in advanced wastewater treatment processes," *Journal WPCF,* 43, No. 7, 1507, July 1971.

Powdered Carbon Improves Activated Sludge Treatment

Process provides alternate to granular activated carbon tertiary treatment of wastewater

C. G. Grieves and **M. K. Stenstrom,** Amoco Oil Co., Naperville, Ill., and **J. D. Walk,** and **J. F. Grutsch,** Standard Oil Co. (Indiana), Chicago, Ill.

USE OF powdered activated carbon is an attractive approach for improving refinery activated sludge effluent. Its use is a viable alternative to granular activated carbon tertiary treatment for meeting proposed 1983 Best Available Technology Economically Available (BATEA) effluent quality standards as required by the Environmental Protection Agency (EPA).

The proposed process involves adding powdered activated carbon to the aeration tank of the activated sludge process, achieving cost effectiveness by operating at a very high sludge age and a low carbon dose. Effective removal of oil and colloidal solids in the pretreatment step is necessary for successful operation.

Effluent quality depends upon both the equilibrium mixed-liquor carbon concentration and the surface area of the carbon. An experimental carbon with a high surface area appears to be several times more effective than the best commercial carbons in achieving an effluent quality standard. Pore size of the activated carbon has no apparent effect upon effluent quality.

In general, the process can be used to meet only the long-term average effluent quality proposed for BATEA. Daily maximum and 30-day maximum variability goals, as presently defined, cannot be met.

The proposed process also enhances nitrification at low temperatures and dampens effects of increased hydraulic flow rate on the activated sludge process. Both phenomena will help to decrease effluent variability.

According to the EPA guidelines for treating refinery wastewaters,[1] the sequence shown in Fig. 1 is recommended for current Best Practical Technology Currently Available standards. For meeting 1983 BATEA goals the guidelines recommend an add-on process using granular carbon adsorption. However, this approach may be both inefficient and very costly. So far as is known, its effectiveness has never been adequately demonstrated. Moreover, preliminary estimates indicate that capital and operating costs for the granular carbon adsorption and

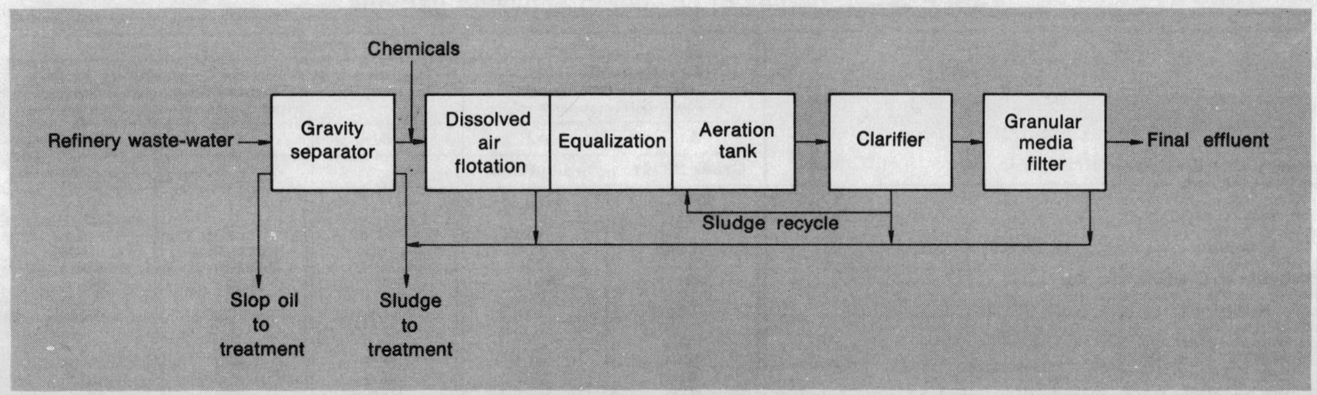

Fig. 1—Simplified refinery Best Practical Technology wastewater treatment system.

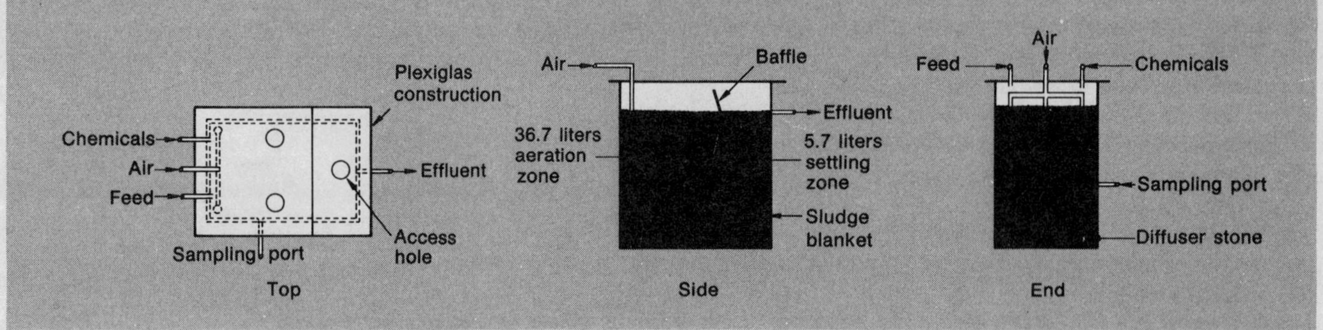

Fig. 2—Activated sludge reactor used in pilot work.

regeneration facilities may equal or exceed those of the entire current activated sludge process.

By contrast, both patents and research studies[2-25] indicate that powdered activated carbon may be a practical and economical substitute for granular carbon. For example, powdered carbon costs only about one-half as much as granular—$0.65/kg versus $1.20/kg.[15] In addition, recent studies have shown that powdered carbon can be added directly to the mixed-liquor in activated sludge aeration tanks.[21, 22, 23, 24] Thus, appropriate alterations in operating procedures may eliminate the need for regeneration by making it economically feasible to discard the spent carbon with the waste sludge.

TABLE 1—Pilot plant operating conditions

Aeration Zone Volume	36.7	liters
Settling Zone Volume	5.7	liters
Nominal Flow Rate	2.45	liters/hr.
Nominal Hydraulic Retention Time	15.0	hours
Nominal Settling Time	2.33	hours
Air Flow Rate	300	liters/hr.
pH	6—8.5	
Caustic Addition Rate	0.12—0.30	liters/hr.
Phosphorous Added to Feed	3	mg/liter
Temperature	Ambient (4—31°C)	

Analytical Work

Frequency	Analysis Performed[27, 28]
Daily	Influent and mixed-liquor pH, temperature, influent flow rate, caustic addition rate. Carbon addition and sludge wastage.
3 Times a Week	Influent and effluent total and volatile suspended solids, soluble organic carbon, soluble chemical oxygen demand, soluble ammonia nitrogen, and soluble phenolics. Mixed-liquor suspended solids and mixed-liquor volatile suspended solids. Sludge volume index.
Once a Week	Material balances to calculate quantity of sludge to be wasted to maintain desired sludge age.

In general, the cost effectiveness of a powdered carbon process increases with the concentration of carbon maintained in the mixed-liquor. A mass balance of such a process is represented by the following equation:

$$C = \frac{C_i \, \Theta_c}{\Theta_h} \tag{1}$$

where

C = Equilibrium mixed-liquor carbon concentration (mg/1)
C_i = Influent carbon concentration (mg/1)
Θ_c = Sludge age (days)
Θ_h = Hydraulic retention time in the aeration tank (days)

Equation 1 shows that the equilibrium mixed-liquor carbon concentration is proportional to the product of the influent carbon concentration (carbon dose) and the sludge age. Thus, equilibrium carbon concentration can be increased by increasing the carbon dose, or the sludge age or both. Therefore, to keep carbon costs to a minimum, it is desirable to operate at as high a sludge age as possible and not at an excessively long hydraulic retention time.

A possible drawback to operation at a high sludge age is the increased risk that toxic, inhibitory, or inert materials will build up in the aeration tank. For example, a build-up of oily solids could reduce the oxygen transfer efficiency and inhibit both the nitrifying and organic carbon utilizing organisms. The dissolved oxygen concentration in the mixed-liquor could also become too low for

TABLE 2—Properties of powdered activated carbons

	Carbon Designation				
	Experimental Amoco High Surface Area—		Commercially Available Conventional Surface Area Carbons		
	A1	A2	B	C	D
Property	Grade PX-21	Grade PX-23			
Surface Area					
BET, m^2/g	3099	3148	717	514	532
Pore Volume, cc/g					
>15 A° Radius	0.16	0.43	0.28	0.38	0.03
<15 A° Radius	1.45	1.60	0.51	0.11—0.42	0.25
Iodine Number	3349	3375	1790	920	888
Methylene Blue Adsorption, mg/g	586	550	100	83	50
Phenol Number	12.8	12.6	34.1	22.9	23.8
Bulk Density, g/cc	0.298	0.228	0.610	0.576	0.484
Screen Analysis					
Passes 100 Mesh, Wt. %	98.4	99.1	99.2	100.0	100.0
Passes 200 Mesh, Wt. %	92.7	93.4	86.7	94.4	97.9
Passes 325 Mesh, Wt. %	84.1	80.8	60.6	68.3	91.8
Molasses Number	10	205	103	85	0

effective nitrification, and the final clarifiers could become overloaded. Therefore, it is desirable in the pretreatment step to remove as much solid material as possible from the wastewater before it enters the aeration tank.

To evaluate the effects of such variables in a process using powdered carbon, an extensive 15-month four-phase pilot plant study was carried out at Amoco Oil Company's Texas City refinery. Pilot plants operating in parallel with the refinery activated sludge process facility were fed the same wastewater for treatment. Specific variables investigated were:

- Carbon type, including surface area and pore volume
- Carbon addition rate
- Sludge age
- Pretreatment of feed to remove oil and solids.

EXPERIMENTAL EQUIPMENT

Fig. 2 shows the configuration of the pilot plants. Each had a volume of 42 liters, and as many as eight units were operated in parallel during portions of the study. They were housed in a rain-tight enclosure but were neither heated nor cooled. Thus, the temperature of the mixed-liquor varied from 4° C to 31° C.

Operating conditions and analytical procedures are summarized in Table 1. The pH was checked daily and controlled by addition of caustic at a constant rate. Dibasic potassium phosphate, K_2HPO_4, was added to satisfy the phosphorus requirement of the microorganisms.

The wastewater feed, a slipstream from the pressure filters of the refinery treatment plant, was passed through a pilot gravity sand filter before being fed to the pilot plants.

Table 2 summarizes the characteristics of the five powdered carbons evaluated. Amoco's experimental high-surface-area carbons are designated as A1 and A2, PX-21 and PX-23, respectively. Those designated as B, C, and D are commercially available carbons having a much lower surface area. Carbon A2 (PX-23) has the highest pore volume.

Effectiveness was judged on the basis of effluent standards proposed for a BATEA facility[1] (Table 3).

TABLE 3

	Concentration mg/liter
Total Organic Carbon (TOC)	15
Chemical Oxygen Demand (COD)	24
Ammonia (NH_3-N)	6.3
Phenolics	0.02

These standards are for a Class C refinery and are based on the guideline effluent flow rate of 0.46 m^3/m^3 of crude thruput per stream day (19 gal/bbl.). Because the BATEA treatment sequence will undoubtedly result in very low concentrations of effluent suspended solids, only the soluble components of the effluent were measured.

To obtain high sludge ages, effluent suspended solids were allowed to settle in 30-gallon plastic containers and then were returned to the pilot plants periodically. At any given sludge age, all plants were allowed to reach steady-state operation over an extended period of time. Then performance data were taken over a 30-day period.

RESULTS

The four phases of the study were carried out in sequence, with the design of succeeding phases based on the results of the preceding ones (Table 4).

TABLE 4

Phase	Objective
I	Effect of carbon type at an addition rate of 100 mg/liter and a sludge age of 20 days with prefiltered feed.
II	Effect of carbon type at an addition rate of 200 mg/liter and a sludge age of 20 days with prefiltered feed.
III	Effect of increasing sludge age to 60 days and reducing carbon addition rate to 25 mg/liter with unfiltered and prefiltered feed.
IV	Effect of further increasing sludge age to 150 days while reducing carbon addition rate to 10 mg/liter.

Phases I and II. The results of Phases I and II (Table 5) indicate that powdered activated carbon significantly enhances the performance of a refinery activated sludge process. Improvement in the quality of the effluents from carbon-fed plants ranged from 65% for soluble organic carbon up to 95% for phenolics. At the 200 mg/liter addition rate, the results usually satisfied the BATEA effluent quality goals. The high surface area carbon A1 was significantly more effective than the other three. The commercially available carbon B produced slightly better effluent than carbon C, which would be expected if efficiency is proportional to surface area. Because nitrification was essentially complete in the control unit, carbon addition could not improve ammonia conversion. Carbon D, which is derived from wood charcoal and has a

TABLE 5—Phases I and II—Effect of carbon type and addition rate on effluent quality* 50% probability data during 30 days of steady-state operation sludge age = 20 days

	Concentration, mg/liter					
		Pilot Plant Effluent				
Component	Filtered Influent	No Carbon	Carbon A1	Carbon B	Carbon C	Carbon D
Phase I: Carbon Addition Rate = 100 mg/liter Equil. Mixed-Liquor Temp = 31°C, Carbon Conc = 3200 mg/liter						
SOC	72.0	22.0	12.5	17.5	18.5	23.0
SCOD	230	73	28.5	48	44	65
NH_3-N	25.8	0.5	0.2	0.5	0.5	0.5
Phenolics	4.35	0.018	0.003	0.010	0.010	0.017
Phase II: Carbon Addition Rate = 200 mg/liter Equil. Mixed-Liquor Temp = 25°C, Carbon Conc = 6400 mg/liter						
SOC	70.0	26.5	9	13.5	15.5	.....
SCOD	230	58	17	24	28	.....
NH_3-N	25.4	0.2	0.2	0.2	0.1	.....
Phenolics	4.06	0.020	0.001	0.001	0.003	.....

*BATEA effluent standards in mg/liter are:

Soluble Organic Carbon (SOC)	15
Soluble COD (SCOD)	24
Ammonia Nitrogen (NH_3-N)	6.3
Phenolics	0.02

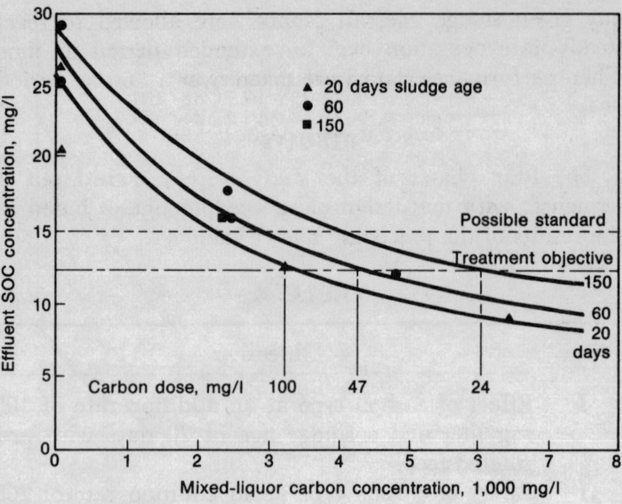

Fig. 3—Effect of mixed-liquor carbon concentration on effluent soluble organic carbon.

significantly lower pore volume than the others, performed so poorly in Phase I that it was dropped from further consideration. The performance of carbon A1 at 100 mg/liter dose was about as effective as carbon B at 200 mg/liter, or about twice as effective as the best commercially available carbon tested.

Phase III. Table 6 shows the effects of sludge age and feed filtration upon performance. The plant with filtered feed performed better than one with unfiltered feed, and a sludge age of 60 days was better than one of 20 days. No deterioration in the settling characteristics of the mixed-liquor suspended solids was observed at this higher sludge age.

TABLE 6—Phase III—Effect of carbon type and addition rate, sludge age, and influent pretreatment on effluent quality 50% probability data during 30 days of steady-state operation equil. mixed-liquor temp = 14° C, carbon conc. = 2400 mg/liter

Influent Pretreatment	Carbon Type	Carbon Addition Rate, mg/liter	SOC	SCOD	NH_3-N	Phenolics
			Influent Concentration, mg/liter			
Filtered Feed	—	—	73.5	294.5	19.3	3.95
			Effluent Concentration, mg/liter			
			Sludge Age = 20 Days			
Unfiltered	—	—	32.0	103.5	12.1	0.027
Filtered	—	—	29.0	83.0	14.5	0.027
			Sludge Age = 60 Days			
Filtered	—	—	25.0	65.9	5.1	0.019
Filtered	B	100	16.0	40.3	0.2	0.001
Filtered	A1	50	12.0	27.5	0.1	0.002
Filtered	A1	25	16.0	50.3	0.4	0.006
Filtered	A2	50	13.0	31.0	1.8	0.004

At a sludge age of 20 days the plant with filtered feed performed marginally better than the one with unfiltered feed. Undoubtedly, greater differences in effluent quality would have been observed in a plant operated at a sludge age of 60 days with unfiltered feed. (Not recorded in these data, however, is the complete failure of the plant fed unfiltered feed shortly after cessation of data gathering for this steady-state period.)

Table 6 also shows how pore size and surface area affect the performance of the carbons. Carbons A1 and A2 have approximately the same surface area, but carbon A2 has much larger pores. Yet, at an equivalent addition rate of 50 mg/liter, both carbons showed about the same performance. Thus, large pore diameters are not required for effective treatment of this refinery wastewater. Moreover, plants fed 50 mg/liter of either A1 or A2 performed much better than the plant fed 100 mg/liter of carbon B. In fact, these high-surface-area carbons are between two and four times more effective than carbon B in enhancing SOC and soluble COD removal.

A comparison of the data in Tables 5 and 6 shows that a low carbon dose and a high sludge age enhance an activated sludge process almost as much as do a high carbon dose and a low sludge age.

It is possible that the difference in performance is solely due to difference in temperature between the phases—mean operating temperature during Phase III was only 14° C, whereas during Phases I and II temperature averaged 31° C and 25° C, respectively.

Also observed during the lower operating temperature of Phase III was an increase in the ammonia removal efficiency of the carbon-fed pilot plants. This phenomenon was unexpected because activated carbon does not normally adsorb ammonia. Possibly, the increased removal rate is due to the adsorption of potentially toxic or inhibitory organic materials which would reduce the rate of nitrification if left in solution. The control plant in Phases I and II had little difficulty in achieving full nitrification, perhaps because of the higher temperature.

Phase IV. As shown in Table 7, Phase IV was designed to push the activated sludge system to the limit by increasing sludge age to 150 days and decreasing carbon addition to 10 mg/liter. Further, in one of the plants, hydraulic retention time was reduced to 7.5 hours, compared with 15 hours in the other plants.

Despite similarities in influent quality during all four

TABLE 7—Phase IV—Effect of high sludge age, low carbon addition rate, and decreased hydraulic retention time on effluent quality* 50% probability data during 30 days of steady-state operation, equil. mixed-liquor temp = 27° C

Carbon Type	Carbon Addition Rate, mg/liter	Sludge Age, days	Hydraulic Retention Time, hr	Equil. Mixed Liquor Carbon Conc, mg/liter	Effluent SOC, mg/liter	Effluent SCOD, mg/liter	Effluent NH_3-N, mg/liter	Effluent Phenolics, mg/liter
—	—	60	15	—	29	99	0.1	0.018
B	25	60	15	2400	22	64	0.1	0.010
A1	25	60	15	2400	18	52	0.1	0.010
A1	25	60	7.5	4800	17	46	0.3	0.010
A1	10	150	15	2400	16	49	0.1	0.010

*Filtered influent contained 78 mg/liter SOC, 270 mg/liter SCOD, 29 mg/liter NH_3-N, and 3.25 mg/liter phenolics.

phases, during Phase IV the effluent SOC and COD of the control increased by about 30-35% over that observed during the first three phases, despite a mean temperature of 27° C (*c.f.* 14° C during Phase III). All pilot plants essentially nitrified completely.

Remarkably, however, the plant with 10 mg/liter of high surface area carbon A1 at a sludge age of 150 days produced an effluent whose soluble organic carbon concentration was 50% lower than that of the control reactor and slightly lower than that of all of the other pilot plants. The plant dosed with 25 mg/liter carbon A1, with one-half the hydraulic capacity of the other plants, produced the second best effluent.

The outstanding performance at a sludge age of 150 days indicates that refinery activated sludge processes can be operated with very little added carbon. The dose may be low enough so that the carbon need not be regenerated but be discarded with the waste activated sludge. At a very high sludge age, there will be smaller quantities of waste sludge to be disposed of.

The data in Table 7 also indicate that powdered carbon can be used to increase the hydraulic capacity of an activated sludge plant, as proposed by others,[13] or to increase the effluent quality of an overloaded plant. The carbon-fed plant that operated at one-half the hydraulic retention time of the control produced an effluent 50% better than that of the control. Experience with pilot activated sludge plants operated at several of Amoco's other refineries has shown that conventional activated sludge processes cannot be operated successfully with a hydraulic residence time of only 7½ hours.

STATUS

The data from Phase IV indicate that the limits of the powdered carbon enhanced activated sludge process have not been reached. In addition, more data are needed before economic studies can be made to weigh the possible options for achieving a given effluent quality:

- High fresh carbon dose at moderate sludge age (20-60 days) with regeneration of spent carbon;

- Low fresh carbon dose at high sludge age (60-150 days) with no regeneration of spent carbon.

Cost analyses should be made for each of these extreme options, and several intermediate ones, and compared with those for tertiary treatment with granular carbon technology.

Fig. 3 shows the qualitative curves this pilot study has generated. Of course, the one for the 150-day sludge age is purely speculative because only one data point exists. However, the trend of the data does show that effluent quality is a function of mixed-liquor carbon concentration. The curves are probably asymptotic to a residual organic carbon concentration, but over the range investigated an increase in mixed-liquor carbon concentration causes a decrease in effluent soluble organic carbon. Furthermore, the relationship between effluent quality, sludge age and carbon dose is clearly non-linear. For example, to achieve an effluent quality of 12.5 mg/liter of soluble organic carbon, the three options are: 100 mg/liter of carbon at a sludge age of 20 days; 47 mg/liter of carbon at a sludge age of 60 days; 24 mg/liter

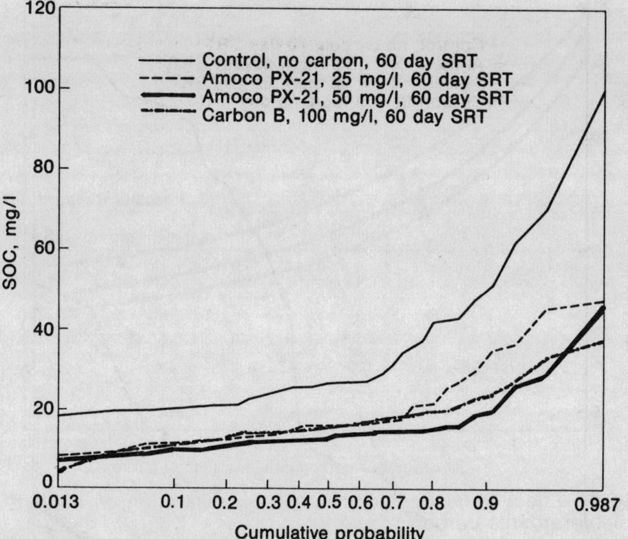

Fig. 4—Soluble organic carbon—Phase III.

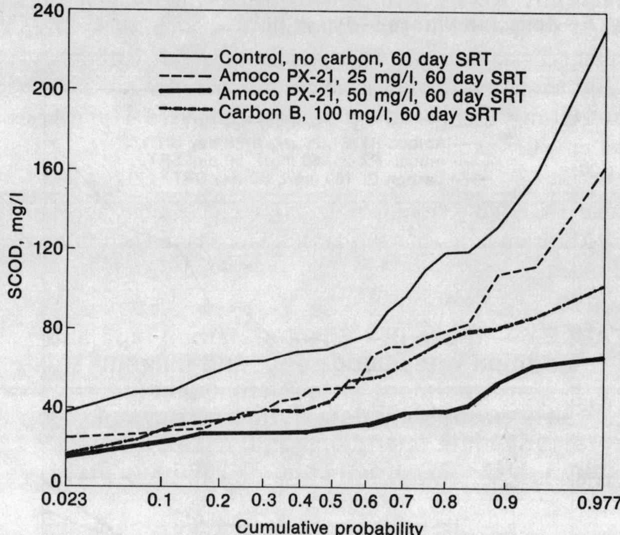

Fig. 5—Soluble chemical oxygen demand—Phase III.

of carbon at a sludge age of 150 days. If the relationship were linear, the values calculated from a base case of 100 mg/liter at a 20-day sludge age would be 33 mg/liter and 13 mg/liter at 60 days and 150 days, respectively.

Apparently, the process loses effectiveness because of incomplete microbial regeneration. Microbial regeneration of the spent carbon is probably not as effective as using fresh carbon; some materials adsorbed by the carbon are undoubtedly non-biodegradable, even after 150 days of contact with microorganisms in the pilot plant. The ability to retain significant effectiveness even at 150 days is the key to cost effective high sludge age operation with powdered activated carbon. Of course, there may be other reasons why carbon loses effectiveness at high sludge age, such as production of cell lysis products which are then adsorbed by the carbon.

EFFLUENT VARIABILITY

Variation in effluent quality over a 30-day (or longer) period is extremely important. The EPA[1] has set the daily maximum variability equivalent to the 99% probability value and the 30-day maximum variability to

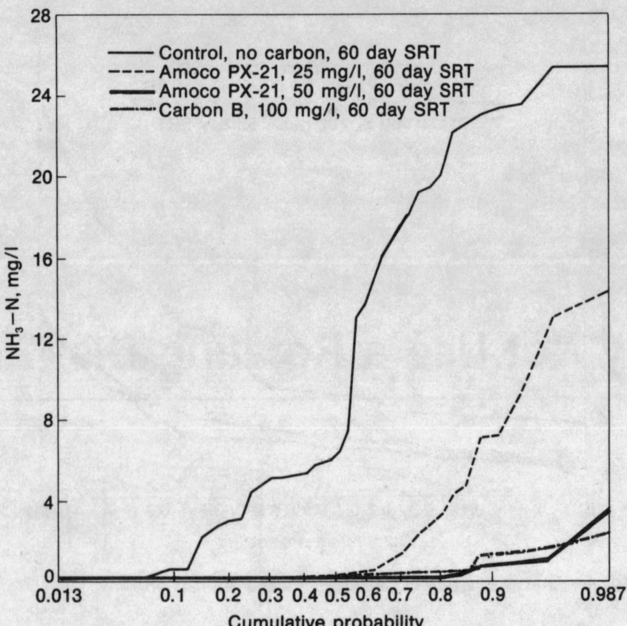

Fig. 6—Ammonia-Nitrogen—Phase III.

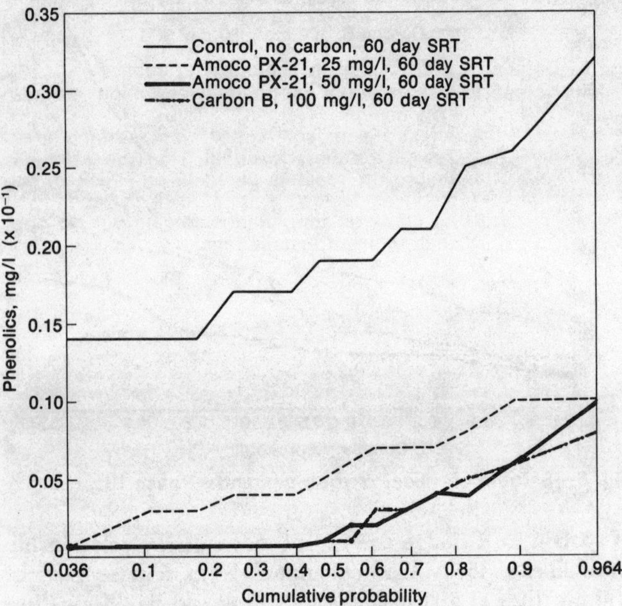

Fig. 7—Phenolics—Phase III.

the 98% level. For BATEA the daily maximum variability factors for TOC, COD, NH_3-N, and phenolics are proposed at 1.6, 2.0, 2.0, and 2.4, respectively. The 30-day maximum values are 1.3, 1.6, 1.5, and 1.7, respectively.

TABLE 8—Phase III—BATEA guideline and actual variability factors for pilot plant fed 25 mg/liter of carbon A1

Parameter	BAT Guideline Variability Factor		Actual Variability Factor	
	Daily Max.	30 Day Max.	Daily Max.	30 Day Max.
Soluble Organic Carbon.......	1.6	1.3	2.8	2.8
Soluble COD...............	2.0	1.6	7.5	7.5
NH_3-N..................	2.0	1.5	2.1	2.0
Phenolics..................	2.4	1.7	5.0	5.0

Figs. 4, 5, 6 and 7 show probability data for the 30-day operating periods during Phase III. Table 8 shows the daily maximum (99% probability) and 30-day maximum (98% probability) variability factors calculated from these figures for the plant fed with 25 mg/liter of Carbon A1. The EPA guideline values are also given. The actual variability factor was calculated as the 99% (or 98%) probability value divided by the target quality value. In general, the variability in effluent quality was higher than the guideline values.

It is important to note that the proposed guideline variability factors are unrealistic. The data base used by EPA[1] for their production was obtained from limited pilot studies. In addition, BPTCA 30-day maximum (98% probability) values were used as the BATEA 30-day maximum values. Variability factors will undoubtedly have to be amended before BATEA *goals* become BATEA *standards.*

ACKNOWLEDGMENT

Presented at The Joint EPA-API-NPRA-UT Second Open Forum on Management of Petroleum Refinery Wastewater, Wednesday, June 8, 1977, University of Tulsa, Tulsa, Oklahoma.

LITERATURE CITED

[1] U.S. Environmental Protection Agency, *(Draft) Development Document for Effluent Limitations, Guidelines and New Source Performance Standards for the Petroleum Refining Point Source Category,* USEPA, Washington, D.C., 20460 (April, 1974).
[2] Derleth, C. P., U.S. Patent 1,617,014, Feb. 8, 1927.
[3] Statham, N., U.S. Patent 2,059,286, Nov. 3, 1936.
[4] Robertaccio, F. L., *et al,* "Treatment of Organic Chemicals Plant Wastewater with the DuPont PACT Process." Presented at AIChE National Meeting, Dallas, Texas, Feb. 20-23, 1972.
[5] Adams, A. D., "Improving Activated Sludge Treatment with Powdered Activated Carbon." *Proc. 28th Annual Purdue Industrial Waste Conference,* 1972.
[6] Robertaccio, F. L., "Powdered Activated Carbon Addition to Biological Reactors." *Proc. 6th Mid Atlantic Industrial Waste Conference,* University of Delaware, Newark 1973.
[7] Flynn, B. P., "Finding a Home for the Carbon Aerator (Powdered) or Column (Granular)." *Proc. 31st Annual Purdue Industrial Waste Conference,* 1976.
[8] Scaramelli, A. B., and DiGiano, F. A., "Upgrading the Activated Sludge System by Addition of Powdered Activated Carbon." *Water and Sewage Works,* 120: 9, 90, 1970.
[9] Hals, O., and Benedek, A., "Simultaneous Biological Treatment and Activated Carbon Adsorption." Pres. 46th Annual Water Pollution Control Federation Conference, Cleveland, 1973.
[10] Kalinske, A. A., "Enhancement of Biological Oxidation of Organic Waste Using Activated Carbon in Microbial Suspensions." *Water and Sewage Works. 115:* 7, 62, 1972.
[11] Koppe, P., *et al,* "The Biochemical Oxidation of a Slowly Degradable Substance in the Presence of Activated Carbon: Biocarbon Unit." *Gesundeits Ingenieur* (Ger) *95:* 247, 1974.
[12] Adams, A. D., "Improving Activated Sludge Treatment with Powdered Activated Carbon." *Proc. 6th Mid Atlantic Industrial Waste Conference,* University of Delaware, Newark, 1973.
[13] Ferguson, J. F., *et al,* "Powdered Activated Carbon-Biological Treatment: Low Detention Time Process." Paper presented at the 31st Annual Industrial Waste Conference: Purdue University, Lafayette, Indiana, 1976.
[14] DeWalle, F. B., and Chian, E. S. K., "Biological Regeneration of Powdered Activated Carbon Added to Sludge Units." *Water Research 11:* 439, 1977.
[15] DeWalle, F. B., Chian, E. S. K., Small, E. M., "Organic Matter Removal by Powdered Activated Carbon Added to Activated Sludge." *Journal Water Pollution Control Fed. 49:* 593. 1977.
[16] Anon, "Chemenator" Chemical Engineering *84:* 1, 35. 1977.
[17] DeJohn, P. B., "Carbon from Lignite or Coal: Which is Better?" *Chemical Engineering* 82:9, 13. 1975.
[18] Hutton, D. G., and Robertaccio, F. L., U.S. Patent 3,904,518, September 9, 1975.
[19] DeJohn, P. B., and Adams, A. D., "Treatment of Oil Refinery Wastewaters with Powdered Activated Carbon." Pres. at the 30th Annual Purdue Industrial Waste Conference. 1975.
[20] Rizzo, J. A., "Use of Powdered Activated Carbon in an Activated Sludge System." *Proceedings of the Open Forum on Management of Petroleum Refinery Wastewaters,* Jan. 26-29, 1976, EPA, API, NPRA, University of Tulsa, at the University of Tulsa, Tulsa, Oklahoma.
[21] Stenstrom, M. K., and Grieves, C. G., "Enhancement of Oil Refinery Activated Sludge by Addition of Powdered Activated Carbon." Pres. at 32nd Annual Purdue Industrial Waste Conference, 1977.
[22] Grieves, C. G., et al, "Effluent Quality Improvement by Powdered Activated Carbon in Refining Activated Sludge Processes." Pres. at the 42nd Mid-year Refining Meeting, API. Chicago, Illinois, May 9-12, 1977.
[23] Thibault, G. T., Tracy, K. D., Wilkinson, J. B., "Evaluation of Powdered Activated Carbon Treatment for Improving Activated Sludge Performance." Pres. at the 42nd Mid-year Refining Meeting, API. Chicago, Illinois, May 9-12, 1977.
[24] Crame, L. W., "Pilot Studies on Enhancement of the Refinery Activated Sludge Process." Pres. at the 42nd Mid-year Refining Meeting, API. Chicago, Illinois, May 9-12, 1977.
[25] Kim, B. R., Snoeyink, V. L., Saunders, F. M., "Influence of Activated Sludge CRT on Adsorption." *Environ. Engr. Div., ASCE, 102:* 55. 1976.
[26] Grutsch, J. F., and Mallatt, R. C., "Optimize the Effluent System." *Hydrocarbon Processing. 55:* 3: 105-112.
[27] U.S. Environmental Protection Agency, *Methods for Chemical Analysis of Water and Wastes,* USEPA, Washington, D.C., 1974.
[28] American Public Health Association *et al, Standard Methods for the Examination of Water and Wastewater,* A.P.H.A. *et al,* 14th Ed., New York. 1975.

Why Not Use a Rotating Disk?

Survey reports problems/progress in this biological treatment system

H. E. Knowlton,
Chevron Research Co., Richmond, Calif.

ROTATING BIOLOGICAL treating units (biodisks or discs) offer U.S. refiners a fresh alternative to treating waste waters. Long a part of the European scene, especially for biologically treating municipal wastes, biodisks have been selected many times for use by U.S. refiners in recent years. A major attraction for some refiners is the relatively low operating attention required by disks. Because of this increasing U.S. interest it is appropriate to review the concept and its application in the U.S.

HOW BIODISKS WORK

Waste water flows into the tank (Fig. 1) as the disk rotates into the water at 1-2 rpm. Microorganisms growing on the disks contact and consume pollutants in the water and gain life-sustaining oxygen as the disks rotate back into the air above. The disk rotation also maintains a dissolved oxygen content in the waste water. After repeated contacts with the biomass on the disks, the waste water flows out the far side of the tank to another disk/tank or to subsequent treating steps.

The amount of surface area required in a biodisk system increases with increasing BOD, increased volume of flow and more stringent requirements for effluent quality. An uncovered disk is shown in Figure 2.

Some advantages of the biodisk system include:

- Minimum operator attention
- Low energy consumption—up to 40 percent of that used by activated sludge for the same application
- Low maintenance costs
- Small space requirements
- Compatibility with existing systems
- Ready expandability
- Odor-free operation.

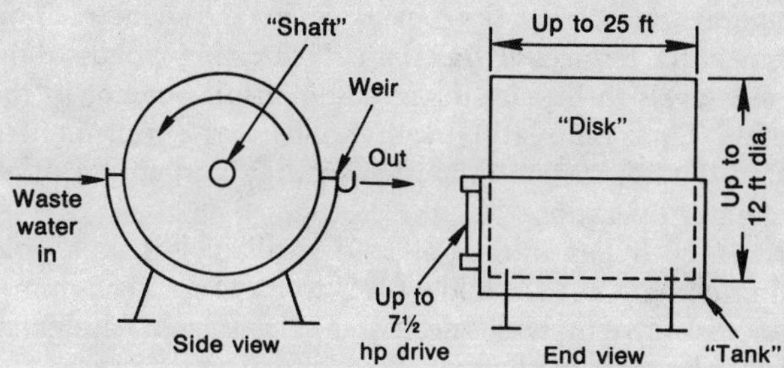

Fig. 1—Details of a disk used in a biodisk system.
Notes: 1. Disk material is polyethylene high surface area design
2. About 100 m sq. ft. per shaft surface area
3. Typical rotation is about 1½ rpm
4. Biogrowth reaches up to ¼ in. thick or as much as ½ lb./sq. ft. of disk surface
5. As many as four stages can be put on one shaft in small installations.

Fig. 2—A biodisk pack without cover.

U.S. USAGE

Four refineries have at least several months experience with biodisks (Table 1) while at least five others have limited experience or are in the process of installing disks. Those with significant experience vary in size from 2 to 18 shafts, depending on the amount of water to be treated. Two refineries (A and C) are using the disks as the sole treatment process.

Chevron experience. Biodisks have been selected for use at two of Chevron's six refineries (Table 2). The decision in each case was based on several factors:

- Availability of ponds
- Winter pond temperature

- Land availability and cost
- BOD content of waste water.

Climate is an important item. In warm climates where winter ambient temperatures are reasonable, existing aerated ponds with some modification are the economic choice. This applied at refineries 1, 2 and 3. However, even though refinery 4 is in a warm climate, there were no existing ponds and land is high priced. The choice then was between a biodisk system or activated sludge. Activated sludge was the economic choice because the high design BOD (500 ppm) of the waste water required relatively large amounts of biodisk surface area per gpm of waste.

Refineries 5 and 6 are in a cold climate where ambient temperatures are as low as 0°F in the winter. This significantly reduces bioactivity.[2] Existing ponds allow permit levels to be met in the summer-fall but not in the winter. Thus biotreating with ponds was supplemented with a biodisk system following deoiling and equalization for winter operation.

Refinery 6 has minimum land available, is in a cold winter climate but has a low BOD to the disks (165 ppm). A biodisk system was selected over activated sludge as the sole biotreat method.

Caution is required if investment and operating costs for a biological treating system are to be held to a reasonable level. The most important of these is source control. Flow volume and contaminants must be controlled at the source. No biological treating system is big enough to allow the operator to bombard it with acid, oil, caustic and higher-than-design water rates. No knobs are available to control these contaminants and flows. However, monitoring procedures can be installed upstream and operating practices can be changed—contaminant control devices such as mechanical seals and flowby sampling valves can be installed, washdown hoses can be shut off when not in use, good water can be reused and other measures taken when "good housekeeping" is properly emphasized by management leaders.

TABLE 1

Refinery	Shafts	M-Ft² of Surface	Treated, gpm	Startup Date	Only Biotreat Equipment
A	2	0.2	150	1975	Yes
B	8	0.2	1000	1975	Ponds Also
C	18	1.8	3000	1976	Yes
D	12	1.2	3000	1977	Ponds Also

Notes: 1. Styrofoam Disk Units Used at B—No Longer Offered by Manufacturer—A, C, and D Use Polyethylene Disk Material.
2. Refinery Size Range—23 M to > 150 M Bbl/Day.
3. Refinery Type—Cracking Type B, C, D; CU-CR Only A.
4. Million ft.² of surface.

TABLE 2

Refinery	System Selected	Climate	Comments
1	Aerated Ponds	Warm	Existing
2	Aerated Ponds	Warm	Existing
3	Aerated Ponds	Warm	Existing—Add Diffused Aeration to Deep Pond
4	Activated Sludge	Warm	High Strength Waste (500 ppm BOD Waste)
5	Biodisk	Cold	Supplement Existing Ponds for Winter Operation
6	Biodisk	Cold	Minimum Land Available (165 ppm BOD Waste)

Note: 1. Selections represent minimum company additional investment to meet 1977 EPA guidelines.
2. Warm means aerated pond temperature is high enough so that bioactivity can be obtained. Cold means ponds not feasible for winter operation.

OPERATING EXPERIENCE

A number of factors affect performance of biological disks. Experience in the operation of four systems (Table 3) suggests recommendations for use in other installations.

Operator attention. The plant operator usually walks through the biodisk system daily to make sure that disks are rotating. Comments on time spent indicate about 15 minutes per day. The operator should look daily at the biogrowth—its color can be indicative of source control problems. For instance, a white biogrowth indicates excess sulfides in feed.

Typical maintenance. Consists of greasing the drives once a month. However, after two years of operation, one refiner had to replace carbon steel tie rod nuts with stainless nuts and reposition the disks. Because of this experience, they check the disks on a quarterly basis and align and tighten the tie rods as necessary to maintain smooth, even rotation. This appears to be a reasonable course to follow for any user. Another refiner had his biodisk replaced by the manufacturer after the plastic tore loose from the drive shaft. Whether this was due to manufacturing flaws or improper operation is not known.

Shock load of BOD. No refiner has observed and documented the effect of a shock BOD load (rapidly increased BOD concentration) on the operation of a biodisk. However, information from a chemical plant operation using a biodisk system showed that a rapid sixfold increase in feed BOD raised the effluent BOD from 2-5 ppm up to 15-20 ppm; 24 hours after feed quality was again normal, the effluent BOD was back at 2-5 ppm.

Toxic materials. Three refiners have had experience with toxic materials or spills. One refinery operator put several weeks' accumulation of caustic water bottoms into the oily drain one day. This completely stripped the growth from the disks. It took five to six days to recover. Today, the caustic water bottoms is drained daily; and the resulting pH change of ½ to 1 number does not cause any observable change in effluent quality.

A second refinery was operating with a 45°F waste water to the biodisk when a small caustic spill occurred. The biomass turned white, and the effluent went off specification on phenol even though it still met BOD. Waste water routing was changed to bring feed temperature above 60°F; the disks regained a growth in a few days and full growth in 10 days. Phenol level in effluent was under control in 15 days in this combined disk-pond biotreat system.

The third refiner's experience was that with 10 ppm sulfide in the feed to the biodisk a white growth developed on the first two of four stages. This white growth interfered with BOD removal and settling of the biomass. The sulfide level was reduced to about 1 ppm and the white growth disappeared. The first two biodisk stages recovered fully in one to two weeks time.

High oil content feed. Two refiners have experienced high oil content feed. The first has only gravity separa-

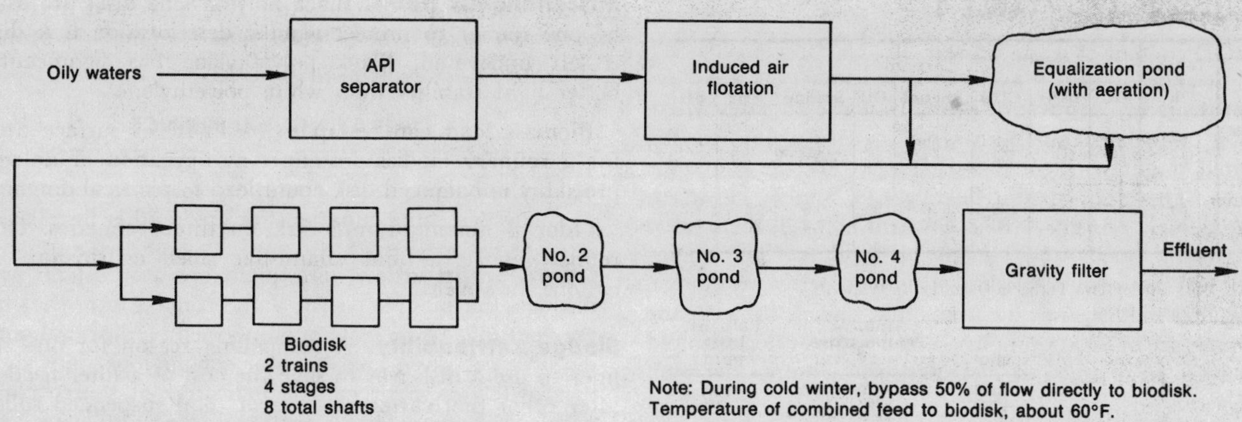

Fig. 3—Refinery B effluent system.

Fig. 4—Refinery B biodisk system.

tion (API-CPI) with no equalization ahead of the biodisks. When an oil skimmer failed, about 1 percent oil content water was fed to the disks for two days. Effluent went off specification on phenol and oil. After correcting the oil in feed, in four to five days the effluent was back on specification—both oil and phenol control was regained at about the same time. No other effects were noticed.

The second refiner fed 100 to 200 ppm oil content water to the disks for extended periods while correcting upstream oil removal problems. Most of this oil carried through the biodisk and caused interference with sludge settleability in his clarifier. This refiner does not have filters after his clarifier. Effluent was marginal during this period on total suspended solids.

Startup time needed. The four refiners report widely varying times required to obtain a steady-state disk growth—anywhere from one to more than seven weeks (Table 3).

Refinery A has the quickest startup time of the four. This probably results from putting warm (100° F) virgin feed water to the disks. The disks are not seeded.

Refinery B had to restart disk growth after losing the biomass due to cold water temperature (45° F) and a suspected caustic dump. When half the waste water

TABLE 3

Refinery	Equalization Before Disks, Days	Horsepower[1] in Equalization	Winter Water Temp. to Disks, °F	Startup Time,[2] Weeks
A	None	None	100	1
B	5	20	50	4-6
C	½	None	68	2
D	3	300	45-60	7+

[1]Horsepower aeration in ponds ahead of disks.
[2]Growth at steady state on the disks.

was bypassed around the equalization pond so that the combined temperature to the disk was more than 60° F, a good growth was obtained in a few days and a steady state growth in 10 days. This compares with four to six weeks in the previous startup.

Refinery C takes about twice as long to achieve startup as does Refinery A. This is probably because water temperature is appreciably lower—about 70° F versus 100° F. The one-half day equalization (no aeration) is a necessary feature which provides control of pH swings, toxics and oil skimming. Refinery D growth takes longer than seven weeks probably due to low water temperature to the biodisks and the low BOD in feed to the disk both caused by thorough aeration prior to entering the biodisk system.

The disk biomass performs the same function that bio-

TABLE 4

	Temperature, °F	ppm DO	BOD	COD	Phenol	Oil	Sulfide	NH_3	pH
Summer									
In.....	80	4.5	20	36	0.023	15	....	3.5	...
Out....	82	6.1	12	26	0.013	11	....	1.8	...
Winter									
In.....	53	3.0	55	120	2.70	25	0.6	26	7.3
Out....	53	4.6	20	80	0.12	15	0.2	22	7.4

Notes:

1. Only One of Two Trains in Operation in Winter.
2. Effluent Items:

	gpm	Ambient Temperature, °F	Effluent, BOD, ppm
Summer...........	1200	70	<10
Winter............	1700	34	<10

3. Aerated Pond Does Most of BOD Reduction in Summer.

TABLE 5

	Effluent Temp., °F	ppm BOD	Phenol	Oil	TSS
Summer August 1976.......	72	11.6	0.03	2.5	7
Winter December 1976....	38	7.4	0.08	4.2	5

Notes: 1. Average of monthly data.
2. Volume of effluent for summer and winter months above was about 650 gpm.

solids recycle does in an activated sludge system. In both systems waste water is contacted with large masses of biosolids generated from some days of previous operation. In the activated sludge process, the biosolids are recycled. In a disk system, the large masses of biosolids accumulate on the disks. In both systems, the mass of biosolids used is many time that which exists in an aerated lagoon.

To encourage maximum biogrowth on the disks, one should run at maximum feasible temperature (80-100° F) and avoid aeration of the biodisk feedstock such that maximum biogrowth will take place on the disk surface. Any growth that occurs prior to the disk reduces the potential for biomass growth. Also aeration ahead of the biodisk can generate fine biosolids that are not removed on the biodisk and don't settle in the clarifier.

Startup after shutdown. When Refinery A shuts down a biodisk the tank is immediately drained, except for short shutdowns when the shaft is kept rotating. If the tank is not drained biogrowth proceeds on the submerged portion such that biogrowth can accumulate on one side of the shaft; in fact, up to as much as 25,000 pounds. Also, if the tank is not drained the exposed biosolids can dry out causing serious unbalance around the shaft. Attempting to rotate an unbalanced disk can lead to tearing the plastic or causing heavy wave action in the tank so that flow over weirs is out of control.

Miscellaneous items. Black polyethylene disks are used by one refiner to protect against deterioration if a disk is left uncovered. Black polyethylene has significantly better light stability than white polyethylene.

Biomass load can be up to ½ lb/ft^2 of surface area for a refinery biodisk system. Any operation of an appreciably unbalanced disk could lead to physical damage.

Odor is absent around disks in three refineries. One refiner reports a faint salami-like smell nearby on occasions, however.

Sludge settleability. A compelling reason for one refiner to use a disk was to save the cost of a filter needed after other biotreatments to meet total suspended solids (TTS) specifications on effluent. Studies indicated that the large, quick settling disk solids did not need a filter. Furthermore, Refinery A uses only a 24-hour settling pond, B uses only a pond and C and D use only a clarifier after the biodisk.

Oil removal capability. Refinery A reports that typical operation reduces the oil content about 50 percent (from 15 to 7.5 ppm) when a 10-20 ppm oil content feed is processed. With large oil content feeds (100 ppm) there is little reduction. This compares with reported reduction of from 25 to 40 percent at Refinery B when feeding a water containing about 15-25 ppm oil (Tables 5 and 6).

Summer-winter operation. The greatest amount of experience has been accumulated at Refinery B, a 45 Mbpd plant with both cracking and coking facilities. The effluent treatment system there consists of gravity oil separation, induced air flotation for further oil removal, a pond for equalization, the biodisk system and 10 days residence in ponds followed by a gravity bed filter (Figs. 3 and 4). Half the feed is bypassed around the equalization pond to insure an adequately heated water (60° F).

Contaminants to the biodisk are consistently higher in winter than in summer operation (Table 4) due to higher temperatures in the equalization pond in summer which causes a higher bioactivity. However, both winter and summer biodisk operation is at a good activity level.

Typical final effluent shows that the over-all system works well in both winter and summer (Table 5). Effluent BOD levels are lower in winter than in summer. Over-all the effluent contaminant level indicates a thoroughly modern waste water treating system.

ACKNOWLEDGMENT

We wish to acknowledge cooperation of various refineries for details of operating experiences, photographs and comments given by two manufacturers of biodisk systems and assistance of Chevron Research Process Design. Adapted from a paper presented to the NPRA Annual Meeting, March 1977, San Francisco, Calif.

LITERATURE CITED

[1] Anon., "Biodisk improves effluent water treating operation," *Oil and Gas Jn.*, Feb. 23, 1976, p 126.
[2] Antonie, R. A., "Evaluation of a rotating disk waste water treatment plant," *JWPCF*, Vol. 46, No. 3, Mar. 1974, p 498.

Polishing Biological Plant Effluents

Pilot plant studies identify best method

William A. Parsons,
Arthur G. McKee & Co., Cleveland, Ohio

ACTIVATED SLUDGE PLANTS and aerated lagoons may need effluent polishing to capture excessive suspended solids from the clarifier overflow. National Pollutant Discharge Elimination System (NPDES) permit specifications in respect to "maximum day" and "30-day average" may be difficult to attain at biological treatment plants employing sedimentation for effluent clarification because biological solids generated in wastewater treatment processes such as aerated lagoons and activated sludge tend to be light and fragile. Light solids resist separation by sedimentation. Fragile solids may disintegrate under turbulence and therefore resist separation by sedimentation or straining. In addition, the performance of sedimentation units is affected by culture transformations, atmospheric turbulence and temperature conditions.

Compliance with stringent limitations in respect to maximum day can be particularly difficult because settling processes are vulnerable to transient situations involving dispersed suspended solids. In such cases, effluent polishing by mechanism other than gravity sedimentation may be required.

Pilot plant studies evaluated dissolved air flotation and dual media pressure filtration for removing excessive suspended solids from clarifier overflow.

Lamella-equipped flotation unit with supplemental polymer coagulant was highly effective for the removal of substantial excesses of suspended solids from activated sludge effluent. Effluents in the range of 25 to 30 mg/l were attained at unit hydraulic loadings of 5 gpm/sq. ft. and unit solids loadings of 13.9 lbs./day sq. ft. The air-to-solids ratio was 0.028. The suspended solids concentration of the float ranged from 5,500 to 7,000 mg/l. The unit was found to be ineffective under the prevailing experimental conditions, for fine tune polishing below 25 mg/l suspended solids.

Dual media pressure filtration can be of limited effectiveness for the clarification of fragile floc-activated sludge effluents. A filter operated at a feed rate of 2.83 gpm/sq. ft. yielded effluents in the range of 20 to 35 mg/l from feeds ranging from 60 to 79 mg/l suspended solids. Its capacity with supplementary polymer feed was 0.22 lbs. suspended solids per cubic foot of anthracite per regeneration. Filtrate quality deteriorated with increased feed concentrations. Thus granular media filtration was indicated to be effective for removal of dribbles of excess suspended solids but was ineffective for more substantial bumps of suspended solids. The backwash volume from the filter was 8 to 12 times the volume of float from the flotation unit.

TEST METHOD

Effluent polishing studies were made on a blend of activated sludge mixed liquor and settled activated sludge plant effluent. The mixed liquor suspended solids concentration was near 4,000 mg/l and the suspended solids concentration of the clarifier overflow was about 35 mg/l. Polymer (Primafloc C-3, Rohm & Haas Co.) was added to the clarifier feed in the amount of 0.005 pound per pound mixed liquor suspended solids to assist with clarification. The Sludge Volume Index was approximately 170 ml/gram and the wastewater temperature ranged from 18 to 22° C.

The dissolved air flotation studies were performed with a FAVAIR (Permutit Co.) Mark II, Model OB-10 pilot plant unit. The truck-trailer mounted unit is 15 feet long, 8 inches wide and 4 feet deep. The 8-inch width is separated by a single lamella to provide two 4-inch channels. The unit is actually a 10-square-foot element of a full-size, operational system which consists of an expanded nest of 4-inch channels. The unit is equipped with air pressurization tank, flocculation tank, chemical feed systems—plus associated mechanical components and instrumentation. Variables evaluated were polymer dosage, hydraulic loading, solids loading, recycle rate, air rate, effluent suspended solids and float suspended solids.

The dual media pressure filtration studies were performed with a vertical filter having a diameter of 3 feet

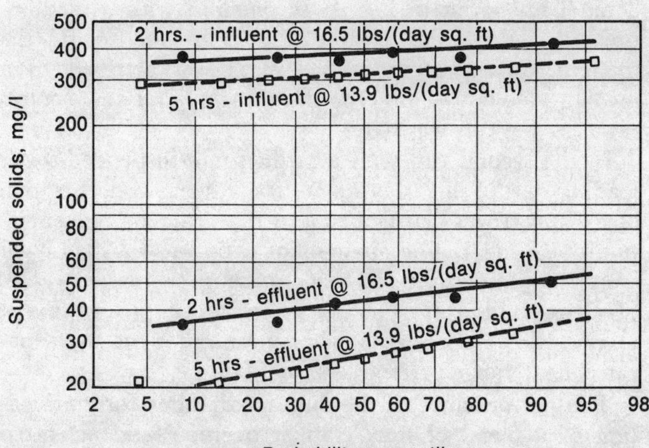

Fig. 1—Effluent polishing performance of dissolved air flotation system at unit hydraulic loading of 5 gpm/sq. ft.

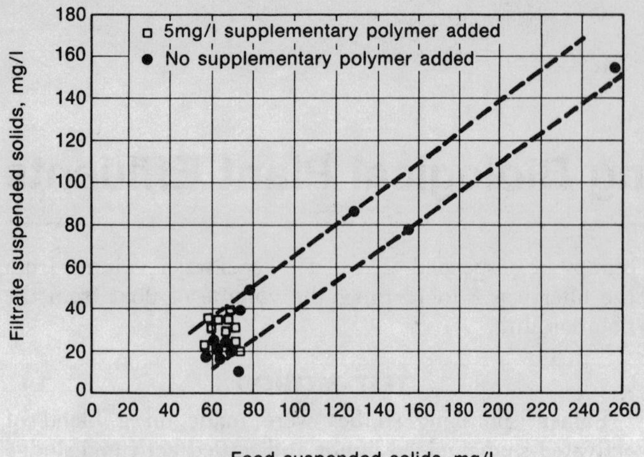

Fig. 2—Effluent polishing performance of dual media filter system at unit filtration rate of 2.83 gpm/sq. ft.

and a side wall depth of 6 feet and commercial type hydraulic distribution systems for influent and backwash. The pressure filter was packed with 18 inches of 2 to 4 mm sand as the under layer and 18 inches of 1.2 to 2.4 mm anthracite as the upper layer.

The filter was operated downflow at 20 gpm (2.83 gpm/sq. ft.) to turbidity breakthrough or 5 psi pressure increase, whichever occurred first. The backwash procedure consisted of a 3-minute air wash followed by a 120 gpm backwash with tap water for 5 minutes. Variables evaluated were solids loading, polymer dosage, filtrate turbidity and filtrate-suspended solids.

RESULTS

The performance of the dissolved air flotation pilot plant for clarification of a blend of activated sludge mixed liquor with final clarifier overflow was evaluated during more than a dozen experimental runs. The preliminary run provided for refinement of operational parameters relative to aeration and recycle and established that supplemental polymer feed was necessary for effective clarification.

The steady state results for two experimental runs at a unit hydraulic loading of 5 gpm/sq. ft. (3.6 gpm feed + 1.4 gpm recycle) with supplemental polymer feed of 3 mg/l are given in Fig. 1. In one run with a median-unit-suspended-solids loading of 16.5 lbs./day sq. ft., the influent concentration of 380 mg/l was clarified to a median value of 42 mg/l in the effluent. The air-to-solids ratio for the run was 0.024.

In the second run with a median-unit-suspended-solids loading of 13.9 lbs./day sq. ft., the influent concentration of 320 mg/l was clarified to a median effluent concentration of 26 mg/l. The air-to-solids ratio was 0.028. Lower values of unit-suspended-solids loading or higher values of air-to-solids ratio did not materially improve effluent clarity. The suspended solids concentration of the separated float ranged from 5,500 to 7,000 mg/l.

The performance of the dual media filter for clarification of a blend of final clarifier overflow and activated sludge mixed liquor was evaluated over a two-week period. The results of operation with and without supplemental polymer feed are summarized in Fig. 2. The results suggested that the suspended solids level in the filtrate was functionally dependent upon the suspended solids level in the feed. Filtrate-suspended solids levels of from 20 to 40 mg/l were generally obtained from feed-suspended solids levels of from 60 to 70 mg/l—irrespective of 5 mg/l of supplemental polymer addition which was indicated to be optimum by jar test. Filtrate-suspended solids increased markedly with increased suspended solids in the feed. Supplemental polymer addition was indicated to increase the capacity of the filter in terms of suspended solids retained prior to breakthrough. Without polymer addition, the median capacity of the filter was 0.14 pounds suspended solids per cubic foot of anthracite per regeneration. With 5 mg/l supplemental polymer addition, the median capacity was 0.22 pounds suspended solids per cubic foot of anthracite per regeneration.

The results indicate that dissolved air flotation and dual media filtration possess different performance attributes for polishing of biological treatment plant effluents. Polymer-assisted dissolved air flotation provided relatively consisten effluent clarity in the 25 mg/l suspended solids range irrespective of feed concentration within unit loading constraints of 5 gpm/sq. ft. (feed + recycle) and 13.9 lbs./day sq. ft. suspended solids. A design factor of 0.67 was recommended to cover contingencies in the translation of the pilot plant results of a prototype system. The collection of float at 5,500 to 7,000 mg/1 suspended solids concentration was an asset in that such concentrations are normally acceptable to waste-activated sludge dewatering systems. It is expected that somewhat higher float concentrations would be obtainable from prototype units with optimized flight speed.

Dual media filtration exhibited a performance pattern that suggested that effluent clarity was functionally dependent upon influent-suspended solids concentration. The highest percentage removals were obtained at low feed-suspended solids concentrations. With feed-suspended solids concentrations in the 60 to 70 mg/l range, the average percentage removal was 61 percent. The feed of supplemental polymer increased the capacity of the filter to 0.22 lbs. suspended solids per cubic foot of anthracite per cycle, but did not improve performance in respect to effluent clarity. The feed of metal salt coagulant would be expected to improve performance in respect to effluent clarity but could complicate disposal of backwash fines. The backwash volume from the filter was 8 to 12 times the volume of float from the flotation unit. In general, the recycle of backwash fines to process is not recommended. Therefore, some means of concentrating the backwash fines may be indicated to accommodate sludge disposal operations.

Thus the dissolved air flotation unit was indicated as the more reliable effluent polishing device for effluent clarity in the range of 25 to 30 mg/l suspended solids, because of the capability of accommodating process bumps of 300 to 400 mg/l suspended solids. For sublime effluent clarity, a two-stage process consisting of a flotation unit followed by a filter may be worthy of consideration. The flotation unit would provide a uniformly low suspended-solids feed to the filter and would process filter backwash for discharge to sludge disposal.

Mutant Bacteria Aid Exxon Waste System

Controlled tests prove effectiveness of mutant bacteria in enhancing operation of well-designed and efficiently operated refinery biological system

K. D. Tracy, Environmental Dynamics Inc., Greenville, S.C., and **T. G. Zitrides,** Polybac Corp., Allentown, Pa.

MUTANT BACTERIA additions speed restart of and enhance efficiency of refinery biological waste treatment systems according to results of controlled tests. Bacterial additives have been known to be useful for solving operating problems created by toxic or inhibitory influents or shock loads. Experience has shown the advantage of using mutant bacteria under such conditions, but conditions of application generally have been difficult to isolate and quantify. One recent exception was a series of long-term controlled tests that clearly demonstrate the advantages of such an additive on system performance and stability.

TEST SYSTEM

The refinery where the tests were conducted had a segregated sewer system and separate treatment facilities for oily wastewaters and stripped sour water. The latter treatment system, used in the additive evaluations, consisted of equalization followed by chemical flocculation and activated sludge (Fig. 1). The activated sludge unit has two parallel treatment trains each consisting of an aeration basin, clarifier, and sludge recycle system. The clarifier effluents were mixed with

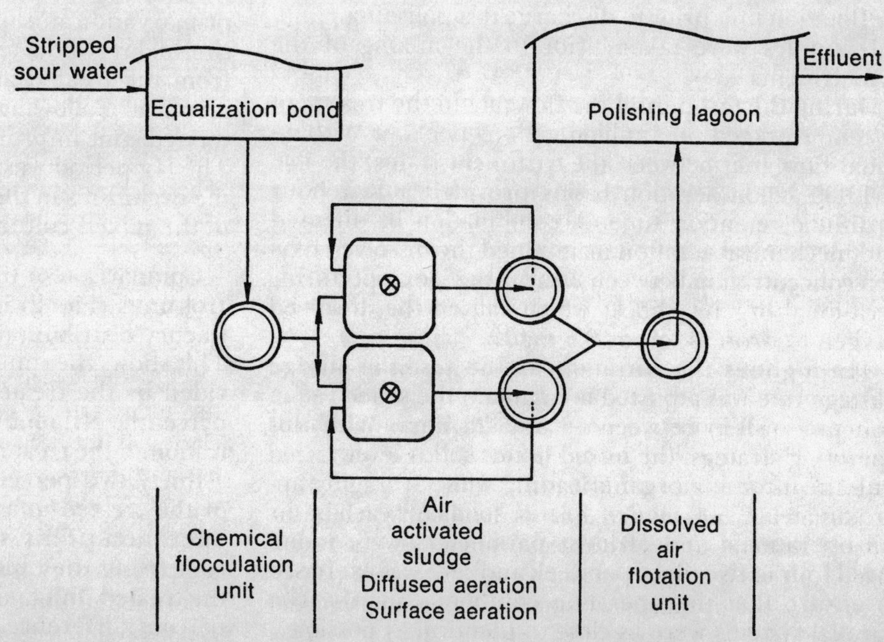

Fig. 1—Refinery wastewater treatment system.

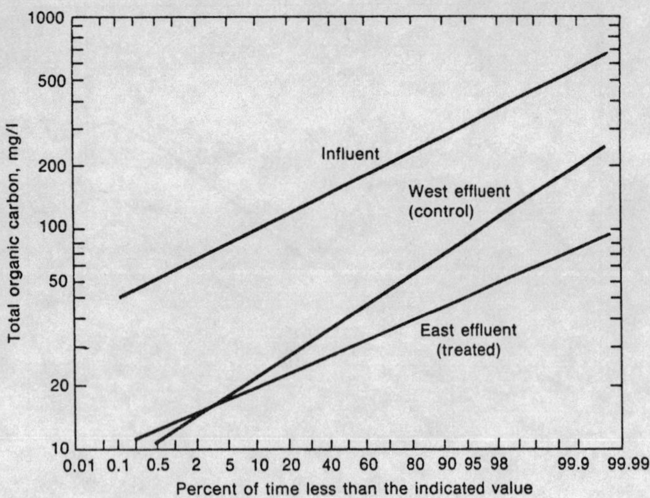

Fig. 2—Effect of additive on activated sludge performance.

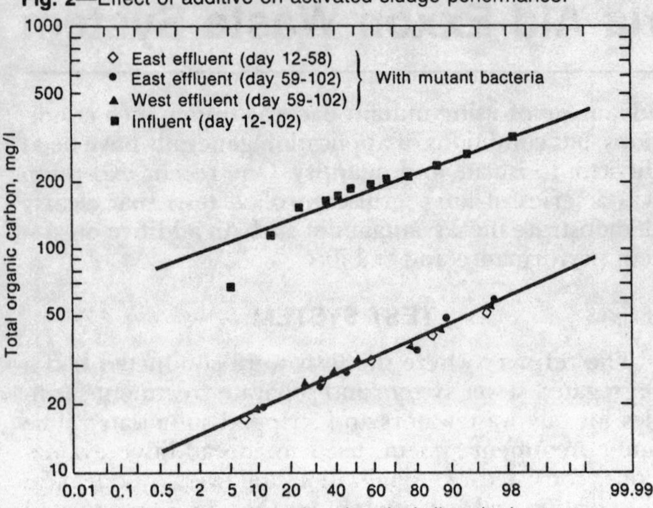

Fig. 3—Activated sludge performance with additive.

the oily-water stream and further treated in a dissolved air flotation unit prior to discharge to a polishing pond. All samples were taken prior to the mixing of the two effluents.

During the test period the flowrate to the treatment system averaged one million gallons per day. With an equal flow split between the treatment trains, the two 264,000-gallon aeration basins provided a 12.7 hour hydraulic retention time. A combination of diffused and mechanical aeration maintained the dissolved oxygen concentration between 2 and 4 mg/l, except during occasional organic shocks which caused the dissolved oxygen to drop as low as 0.5 mg/l.

Throughout the duration of the test the sludge wastage rate was adjusted to maintain the solids retention time (SRT) between 18 and 22 days. With this operating strategy, the mixed liquor solids level varied with the incoming organic loading while averaging approximately 2,500 mg/l of volatile solids for each basin. All operational and effluent parameters were monitored four to five times per week and care was exercised to ensure that the operating conditions for the two parallel systems were as close to identical as possible.

TEST PROGRAM

The test program was initiated following a major refinery turnaround. During the turnaround all major refinery process units were shut down and the wastewater was generated largely by cleaning operations. By the week prior to the program the influent wastewater was very low in organic content and microbial activity had approached endogenous levels. These conditions provided the opportunity to test the additive under conditions approaching treatment system start-up. During previous turnarounds the activated sludge system had been fed acetic acid to maintain microbial activity. This practice, it had been observed by plant operations personnel, did little to aid in treatment system restarts.

Following the refinery turnaround, the east aeration basin of the parallel-train system was seeded with the mutant bacterial additive (PHENOBAC Mutant Bacterial Hydrocarbon Degrader),[4] while the west served as a control. The addition program consisted of 12.5 pounds on the first day followed by 2.5 pounds per day on each of ten succeeding days. The product used is a mixture of microorganisms specifically adapted to the degradation of organics found in the refinery wastewaters. The strains in the formulation are selected from nature for activity on hydrocarbons, adapted to the typical constituents of refinery wastes and then mutated to fix the adaptation. The preserved microoganisms, in the form of free-flowing granules, were hydrated with warm water for one hour prior to addition.

After eight weeks of parallel testing a shock load caused complete failure of the control unit while the treated unit suffered no ill effects. This provided the opportunity to evaluate the bacterial additive in upset recovery, and the product was added to the previously untreated west cell.

PARALLEL TEST

Total organic carbon (TOC) levels were used as the primary indicator of system performance. For the first twelve days of the test program the effluent TOC levels from the treated and the control unit were virtually identical. Following this lag, the performance of the treated unit improved steadily relative to the control. The lag period was attributed to the time required for the organisms in the inoculum to reproduce and adapt to the mixed culture conditions.

Comparison of the performance of treated and control units (Fig. 2) is illustrated by comparing the frequency distribution of feed and effluent TOC values. This shows the continuing enhanced performance provided by the treated system. Comparison of the fifty percentile effluent TOC values shows a 28 mg/l concentration in the treated unit versus 41 mg/l in the control, a thirty two percent improvement. The flatter slope of the treated-unit plot is indicative of a more stable operation. This stability is more apparent when comparing the ninety percentile values of 45 mg/l for the treated unit and 78 mg/l for the control, a forty-two percent difference.

A variety of other water quality parameters were monitored during the test period. The effluent phenol averaged 0.11 mg/l for the treated unit and 0.16 mg/l for the control. Ammonia concentrations averaged 93 and 104 mg/l, respectively. Although the differences did not prove to be statistically significant, the treated unit provided the better effluent quality for both parameters. Suspended solids concentrations were not monitored in the clarifier effluents, and the sludge volume indices were excellent for both units. No conclusion, therefore, can be made regarding the effect of the additive on settling properties in this test. Foaming, on the other hand, was markedly reduced in the treated unit. In fact, a severe foaming problem caused a failure in the untreated cell and led to termination of the parallel test. Performance of the treated unit was unaffected during this incident.

Throughout the parallel-test phase of the study care had been taken to maintain identical operating conditions in the two treatment trains; however, to ensure that improvement in performance was not a result of inherent differences between the two process trains, mutant bacteria were added to the control cell. The resulting performance data (Fig. 3) show that effluent quality of the former control unit now matched the treated unit. The effluent TOC levels from the two treated cells during this phase of study followed the same probability distribution as the treated unit in the parallel test, confirming the performance improvement.

RESISTANCE TO SHOCK

During the five-month study, there were several opportunities to test the upset resistance of treated systems. First, the parallel test was terminated because of a change in influent quality as a result of an operational change in the sour water stripper. TOC data (Fig. 4) show a steady decline in effluent quality of the control with no appreciable effect in the treated unit. Deteriorating performance of the control unit endangered the overall effluent quality which necessitated stopping flow to the unit. Following shutdown of the control system, the treated cell was able to handle the entire plant flow for two days while maintaining a seventy-one percent TOC reduction.

To speed recovery after this upset, mutant bacteria were added to the control cell and sludge was wasted from the previously-treated cell to the control cell. TOC removal, at reduced flow rates, reached normal levels within two days. Flow to the unit was then increased to normal rates over a period of three days without reducing treatment efficiency.

After both trains had been treated, resistance to shock loads was confirmed by a toxic shock followed by a large organic shock. Fig. 5 shows a slight increase in effluent TOC levels following the toxic shock on day 104; but, recovery was very rapid and, normal TOC levels were restored within two days. The subsequent organic shock, commencing on day 110 was assimilated with no increase in effluent TOC.

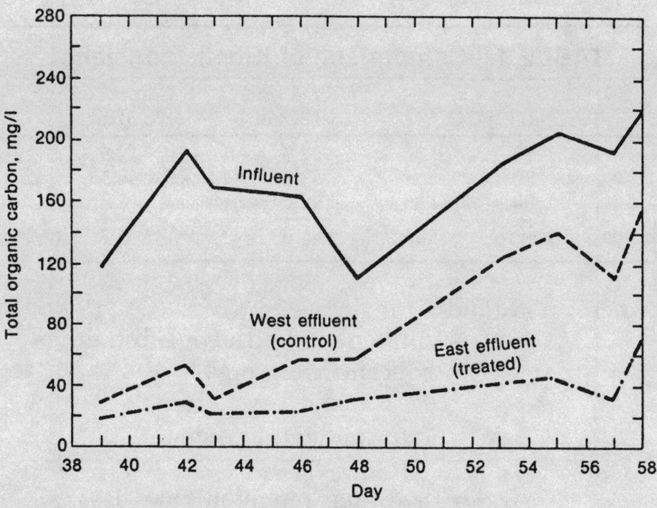

Fig. 4—Additive increases resistance to upsets.

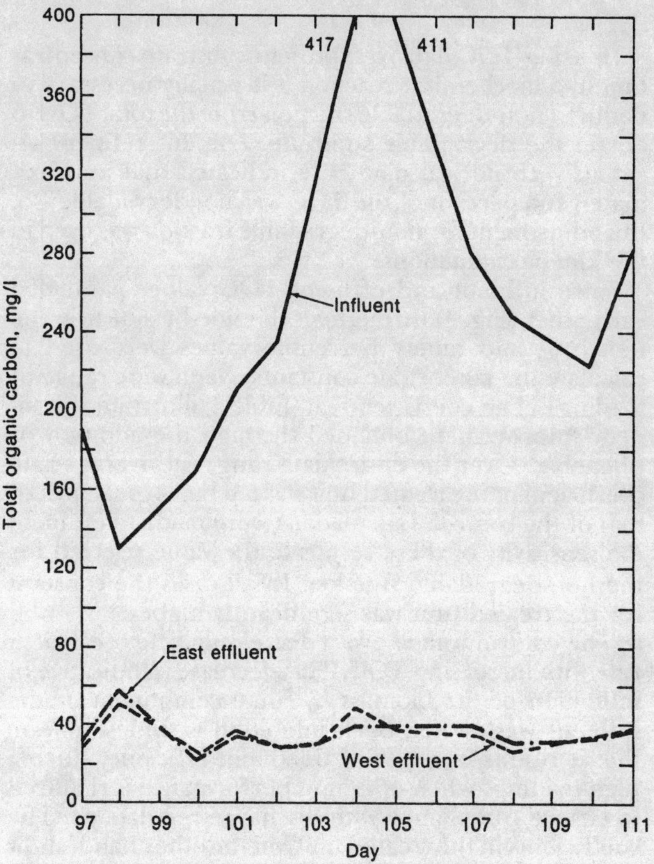

Fig. 5—Increased shock resistance in treated units.

EFFECT ON PROCESS KINETICS

Process kinetics for the treated and untreated cases were defined to assess additive effects for conditions other than those which prevailed during the test. In evaluating the kinetics the following mathematical model was used:

$$S_e = \frac{S_o}{1 \cdot KX\,\Theta_h}$$

TABLE 1—Comparison of Kinetic constants

Percentile TOC	Treated unit coefficient	Control unit coefficient
10	32.6	12.6
50	53.8	6.5
90	42.6	3.9
Average	43.0	7.7

where: S_o and
S_e = influent and effluent substrate concentration, mg/l

K = rate constant, l/g—day

Θ_h = hydraulic retention time, days.

X = MLSS concentration, g/l

In using TOC data to represent substrate concentration in a biochemical reaction it is usually necessary to deduct a non-degradable fraction from the total TOC to obtain the degradable substrate. For this refinery an extensive historical data base indicated that approximately ten percent of the TOC was nondegradable; so, this adjustment for non-degradable fraction was used in the kinetic evaluations.

Since influent and effluent TOC values paralleled each other (Fig. 4) throughout the side-by-side test, the ten, fifty, and ninety percentile values were used to calculate the kinetic rate constants over a wide range of loadings. The coefficients in Table 1 illustrate the kinetic improvement obtained through the addition of phenobac. Over the entire data range the average rate coefficient of the treated unit was 5.6 times greater than that of the control. Calculations were made to evaluate the sensitivity of these results to the value selected for the non-degradable fraction. In all cases the constant for the treated unit was significantly higher.

The control unit shows a decreasing rate coefficient rate with increasing TOC. This decrease is indicative of inhibition of the biomass by some component of the influent wastewater. Such inhibition is undesirable in that it results in a loss of treatment efficiency during high loadings when optimum performance is required to ensure compliance with discharge regulations. The coefficients in the treated unit, on the other hand, show no inhibitory effects. The differences in kinetic coefficients at various loadings account for the greater resistance of treated systems for substrate shocks. During substrate shocks, non-treated systems are subject to kinetic limitation and deterioration of effluent quality.

Additive effectiveness was considered surprising[1] since it seems to violate the "ubiquity principle": a microbial population will naturally and spontaneously develop which is best suited to metabolize the available food supply in a given environment. These findings are attributed to the fact that the organisms investigated are mutants. Although they are able to grow and reproduce in an environment of refinery activated sludge, they would not necessarily develop of their own accord because of the specialized conditions necessary to mutation. In addition, recent work[2] has shown that the "biodegradability" of various organic substrates is highly dependent on the starting inoculum. This is particularly true of compounds which might be expected in refinery or petrochemical effluents. Thus, inoculation of an existing biomass can assist the indigenous biology to assimilate "new" substrates which enter the plant under variable conditions. Several interpretations of the results have been proposed,[3] but whatever the explanation this mutant bacteria formulation has proven useful for starting up and improving routine operation as well as aiding in upset recovery.

The additive has been shown to improve process efficiency and stability in a properly designed and operated activated sludge system. This approach provides the operator of such a treatment system with an additional tool for coping with widely varying influent wastewater quantity and quality. The application of the technique to refinery wastewaters can also result in a cost savings through reduced usage of chemicals such as antifoams.

In addition to improving performance of adequately-sized systems, the bacterial additive can aid in bringing overloaded systems into compliance. Increased production levels in many industries have resulted in wastewater loads which exceed the design capacity of existing treatment systems. Biomass augmentation offers an alternative to large capital investment in some cases, or, a means of maintaining permit compliance while additional facilities are being constructed.

LITERATURE CITED

[1]Tracy, K. D. and Shah, P. S., "Application of Bacterial Additives to Refinery Activated Sludge," 50th Annual Meeting, California Water Pollution Control Association, Sacramento, CA, April 21, 1978.

[2]Haller, Helen D., "Degradation of Mono-substituted Benzoates and Phenols by Wastewater,; *Journal Water Pollution Control*, Dec. 1978.

[3]McDowell, Curtis S., "Biomass Engineering Improves Wastewater Treatment System Performance," 39th Annual Meeting International Water Conference, Pittsburgh, PA, Oct. 31-Nov. 2, 1978.

[4]POLYBAC Corp., 505 Park Ave., N.Y.

Use Gravity Belt Filtration for Sludge Disposal

Application shows promise

George W. Grove, Exxon Research and Engineering Co., Florham Park, N. J.

USE OF GRAVITY belt filtration to concentrate refinery and petrochemical wastewater sludges offers advantages over other systems. Long used in Europe for municipal sludges, their application to oily refinery sludges is yet to be proven commercially.

Wastewater treatment processes such as dissolved air flotation, filtration or biological systems result in dilute slurries of 0.1-5 wt% oily solids that must be concentrated to reduce volume to minimize cost of disposal.

The process sequence for a typical refinery or chemical wastewater plant is based on pretreatment by passing suspended solids and oils through sand filters or dissolved air flotation (DAF) to reduce contaminant levels prior to biological oxidation. Final polishing filters may be used following the biox unit depending on effluent quality requirements.

Oily solids from pretreatment and biological sludges are concentrated and disposed in three steps. A typical sludge concentration sequence might include gravity settling, centrifugation and incineration. These generally account for 40 percent of total investment and 60 percent of the operating cost of the total wastewater treating plant.

One of several processes used here is mechanical dewatering for which critical technology is needed.

Three general types of sludge concentration steps are (Table 1):

TABLE 1

FIRST STEP	SECOND STEP	THIRD STEP
Gravity settling	Filtration Vacuum	Landfill
Flotation	Pressure Centrifugation Scroll Disc Imperforate basket	Sludge farming Sand lime Incineration
	Gravity belt filtration	

• Gravity settling or DAF utilizes differences in density to concentrate the sludge to approximately 2 to 5 weight percent oily solids.

• Mechanical dewatering by vacuum or pressure filtration, centrifugation or gravity belt press filtration is

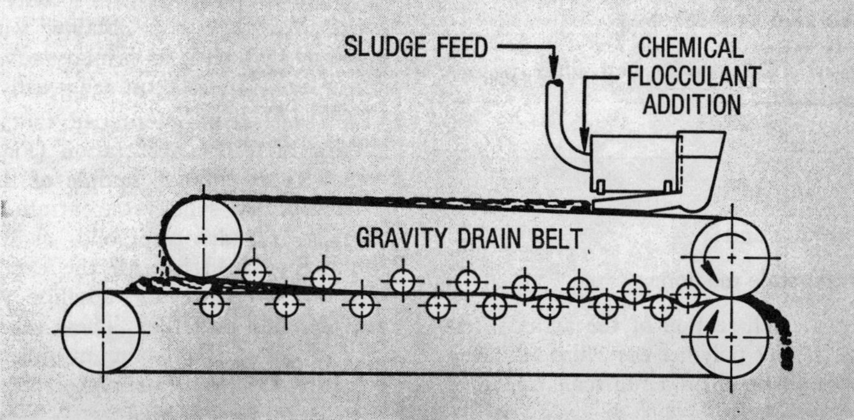

Fig. 1—Gravity belt filter press.

applied. Mechanical dewatering concentrates the sludge to about 10 to as high as 30 to 40 weight percent oily solids.

• The sludge is incinerated utilizing heat to reduce volume by evaporating water and burning the volatiles. This creates an inert ash and requires high investment and operating costs.

Each step is increasingly higher in investment and operating cost per ton of sludge handled. Therefore it is important to be sure that sludge solids concentration is maximized in each step. The area of mechanical dewatering offers the best incentive for improvement. Vacuum or pressure filters have not been used extensively for refinery sludges because filter surfaces blind rapidly. This necessitates adding additional inert solids to increase porosity and results in increased solids for disposal by factor of two to three.

Scroll and disc centrifuges have been used more extensively for petroleum and petrochemical sludges. These are high-speed, precision machines. The grit in the sludge can cause rapid erosion and in some cases plugging. This results in frequent downtime for maintenance and cleaning. Imperforate basket centrifuges which are relatively new for sludge dewatering are probably preferred centrifuges for concentrating wastewater sludges. For most service they are less subject to abrasion and plugging than scroll and disc centrifuges and provide a solids recovery of 85 to 90 percent.

Gravity belt filter presses dewater sludge in three basic steps (Fig. 1):

• Polymers are added to flocculate the fine sludge particles

• Flocced feed is poured onto a coarse screen where the clear water readily drains away

• Cake is pressed to remove additional water.

Polymer is added to the sludge in a rotating drum where the floc is formed. Water then drains from the sludge on the top belt. At the end of the top belt, sludge falls to the bottom belt and is pressed between the two screens. The rate at which pressure is applied is very important since too rapid application forces the sludge through the screen. However, gradual application lets additional water drain from the sludge and solids concentration in the cake increases.

Tests on several full scale gravity belt filter presses fixed four parameters:

• Polymer dosage and cost

• Solids capture or filtrate clarity

• Cake solids concentration and

• Production rate of full scale machine.

A simple test using a small section of the actual press screen can be used to predict polymer cost, filtrate clarity and cake concentration obtained on a full scale machine.

Polymer costs are predicted to be in the range of $2-$10 per ton of dry solids. High polymer costs result from dilute

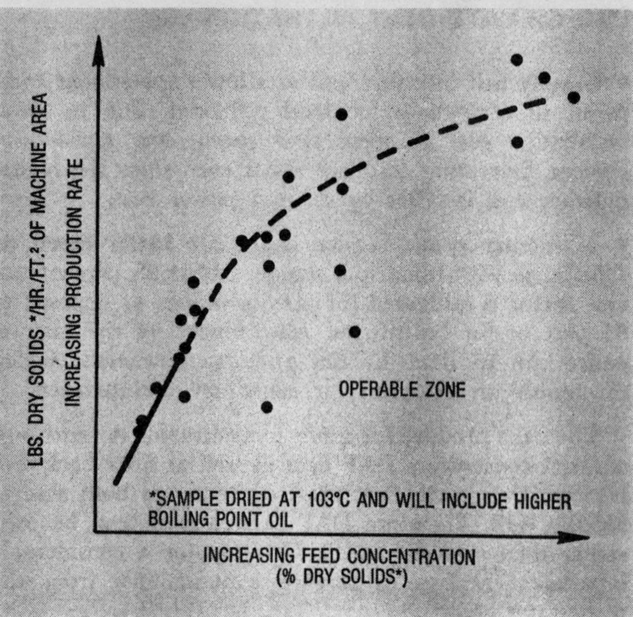

Fig. 2—Effect of feed concentration on production rate.

feeds and can generally be reduced by gravity settling or dissolved air flotation ahead of the press. Combining sludges can result in reducing polymer cost 10 to 25 percent compared to handling the sludges separately. This is probably due to neutralization of electrostatic charges in the feed. Although combining sludges may not always result in lower cost, it should never result in increased cost. Gravity belt filter presses use the normal water treating high molecular weight anionic and cationic polymers. The flocculation of sludge does not appear to be very sensitive to the amount of polymer used. There is an optimum dosage from a cost standpoint, but operation is not adversely effected by minor changes in sludge composition and minimum operator attention is required.

Solids capture is very high for gravity belt filter presses compared to centrifuges. Filtrate plus wash water, from biological sludges has contained as low as 100 ppm which is a solids capture of about 99.7 percent. Pretreatment sludges usually contain higher amounts of oil for which solids capture is about 97-98 percent compared to 86-92 percent using imperforate basket centrifuges.

Cake concentrations with gravity belt filter presses are two to three times that obtained with imperforate basket centrifuges which is its primary advantage since incineration or final disposal cost are greatly reduced.

Full scale tests determined that production rate is a function of feed concentration (Fig. 2) based on drying a small representative sample of the feed at 103° C to determine dry solids concentration. This "dry solids" therefore includes true solids as well as higher boiling hydrocarbons. Although the experimental relationship (Fig. 2) is typical, the absolute values vary from one press design to another. These machines will operate at any rate up to the maximum and rate is independent of the ratio of oil to solids in the feed.

Gravity belt filtration offers several advantages over the previously preferred mechanical dewatering device, the imperforate basket centrifuge:

- Gravity belt filtration indicates lower investment compared to centrifuges for most refineries, due to lower installation cost of these slow speed, low horsepower devices. Operating costs are about even since the higher polymer cost is offset by reduced power cost.

- Maintenance and service factor are better based on experience with municipal sludge. About 95 percent service factor is estimated for gravity presses as opposed to 85 percent for centrifuges. Also, much of the maintenance can be done by the operators whereas skilled machinists are required for centrifuge maintenance.

- The cake produced is more concentrated. A centrifuge will not concentrate DAF float as well as filter backwash whereas the gravity belt press concentrates both sludges equally well. Therefore DAF pretreatment can be used rather than granular media filtration for a considerable investment savings to remove contaminants from the wastewater.

- Dryer cake reduces the amount of water to be evaporated in an incinerator resulting in a considerable fuel savings. If the sludge is to be hauled offsites for disposal, hauling cost savings will be directly proportional to the volume reduction in the amount of sludge.

LANDFILL

Landfill has been widely used for disposal of many solid and semi-solid wastes including oily sludges. However this approach is becoming increasingly costly and frequently only postpones the problem. Stricter regulations on licensing, control of leachate and cover material are increasing landfill costs. Even more important buried oily waste will remain unchanged for hundreds of years. Potential problems of leaching to subsurface water or any future use of the land must consider that it had been a landfill site.

Incineration prior to landfill is another possibility where gravity belt filtration should reduce equipment size and fuel cost since the sludge contains less water. Also, it can be furrowed with the rakes of a multiple hearth furnace which will be lower cost than fluid bed incinerators used for liquid sludges for many smaller installations. In spite of these improvements, incineration will still result in high investment and operating cost.

Also, energy and environmental problems from incineration must be considered. Sludges will still be 70-90 percent water which will require about 3,000 Btu/lb. to evaporate when stack losses, radiation, etc., are considered. With rising fuel costs this will become increasingly expensive. Futhermore, incineration may create air pollution problems. Finally incinerator ash will still contain the heavy metals which may leach to the soil.

Sludge farming of oily wastes although practiced for many years is not technically well defined with respect to short or long term environmental effects. Such data is required to convince regulatory agencies of the environmental soundness of the practice.

TABLE 2

Location	Approximate % oil in soil	Oil degraded	Oil degraded bbl./acre/yr. (8 mo. yr.)	Fert.	Aerated
Continental Oil	1	....	32—65	yes	yes
C—Esso Europe...	1	0.11 %/mo.	44.5	yes	?
B—Esso Europe...	2.1	2.2 %/mo.	2,555?	yes	1/mo.
A—Esso Europe...	4.2	0.98 %/mo.	372	yes	?
U.C. Oak Ridge	20	1.0 %/mo.	800	yes + H_2O	1/wk.
Shell.........	20	1.2 %/mo.	1,161	yes	2/mo.
Shell.........	20	0.6 %/mo.	580	no	2/mo.
Exxon Bayway.	50	1.96 lb./day	1,000	?	?
Exxon Bayway.	8	0.59 lb./day	....	....	....
Exxon Bayway.	5	0.17 lb./day	....	....	....

TABLE 3

	Sludge farming	Incineration	Factor
EUROPE			
Esso (1974).....................	$7/m³	$40/m³	5.7
Mobil (1974)....................	$7/m³	$35/40 m³	5-5.7
U.S.			
Continental (1974) 20% oil.......	$3.35/bbl.	$19/bbl.	5.7
Shell (1972) 33% oil.............	$3/bbl.	$11-$26/bbl. (est).	3.7-8.7

Data sources:
Continental Oil—Report to October 1974 API Division of Refining Meeting.
U.C. Oak Ridge—Disposal of Oil Waste by Microbial Assimilation by H. C. Francke and F. E. Clark. Prepared for the U.S. Atomic Energy Commission under U.S. Government Contract 2-7405 eng 26.
Shell—Oily Waste Disposal by Soil Cultivation Process by C. Buford Kincannon. EPA R2-72-110 December 1972.
Mobil—Report to October 1974 API Division of Refining Meeting.

Review of tests conducted during the last several years indicates sludge farming has several advantages compared to incineration, primarily lower cost and no air pollution.

Also, penetration of oil and heavy metals to subsurface water is not indicated to be a problem. Further work is needed in this area to determine the long-term effect of long-term buildup of heavy metals or undegraded hydrocarbons.

Degradation of hydrocarbons by living organisms requires certain criteria be met for microbiological growth. First, soil must be mixed with the sludge to provide contact with the bacteria which are normally present in the soil. Next, the moisture content, temperature, essential nutrients as well as oxygen must be present. Aside from nitrogen (which is provided by fertilization) and oxygen (which is provided by tilling) the other requirements are generally present in sufficient quantity.

Degradation rates from various sludge farming experiments (Table 2) show that oil degradation rate is a function of oil concentration in the soil; fertilization, especially nitrogen, increases degradation rates; and 800-1,000 barrels of pure oil per acre should be degraded in an eight-month growing season.

The cost of sludge farming is one-fifth to one-sixth that of incineration (Table 3). In view of this cost incentive and preliminary indications of environmental problems sludge farming should be given more consideration.

Part III.

Solids Control

How Bayer Incinerates Wastes

This large chemical waste disposal system handles 25,000 tons per year of special wastes and produces 140,000 tons of steam while meeting environmental rules

H. W. Fabian, P. Reher and **M. Schoen,** Bayer AG, Leverkusen, Germany

AFTER APPROXIMATELY one year of trials the No. II incinerator system at Bayer's Leverkusen plant was commissioned for regular service in November 1977. The construction of this system with downstream flue-gas scrubber had become necessary as a result of the increasing volume of waste and the additional requirements imposed for the disposal of "special waste."

The system's capacity allows the incineration of approximately 25,000 metric tons of special wastes per annum and produces approximately 140,000 tons of steam. It consists primarily of the waste bunker building, the rotary kiln with secondary combustion chamber including all feed devices for solid, paste and liquid waste, the heat recovery boiler, the electrostatic filter and the flue-gas scrubber which in turn consists of the injection cooler, two rotary scrubbers and one jet scrubber.

The purified flue-gas is heated in two heat exchangers and discharged by a suction blower via a stack of 100 m heighth. The stack is fitted with sensors and apparatus for the continuous measurement and recording of the type and quantity of the emissions. The scrubbing water is transferred to the company's own waste water treatment plant. Solid ash and dust are deposited in licensed disposal sites.

The system requires an operator staff of 35 persons including the instrumentation and maintenance personnel. The acquisition cost was DM 27 million. The system concept was developed in collaboration with L.&C. Steinmueller, D-5270 Gummersbach. Steinmueller was also the design and construction contractor. The operating cost is DM 10 million per annum. The average incineration cost is DM 400 per ton of waste.

Background. In 1973 the construction of a second incineration system with downstream flue-gas scrubber was decided as a result of the increasing volume of "special waste" generated by the plants of Bayer AG.

It was planned to use this system for the disposal of all "special waste" as defined in § 2.2 of the Waste Disposal Act (Abfallbeseitigungsgesetz) dated June 7, 1972, and in accordance with the Federal Anti-Pollution Act (Bundesimmissionsschutzgesetz—BImSchG) dated March 15, 1974, and the Technical Directive on the Prevention of

Fig. 1—Over-all view of the No. II incinerator system with flue-gas scrubber, seen from the west.

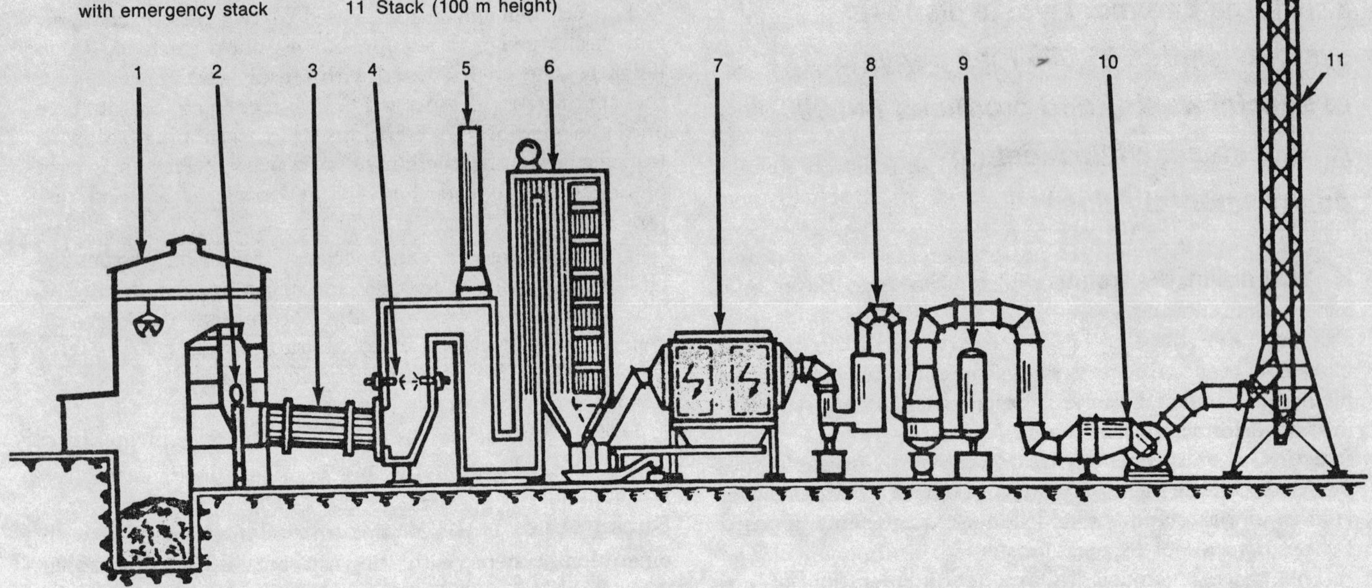

Fig. 2—Process schematic of the No. II incinerator system.

Air Pollution (Technische Anleitung zur Reinhaltung der Luft—TA—Luft) dated Aug. 28, 1974.

The following requirements were imposed on the design and operation of this system and the clean incineration of the "special waste" involved in accordance with the statutes:

1. Incineration must always be preceded by careful analysis and classification of the waste in order to determine the best suited treatment method.
2. The design, process technology and mode of operation of the incinerator system must be suitable for adaptation to the specific waste problems encountered, i.e., the system must be designed so that any change in the quantity and composition of the waste will not cause any malfunction in the process.
3. The design of the combustion spaces and heat recovery systems as well as the materials used for them must minimize the flue gas damage.
4. Finally, the flue-gas scrubbers must be designed so that removal of the pollutants from the flue gas is ensured regardless of quantity and ratio between the individual flue gas components.

SPECIAL WASTE

In order to facilitate special waste handling and to allow a certain classification which is also required for intermediate storage, this waste type was roughly classified into four categories on the basis of the chemical properties:

1. Halogen, nitrogen, sulfur, phosphorous and silicon-containing production waste,
2. Toxic, spontaneously igniting, spontaneously decomposing and spontaneously polymerizing waste as well as waste generating strong odors,
3. Waste of high reactivity which may react spontaneously with other waste or have a strong corrosive effect, and
4. Miscellaneous waste consisting of minor quantities of highly different waste whose composition is mostly unknown.

Classification under physical aspects would have been possible on the basis of the state of aggregation (solid, liquid) or on the basis of their calorific value which varies from 0 to 40,000 kJ/kg. However, in practical operations such a classification would not have been feasible since the chemical properties dominate.

INCINERATION PROBLEMS

In addition, the concept for the No. II incinerator system had been influenced by some basic considerations arising from the specific incineration problems of some waste types. As an example for these problems, the disposal of chlorine-, phosphorous- and silicon-containing waste and the heavy-metal-containing ash generated upon incineration shall be explained in more detail.[1]

Incineration of chlorine-containing waste. In addition to CO_2 and H_2O the incineration of chlorine-containing waste primarily produces HC1. The concentration of the generated HCl can be estimated fairly rapidly if the chlorine content of the waste and the excess air number are known. Thus it can be shown that the limit of 100 mg/nm[3] HC1 in the exhaust air, which the German air pollution directive requires, will be reached and exceeded even when the waste has a very low chlorine content. For instance, the flue gases of waste containing

0.2% by weight chlorine, incinerated with 50% excess air, i.e., at $\lambda = 1.5$, already contains 160 mg HCl/nm^3. For waste whose chlorine content may be 60% by weight and more, the HCl content in the flue gas will be in the order of several grammes HCl/nm^3. The possibility that major quantities of HCl may be generated requires very careful selection of materials and imposes maximum requirements on the efficiency of the flue-gas scrubbers.

Apparently caused by HCl and supported by other aggressive volatile flue-gas components, corrosion damage was also temporarily encountered on system components which, due to their composition of materials, should have resisted this corrosive attack. Some parts made of Hastelloy C in the injection cooler, in the rotary scrubbers and in the droplet traps were completely destroyed after only a few thousand operating hours. Similar damage was also encountered in the heat exchanger and in the suction blower. These damage and malfunction problems have now been corrected through design changes and through the use of improved materials.

In the combustion of chlorinated hydrocarbons the formation of elementary chlorine through oxidation of HCl cannot be theoretically excluded. Chlorine represents a certain danger to metallic system components. The thermodynamic data of this oxidation process, which is referred to as the Deacon reaction, show that chlorine formation is strongly reduced with increasing temperature and increasing water content, but is facilitated with decreasing temperature and increasing HCl and oxygen concentration. At the usual combustion temperatures between 1,200 and 1,300°C, however, the theoretically possible chlorine quantity is only 10 mg Cl_2/nm^3 which is negligible.

Considerably more significant, however, is the question as to the change of equilibrium taking place as the flue gases pass through the ducts of the heat recovery boiler until they have cooled to approx. 300°C. In this temperature range the equilibrium concentration of chlorine is so high that the corrosion of the boiler and system components must be feared. It is a known fact that carbon as well as iron and copper oxides have a catalytic effect and accelerate the equilibrium. In fact, a concentration of 200 mg Cl_2/nm^3 has been measured in the flue gas during the incineration of copper-containing chlorinated hydrocarbons. The combustion space temperature was 1,100°C, and the flue gases passed the temperature drop from 1,100°C to 600°C in approximately 2 seconds. In addition, the catalytic effect of dust from the electrostatic filter on chlorine formation was studied under laboratory conditions. For these studies a mixture of 90% by volume of air, 9% by volume of water vapor and 1% by volume of HCl was passed through a reaction tube at 500°C filled with glass wool and dust from the electrostatic filter, and the chlorine content determined at the outlet. After a residence period of 15 seconds, chlorine contents up to 800 mg Cl_2/nm^3 were determined.

Similar reactions take place in the incineration of bromine and iodine-containing waste. As a result of the lower electron negativities of bromine and iodine, the unfavorable effect arises that the formation of the corresponding halogenated hydrocarbons will shift in favor of free halogens so that increasing quantities of free bromine and iodine are produced.

In order to avoid the problems in systems technology and apparatus involved in the possible occurrence of free halogens in the flue gas, it has been found advantageous to incinerate halogen-containing waste together with sulfur-containing waste. Apparently the SO_2 produced during this process will reduce the halogens present to the corresponding halogenides to a degree where not even traces of halogen can be identified by analysis. Therefore, the prerequisite for stopping the halogens in the flue-gas flow is to maintain a constant waste mix and an adequately great SO_2 concentration. As a result of this mode of operation it has been possible so far to avoid any corrosion damage by halogens.

Fig. 3—Bunker building for solid waste; delivery in tipping truck.

Incineration of polychlorinated compounds. Certain polychlorinated compounds may appear problematic from the point of view of incineration in that they are characterized by their increased thermal stability. Consequently, their complete incineration may require higher temperatures and longer residence times in the combustion zone. Our own studies in this field have shown that, at temperatures above 1200°C and minimum residence times of 1 second, it was impossible to identify any chlorinated compounds in the flue gas, the lower identification limit of the analysis being approximately 0.003 μg/nm^3.

Incineration of organic phosphorous compounds. The primary combustion product of organic phosphorous compounds is gaseous phosphorous pentoxide. This compound has extremely unfavorable chemical and physical properties for the process of an incinerator system. Phosphorous pentoxide is fairly volatile, its vapor pressure being 10^{-2} Torr at 300°C which is equivalent to 170 mg P_4O_{10} per nm^3, its melting point is 569° C and its boiling point 591°C. Being the anhydride of a strong tribasic acid, phosphorous pentoxide violently reacts with water even at elevated temperatures, forming metaphosphoric acid. Phosphates of varied basicity are formed in the presence of basic oxides or salts. At temperatures above 600°C reactions with airborne dust deposited on heat exchanger surfaces and on the internals of the electrostatic filters take place with the result that these loose deposits turn into sticky, firmly adhering and difficult to remove incrustations. This will result in increased pressure drop and reduced efficiency of the heat recovery boiler

Fig. 4—Incineration of liquid waste from settling tanks.

and electrostatic air filters. In addition, reactions with the boiler walls cannot be excluded. The gaseous P_4O_{10} penetrates the dust deposits and reacts with the protective iron oxide layers on the boiler wall, forming iron phosphates. In addition, it has been observed that, due to the very small temperature interval between condensation point and melting point ($\Delta T = 22° C$), when the phosphorous oxide-containing flue gases cool, it will easily happen that very fine aerosols are produced which are only highly inadequately separated by conventional dust traps. Many experiments and design modifications were necessary in order to find a solution for these problems. In the meantime the process technology and the plant design have been adapted to the point where no emissions are observed even under surging P_4O_{10} loads.

Incineration of organic silicon compounds. Extremely fine silicon dioxide particles are produced during the incineration of organic silicon compounds. SiO_2 has a very high melting point (1713° C) and high electrical resistance (10^{12}-10^{13} Ω/cm). Both are adverse properties from the point of view of optimum electrostatic dust separation. As a result of the very high melting point the SiO_2 aerosols from the primary process will not agglomerate into larger particles, and the separation efficiency of the electrostatic filter is correspondingly poor. Due to the high electrical resistance, the electrical charge is transferred very slowly from the deposits to the separation electrodes. The SiO_2 deposits have the effect of a dielectric, impeding the electron flux between spray discharge electrodes and separation electrodes which is necessary for dust separation.

It is found in day-to-day operations that during the incineration of silicon-containing compounds the separation efficiency will drop below 30 percent of the usual value in spite of increased field strength. In these cases the perforated plates, the spray discharge wires and the separation plates of the electric dust filters are subsequently covered with dust deposits up to 10 mm thickness. Normal vibrating devices are inadequate for the removal of this type of dust accumulation. Ninety-nine percent of this dust consists of SiO_2; its density has been determined at 0.05 g/cm^3. Consequently, the incineration of organic silicon compounds usually requires extensive manual cleaning work in the electrostatic filter systems.

This is another case where the process was optimized through systematic experiments and some redesign to the degree where malfunctions of the type described above no longer occur.

Ash with heavy-metal content. One problem of special significance is the heavy-metal content of slag and ash. In addition to metal compounds which are insoluble in water, a great number of soluble metal compounds in the form of oxides, chlorides or sulfates are produced during the incineration of chemical waste. These compounds, contained in the slag and ash, are easily eluted when the slag or ash is flushed from the system, and thus may get into the waste water. The prevention of this contamination requires additional methods to reduce the heavy-metal content of scrubbing and slag water. One such method, for instance, is the precipitation of heavy-metal ions with milk of lime, forming hydroxides which are almost insoluble in water and can be filtered.

Another problem is the volatility of certain metal compounds. Some metal oxides and chlorides have such a high vapor pressure at temperatures from only 150 to 350°C that they will pass the electrostatic filters and, unless there is a flue-gas scrubber, would condense only after further cooling and discharge into the atmosphere. These include predominantly the oxides of arsenic, selenium and phosphorous as well as the chlorides of antimony, arsenic, iron, mercury, tin, cadmium, bismuth, zinc and tantalum. Since some of these compounds are highly toxic, the flue-gas scrubbers must be extremely efficient. For instance, a mercury content of only 0.1% by weight in waste would be sufficient to cause an emission of approximately 120 mg $HgCl_2$/nm^3 in the exhaust gas. Or, during the incineration of waste containing 0.1% by weight of $CdCl_2$ and provided that all of the easily soluble cadmium chloride is dissolved in the slag water, the waste water, based on a realistic waste water ratio of 2 m^3 per ton of waste incinerated, would contain approximately 500 mg $CdCl_2$/liter!

WASTE ACCEPTANCE

Waste generated in Bayer AG plants is first reported by the operating departments to the anti-pollution and waste disposal department, using an incineration permit accompanied by a 5 kg sample. This incineration permit includes data on the quantity, the precise designation and characteristic components of the waste to be incinerated. Other important information for waste disposal include the type of packaging and/or delivery to the plant, the date of delivery, the state of aggregation of the waste, its combustibility, the flash point, the ignition temperature, the calorific data, its behavior upon heating, its boiling point, gas generation, reactivity with water, toxicity, any odors generated and the hazard class according to the Flammable Liquid Directive (VbF). Within the waste disposal department the waste acceptance and inspection laboratory group, on the basis of this information and if necessary its own laboratory studies, determines the technically and economically optimum waste disposal method (incineration with or without flue-gas scrubbing, incineration on the high seas, surface or underground disposal), completes the study by assigning an inspection number and communicates the result to the operating department concerned. If the waste disposal department proposes the disposal of waste by incineration, the waste will be delivered to the plant together with a waste way bill which lists predominantly the quantity, number and type of containers, the inspection number and the materials classification code. This materials classification code is important for computing the incineration cost. Practical experience has shown that a rough classification of combustible waste into eight materials classes is best. The incineration cost per materials class varies from DM 165/ton for liquid non-chlorinated waste to DM 885/ton for waste in small containers of highly variable content.

On the basis of the materials classification code the incoming waste is channelled to certain collection points by a control center. Solid halogen-, sulfur- and/or phosphorous-containing waste is collected in a bunker building. This bunker building includes four chambers having a capacity of 200 m^3 each. On the basis of their basicity, neutral, basic or acid wastes are segregated and collected in a certain chamber each. The fourth chamber is available for special waste types. Liquid halogen-, sulfur- and/or phosphorous-containing waste is either incinerated directly from settling tanks (maximum capacity 10 m^3) or collected in larger tanks (capacity approx. 400 m^3) for intermediate storage. The method used is decided on the basis of the quantities generated and the chemical compatibility of the different waste types. Highly reactive, decomposing and/or generally difficult to handle waste is delivered in combustible plastic drums whenever possible (capacity 60 to 100 liters), handled via a separate drum elevator and dropped with drum into the rotary kiln for incineration.

NO. II INCINERATOR SYSTEM

The process concept for the No. II incinerator system was developed on the basis of the specific waste situation prevailing at Bayer's Leverkusen plant with due consideration to the theoretical studies above. The practical experience acquired in several years of operation on No. I incinerator system proved highly advantageous for the detailed planning effort.[2]

One new feature of this process is the multi-stage flue-gas scrubber system. Other design modifications included the configuration of the ash bunkers, the lay out and design of the afterburner chamber and the design of the heat recovery boiler.

The process schematic of No. II incinerator system is shown in Fig. 2. The functions of the individual system components are described below.

The most important technical specifications of this system have been compiled in the following table:

Fig. 5—Rotary kiln with upright, 2-duct afterburner chamber.

Bunker building		
Waste acceptance	4 chambers	200 m^3 capacity per chamber
Waste charging system	1 grapple	0.8 m^3 capacity
	1 plate conveyor	3-4 ton/hour
	2 slide gates	approx. 2-minute cycles
	1 charging ram	
	1 drum elevator	capacity 2-3 ton/hour
Rotary kiln		length 13.5 m diameter 3.5 m maximum thermal load 15 Gcal/hr (approx. 60 GJ/hr)
Solid waste incineration (see above)		
Liquid waste incineration	various settling tanks	capacity 3-5 m^3
	tank farm	several tanks
	1 sustaining burner	capacity 300 kg/hr
Sludge incineration	1 injector	
Combustion air	1 blower	capacity 37,000 m^3/hr
Cooling air	1 blower	capacity 5,400 m^3/hr
Flue-gas outlet temperature		1,100 to 1,200° C
Afterburner chamber		
Secondary combustion	various settling tanks	capacity 1-5 m^3
	tank farm	several tanks
	3 liquid burners	capacity 300 kg/hr per burner
	1 booster burner	capacity 50 kg/hr
Residence time		2-4 secs.
Flue-gas flow rate		40,000-50,000 nm^3/hr, max. thermal load 20 Gcal/hr (approx. 80 GJ/hr)
Flue-gas outlet temperature		1,200-1,300° C
Heat recovery boiler		
Heat recovery	4 ducts	20 ton steam/hr, 38 bars, 350° C
Flue-gas outlet temperature		300° C
Electrostatic filters		
Dry dust removal		max. 20 kg dust/hr
Flue-gas scrubber		
Cooling and HCl separation	1 injection cooler	6-8 m^3 water/hr
Residual dust separation	2 rotary scrubbers	5-25 m^3 water/hr
Alkaline scrubber	1 jet scrubber	2-12 m^3 water/hr
Sulfite oxidation	2 columns	
Exhaust air outlet temperature		60° C
Waste-air heating	2 heat exchangers	
Suction blower	1 ducted-impeller blower	capacity 60,000 m^3/hr, drive rating 335 kW
Stack	insulated steel tube	height 100 m
	instrumentation and recording devices	waste-air outlet velocity 12-15 m/sec.

CHARGING SYSTEMS

Solid waste charging. Solid waste is predominantly delivered in tipping lorries and collected in a sealed bunker building. Bulky materials are crushed in an impact mill. The charging hopper located above the rotary kiln inlet is charged by a grapple which is controlled from a fully air-conditioned operator cab which is sealed against the bunker space, using TV cameras and TV screens in a partially automatic, partially manual operation. The rotary kiln inlet is sealed against the bunker space by a lock fitted with two sliding gates. When the inclined sliding gate (in the drop chute of the rotary kiln inlet) is closed, a horizontal sliding gate located in the charging hopper will open.

A continuously operating plate conveyor belt located below moves the waste into the drop chute and deposits it on the inclined sliding gate. After a certain cycle time the horizontal sliding gate will open and the waste will drop down an inclined chute into the rotary kiln inlet. Any waste remaining on this inclined chute is pushed down by a ram.

Liquid waste charging. Liquid waste is delivered in sealed settling tanks having a capacity up to 10 m^3. Part of the liquid waste is incinerated directly from these settling tanks while another part is collected for intermediate storage in larger tanks (400 m^3 capacity) prior to incineration.

The transport of this liquid waste to the individual incineration points is accomplished via pumps or under nitrogen pressure (3 bars). Waste oil containing solids is first purified in a screening machine. The screen residue with a particle size of over 2 mm is moved by a compressed-air conveyor system (Gulliver gun) via a pipe line directly into the rotary kiln. Some of the waste delivered in settling tanks requires liquefaction through heating prior to combustion so that it can be pumped and atomized in the burners.

Steam outlets at up to 30 bars pressure are available for this purpose. After removal of the waste from the settling tanks through pressure, the tanks are de-pressurized via gas lines terminating in the afterburner chamber via intermediate safety devices. Waste liquids of low calorific value are atomized together with waste having high calorific value in multiple-fuel burners and incinerated. Atomization is accomplished either with steam or with compressed air.

These burners are located in the face side of the rotary kiln (1 burner) and in the wall of the afterburner chamber (3 burners, 1 booster burner). Burner monitoring is accomplished via a temperature sensor system. When the temperature drops below 1050°C a booster burner fired with natural gas is ignited, supporting the main burners. If the temperature continues to drop below 950°C, the fuel feed is interrupted by quick-action valves.

Charging of problematic waste in small containers. A separate system is available for charging the system with waste in small containers (cans, drums) which consist predominantly of an elevator, a roller conveyor, a lock with two sliding gates and a tilting table from which the small containers are dropped into the rotary kiln. This charging system is interlocked by the control system so that these small containers can be dropped into the rotary kiln only when the inclined sliding gate is closed. This insures that there will not be any interference between the drum charging system and the solid waste charging system.

Rotary kiln, afterburner chamber. The rotary kiln has a length of 13.5 m and an outer diameter of 3.50 m. The ceramic refractory liner has a thickness of 0.25 m. Consequently, its inner diameter is approximately 3 m. The rotary kiln is inclined 3 percent toward the outlet end.

The average rotating speed is approximately 9 revolutions per hour. It is infinitely variable between 7 and 24 revolutions per hour. The residence period of the waste in the rotary kiln is approximately 1½ hour. The internal temperatures reach 1,200 to 1,300°C. The hottest zone is located in the 7 to 11 m range. This is the zone where the strongest corrosion and erosion attack takes place. It is attempted to keep this attack within justifiable limits through the selection of extremely high-quality refractory materials and the continuous presence of a protective slag layer. At the discharge from the rotary kiln the slag drops into a water-filled plate conveyor slag removal system. The slag is free of any pollutants and can be deposited at a disposal site without any danger. The combustion air at a rate of approx. 40,000 nm^3/h is aspired from the bunker building and injected into the rotary kiln inlet below the solid waste charging system. The gross thermal rating of this rotary kiln is approx. 60 GJ (15 Gcal).

A circular, 2-duct, upright afterburner chamber is placed immediately downstream of the rotary kiln. Its height is approx. 9 m and its inner diameter 4.10 m. Both ducts are provided with a liner of high-temperature ceramic refractories. At approximately half the height of the first duct three liquid burners with an average thruput of approximately 300 kg/h per burner are distributed over the circumference.

In continuous operation this arrangement will achieve intensive admixture of the flue gases, combustion temperatures up to a maximum of 1300°C and a residence period of 2 to 4 seconds. Secondary air nozzles are installed above these burners for safety reasons through which additional combustion air can be injected. The entire inner space of the rotary kiln can be inspected from a port located in the base of the first afterburner duct. This has special advantages for the continuous inspection of the rotary kiln refractory liner and the protective slag layer in the kiln.

The second duct serves as an abatement chamber, and its bottom end has an ash collection space. This space can be cleared at certain time intervals through a port of several square meters area which is closed by four bulkheads; a front end loader is used for ash removal. Continuously operating conveyor devices have not been found feasible at this point. The afterburner chamber is designed for an over-all thermal rating of 80 GJ (20 Gcal).

Heat recovery boiler. The flue gases departing from the afterburner chamber at approx. 1200°C are cooled to about 300°C in the heat recovery boiler, generating approx. 20 tons of steam per hour at 38 bars and 350°C. This heat recovery boiler is an upright 4-duct boiler type. The first two ducts are empty ducts with radiative heating

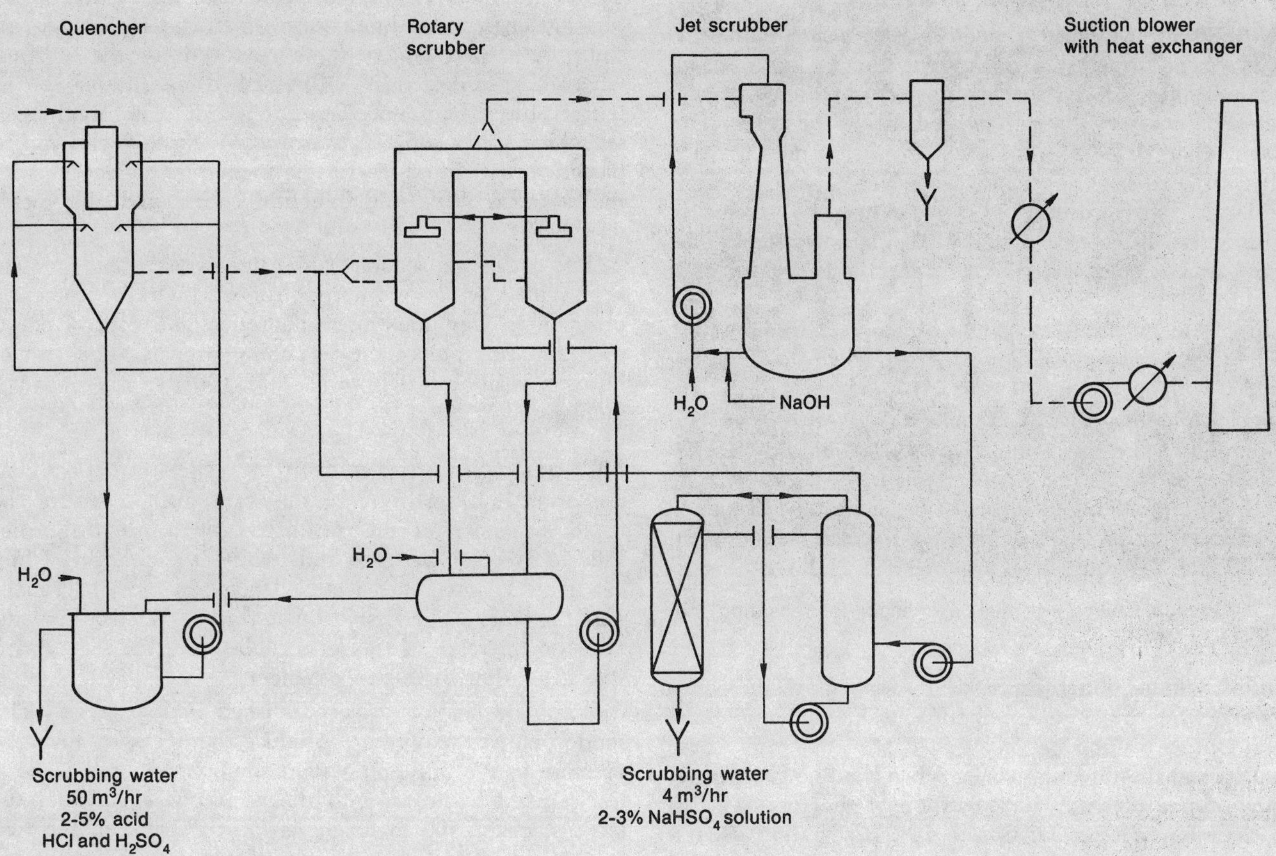

Fig. 6—Schematic of the flue-gas scrubber system.

surfaces. The last two ducts contain superheater, evaporator and economizer.

These boiler ducts have been made of finned tubing cooled by natural circulation. The contact heating surfaces have been welded from square tubing, forming smooth elements that are easy to clean with soot blowers, and connected to a forced circulation cooling system. The rotary kiln charging system and the heat exchangers used for heating the scrubbed flue gases are also installed in this cooling circuit.

Flue-gas scrubbers. In addition to CO_2 and H_2O the combustion of chemical waste produces predominantly SO_2, HCl, NO_x and dust as well as minor quantities of HF, SO_3 and P_4O_{10} whose quantities in the flue gas have been limited under the Anti Air Pollution Directive (TA-Luft) so that most of them must be scrubbed from the flue gas.

The concept definition for the flue-gas scrubbing system was based on the assumption that the flue gases will contain an average of 10 g HCl/nm^3, 2.5 g SO_2/nm^3 and 2 g dust/nm^3, and that multiples of these concentrations may easily occur under peak loads. The quantities of HF, NO_x, SO_3 and P_4O_{10} were considered to be of minor significance.

The dry portion of the combination dry-wet flue-gas scrubbing system consists of the electrostatic filter and the 3-stage wet portion of the injection cooler (quench), the two rotary scrubbers in a parallel configuration and the jet scrubber.[3]

The majority of the dust is precipitated from the flue gases at approx. 300°C in the electrostatic filter. The filter efficiency is approx. 90-95%. The precipitated dust is removed by a scraping conveyor belt, collected in a covered tipping lorry and deposited at a solid-waste disposal site. This procedure has the advantage that any heavy metals contained in the dust and/or precipitated together with the dust cannot get into the subsequent wet scrubbing system where it would be dissolved in the form of chlorides or sulfates which could interfere with the biological treatment process of the liquid-waste treatment plant.

In the injection cooler the flue gases arriving from the electrostatic filter are cooled from 300°C to a level of 60-70°C and, at the same time, the majority of the HCl is separated. The cooling and scrubbing water required for this purpose is re-circulated. The evaporation losses are compensated by 6-8 m^3/hr make-up water. The design of this section of the system and the materials used have been selected in accordance with the enormous thermal and corrosive loading of this section. The HCl concentration of the departing scrubbing water varies from 2-5 percent.

From the injection cooler the flue gas passes into two parallel rotary scrubbers. In these scrubbers the flue gases pass through a dense cloud of very fine water droplets generated by two rapidly rotating paddle-wheels.

These rotary scrubbers have three functions: (1) separation of any residual dust and extremely fine dust as well as volatile metal compounds; (2) separation of sulfuric acid aerosols which may be generated in considerable

Fig. 7—Suction blower with heat exchanger for reheating the purified flue gases.

quantities in the injection cooler when the flue gases have certain compositions; (3) separation of the total dust if the electrostatic filter fails or if dust penetration occurs under excessive loads.

The scrubbing water is collected in a manifold and pumped out from this point. 2-5 m^3/hr fresh water is fed into this manifold in order to prevent any sludge accumulation. The overflow is passed to the collection point for transport to the liquid-waste treatment plant. Compared to other comparable scrubbers this rotary scrubber has the advantage of being insensitive to gas flow rate fluctuations, ensuring a high solids concentration (up to 200 g/liter) and having a lesser pressure drop (approx. 50 mm WC).

The latter effect was decisive for the decision in favor of this scrubber type. It meant that a flue-gas suction blower of lower pressure (400 mm WC), slower speed and lesser noise pollution could be installed.

Downstream of the rotary scrubbers the jet scrubber, using a weakly alkaline medium, is the third scrubbing stage. In this scrubber especially the SO_2 in addition to residual HCl, chlorine and NO_x are removed. In principle this scrubber is an over-dimensioned water jet pump. The scrubbing water is injected in the direction of the gas flow at a pressure of 3-6 bars. The downward velocity of the scrubbing water is 25-35 m/sec, that of the flue gas 10-20 m/sec.

As a result the scrubber automatically aspires the flue gas. The small pressure gain is adequate to overcome the internal drag. The scrubbing water is collected in a manifold and re-circulated. Scrubbing water losses are compensated by approx. 2 m^3 fresh make-up water per hour. The scrubbing water pH value is maintained constant at a certain adjustable value by means of a measuring and control device. The pH value of the scrubbing water is adjusted by adding sodium hydroxide solution.

The Na_2SO_3 and $NaHSO_3$ produced through reaction of SO_2 with sodium hydroxide in this jet scrubber must be oxidized into Na_2SO_4 prior to discharging the scrubbing water into the liquid-waste treatment facility. A specially developed 2-stage facility is used for this purpose where the oxidation is accomplished instantaneously at temperatures between 70 and 90°C. The discharged scrubbing water contains approx. 5% Na_2SO_4 and can be passed to the liquid-waste treatment plant together with the overflow from the injection cooler and the rotary scrubbers.

The entire wet section of the flue-gas scrubbing system is provided with an internal rubber liner for corrosion protection. Numerous instrumentation and control points monitor and control the scrubbing process. The cost of flue-gas scrubbing is approx. DM 100 per ton of waste incinerated.

Flue-gas drying and heating. In order to prevent condensation and corrosion in the stack and to ensure the rapid dispersion of the purified waste air in the atmosphere, two droplet traps and two heat exchangers (flue-gas heaters) have been installed between the jet scrubber and the stack. The first droplet trap is located immediately downstream of the jet scrubber and the second upstream of the first heat exchanger.

Waste-air heating is accomplished in two stages. The suction blower required to produce the necessary negative pressure in the different system components and to move the purified waste air into the atmosphere via an insulated stack of 100 m height is located between the two heat exchangers. The hot water required for the heat exchangers is obtained from the natural circulation of the heat recovery boiler.

This flue gas reheating process requires approx. 5 tons hot-water steam per hour, equivalent to 35,000 tons steam per annum at a cost of approx. DM 0.5 million for an average of 7,000 hours of operation per annum. If this reheating system were eliminated, the incineration cost could be reduced by 5-10%.

Efficiency. The energy efficiency of this system is computed as follows: under normal load approx. 1.8 tons of solid waste and 1.2 tons of liquid waste as well as approx. 0.6 tons of small-container waste are incinerated per hour. This waste has an over-all calorific value of approx. 83.7 GJ (20 Gcal). Out of this quantity, 41 GJ/hr (9.8 Gcal/hr), equivalent to 49%, is consumed for steam generation, and 9.2 GJ/hr (2.2 Gcal/hr), equivalent to 11%, is required for re-heating the purified waste gases. The remaining 40% includes 29.4% over-all heat losses (radiative losses on the rotary kiln, the afterburner chamber and the boiler, heat losses in the injection cooler and heat removal via the flue gases), and 10.6% for the operational uses of the plant.

ANTI-POLLUTION REQUIREMENTS

The license granted by the Düsseldorf Licensing Authority on Dec. 31, 1974, established the following maximum emission values for the No. II incinerator system:

SO_2	1,250 mg/nm³ waste air
NO_x	1,000 " " "
HCl	100 " " "
HF	10 " " "
Dust	100 " " "
O_2	11% by volume

Fig. 8—Control center for the No. II incinerator system.

These components must be measured continuously for anti-pollution monitoring. For some components the monitoring is accomplished by means of continuously recording apparatus.

Systematic flue gas analysis under different loads and modes of operation was accomplished during the period from 6,000 to 8,000 hours of operation in order to study the efficiency of the flue-gas scrubbing system. The following results have been achieved:

Flue-gas component	Normal load (in mg/nm³)	Peak load (in mg/nm³)
SO_2	300	800
NO_x	400	up to 1000 and more
HCl	50	100
HF	3	5
Dust	20	100

On the whole the anti-pollution limits are maintained. They are occasionally exceeded with respect to NO_x. This has not been due to deficiencies in the apparatus or in the process, but to the fact that the physical and chemical properties of nitrogen oxides render any absorption very difficult. To this date no scrubbing process has been developed where separation efficiencies are achieved which are comparable to those achieved when scrubbing SO_2 and HCl.

In addition, the license requires all waste water and scrubbing water to be treated in a biological liquid-waste treatment plant.

Under normal operating conditions the afterburner chamber must be operated at a minimum temperature of 1000°C and a minimum residence period of 0.3 secs. The minimum temperature must be increased to 1200°C for the incineration of polychlorinated compounds.

The afterburner chamber temperature must be monitored by continuously recording apparatus. If the temperature drops below 1000°C, the charging system of the rotary kiln must be stopped.

ACQUISITION COST, PLANNING AND ERECTION TIME

The investment to acquire a capacity of 25,000 tons per annum of chemical waste amounted to DM 27 million. The system concept was developed jointly with L. & C. Steinmüller GmbH, D-5270 Gummersbach, who were also the designers and contractors for the construction of the facility. The planning period extended over two years. Erection required one year. During the three years of planning and construction this project secured approximately 100 jobs. The operating cost is DM 10 million per annum. Consequently, the average incineration cost is approx. DM 400 per ton of waste.

After trial operation of approx. 7,000 operating hours the No. II incinerator system of Bayer AG's Leverkusen plant was commissioned for regular service in November 1977. This system has passed the performance tests, achieving the specified operating parameters and has demonstrated its efficiency. It was presented to the public in a brief ceremony on Jan. 18, 1978.[4]

The trend of the waste situation during the next one or two decades can be estimated on the basis of two facts:

1. The Leverkusen plant has reached its maximum expansion level, and

2. Low-waste technologies are used with preference when new products are marketed.

The conclusion may be drawn that the amount of production waste of the type described in this article will not increase significantly in the short run. Therefore, the construction of a third incinerator system is not contemplated.

POSTSCRIPT

One important item in the planning concept was the company requirement for a maximum of continuous mixing and transport of the different waste materials with respect to their calorific value and quantity in order to create uniform thermal conditions in the combustion spaces. In this respect a problem was created by the so-called small waste containers, cans or drums of 60 to 100-liter capacity and having a great variety of contents.

These small containers are filled with solid, paste and liquid production or laboratory waste in pure or mixed form whose calorific value fluctuates from 0 to 10,000 kcal/kg. The composition of these small container wastes may vary greatly. Often the contents are toxic, frequently they are not admixable, some of them will burn very slowly, but most of them spontaneously, and they sometimes react with great sensitivity to air and humidity.

The general practice of dropping the small containers into the rotary kiln without emptying them has process disadvantages. Occasionally there will be deflagrations with strong soot generation and excessive thermal and mechanical loading of the kiln refractories resulting from this practice.

In order to prevent these defects, paste and liquid small-container contents should be pressed into liquid nitrogen, granulated under nitrogen shielding, mixed and removed from the nitrogen bath into the rotary kiln by a conveyor system.[5]

This type of system would have been an elegant solution for all process problems arising in conjunction with these small-container wastes. Unfortunately, it would have caused some difficulties which were not accurately predictable and/or could not be exactly determined through preliminary studies:

1. The freezing characteristics of different organic wastes and waste mixtures vary greatly. One special disadvantage is that certain groups of materials will form continuous layers of incrustations floating on the nitrogen surface or adhering to the inner walls of the containers which do not settle on the bottom so that they cannot be removed and will gradually clog the system. Other groups of materials will freeze into minute crystals suspended in the liquid nitrogen bath, gradually enriching and forming a thick paste of crystals which finally agglomerates and blocks the conveyor system.
2. The small-container waste is mostly highly heterogeneous. Plastic film, plastic beakers, leather and rubber gauntlets as well as glass bottles are frequent accompanying products. Where these materials get into the liquid nitrogen bath, they will quickly clog the granulator and the transport screw. The necessary repairs are time-consuming and require many man-hours and, moreover, they are not entirely free of danger since reactive mixtures or explosive vapors/air mixtures can form during thawing, or toxic substances may be released.

Since these difficulties involved in retention of the concept could have been resolved only at additional enormous effort in terms of research and process development, if at all, it was attempted to optimize the combustion conditions through design modifications in the incinerator system itself and through process modifications so that even these problematic waste types could be properly disposed of without the disadvantages named above. These experiments which might be the subject of a later article have been successful every step of the way so that these problems may now be considered resolved.

This example was given to indicate that the apparently direct approach will not always yield the desired success. It may be possible that, in view of the increasing complexity of the problems, detours will become increasingly unavoidable which will be associated with high cost and will be very time consuming.

LITERATURE CITED

[1] Fabian, H. W. and M. Schön, Thermal Waste Treatment (Thermische Behandlung der Abfallstoffe), Paper presented at the first Special Waste Seminar of the Management Information Centre, Verlag Moderne Industrie Publishers, Munich, 29/30 September 1977.

[2] Fabian, H. W., and H. O., Weber, The New Waste Incineration System of Farbenfabriken Bayer AG, Leverkusen (Die neue Rückstandverbrennungsanlage der Farbenfabriken Bayer AG, Leverkusen), published in Brenstoff-Wärme-Kraft (BWK), Volume 19 (1967), No. 10, pp 484-486.

[3] Reimer, H., Flue-Gas Scrubbing in Waste Incinerator Systems Rauchgaswäsche nach Müllverbrennungsanlagen), published in VGB Kraftwerkstechnik, Volume 53 (1973), No. 11, pp 735-742.

[4] Fabian, H. W., Incineration of Chemical Waste (Verbrennung von Chemieabfällen), published in Umwelt No. 3/78, pp 153-158.

[5] Gromotka, H., New Processes for the Thermal Treatment of Polluting Industrial Waste (Neue Verfahren für die thermische Behandlung von umweltschädlichen Industrieabfällen), published in Energie und Technik, 1974, No. 6, pp 131-134.

Incinerate Refinery Waste in a Fluid Bed

K. P. Becker, Dorr-Oliver Companies, Weisbaden, W. Germany, and **Clarence J. Wall,** Dorr-Oliver Inc., Stamford, Conn.

FLUID BED INCINERATION offers a practical, economical and environmentally compatible method for disposing of refinery waste sludge. In conjunction with other systems for concentrating waste, fluid bed incineration can transform sludges into suitable streams for disposal through acceptable environmental techniques.

Design alternatives offer equipment selections that provide streams acceptable to environmental regulation while maintaining operable conditions. Design parameters include size, bed type, operating temperature, heat recovery and stream contaminations.

Operating problems incurred in several HPI installations center on selection of proper bed temperature and composition, heat exchange and waste water disposal. These are normally soluble by simple operating controls.

PROCESS DESCRIPTION

Fluid bed incineration of waste uses technology learned from fluid bed operations through years of experience in chemical, metallurgical and oil industry operations such as fluid catalytic cracking. Incinerator design is based on one fundamental arrangement with various alternatives for heat utilization.

A vertically oriented vessel is divided into three parts: windbox at the base; fluid bed middle section, and reactor top zone. Air for combustion and fluidization is supplied directly to the windbox (Fig. 1) while dewatered sludge solids and/or fluids are introduced into the fluid bed by pumps or screw feeders. Auxiliary fuel, oil or gas as required, is introduced into the fluid bed and burned along with organic sludge components to provide complete destruction of waste solids. Sludge water is converted to steam and emitted through the reactor with gases from combustion of organics and fuel and suspended fine inert ash solids. This arrangement is essentially the same regardless of what arrangements are provided for heat utilization or makeup.

Three basic arrangements are normally employed for heat recovery or additional supply: cold windbox (Fig. 2); hot windbox (Fig. 3), and heat recovery system (Fig. 4).

No heat is added or recovered in cold windbox design such as that normally used when high heat value sludges must be handled. Reactor gases are cleaned and cooled by contact with a water spray to about 120° F and a particulate solids content normally less than 0.05 grains per standard cubic foot (dry).

Heat is added to combustion air in a hot windbox system by recovery from reactor effluent gases. Fresh windbox air temperature is increased to about 800-1,200° F while incinerator off gases are cooled to about 1,000° F such that over-all thermal efficiency is improved enough to cut water evaporation requirements to about 2,500-2,600 Btu's per pound compared with 3,700-4,100 Btu's per pound where no gas heat exchange is used.

Additional heat to insure complete combustion of waste sludge may be added in either cold or hot windbox arrangements by introducing auxiliary fuel, oil or gas, to the fluid bed as required.

Heat recovery may be accomplished by use of a waste heat boiler. Under this arrangement, windbox air is not preheated but all available heat is used for generating steam at 350 psig and 437° F superheat for use for unit drives. Any excess steam not needed for sludge treatment is used for heating or power generation or for driving plant rotary equipment.

Incinerated gases are cooled to about 650° F after which they are scrubbed for final cleaning and cooling. Sometimes a hot Cottrell precipitator or electrofilter may be used for final gas cleaning.

EFFLUENT STREAMS

Exhaust from gas cooling presents no environmental pollution problem for any known HPI operation (Table 1). Only obnoxious contaminants which are vapors at atmospheric pressure and 120° F or less are possible problems. Other materials can be scrubbed to a level of particulate cleanliness required to meet any expected local rule by arrangement of gas cooler design.

Gas cooler effluent water, however, can be a problem if concentrations of alkaline salts are high. Such problems require special consideration and may result in a need to evaporate water for ultimate disposal. This is a rare circumstance, however, since water can be disposed of normally by lagoon evaporation or through filtration and land fill disposal of filter cake. Effluent water may also

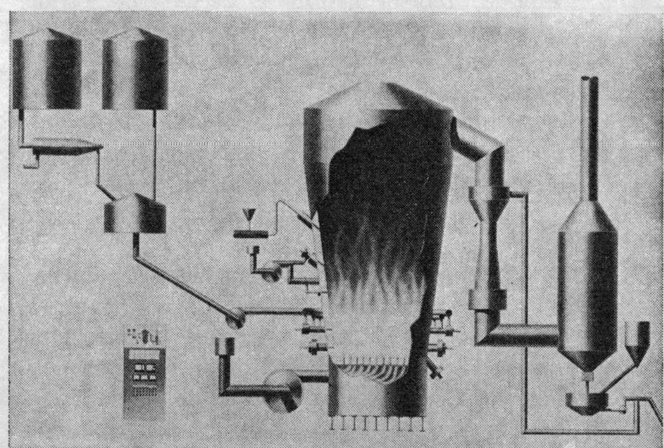

Fig. 1—Fluid bed incinerator for refinery waste.

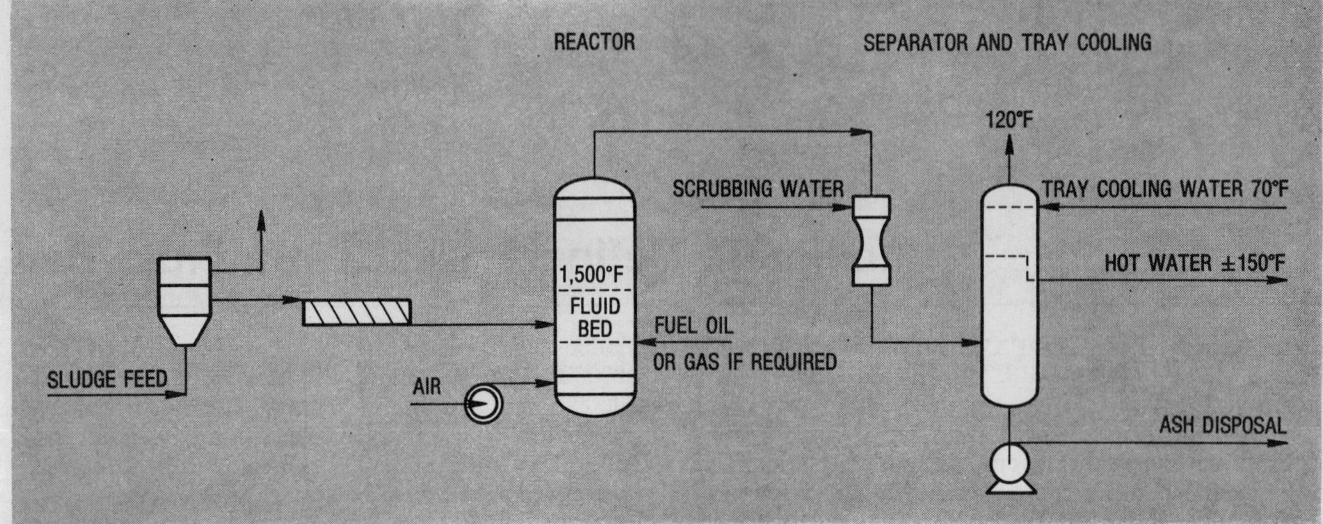

Fig. 2—"Cold windbox" fluid bed incinerator.

be reused when neither chloride nor alkali concentrations prohibit. Under no conditions are combustible materials found in this water.

UNIT DESIGN

Factors to be considered in fluid bed incinerator design include over-all requirements of capacity and heat balance as well as special operating conditions around the bed. Sludge heat content and amount determine, in part, whether auxiliary fuel will be required and whether heat can be recovered for other services. Furthermore, sludge composition can determine bed type to be used because of impurity influences. Normally, bed selection is limited to that material which is readily and cheaply available but which will not cause chemical interaction and operating problems. Silica sand is most widely used; however, in some instances, initial bed type is gradually transformed by products of waste combustion. This is especially true where concentrations of alkalis and other materials result in solid agglomerates that are stable at temperatures below that of normal bed conditions.

Bed selection also determines bed operating temperature. In every case, effluent quality is controllable with respect to vapor and water except where high concentrations of chlorides and/or alkalis cause excessive buildup in effluent cooler water.

Ash formed and removed along with cooler water creates a unique problem of disposal because of possible quality problems. High chloride and/or sulfate concentrations can preclude easy disposal of ash and necessitate special disposal arrangements.

Attention must be given to materials of construction for fluid bed incinerators. Where high concentrations of acids and/or alkalis are present in waste, special consideration must be given to the protection of the mild carbon steel. Normally, a rubber lining between the steel and acid brick is provided or conditions are set to insure that steel wall temperatures are held high enough to prevent condensation of acidic materials. Reactor steel temperature is normally maintained at no less than 450-500° F where high sulfur concentrations are present and at temperatures not below 225° F where hydrogen chloride gases are present. Frequently, protective coatings other than rubber are applied between the steel and the refractory materials for further protection.

PLANT OPERATIONS

Fluid bed incinerators are relatively simple to operate provided proper bed temperatures are selected. General technology for fluid bed operations applied here.[x] Use of proper fluidizing gas flows as well as bed compositions are essential to good operations.[y, z]

Problems. Significant problems in fluid bed incinerators occur in bed composition and gas cooling.

Fluid bed compositional change can result in conditions where bed material becomes tacky and ceases to be fluid.

Various wastes are in themselves chemically quite complex. In addition to normal organics or petroleum hydrocarbons, they may contain sodium, potassium, magnesium, sulfur and phosphorus which may come from the various waste water streams or from various petroleum streams being processed in the refinery. Chemicals may be added to waste for treatment or enter with rainwater run-off, etc., so that elements such as iron and aluminum and compounds like silica dioxide (sand and silt) and clay (aluminum oxide in mixture with silica oxide) may be present in the final waste sludges. For refineries on sea coast estuaries or bays, seawater is a most likely component of waste sludges. Also, ballast water may contain seawater and sulfur may be present in auxiliary oil feed for incineration. Therefore, various elements that might be present in waste sludges include sodium, potassium, magnesium, calcium, phosphorus, sulfur, iron, aluminum, silica, oxygen, nitrogen, carbon, hydrogen and other trace-quality elements found in seawater or waters that are treated in waste water plants.

Such chemicals create incinerator reactions at high temperature that can seriously affect operation through defluidization or carryover. Weak spent caustic for example, is converted into sodium carbonate and sodium sulfate in association with other materials. Sodium carbonate to sodium sulfate ratio which depends upon the amount of sulfur being fed with auxiliary fuel or in other feed streams can create chemical reactions that determine acceptable fluid bed temperature levels. Pure sodium sulfate has a melting point of 1,623° F, and sodium carbonate has a melting point of 1,564° F but combinations

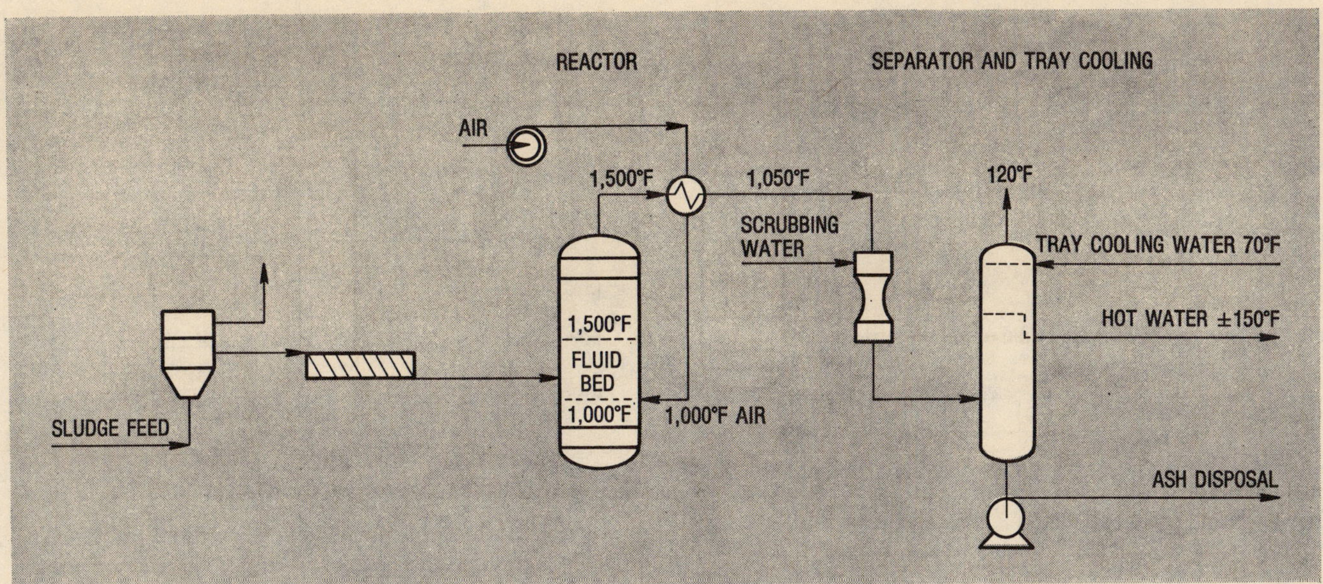

Fig. 3—"Hot windbox" fluid bed incinerator.

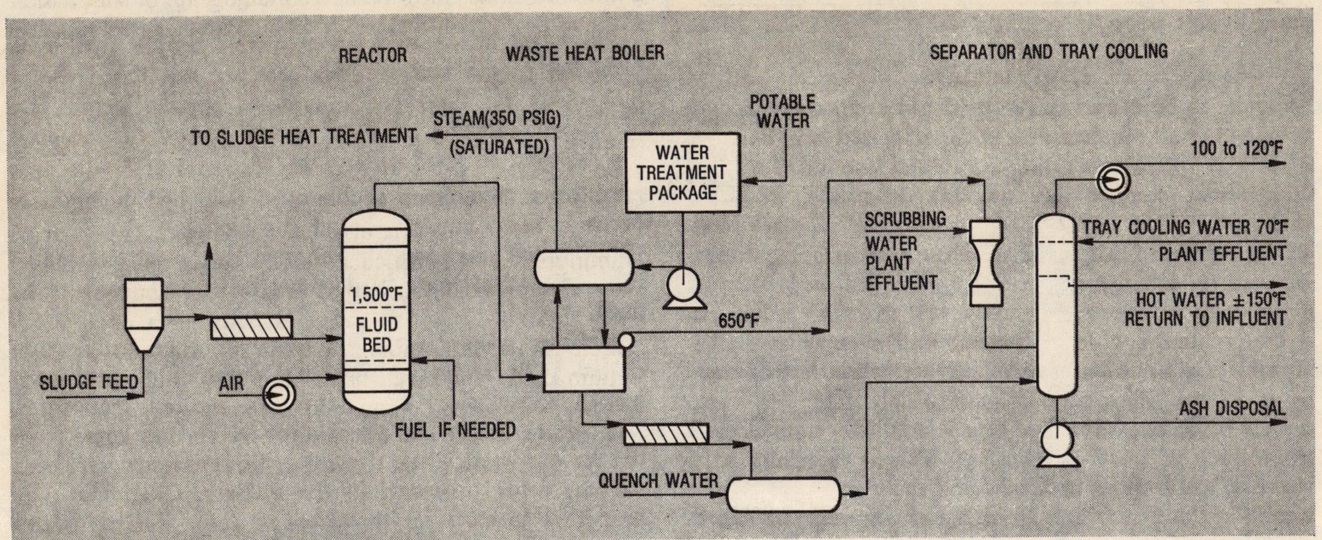

Fig. 4—Heat recovery fluid bed incinerator.

form a eutectic (Fig. 5) with a minimum melting point of 1,552° F at 47 percent sodium sulfate and 53 percent sodium carbonate. Furthermore, all combinations of these materials up to and including approximately 70 percent sodium sulfate create melting point mixtures that are below that of sodium carbonate. When this occurs, fluidization is impaired for any temperature approaching these levels and necessitates selection of operating conditions to avoid partially melting the bed.

Chlorides also can cause operating problems. Seawater can cause bed salting due to conversion of sodium chloride to sodium carbonate in association with sodium sulfates. Under these conditions, bed temperature must be maintained around 1,350-1,380° F for efficient organics combustion with a minimum of excess air while avoiding bed softening. Lowering the bed temperature usually requires an additional use of excess air making for poorer combustion.

Seawater, when evaporated to dryness and heated to 800° F produces sodium chloride, magnesium oxide, magnesium sulfate, potassium chloride and calcium sulfate. Magnesium chloride hydrolyzes to magnesium oxide and hydrogen chloride during final drying. Since sodium chloride has a significant vapor pressure at normal incinerator operating temperatures (1,472-1,652° F), one would expect total vaporization at temperatures 50 degrees or more above theoretical (Fig. 6). In evaporating seawater, calcium sulfate precipitates until calcium contents are very low. Remaining sulfate ends up as magnesium sulfate so that all sulfur should be present as calcium or magnesium sulfate. However, in incineration of seawter, up to 70 percent of all sulfate appears with ash residue as sodium sulfate. With both sodium sulfate and sodium chloride present, a low melting eutectic may be formed (Fig. 7). If this eutectic liquid phase is allowed to accumulate, bed defluidization will take place eventually. With a silica sand bed or with silica dioxide in waste sludges, sodium sulfate reacts with silica to form a very sticky viscous sodium silicate glass which can cause very rapid bed defluidization.

Although fluid bed incineration is possible with limited amounts of chlorides, any substantial amount of more

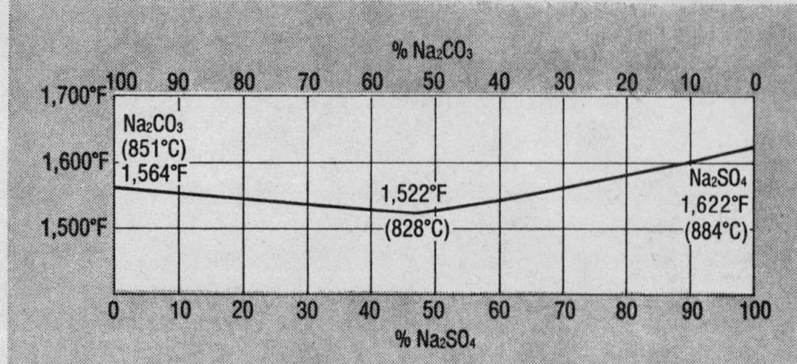

Fig. 5—Sodium carbonate-sodium sulfate system.[1]

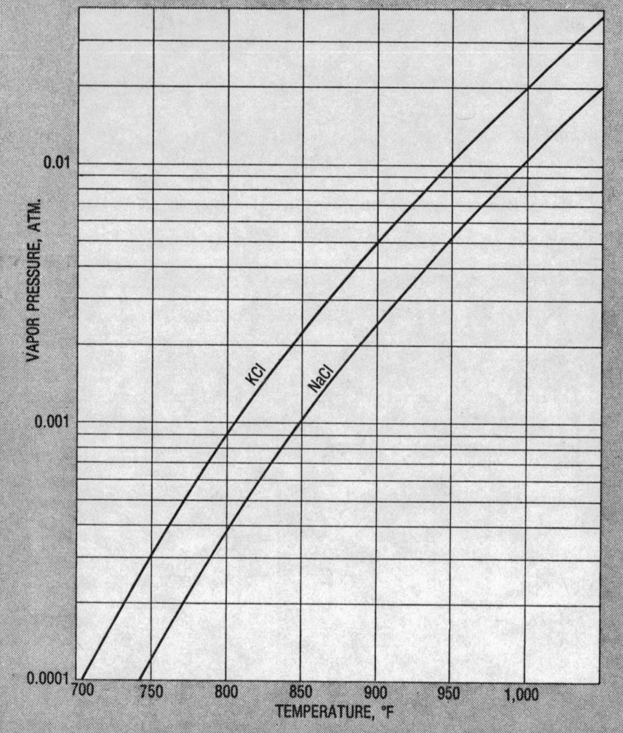

Fig. 6—Vapor pressure of sodium chloride and potassium chloride.[1]

than 500-600 ppm of chloride in feed sludge causes serious problems. Some sludges normally may contain up to 50 percent seawater and at times can be almost totally seawater. Such a condition creates a problem which may be overcome by adding sufficient sulfur to convert all alkali metals to sulfates thus driving off chlorides as hydrogen chloride if temperatures are held below volatilization levels of metal chlorides. In some cases, however, presence of certain elements in waste streams leads to low melting eutectics which cause defluidization. One such element is magnesium. A low melting eutectic (Figs. 8 and 9) for calcium, sodium and magnesium sulfates limits acceptable levels for magnesium.

All problems relating to handling waste containing alkalis and chlorides are solved by selection of a operating temperature below that required for sodium potassium volatilization. Accumulation of alkali metal sulfate is prevented by reacting them with silica dioxide. Alkali metal silicate, a sticky, viscous glass which is produced from decomposition of alkali sulfate, is further reacted with various metal oxides which will form high-melting crystalline silicate compounds when reacted with alkali metal silicate. Some metal oxides that my be used include calcium oxides, aluminum oxide and iron oxide. This

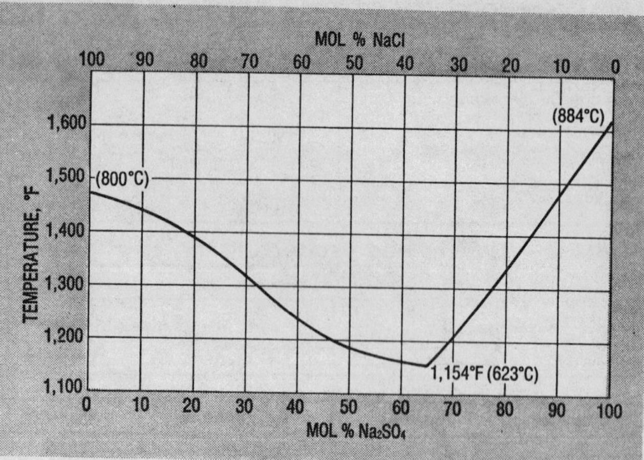

Fig. 7—Sodium chloride-sodium sulfate system.[2]

final end product of metal oxide (sodium oxide-silica dioxide) requires and generally has a high temperature melting point well above that required for vaporizing alkali metal chlorides. To be effective, both silica dioxide and metal oxides must be in a suitable state of subdivision so as to be available to react as indicated. In many cases some required compounds (silica dioxide and metal oxides) are present in waste sludge and may supply all or part of the additive required if they are of sufficient small particle size. In other cases, these compounds have to be added to waste sludge.

Clay is a natural mixture of hydrous aluminum silicates with two to three parts of silica dioxide and one part of aluminum dioxide, which normally occurs in a very fine state of subdivision such that it is a very convenient incinerator additive. Clay provides both silica dioxide required for decomposition of sodium sulfate to produce sodium oxide and silica oxide and also aluminum oxide to react with silica compounds to form a crystalline high-melting-point sodium aluminum silicate. Reactions involved with kaolin clay are as follows:

Dehydration

$$Al_2O_3 \cdot 2SiO_2 \cdot 2H_2O \longrightarrow Al_2O_3 \cdot 2SiO_2 + 2H_2O \quad (1)$$

Reaction with NaCl and H_2O

$$Al_2O_3 \cdot 2SiO_2 + 2NaCl + H_2O \longrightarrow 2HCl + Na_2O \cdot Al_2O_3 \cdot 2SiO_2 \quad (2)$$

This occurs "in flight" and ties up sodium chloride directly as a high-melting-point sodium oxide—aluminum oxide—silica dioxide mixture without going through any other intermediate reactions. Reaction with small amounts of potassium chloride in seawater are similar to that for sodium chloride. By direct reaction between clay and sodium chloride, a complex of sodium sulfate, sodium

TABLE 1

Company and plant location	Design sludge rate		Reactor	
	mtpd	% water	Inside diameter, feet	Height, feet
Mobil Oil Worth (Karlsruhe), Germany	40	95	13	30
Gulf Oil Bertonico (Lodi), Italy......	10	93.5	7.2	28
Gulf Eastern Bantry Bay, Ireland........	5	95	3.5	28
Stanic (Esso) Bari, Italy................	38	87	14	28
Stanic (Esso) Livorno (Leghorn), Italy....	84	83	19.7	28
American Oil Whiting, Ind...............	338	85—95	28	40

TABLE 2—Total mixed feed Amoco, Whiting, Ind.

Sludge source	Total tons/day	Weight, % water	Combustibles, tons per day
Flotation clarifier skimmings	181.4	90	18.1
API separator sludge	36.3	85	5.4
Tank cleanings and misc. oils	63.5	50	31.7
Biological treatment sludge	6.3	70	1.9
Spent caustic	50.8	90	0
Total	338.3	..	57.1

TABLE 3—Typical bed product

Chemical analyses

Water soluble		Water insoluble	
Element	Amount	Component	Amount
Na	5.34%	Al_2O_3	12.62
Cl	150 ppm	Fe_2O_3	6.69
S^{+6}	4.12%	SiO_2	23.73
Ca	810 ppm	Na	1.94
Mg	1 ppm	Ca	15.58
Al	37 ppm	Mg	2.76
K	3,641 ppm	P_2O_5	2.58
Cr	250 ppm	LOI	2.44
		S^{+6}	5.79
Total	9.95%		74.13

Probable compound composition

Water soluble		Water insoluble	
Compound	Weight %	Compound	Weight %
Na_2SO_4	16.45	$CaSO_4$	24.62
K_2SO_4	0.80	$Ca_3(PO_4)_2$	4.77
NaCl	0.02	$Na_2O{\cdot}Fe_2O_3{\cdot}SiO_2$	16.31
$CaSO_4$	0.27	$CaSiO_3$	10.32
		$MgSiO_3$	10.00
		CaO	2.02
		Al_2O_3	10.66
Total	17.54		78.70

oxide, silica dioxide, sodium sulfate and sodium chloride is avoided and only calcium sulfate, magnesium sulfate, magnesium oxide, sodium oxide, aluminum oxide and silica dioxide compounds are formed from seawater or other sodium, chlorine, sulfur, magnesium, calcium, potassium elements present in waste sludges.

Tests show that addition of clay essentially eliminates buildup of any molten salts on bed particles to eliminate bed stickiness. Reaction between sodium salts and clay are essentially 100 percent complete and few insoluble sodium salts are retained. If sodium salts in incinerator feed can be completely reacted with clay while in the reactor, sticky compounds can be avoided in exit gas and possibly a heat exchanger can be considered for additional heat recovery if desirable. If sodium oxide—silica dioxide is formed on bed particles where a sand bed is used or where there is coarse silica in waste feed stream, reaction with clay would be:

$$Na_2O{\cdot}3SiO_2 + Al_2O_3{\cdot}2SiO_2 \longrightarrow Na_2O{\cdot}Al_2O_3{\cdot}2SiO_2 + 3SiO_2 \quad (3)$$

In the presence of sodium sulfate the reaction is:

$$Na_2SO_4 + Al_2O_3{\cdot}2SiO_2 \longrightarrow Na_2O{\cdot}Al_2O_3{\cdot}2SiO_2 + SO_2 + \tfrac{1}{2}O_2 \quad (4)$$

With metal oxides of calcium and iron, there is a possibility of forming some sticky sodium compounds.

If for some reason clay is not used, then the reactor is designed with special gas outlet duct to avoid or minimize duct scale formation. With clay, however, conventional exit gas ducting design is employed.

Selection of an alternate form of bed material may circumvent these problems. Other materials for consideration are iron ore, alumina or that bed which is formed

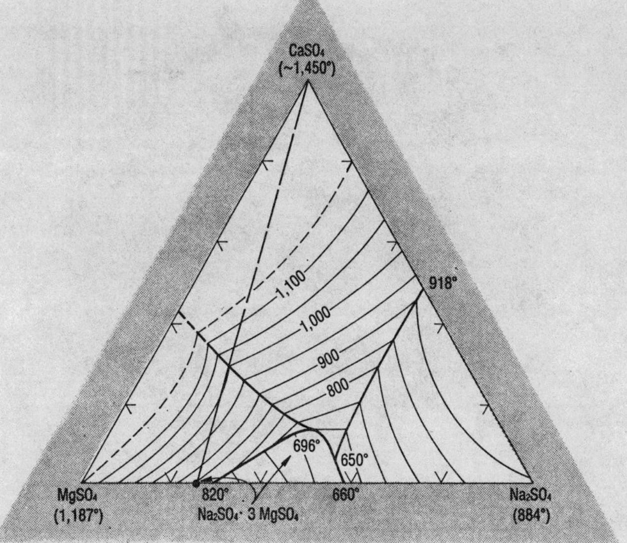

Fig. 8—Sodium sulfate-calcium sulfate-magnesium sulfate system.[3]

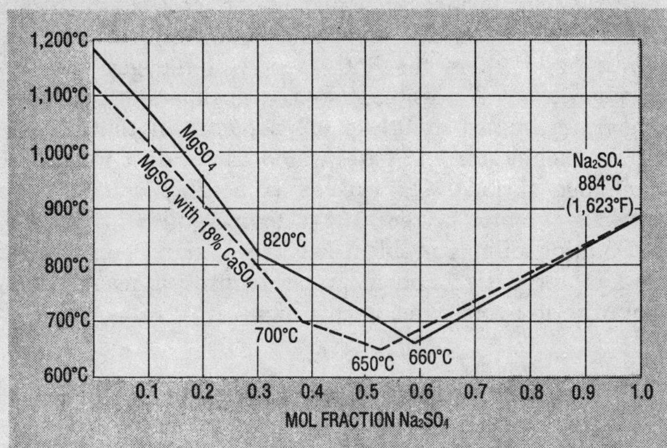

Fig. 9—Magnesium sulfate-sodium sulfate system with zero and 18% calcium sulfate.[1, 4]

from normal operations.

Incinerator feed impurities can change bed composition in normal operations. This occurs when feeding a sludge such as one consisting of API separator sludge, water-oil emulsions from slop oil recovery, tank cleanings, sludges developed in flotation clarifiers in processing sanitary waste water treatment, low-strength caustic and other miscellaneous waste liquids having a typical composition like that shown in Table 2. Low-strength caustic which is too dilute to be recovered must be fed to the incinerator. In addition to hydrocarbons, these sludges also contain some sulfur, iron, aluminum and phosphates as well as alkali elements such as calcium, magnesium and potassium. Accumulation of products from these sludges converted the original silica sand bed by attrition to particles of chemical solids in pellets of nodule form (Fig. 10) composed mainly of aluminum oxide derived from alum used for flocculent in waste water treatment, and also oxides, sulfates, carbonates and iron elements which impart a red color to the pellet. A typical analysis of this bed material is shown in Table 3. A bed temperature of 1,270-1,300° F must be maintained to avoid bed softening or melting.

Exchanger contamination. Ash carryover and vapor-

Fig. 10—Fluid bed product derived from normal plant operation.

Fig. 11—Fluid bed incinerator at the Amoco, Whiting, Ind., plant.

ized compounds sometimes deposit in the transfer line and on exchange surface such that under most severe conditions, flow is seriously impaired. Relatively soft ash deposits build up in the hot gas side of the gas heat exchanger tubes. A possible cause is a small amount of liquid phase (estimated at 0.1 to 0.2 percent) in the fine ash in the incinerator exit gases which causes dust to stick to exchange surface. Carryover is possibly caused by a eutectic mixture present at the temperatures involved.

Solution of this problem lies in selection of exchanger and piping design and in proper control of reactor temperature to avoid eutectic conditions.

PLANT INSTALLATIONS

Six fluid bed incinerators (Table 1) are in operation world-wide for disposal of refinery sludges. In addition, there are a number of other installations in services outside the HPI.

Sludges produced in a three-train liquid waste treatment operation at the Mobil Oil refinery in Karlsruhe, Germany, are processed in combination with API oil separator sludge, flotation clarifier skimmings and waste-activated sludge in a hot windbox incinerator. The sludge solids are 55 percent organic and 45 percent inorganic. They are fed from a surge tank to the fluid bed which is operated with fluidizing and combustion air preheated to 1,000° F by flue gas which is cooled to 1,180° F. The reactor operates at an equilibrium temperature of 1,700° F because of high heat content sludge. Partially cooled incinerator gases from the heat exchanger are scrubbed in the venturi scrubber and cleaned, cooled and exhausted to atmosphere at less than 0.065 grains/scf particulate matter.

The plant operates at 24 to 29 metric tons/day although it was designed for 40 metric tons/day. Reduction results from increased concentrations of oil in sludge compared to 95 percent water sludge for which it was designed.

A cold windbox incinerator is employed at the Gulf Oil plant at Bertonica, Italy, to treat API oil separator sludge, and waste-activated sludge plus dissolved air flotation (DAF) skimmings which are thickened in a common thickener before being fed to the incinerator. Silica sand is used for fluid bed and reactor held at 1,560-1,650° F. Auxiliary heat is supplied by refinery gas burned in the fluid bed.

Another cold windbox type incinerator is used to process sludge containing 83% water, 8% oil, 2% biological solids and 7% inerts (mostly aluminum hydroxide and iron hydroxide) at Esso's Stanic Livorno, Italy, refinery.

The world's largest fluid bed incinerator (Fig. 11) is at the American Oil Co., Whiting, Ind., installation. The unit is designed for a capacity of 338 metric tons of wet sludge of 85 to 95 percent water content. The incinerator is 40 feet high, 20 feet ID at the base, tapering above the bed to 28 feet in diameter at the top. Waste sludge feed consists of several streams including: API separator sludge; water-oil emulsions from slop oil recovery; tank cleaning; sludges developed in flotation clarifiers in processing waste water; low-strength caustic, and other miscellaneous waste liquids. The system is of the hot windbox design with reactor operating at 1,270-1,300° F to avoid softening or melting the bed.

LITERATURE CITED

1 *International Critical Tables*, McGraw Hill.
2 Levin, Robbins, McMurdie, *Phase Diagrams for Ceramics*, Fig. 1773, American Chemical Society, 1964.
3 *Ibid*, Fig. 2923.
4 *Ibid*, Fig. 1118.

Make Fuel from Plastic Wastes

This self-sufficient thermal cracking process recovers over 90 percent residual liquid fuel from plastic wastes

H. Sueyoshi, Mitsui Petrochemical Industries Ltd., and **Yoji Kitaoka,** Mitsui Shipbuilding & Engineering Co., Ltd., Tokyo, Japan

JAPANESE PLASTICS PRODUCTION is estimated by MITI to reach 10 million metric tons by 1975 and about 5.1 million metric tons are expected to be discharged in some form of wastes. Because plastics do not biodegrade, effective and economical methods of treating these wastes must be developed. Such treatment should put plastics into a natural ecological cycle.

Plastic wastes can be divided into two main groups—industrial wastes and city rubbish. Industrial wastes have much higher plastic concentration such as: byproduct and substandard polymers from synthetic resin plants (polyethylene of low molecular weight and atactic polypropylene); scrap from plastics converting plants; one-way plastic containers (bottles); and plastic packaging materials.

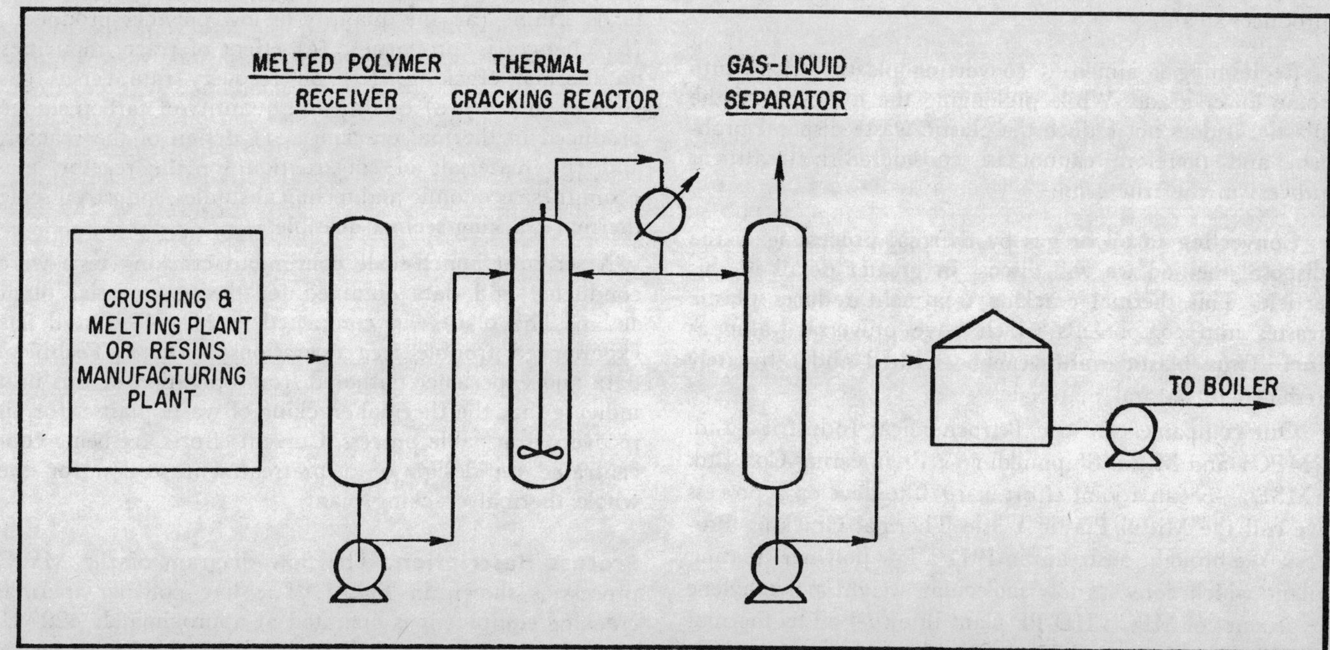

Fig. 1—Schematic of Mitsui's MWC thermal cracking process for plastic wastes

TABLE 1—Typical characteristics and yield of recovered oil from low polymer

Specific Gravity (15/4 °C)	0.77 — 0.79
Viscosity (C.S. 50°C)	1.47
Average Molecular Weight	150
Average Bromine Number	57
Flash point	30°C
Gross Calorific Value (Kcal/kg)	11,000
Oil Yield	90% (weight) of charged plastic waste
Residue Gas Yield*	5% (weight) of charged plastic waste

* Carbon number of 50% of the residue gas is 2 and 3.

TABLE 2—Properties of the low polymer

Density at 150° C	0.76 g/cm^3
200° C	0.74 g/cm^3
Viscosity at 150° C	1000 C.P.
H/C ratio	2.06
Ash	0.068 wt%
Pour point	<115° C

TABLE 3—Composition of gaseous product

Component	Vol%
CH_4	13.7
C_2H_4	9.6
C_2H_6	17.9
C_3H_6	21.2
C_3H_8	18.9
C_4H_8	11.8
C_4H_{10}	7.8
Average molecular weight	38.4

TABLE 4—Properties of discharged oil

Specific gravity 15/4°C	0.7897
Viscosity at 100°C	42.0 C.S.
at 150°C	7.13 C.S.
Pour point	60°C
Flash point	262°C
Ash	0.082 wt%
Average molecular weight	893
Iodine number	14.6

TABLE 5—Composition of liquid products obtained by thermal cracking of waste polystyrene and atactic polypropylene

Polystyrene	%
Lowers	0.06
Benzene	0.15
Toluene	15.66
Ethylbenzene	28.89
Cumene	6.41
n-Propylbenzene	0.08
Styrene	28.50
α-Methylstyrene	13.02
Highers	7.23
Atactic Polypropylene	**%**
C_6^-	16.0
C_7	1.8
C_8	3.3
C_9	22.2
C_{10}	3.2
C_{11}	7.0
C_{12}	3.2
C_{13}	9.8
C_{14}	3.1
C_{15}	3.4
C_{16}	3.8
C_{17}^+	23.2

Since separating plastic wastes mingled in city rubbish is so difficult, we have concentrated on treating industrial plastic wastes. Three types of treatment are currently being developed.

Incineration is the simplest method and can be done so as to recover heat energy. We have been operating successfully such an incinerator to dispose of atactic polypropylene since 1969 at our Chiba Works which produces steam.

Reclaiming is aimed at converting plastic wastes into some lower grade. While prolonging the life cycle of the plastic, it does not reduce the plastic waste disposal problem and therefore cannot be considered a treatment process in the true sense.

Converting to oil or gas by thermal processing is the disposal method we will discuss in greater detail in this article. This thermal cracking treatment reduces plastic wastes into components which have universal value as fuel. Thus, plastic wastes can be treated and ultimately reduced to natural materials.

Our companies—Mitsui Petrochemical Industries, Ltd. (MPC) and Mitsui Shipbuilding & Engineering Co., Ltd. (MSE)—began a joint effort in 1970 to develop a process we call the Mitsui Plastic Waste Thermal Cracking Process. We brought onstream in 1971 a low polymer treating plant which converts low molecular weight polyethylene byproduct of MPC's HD-PE plant into fuel oil by thermal cracking at the rate of 36 tons per day.

Formerly, waste polymer had been flaked and bagged for transport 2 km to an incinerator. Oil recovery seemed more desirable since it could be stored and used as needed for fuel whereas the steam from the incinerator must be used as generated. Also, the economics of the oil process were better.

Initial studies. Before developing the thermal cracking unit, studies and laboratory experiments were conducted to determine (a) the quantity of low polymer produced; (b) changes in properties; (c) effect of trace impurities on thermal cracking; (d) oil recovery rate versus gas generation rate; (e) quality and quantity of carbon sludge produced in thermal cracking; (f) design of the reactor, and (g) materials of construction for the reactor, etc. From these economic and technical studies, industrial scale thermal cracking seemed feasible.

A series of bench-scale continuous cracking tests were conducted and data obtained for the commercial plant design. The plant was completed in May 1971 and has experienced trouble free operations to date. Technical data and experience gathered from operation of this unit indicate that the thermal cracking of waste plastics for oil recovery is a viable process. Current efforts are being concentrated on design of a pretreatment section for the whole thermal cracking plant.

Process description. The flow diagram of the MWC process is shown in Fig. 1. The low polymer thermal cracking equipment is operated at approximately 420° C. Byproduct gas is flared. Product oil is mixed with heavy fuel and used as fuel in the power plant. No single full-time operator need be assigned to this plant as it is fully instrumented and capable of automatically shutting down in case of emergency—power failure, cooling water failure or clogged piping.

Carbon sludge produced in the thermal cracker is continuously removed by an incorporated system and we have not experienced forced shut down due to carbon trouble. Fuel for the thermal cracker is supplied completely from the residual oil make.

Product oil yield is approximately 90 percent based on weight of feed after subtracting fuel requirements of the furnace. Typical specifications of the oil is shown in Table 1 and is good quality as a fuel.

Industrial plastic wastes disposal. This check list must be consulted when considering disposal of industrial plastic wastes by the MWC process:

1. Plastic wastes must be easily collected and separated;
2. Corrosive or poisonous gases—hydrogen cyanide and hydrogen chloride—must not be generated;
3. The yield must be high—80 percent is the minimum, and
4. Foreign materials, e.g., metal and glass used in composite molding must be excluded.

These factors not only create technical problems but are detrimental to good economics of the thermal cracking process.

When the disposal process is applied to plastic crates, carboys, converter shop scraps, etc., collection and sorting are easy, but care must be taken to exclude harmful plastic materials and impurities. In cases where impurities of many kinds and in large quantities have to be removed, the cost of removal may run fairly high.

Beer bottle crates of polyethylene or polypropylene are first crushed into chips, then charged to the hopper of an extruder which functions as the melter and metering pump. The molten material is fed continuously into the reactor. A block diagram of this process is shown in Fig. 2. Experiments have shown that HD-PE is the same as low polymer for cracking conditions and products. Poisonous metallic pigments contained in colored crates cause difficulties and methods for treating these metals are now under study.

Polyvinyl chloride materials are not suitable for the process as the theoretical yield is about 40 percent and

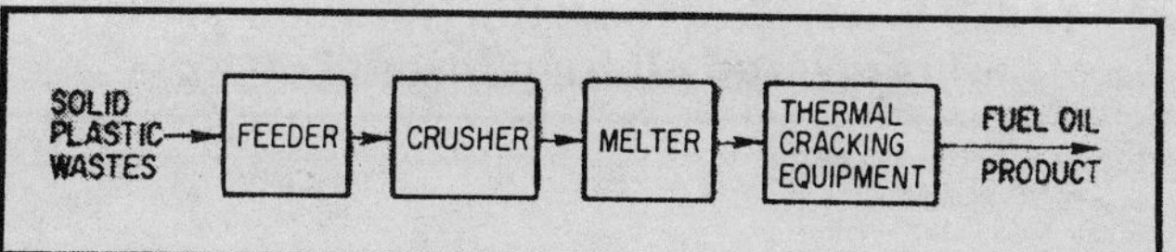

Fig. 2—Block diagram for thermal cracking of industrial plastic wastes

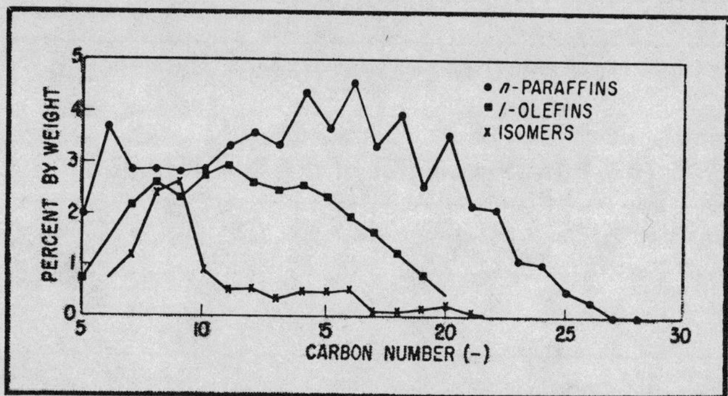

Fig. 3—Gas chromatographic analysis of oily product from cracking

the chlorine content of the PVC remains as organic chlorides in the oil which would corrode the boiler when used as fuel. Also, poisonous metals in PVC, e.g., Cd, Pb, etc., adversely affect the thermal cracking process.

Plastics in city rubbish. Because of the difficulty in separating the plastics by the various types and kinds, the process does not seem to be applicable to city rubbish disposal. However, as the plastics content of rubbish grows to more than 50 percent of total plastic wastes studies are underway to find methods of sorting and disposal of these waste plastics. Small scale tests are being conducted on sorting the plastic wastes by types at the household for pickup and disposal.

Bench scale test results. While the commercial plant has been operated on low polymer (see Table 2), details from some bench scale tests will be of interest. Table 3 shows the composition of the gas product. Table 4 shows properties of the residual oil.

Other plastic wastes have been cracked and Table 5 gives the composition of liquid products obtained from cracking polystyrene and atactic polypropylene.

Cracked oil from low polymer in bench scale tests ranged from C_5 to C_{28}, all straight chain (see Fig. 3). For this run, the *n*-paraffin, 1-olefin and isomers ratio is 6:3:1 and the peaks of carbon atom number distribution curve at C_{15}, C_{11} and C_9 respectively.

Use Land Farming for Oily Waste Disposal

With disposal of large amounts of solid biological and oily wastes, land farming is preferred because of its environmental advantage and reduced use of energy resources as compared to incineration

G. W. Grove, Exxon Research & Engineering Co., Florham Park, N.J.

DESPITE BEST EFFORTS, a certain amount of oily and biological sludge must ultimately be disposed upon the land. Options include:

1. Direct landfill

2. Incineration and landfill of the ash, and

3. Land farming (biodegradation of oil in the soil).

Land farming is the repeated controlled application of oily or biological sludge to a given soil and the promotion of naturally occurring microbial assimilation to convert the hydrocarbons and organic matter to the end products of CO_2, H_2O and increased humus content of the soil.

Land farming of oily refinery sludges has been practiced in the U.S. for at least 25 years but only recently has the process been studied in a scientific manner. A properly operated land farm with treatment of surface water runoff and monitoring of subsurface water is the preferred environmental practice compared to the alternatives.

It is estimated that 7% of oily refinery sludges disposed of in this manner in 1976 will increase to 34% in 1983.

The objectives in refinery waste management are to minimize oil losses and recover oil from waste streams. However, there are large quantities of effluent water to be treated to remove low levels of oil, silt, rust and other contaminants. Typically, this is accomplished by a combination of processes, e.g., gravity settling, air flotation, biological oxidation and granular media filtration. These processes yield sludges containing 1-5% oily solids.

Over the last several years, as refineries have installed a large number of wastewater treating facilities, the sludges have increased tremendously. While these sludges are concentrated by centrifugation, filteration, etc., large amounts of semi-solid oily wastes remain for disposal. Present technology makes further oil recovery from these wastes impractical. Even with total recycling of effluent water, these same sludges would result. In fact, sludge quantities would probably increase since removal of dissolved salts might be required before the water could be recycled.

In comparing the disposal alternatives (mentioned earlier) from an environmental and cost standpoint, the conclusion is that land farming is the preferred disposal technique for many oily and biological petroleum sludges. Landfilling and incineration are conventional methods for municipal sludge disposal and it is not surprising that these were first proposed for waste disposal. At first there was no provision for use of land farming. More recently this has been changing, due in part to efforts by the API and member companies in providing land farming data to regulatory agencies.

This is also true for Europe where CONCAWE are presently studying land farming. Sludge farming is already being done in the U.K., The Netherlands, Sweden, Denmark and France. As in the U.S., the European locations are working with agricultural colleges and government agencies to assure that the process is being carried out in an environmentally sound manner. There are no known cases in the U.S. or Europe where land farming has been discontinued because of environmental hazards.

ALTERNATIVES COMPARED

Landfilling is the least expensive and easist method of disposal. Costs range from $6-12/ton of waste depending upon land value and hauling distance. This corresponds to the common sanitary landfill used for municipal garbage. The wastes remain essentially unchanged and because of frequent soil cover, the site soon becomes exhausted.

Landfilling is satisfactory for many inert wastes. However, for oily and biological refinery wastes (because there is very little oxygen in these sludges) they do not degrade. The use of the land is severely restricted because the layers of oily sludge remain for many years. There is a continued potential for leaching oil and soluble metals from the site. If the oil has been mixed with garbage, the leaching of metals is a real threat since organic acids from the garbage decomposition will solubilize the metals.

Incineration and ash landfill. Fluid bed incinerators are most commonly used for refinery wastes. Environmental problems accompanying fluid bed incineration include:

1. Fine solids in the sludge and attrition fines from the fluid bed are scrubbed from the incinerator stack. Since scrubbers can only remove a portion of the fines, the smallest and environmentally more important particles go out the stack and are distributed over large areas.

2. The ash containing most of the metals (except those escaping through the stack) must still be disposed of on the land.

3. Auxiliary fuel is required to incinerate oily and biological refinery sludges. This fuel might otherwise be put to a more beneficial use.

Fig. 1 shows a typical fluidized bed incinerator scheme. Sludge sprayed above the bed is to cool the combustion zone. The fluidizing/combustion air is preheated to about 1200° F (650° C). All the required heat cannot be added in this manner since higher temperature would accelerate the oxidation rate of the metallic bed support. Additional fuel is added through lances in the bed to maintain a bed temperature of about 1500° F (800° C). Lower bed temperatures cause bed bogging due to sodium salts in the sludges.

A particular incinerator has a bed area of 135 ft.^{2} (12.5 m^2) and was designed for a feed rate 17,600 lbs./hr. (8 m^3/hr.) of sludge. The maximum feed rate that can normally be achieved is about 13,000 lbs./hr. (6 m^3/hr.). This 6 m^3/hr. limit is reached because of the amount of fuel which must be added in the bed to maintain bed temperature. Due to poor distribution of this heat in the bed, fuel added in this manner rapidly overheats the combustion zone. Further increase in combustion zone temperature would damage the refractory in that area.

However, another limit is reached at an even lower feed rate of 8,000 lbs./hr. (4 m^3/hr.). At this feed rate the quantity of fine particles which are not captured by the scrubber exceeds 0.057 gr./ft.^{3} (130 mg/Nm^3). This is a typical incinerator air emission regulation. A solution to the problem of overheating the refractory would be to dewater the sludge. The higher calorific value of the sludge would then require less auxiliary fuel in the bed and sludge could be fed to the incinerator at a faster rate. But dewatering the sludge also increases the concentration of solids in the feed and therefore the amount of solids

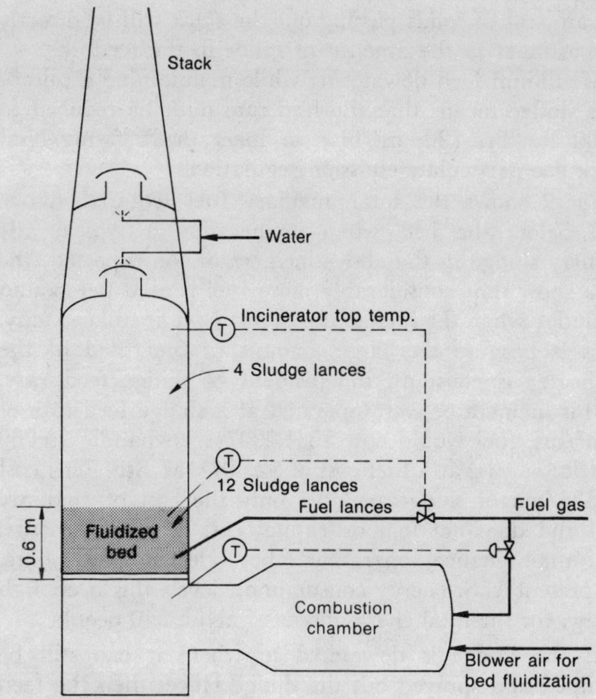

Fig. 1—Typical fluid bed incinerator schematic.

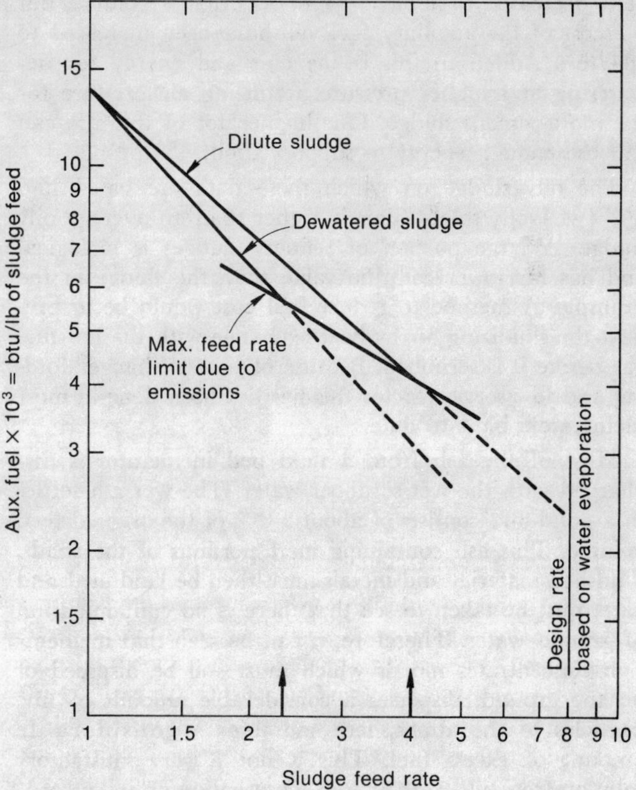

Fig. 2—Sludge feed rate versus auxiliary fuel required.

going out the stack. Unless there is a particle size change, the amount of solids passing out the stack will be directly proportional to the amount of solids in the feed.

Maximum feed dewatering while maintaining a pumpable sludge means that the feed rate must be reduced to 5,500 lbs./hr. (2.5 m^3/hr.) to meet the 130 mg/Nm^3 stack gas particulate emission regulation.

Fig 2 shows the total auxiliary fuel required, above and below the bed when incinerating a typical oily refinery sludge in the above incinerator. As expected, the data show that considerably more fuel is used per pound of sludge when the incinerator is not used at full capacity. This is because the large amount of fuel used in the preheater is constant, independent of sludge feed rate. If this incinerator were operated at a sludge feed rate of 4 m^3/hr., fuel would cost $631,000/yr. to handle 25,000 metric tons/yr. at a fuel cost of $2.80/MM Btu. This cost of $25/ton of sludge includes only the cost of auxiliary fuel and does not include capital cost or other expense, e.g., maintenance, operating labor, electric power, etc. At present U.S. energy consumption levels this is enough energy for the total energy needs of about 540 people.

If the sludge is dewatered to where it can still be pumped and sprayed out the sludge lances then the feed rate must be reduced to meet the emission regulation. There is less sludge, but the particulate concentration in the stack gas remains the same and yearly fuel cost is only reduced to $508,000/yr. The amount of sludge will have been decreased to about 50% of its original volume, but the cost of the auxiliary fuel will have been increased to $34/ton. Added to this is the cost and energy for dewatering and higher pressure atomizing air pressure for the more viscous sludge. One incinerator of this size can still use enough energy to support about 450 people.

The oily sludge on which these data are based had 8% (wt.) oily solids. This is higher than an average oily sludge. A large portion of refinery sludges is biological and has far lower calorific value than the sludge in the example. A method to reduce fuel cost would be to preheat the fluidizing air by heat exchange with the hot flue gas before it is scrubbed. Because of heat exchanger fouling and low service factor this has not been done in most incinerators built to date.

Most of the ash from a fluid bed incinerator is discharged with the wet scrubber water. The wet ash settles in a pond and consists of about 5-8% of the original feed volume. This ash containing inert portions of the solids, fluid bed material and metals must then be landfilled and care must be taken to see that there is no contamination of ground water. Therefore, it can be seen that incineration concentrates metals which must still be disposed of on the ground, disperses a considerable amount of fine particles to the atmosphere and uses a considerable amount of excess fuel. This is not a very satisfactory solution from either an energy conservation or an environmental standpoint.

Land farming. The biodegradation of oily sludges (commonly called land farming) is the repeated application of oily sludge to a given soil and the controlled promotion of naturally occuring microbial assimilation which converts the hydrocarbons to end products of CO_2, H_2O and increases humus content of the soil.

Land farming is the controlled application of a known quantity of waste oil to a given plot of land which has been selected and prepared to receive the sludge. The site is dedicated for the purpose of land farming and like any commercially developed land is not intended for agricultural farming in the future. Once sludge is applied, the conditions of moisture, aeration and supply of nutrients are controlled to promote the microbial assimilation in the soil. Likewise the site is monitored to prevent contamination of surface and subsurface water.

In land farming, a fairly level site is selected to prevent erosion. If necessary, the area is diked and all runoff from sludge and rain is collected for evaporation, treatment or reapplication to keep the land farm site moist. Most soil types are suitable although sand might be added to silty soils to increase the amount of air and water the silty soil can hold. Just as for farming crops, a wide range of soils is suitable.

The sludge is spread on the land generally using sludge trucks with distribution pipes. Subsurface injection may also be done by using plow-like devices which inject the sludge 4-6 inches below the soil surface. This is more commonly done with sewage sludge where odor may be a problem.

After application, some of the water in the sludge is allowed to soak into the soil or evaporate. This makes it easier for the mixing/aeration equipment to maintain traction. Typical farm type disc aerators are commonly used. Aeration allows oxygen to enter the soil for the naturally occurring soil bacteria. It also increases the ability of the soil to hold moisture. From time to time the soil pH and nutrient level are checked and lime or fertilizer added. Lime is added to maintain soil pH above 6.5. This maintains the metals as precipitates and minimizes their being carried away with water runoff. Fertilizer may be added to increase bacterial growth and enable more sludge to be applied to the soil.

Land farming has been practiced by refineries since 1954 and there has been no evidence that the naturally occurring biological degradation diminishes with time. Recent reported costs have been about $12-18/ton of sludge depending upon the land value. Other than use of the land to grow crops for eventual human consumption, there should be no restrictions on future use of land. The land can be used for construction and since the humus content of the soil is increased it is better suited for growing trees and grass.

Thus, it can be seen that landfilling is the lowest cost alternative, but since the oil remains in the soil it presents a continuing potential for ground water pollution and the site cannot be used in the future.

Land farming and incineration are alike in that they convert oily wastes to CO_2, H_2O and ash. However, a sludge incinerator requires the use of our limited supply of energy and natural resources to build. Also, it continues to consume energy and resources to operate. It causes pollution in its manufacture and continues to cause pollution in its operation.

ACKNOWLEDGMENT

Originally presented before the API Midyear Refining Meeting, May 10, 1978, Toronto, Canada.

Part IV.

Air Control

Source/Control of Air Emissions

Where do you expect to find pollution? What can be done about stopping it when it is found? This review is a good check for any plant

H. F. Elkin and **R. A. Constable,**
Sun Oil Co., Philadelphia, Pa.

FACTORS influencing refinery atmospheric emissions include capacity, crude type, processing scheme, control measures, and maintenance and housekeeping practices. The potential sources of specific emissions are shown in Table 1.

In crude separation the use of barometric condensers in vacuum distillation can release noncondensable hydrocarbons to the atmosphere. Regeneration of catalyst in cracking by controlled combustion can release unburned hydrocarbons, carbon monoxide, ammonia, and sulfur oxides although these gases are now commonly rendered less objectionable through the use of CO boilers. Catalyst handling system can also be the source of discharge of catalyst fines, but conventional mechanical or electrical separation equipment has proved successful for control of this material.

The molecular rearrangement processes—alkylation, reforming, polymerization and isomerization—are significant in that they handle volatile hydrocarbons; and tight equipment such as valves and pumps becomes of greater emission control importance than in the heavier oil processes yielding fuel oils, lubes or asphalts.

Pollution aspects of treating operations are concentrated on methods of disposing of the spent chemicals and extracted impurities such as spent acids, spent caustics and hydrogen sulfide. Air agitation and blowing, vessel vents, drains, valves and pump seals are possible sources of loss or leakage of sulfur bearing compounds and hydrocarbons in any of these processes or handling procedures.

A number of control measures are practiced throughout the industry for control of emissions of sulfur, hydrocarbons, particulates, etc. Sulfur, by far the most publicized impurity found in crude oil, exists as sulfides, mercaptans, polysulfides and thiophenes. Crudes vary significantly in sulfur content, thus, the processing scheme must be able to handle the maximum sulfur contained in any crude.

To limit atmospheric emissions of SO_2, a refiner must remove H_2S from fuel gas streams, maximize use of natural gas as a fuel, blend heavier high-sulfur internal fuels with desulfurized distillates and maintain a well operated, high efficiency sulfur recovery plant.

TABLE 1—Potential sources of specific emissions from oil refineries[7]

Oxides of sulfur	Boilers, process heaters, catalytic cracking unit, regenerators, treating units, H_2S flares, decoking operations.
Hydrocarbons	Loading facilities, turnarounds, sampling, storage tanks, waste water separators, blowdown systems, catalyst regenerators, pumps, valves, blind changing, cooling towers, vacuum jets, barometric condensers, air-blowing, high pressure equipment handling volatile hydrocarbons, process heaters, boilers, compressor engines.
Oxides of nitrogen	Process heaters, boilers, compressor engines, catalyst regenerators, flares.
Particulate matter	Catalyst regenerators, boilers, process heaters, decoking operations, incinerators.
Aldehydes	Catalyst regenerators.
Ammonia	Catalyst regenerators.
Odors	Treating units (air-blowing, steam-blowing) drains, tank vents, barometric condenser sumps, waste water separators.
Carbon monoxide	Catalyst regeneration, decoking, compressor engines, incinerators.

Many relatively small sources of sour gases from crude distillation or vacuum distillation units are generated at low pressure and require compression for recovery and treatment if venting to flares is to be avoided.

Nearly all processes generate some gases which contain hydrogen sulfide or other low molecular weight sulfur compounds. However, the major portion of these gases are generated in thermal and catalytical cracking, hydrodesulfurization, reforming and hydrocracking. These gases, after treating, are used as fuel in process heaters and boilers, and result in sulfur oxide emissions if sulfur compounds are not removed. Fortunately, economic techniques for removal of hydrogen sulfide and other acid gases are well established.

Refiners are confronted with the same difficulties in controlling SO_x emissions from burning residual fuel oil in stationary units as other industries. Processes for removing SO_x from stack gases are still in the developmental phase and cannot be considered for refinery applications at this time.

Sulfur dioxide emissions from residual fuel oil combustion in refineries will become more difficult. Customer demand for low sulfur fuel oil is increasing, while the sulfur content of residual fuel oil is increasing as refineries convert more "bottom of the barrel" to other products. The average sulfur content of all residual fuel oil produced in the United States increased from 1.2% in 1950 to 1.75% in 1970, while the yield decreased from 20 to 7%. A recent EPA report[20] stated that only the larger refineries (100,000 + bpcd) can support hydrogen production facilities necessary to desulfurize residual fuels below 1% sulfur.

Sulfur removal from whole crude is not widely accepted economically; however, desulfurization of products and intermediate stocks is widely used. Atmospheric residual material is desulfurized, but not to low levels (0.3%) called for by regulations in certain metropolitan areas. Residual vacuum-tower bottoms is even more difficult to desulfurize. Consensus today is that technology is not sufficiently developed to make this a feasible route to low-sulfur heavy fuels. Caribbean refineries desulfurize vacuum gas oil to 0.3% and back blend with vacuum resid, desulfurized atmospheric bottoms, and/or naturally occurring low-sulfur atmospheric bottoms to satisfy the 1% sulfur fuel demand.

A procedure for removal of sulfur dioxide from catalytic cracking flue gases has not been developed. However, sulfur oxide emissions are reduced (75-80%) when cracking unit feed is desulfurized by hydrogenation. Hydrotreating feedstock, usually initiated to improve yield of high demand products, is expected to increase and should result in a major reduction in emissions of sulfur oxides.

Waste acids and sludges from product acid treatment and other refinery processes such as alkylation, were frequently burned in earlier years. This procedure, which produced large local concentrations of SO_x, largely has been discontinued. Most waste acids and sludges are reprocessed by acid manufacturing units for recovery of acid values.

Sour water strippers are also a potential source of sulfur emissions.

Storage is potentially the most important source of hydrocarbon emissions in the HPI. Vapors can be emitted when storage tanks "breathe," when vapors are displaced during filling, and when liquids evaporate. Pressure tanks, fixed-roof tanks, floating-roof tanks, and conservation tanks provide closed storage. Pressure tanks, such as spherical storage tanks, will withstand pressures exerted by their contents and are themselves a control measure. Fixed-roof tanks often have open vents to the atmosphere. Floating-roof tanks have pan, pontoon, or double-deck floating roofs. Conservation tanks are connected to gas storage systems and have internal flexible diaphragms, or floating plastic blankets.

Waste gases are generally collected and used as fuel or as raw material for further processing. Sudden or unexpected upsets in process units and scheduled shutdowns, however, can produce gas in excess of the capacity of the gas-recovery system. A relief system for disposal of emergency and waste refinery gases normally consists of a manifolded pressure-relieving or blowdown system and a system of flares for the combustion of the excess gases, or both. These systems are also used in evacuation of units during shutdowns and turnarounds.

Smokeless flare combustion can be obtained by the injection of an inert gas to the combustion zone to provide turbulence and inspirate air. A mechanical air-mixing system is ideal, but is not economical in view of the large volume of gases handled. The most commonly used air-inspirating material for an elevated flare is steam.

Rupture disks are used to insure that safety relief valves do not inadvertently leak hydrocarbon vapors to the atmosphere.

A waste-water gathering system for a modern refinery typically includes gathering lines, drain seals, junction boxes, and pipes of vitrified clay or concrete for transmitting waste water from processing units to large basins or ponds used as oil-water separators. These basins, sized to receive all effluent water (sometimes even rain runoff) are constructed as earthen pits, concrete-lined basins, and steel tanks.

Organic compounds found in the system can escape to the atmosphere from openings in the sewer system, channels, vessels, and oil-water separators. The large exposed surface area of these separators can result in large hydrocarbon emissions. The most effective means of emissions control is to cover forebays or primary separator sections. Either fixed roofs or floating roofs are acceptable covers. Since over 80% of the floatable oil layer is removed in the covered sections, only a small amount of oil is contained in effluent water, which flows under concrete curtains to the open afterbays or secondary separator sections. In addition to covering the separator, open sewer lines that carry volatile products can be converted to closed, underground lines with water-seal-type vents. Junction boxes can be vented to vapor recovery facilities, and steam can be used to blanket sewer lines to inhibit formation of explosive mixtures.

Catalyst particles, used to crack heavy fractions to lighter components, become coated with carbon and high-molecular-weight compounds. In regeneration, a controlled amount of air is admitted to burn off the coatings causing the formation of CO and hydrocarbon vapors. These emissions can be controlled by incineration using a waste heat boiler, commonly referred to as a CO boiler.

Leakage from pumps can cause organic emissions from the opening in the cylinder or fluid end through which the connecting rod actuates the piston in a reciprocating pump, or where the drive shaft passes through the impeller casing in a centrifugal pump. Since the contact surfaces between packing and shaft in reciprocal pumps are lubricated by a controlled amount of product leakage to the atmosphere, packing seals are undesirable in applications where the product can cause a pollution problem.

Valves are subject to product leakage from the valve stem as a result of the actions of vibration, heat, pressure, and corrosion or improper maintenance of valve stem packing. Obviously, maintenance is the controlling factor in preventing leakage from valves. An effective schedule of inspection and preventive maintenance can keep leakage at a minimum.

Asphalt, normally obtained from select crude oils by means of vacuum distillation or solvent extraction, is made suitable for paving, roofing, or pipe coating by reaction with air. Effluents from air-blowing include oxygen, nitrogen, water vapor, sulfur compounds, and hydrocarbons in the forms of gases, odors, and aerosols. Control of these emissions is accomplished by gas scrubbing and incineration, singly or in combination.[7] Where removal of most of the potential air pollutants is not feasible by scrubbing alone, the noncondensables must be incinerated.

Emissions from acid treating can be reduced by substituting continuous mechanical mixing for batch-type agitators that employ air-blowing for mixing. Acid regeneration can also be used instead of the hydrolysis-concentration method of acid recovery. Gases emitted during acid-sludge recovery can be vented to caustic scrubbers to remove sulfur dioxide and odorants. Gases from scrubbers can then be vented to a firebox or flare. For new installations, acid treatment can also be replaced by catalytic hydrogenation or by other, more effective processing.

In doctor treating, doctor solution can be steam-stripped to recover hydrocarbons prior to air-blowing for regeneration. Effluent from air-blowing can then be incinerated to destroy hydrocarbon vapors.

In the disposal of spent caustic, entrained hydrocarbons are often removed by stripping with inert gases. The vapors can be vented to a flare or to a furnace firebox.

Proper maintenance of burner equipment is essential to proper control of boilers and process heaters to assure efficiency of combustion and minimum emission of unburned hydrocarbons.

Emissions of hydrocarbons from vacuum jets can be controlled by venting the discharge to blowdown or vapor-recovery systems. They may also be vented to the firebox of a boiler or heater.

Loading into tank trucks, tank cars, and drums can result in hydrocarbon vapor loss by displacement or evaporation. Careful operation to minimize spillage, and vapor collection and recovery equipment will control vapor loss from this operation. The volume and composition of vapors produced during loading operations is greatly influenced by the type of loading or filling employed. Overhead loading may be either splash or submerged type. In splash filling, the outlet of the delivery tube is above the liquid surface during all or most of the loading. In submerged filling, delivery tube outlet is extended to within 6 inches of the bottom and submerged during most loading. Bottom loading is accomplished by connecting a swing-type loading arm or hose at ground level to the underside of tank vehicles. Splash filling generates more turbulence and therefore more hydrocarbon vapors than submerged filling. In bottom loading turbulence is minimized and less vapor is produced.

To control vapor emissions from loading operations, devices can be installed to collect the vapors at the tank vehicle hatch. Bottom loading permits easier collection of displaced vapors because filling and vapor collection lines are independent of each other. The vapor collection line is a flexible hose or swing-type arm connected to a quick-acting valve fitting on the vehicle dome.

Vapors collected during loading may be used as fuel recovered as product. When fired heaters or boilers are available, displaced vapors may be directed through a drip pot to a small vapor holder, which is gas-blanketed to prevent formation of explosive mixtures. The vapors are drawn from this holder by a compressor and discharged to the fuel gas system. If loading facilities are near the refinery, a vapor line can be connected from loading to an existing vapor recovery system through a regulator valve. If not, packaged units may be used to recover vapors.

Odors associated with minor releases of unusual compounds are probably the most perplexing problems associated with refinery operations. Some malodorous compounds are perceptible at concentrations a fraction of a part per billion. As a result, extremely small concentrations, well below toxic or harmful levels, cause complaints from nearby residents. Since the volume is too small to be a significant weight emission and the compounds are usually destroyed by oxidation after a short time in the atmosphere, odor control is largely a local nuisance.

Odorous gases—hydrogen sulfide, mercaptans, phenolic compounds, naphthenic acids, organic sulfides, nitrogen bases, aldehydes and ammonia—can be released from process steam condensates, drain liquids, barometric condenser sumps, sour volatile product tankage, and spent caustic solutions from treating operations.

Odorous compounds in steam condensers can be removed by stripping with air, flue gas, or steam and the gases burned in furnaces or boilers. Drain liquids can be collected in closed storage systems and recycled to the process. Barometric condensers can be replaced by more modern surface condensers and noncondensables may be burned in process heaters or separate incinerators.

Spent caustic can be degasified, neutralized with flue gas and/or stripped before disposal. Sulfides can also be removed from sour process water and spent caustic solutions by air oxidation to thiosulfates and sulfates.

Increased use of hydrogen treating is simplifying the nuisance problem of malodorous compounds. Hydrogen treating destroys most foul compounds and converts them to more desirable products and hydrogen sulfide.

A major problem in odor control is prevention of even extremely small leaks from pumps, valves and compressors. Good maintenance practices and good housekeeping must be practiced. Also, if a refinery is to be maintained essentially odor-free, development of improved equipment may be required.

Major sources of particulate matter are catalyst regenerators, airblown asphalt stills, and sludge burners. Lesser sources include fired heaters, boilers and emergency flares. From 100,000 to 300,000 cfm of hot, dusty gases are vented from a large catalytic cracker regenerator. Dust collectors, as well as carbon monoxide waste heat boilers are often used to control air pollution. In typical installations 2 or 3-stage cyclones are located in regenerator vessels of FCC units for catalyst recovery. In some cases, external cyclones are installed to reduce the particulate content of flue gases leaving regenerators. Catalyst dust losses from these regenerators ranges from about 100 to 1,000 pounds per hour depending on unit size, age and design.[15]

Electrostatic precipitators are widely used to collect fine particles from regenerator exit gases. Some refiners have reported catalyst dust losses as low as 30-60 pounds per hour although typical installations have higher emission rates. One particular problem in recovering FCC catalyst dust via electrostatic precipitators is high resistivity of particles and extremely small particle size. Most installations must use ammonia injection to properly condition dust and lower its resistivity.

CO sources include catalyst regenerators, coking operations, blanketing gas generators, flares, boilers and process heaters. Only moving-bed catalyst regenerators and fluid cokers emit significant amounts of CO. Use of waste heat boilers to burn CO to CO_2 adequately controls CO emissions from these units.

A fluid-coking unit, like a FCC unit, has a bed of fluidized solids to transfer heat to partially vaporized feed material. Solid coke particles, products of the cracking reaction, are burned under conditions of limited air making the flue gas very rich in CO. This gas can be burned in CO boilers like those used for catalyst regenerator flue gas.[16]

NO_x emissions result from combustion of fuel in process heaters and boilers and from internal combustion engines used to drive compressors and electric generators. NO_x is also released from catalytic-cracking regenerator and from CO boilers. Combustion modifications such as low excess air, flue gas recycle, two-stage combustion, and steam or water injection may be applicable to both boilers and refinery heaters.

ACKNOWLEDGMENT

Extracted and condensed from a paper presented at the Air Technical Forum, national AIChE meeting, Dallas, Texas, February 1972.

LITERATURE CITED

1 API News Release, Jan. 10, 1971.
2 API, Division of Statistics.
3 A. C. Stern, ed., Air Pollution Volume III, 2nd Edition 1968, Academic Press, New York, p. 100.
4 "Emissions to the Atmosphere from Petroleum Refineries in Los Angeles County," Joint District, Federal, and State Project, Final Report No. 9. Los Angeles County Air Pollution Control District, Los Angeles, 1958.
5 S. S. Griswold, et. al., Proc. National Conference Air Pollution Washington, D.C. 1958, pp. 140-146. Washington, D.C.
6 "Air Pollution Data for Los Angeles County," LAAPCD, 1967.
7 "Atmospheric Emissions from Petroleum Refineries; a Guide for Measurement and Control," P.H.S. Publ. No. 763 U.S. Division Air Pollution, Washington, D.C., 1960.
8 E R. Stephens, E. F. Darley, and F. R. Burleson, Proc. American Petroleum Institute, Section III 47 (1967).
9 "Air Pollution and Smog," *Air Pollution Found,* (Los Angeles) (1960).
10 W. L. Faith, *Ind. Wastes* 5, No. 4 (1960).
11 "Control Techniques for Sulfur Oxide Air Pollutants," U.S.D.H.E.W., AP-52, January 1969.
12 "Control Techniques for Hydrocarbon and Organic Solvent Emissions from Stationary Sources," U.S.D.H.E.W, AP-68, March 1970.
13 "Profile of Air Pollution Control in Los Angeles County," Air Pollution Control District L.A. County, January 1969.
14 "Waste Management and Control," Committee on Pollution, (Washington National Academy of Sciences Publication 1400, 1966) p. 11.
15 "Control Techniques for Particulate Air Pollutants," U.S.D.H.E.W., AP-51, January 1969.
16 "Control Techniques for Carbon Monoxide Emissions from Stationary Sources," U.S.D.H.E.W., AP-65, March 1970.
17 "Control Techniques for Nitrogen Oxides from Stationary Sources," U.S.D.H.E.W., AP-67, March 1970.
18 "Environmental Conservation—The Oil & Gas Industries," forthcoming report from the National Petroleum Council.
19 "Manual on Disposal of Refinery Wastes," Vol. II, American Petroleum Industries.
20 "Air Pollution from Fossil Fuel Combustion in Stationary Sources in the United States," EPA, PAC/70.4/07. (1971).

Indexing terms: Air-8,5, Control-2,7,10, Emissions-1,6, Hydrocarbons-3,5, Nitrogen oxides-3,5, Odor-3, Particulates-3,5, Pollution-8,3, Sulfur oxides-3,5.

Blanket Tanks for Gas Control

Improved approach saves energy while avoiding pollution

D. M. Gammell, Engineering Consultant, Arcadia, Calif.

GAS BLANKETING TANKS and loading racks for emission control is effective when using improved techniques. Safety is greatly enhanced, operator attention is minimized and problems previously associated with blanketing are minimized with instrumentation methods now available. Vapors previously lost to the atmosphere are recovered for use and energy is conserved over all.

Improved practices now in operation are being evaluated by field tests to establish empirical design criteria for future applications.

SYSTEM CONCEPT

Natural gas is used for blanketing in a system (Fig. 1) installed experimentally in a southern California refinery and in a similar system installed for recovering vapor from tankage only. Net vapor recovery amounts to about 3 percent of the refinery's fuel demand, although 25 tanks and 2 gasoline truck top-loading racks are involved. This level of usage is established under some of the most stringent conditions in the United States where the average ambient temperature change between day and night is maximum.

The system is less expensive to install and operate than commercial vapor recovery units and eliminates surveillance of vapor collection and emission rates. Net vapor recovered may either be burned as fuel or traded to a utility or fuel consumer.

Tank inhalations actuate self-contained regulators, usually at 0.25 ounces per square inch (osi) vacuum, admitting blanket gas into connected tank ullages before tank vacuum relief valves actuate, usually at 0.5 osi vacuum. One regulator can be used for several adjacent tanks if interconnecting pipe sizes are adequate. Sour crude oil or light straight run distillate vapors are prevented from contaminating finished products or sweet

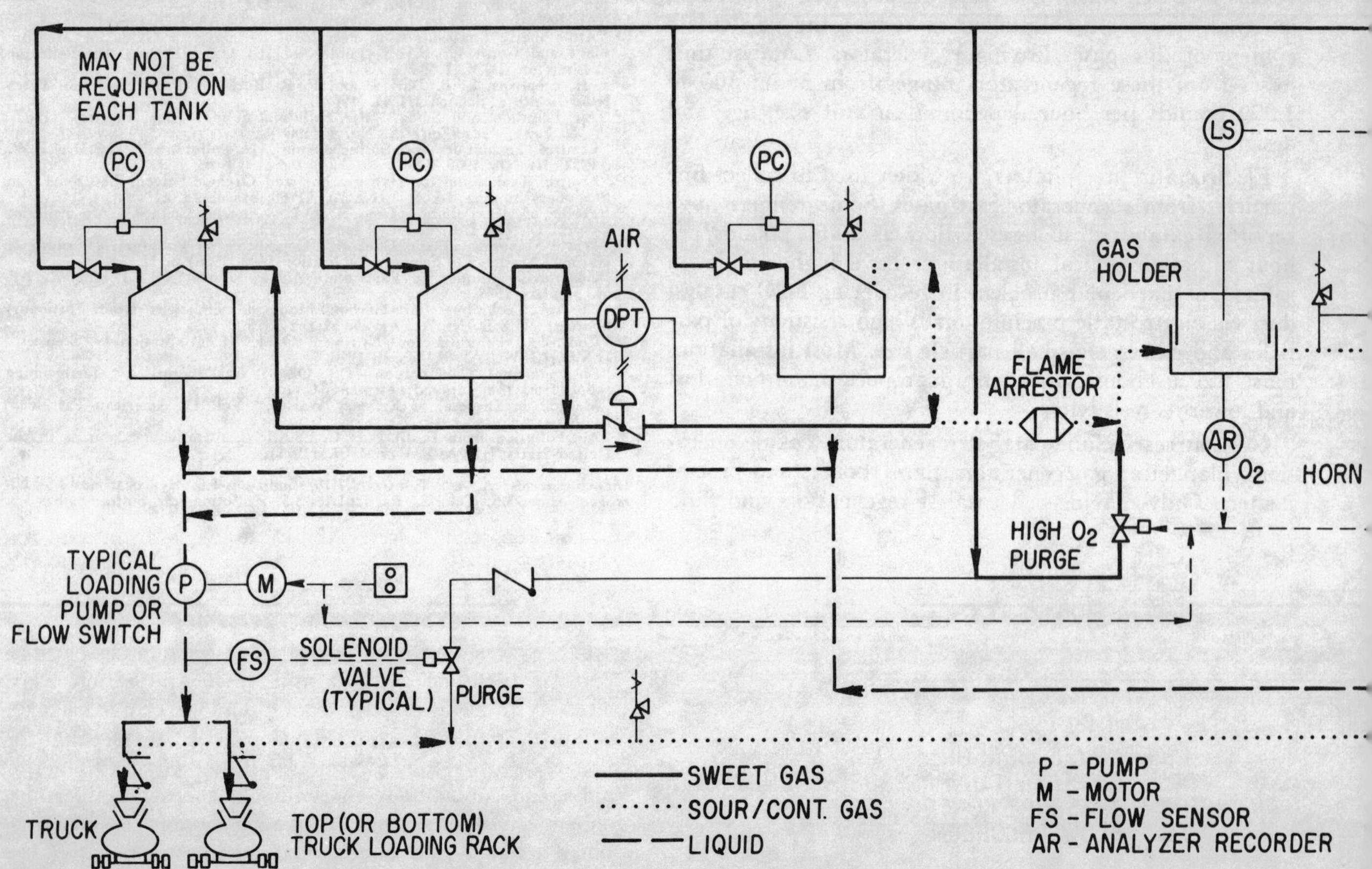

Fig. 1—Gas blanketing system for recovery of fuel.

intermediary storages by a normally closed butterfly valve which opens fully when the upstream pressure exceeds that downstream by 0.1 inch of water.

Tank exhalations breathe into gas holders by the very small difference between the weight of the tank cone roofs (0.5 osi) and that of the gas holder diaphragms (0.2 osi). This pressure difference motivates net exhalations into the gas holder, even through the differential check valve which prevents intertank contaminations and the flame arrester, without lifting the pressure relief valves on the storage tanks, which are usually set at 0.5 osi. A fan, and/or a heavier roof than one of 3/16 inch steel plate, would provide a greater pressure differential, but such has not been found to be necessary in most cases.

A variable volume gas holder is provided so that the compressor and evaporative condenser need not operate continuously, but only to empty the holder as far as tankage is concerned. When the gas holder diaphragm reaches almost full capacity, its level switch starts the compressor and condenser. Surplus holder volume is provided for the special case where net tank exhalations into the holder plus loading rack vapors exceed the compressor capacity. When the holder is evacuated and the diaphragm drops to a low capacity, the level switch stops the compressor and condenser unless the loading rack is in operation. Compression is necessary only to enter the refinery fuel gas system, usually about 40 psig. Any con-

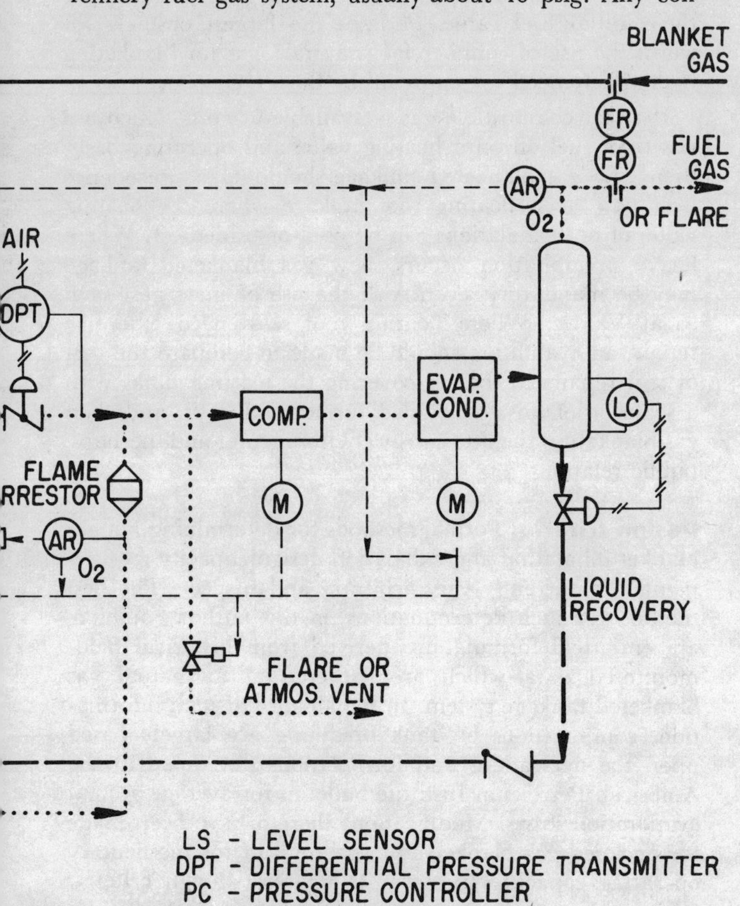

TABLE 1

Circumstance	Floaters	Blanket (Fig. 1)	Blanket with a commercial package
No blanket gas economically available	Yes	No	Rack only
Inert blanket gas economically available	No	Yes	No
Combustible gas but no dispositions economically available	Maybe	No	Maybe
Combustible gas and dispositions economically available	No	Yes	No

densation is returned to storage. Flowmeters are provided to record blanket gas usage and the net gain realized in vapor recovery.

All tanks in the vapor recovery system must be airtight at operating pressures and tank PSVs must operate properly without leaking. An oxygen analyzer continuously records oxygen concentrations in the gas holder, where no oxygen should ever be present.

Loading racks. Fig. 1 also illustrates how truck, rail or ship loading systems can be safely gas blanketed in conjunction with either associated or unassociated tankage to essentially eliminate atmospheric vapor emissions during loading operations. The same compressor and condenser can be utilized if its size is adequate for both tank and loading rack loads.

Because empty vehicle compartments may contain some air, blanket gas is purged through the vapor collection header to the compressor whenever the loading rack is in operation. If oxygen concentrations above some predetermined level occur, an oxygen analyzer actuates a warning to the loading rack attendant who may then elect to do nothing for 10 or 20 seconds in hopes that the warning will cease; stop the compressor and condenser and vent loading rack vapors to flare or atmosphere until satisfactory oxygen levels are reached, (the compressor and condenser will automatically start again and the vent automatically close) or switch off the loading pumps that feed the rack.

A compressor shutdown signal from the loading rack overrides that from the level switch on the gas holder, which is another reason for surplus volume in the holder. Safety relief to atmosphere is provided to prevent pressure build-ups which could structurally damage vehicle loading compartments, loading connections, or etc.

A normally closed butterfly check valve prevents loading rack vapors from entering the gas holder, so that oxygen recorded there is known to come from tankage only.

SYSTEM IMPROVEMENTS

Oil refiners have been averse to gas blanketing because sensitive operating controls and diligent attention were necessary to avoid structural tank damage or gas leakage, explosive conditions could develop with devastating results, tank turnarounds were difficult and vapors from one tank could contaminate the liquid in another tank. Some refiners have phased out blanketing systems for floating roofed tankage and commercial loading rack vapor recovery packages. However, since both of these installations have vapor emission factors above zero, a second look has been taken at the totally enclosed system outlined above.

First of all, gas blanketed tankage is basically much safer than fixed roof tankage without blanket gas. This is true for crude oil, intermediate refinery products and finished products. A hydrocarbon gas mixture without hydrogen which is too rich to burn is about 40 times easier to maintain than one that is too lean to burn. Very reliable, sensitive, self-calibrating oxygen analyzer-controllers have been strategically located in the system illustrated by Fig. 1 to continuously record and control (in two cases) oxygen concentrations.

When oxygen concentration in the gas holder reaches only 10% of the upper explosive limit (UEL), for instance, blanket gas automatically purges the holder until less than that concentration is obtained. If the holder fills before the oxygen concentration is diluted sufficiently, it will be relieved to atmosphere. If oxygen concentration in the loading rack vapor collection header reaches 30% of the UEL, an alarm is actuated for the attendant and, simultaneously, a greater than normal blanket gas purge rate is automatically admitted to the header. A third oxygen analyzer is located after condensations are removed because such removals increase oxygen concentrations in the gas phase. This analyzer-recorder backs up the other two and warns operators of concentrations over 50% of the UEL.

Task pressure-vacuum relief valves do need to be inspected periodically for tight, but not sticky, seat rings-to-knife edges, as should those on any fixed roof tank, and the rim vents and bleeder vents on any floating roof tank. It is perhaps more imperative that tankage in gas blanketed service be generally air tight. Hand-operated butterfly dampers may be installed at each tank vapor collection roof flange to prevent blanket gas losses during tank gaging and sampling operations, but such is not required by code. Modern instrumentation (Fig. 1) has been found to be reliably responsive, including that which prevents contaminations between tanks.

Turnaround procedures have been formalized to prevent a tank from ever containing a combustible gas mixture while it is selectively taken out of and placed into service in the vapor recovery system.

APPLICATIONS

New tankage and tankage-loading installations should consider the gas blanket alternative in view of current ambient air quality regulations, air pollution rules and the availability and disposition of blanket gas.

Gases for blanketing may come from a number of sources; but they should be sweet and not otherwise detrimental to the liquid in the tanks:

- **Combustible blanket gases** (net exhalations to fuel gas)
  Commercial gas from gas utility companies
  Natural gas from private fields
  Treated refinery gas having no hydrogen
  Vaporized liquid petroleum gas (LPG) when absorption not critical.

- **Inert blanket gases** (net exhalations to flare)
  Compressed flue gases
  Nitrogen from inert gas generators
  Bottled nitrogen or CO_2
  Compressed CO_2 exhausts having no hydrogen.

Combustible gases offer the advantage of total heat recovery from net exhalations, which include hydrocarbon vaporizations from the stored and loaded liquids. Inert gases offer the advantage of easier and safer shutdowns for entering into the tanks. The use of combustible gases is usually more economical if such gases are available.

From the oil refiner's point of view, the more blanket gas at profitable fuel oil equivalent prices that is burned necessarily for processing, the more oil there is available to process. From the bulk oil terminal operator's point of view, where fuel gas may not be needed or may not be easy to trade off to a local user, the most economical system that satisfies the environment is best. From the public utility commission's point of view, commercial gas should not be used if any other gas is available. From the air pollution agencies' points of view, the best system is that which reliably emits nothing to the atmosphere.

On balance, the ideal solution is dependent upon the owner's need for fuel, or his ability to trade-off net exhalations, and the long term availability and cost of the blanket gas (Table 1).

Net exhalations from inert gas blankets would normally be flared, but flaring would not normally be economical for net exhalations from combustible gases because of their initial fuel value. Perhaps the largest obstacle towards the use of commercial (natural) gas for blanketing is the likely need for approval by the PUC.

If sweet combustible gas is available at costs which are less than fuel oil on a heating value and operating basis, then new gas blanketed tankage should take precedence over new open floating roof tankage where the heating value of net exhalations can be used or traded-off. Where heavy precipitation occurs, new gas blanketed tankage may be mandatory, even with the use of inert gas, such as at Valdez. Where floating roof seals need extensive repair, an evaluation should be made to compare the cost of seal repairs to that of covering the existing tanks with a sealed roof on a scheduled turnaround basis, and then gas blanketing them to satisfy environmental and, perhaps, public relations.

Design criteria. Formal methods for determining net gas blanket inhalation and exhalation design capacity requirements are currently very arbitrary and suspect. The best method for such determinations, in this author's opinion, are empirical formulations derived from historical field-monitored data which are gathered at an actual gas blanketed tankage system. In the absence of such information, computations on tank breathing are largely based upon the derivations and formulations developed in the American Petroleum Institute bulletins for average annual evaporation losses. Modifications thereto have been made for enclosed gas blanket atmospheres and for momentary operating capacities in order to establish design criteria. Reasonableness checks have been made by independent approaches that substantiate the conservativeness of those derivatives, and the sensitivity of some variables that inherently are assigned arbitrary values have been evaluated.

Tankage. Net exhalations from a vapor recovery system containing several tanks is the concurrent result of individual breathing, V_b, and working, V_w, exhalations. Since tank breathing results from ambient temperature changes, all tanks tend to breathe in and out together. Working exhalations, however, result from independent tank filling rates less emptying rates, expressed as V_w' and V_w'' in cubic feet per minute (cfm), respectively. The net effect, Q, in cfm is:

$$Q = \Sigma \left[(V_b + V_w' - V_w'') \text{ for each tank} \right]$$

The amount of hydrocarbon vapor in Q would be something less than that from cumulative concurrent true vapor pressures in each tank by the amount of non-equilibrium stratification in the tanks.

Breathing exhalations can be approximated by reducing the API formula for average annual breathing losses from cone roof tanks to atmosphere to an instantaneous rate of vapors plus blanket gas, using perfect gas laws and considering an average daily quantity to occur evenly for 8 of 24 hours exclusively. Maximum monthly average daily temperature changes should be determined from data published by the nearest weather bureau. Exhalations are assumed to be 15° F warmer than bulk liquid temperatures due to their residence time in steel tank ullages, vapor recovery piping and the gas holder.

$$V_b = 3.5 \times 10^{-5} \frac{\rho_L (475 + T_3) P (T_2)^{0.5} (D)^{1.73} (H)^{0.51}}{M_v (p)^{0.32} (P\text{-}p)^{0.68}} (F)\ (C)$$

Working losses are simple volume displacements in and out of each tank, unless tank outages cause V_w'' to exceed exhalations $V_b + V_w'$, which results in net inhalations for the system. In this special case, simple outage volume displacement is offset by vaporization to accommodate Raoult's Law, and:

$$V_w'' = \frac{(G)(P\text{-}p)}{7.48(P)}$$

in cfm into tanks.

Loading racks. Total vapor loads from loading rack operations include direct volume displacements plus whatever entrainment and vaporization occur during loading operations. The rate of vaporization in each compartment being filled depends upon:

- Vapor pressure of the liquid
- Total absolute pressure of the compartment
- Amount of vapors already in the empty compartment
- Loading method.

Petroleum products are loaded into vehicle compartments by top (splash) methods, bottom (submerged) methods or combinations thereof. Truck and trailer loading racks may utilize any one of these three methods. Rail cars are usually splash loaded and marine vessels are submerged loaded.

If the compartments to be loaded are already saturated with hydrocarbon vapors, the rate of vapors expelled from the loading rack will only equal the rate of liquid displacement, no matter what loading method is used, disallowing for liquid entrainments. If the compartments are not saturated, however, the rate of expelled vapors will equal the rate of liquid displacement plus a rate depending upon the factors listed above. Submerged loading will cause relatively slight increases, while splash loading causes significant increases, depending upon the degree of unsaturation at the start of loading.[2] A 1973 publication claims that trucks average about 30% saturation at the start of loading.[3] With vapor recovery systems being required at unloading points since then, a 50% saturation was assumed for designing the system illustrated by Fig. 1.

The capacity of a combined tankage and loading rack system is limited by the compressor capacity. The minimum time for a truck and trailer, rail car or ship to be at the loading rack can be expressed as loading time, t_L, plus set-up time, t_s. The minimum set-up time allowed by the capacity of the vapor recovery system, without overloading it, is:

$$t_s\ (minimum) = \frac{V_L t_L}{R\text{-}Q} - t_L$$

wher V_L = cfm of emissions from the loading rack, including purge gas and t_L = time in minutes to load the vehicle. The higher the rack loading pump capacity, the more set-up time there is available. The number of vehicle loadings per hour, N, becomes:

$$N = \frac{60\ (R\text{-}Q)}{V_L t_L}$$

NOMENCLATURE

C = Small tank correction factor[1]
D = Diameter of tank, feet
F = Paint factor[1]
G = Liquid rate into or out of tank, gpm
H = Average annual outage of liquid in tank, feet
M_v = Molecular weight of vaporized tankage liquid
N = Number of vehicle loading set-ups per hour
P = Average annual atmospheric pressure, psia
p = True vapor pressure in psia of bulk liquid at temp. T_3
Q = Momentary net exhalation rate from tankage, cfm
R = Actual compressor capacity, cfm
T_2 = Largest daily temperature change °F in 30-yr. monthly averages
T_3 = Temperature of bulk liquid in storage, °F
t_L = Time to fill vehicle compartments, minutes
t_s = Time to set up a vehicle rig, minutes
V_b = Tank breathing exhalations, cfm
V_L = Loading rack vapors in cfm, including purge gas rates
V_w = Tank working losses, cfm
V_w' = Tank working exhalations, cfm
V_w'' = Tank working inhalations, cfm
ρ_L = Specific gravity of liquid as stored

LITERATURE CITED

1 API Bulletin 2518 on Evaporation Loss from Fixed-Roof Tanks dated 6/62.
2 Air Pollution Engineering Manual, U.S. Department of H.E.W., Public Health Service Publication No. 999-AP-40.
3 Hydrocarbon Emissions from Refineries, API Publication No. 928 dated 7/73.
4 API Bulletin 2513 on Evaporation Loss in the Petroleum Industry—Cause and Control, dated 2/59.

Pick the Right Seal

Emission losses from floating roof seals are less than estimates and vary widely according to seal type

R. E. Hills and **D. K. Neely,** Pittsburgh-Des Moines Steel Co., Pittsburgh, Pa.

EMISSIONS from floating roof tanks vary markedly with seal design and ambient conditions. Furthermore, generally used procedures for predicting emissions are unsatisfactory and usually result in predictions that are several times higher than actual. The consequence of this condition is alarming since longstanding prediction methods are often used for regulatory standards. In addition, actual emissions are capable of significant improvement by the proper selection of seal type.

Test procedures used in evaluating seal leakage are critical. Not only is the selection of procedure critical but the technique used in carrying out the procedure is also critical. Actual results are valid only when the procedure is well chosen and executed.

Emission levels from floating roof tanks have received considerable scrutiny. Numerous studies by the American Petroleum Institute (API), the Western Oil and Gas Association (WOGA), and others have been sponsored in the light of regulatory standards being written for the design and operation of such tankage. Major projects sponsored by WOGA[1] and Sohio[2] collected data which was submitted to the California Air Resources Board (CARB). These projects involved testing in a twenty-foot diameter test tank equipped with a floating roof and either a resilient foam type seal (Fig. 1) or a mechanical shoe type seal (Fig. 2). In this testing, the foam seal was mounted above the product surface. Thus,

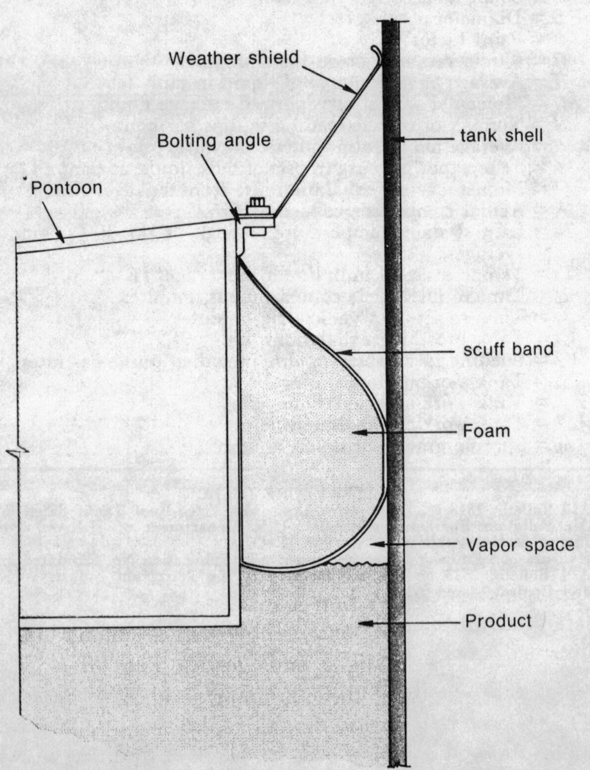

Fig. 1—Resilient foam-type seal.

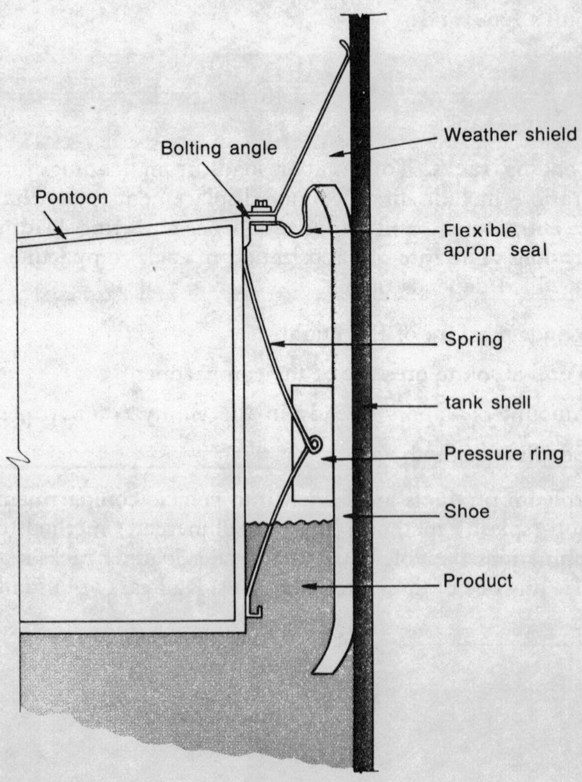

Fig. 2—Mechanical shoe-type seal.

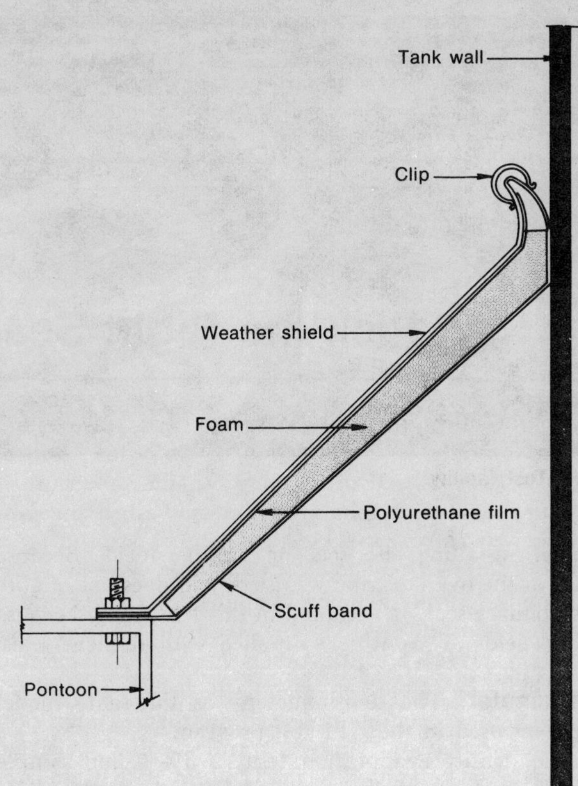

Fig. 3—Secondary-type seal.

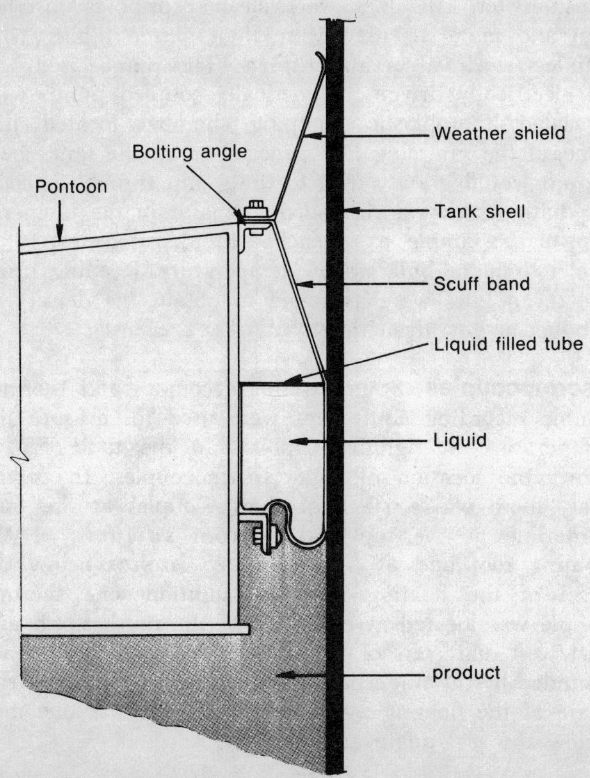

Fig. 4—Liquid-type seal.

a vapor space existed beneath the seal which was open entirely around the circumference of the tank. From the information presented before CARB, it was concluded that resilient seals were more susceptible to the effects of wind than shoe seals.

To understand the reasoning behind this conclusion, wind must be considered to flow across the open top of the tank. As the wind flows, an increase in pressure occurs above the seal on the downwind side of the tank. Under the influence of this increased pressure, the wind passes through any gaps between the seal and the tank shell and into the vapor space beneath the seal. It is then free to move circumferentially around the tank where it tends toward saturation. When the near saturated air reaches the upwind side of the tank it exits through any gaps and an emission results.

With recognition of this process, secondary seals (Fig. 3) were seen as effective means of reducing emissions. They would interfere with the flow of such wind into the space occupied by the seal. Secondary seals are being accepted by regulators and tank owners alike.

A review of this reasoning on the effect of wind, and in particular a recognition that the vapor space is the offending element, leads one to another conclusion; eliminating the vapor space beneath the seal reduces emissions. This may be accomplished by using a liquid-type seal (Fig. 4). However, no test results have been reported on this particular design. Therefore this study was undertaken to establish actual results on emissions using a liquid-type seal.

ESTABLISHED TECHNOLOGY

The API has been considered the authority with regard to emissions from petroleum storage tanks. API has published a series of bulletins which are used to estimate the emissions from different types of tankage. These bulletins were written several decades ago and were intended principally to be used in the economic comparison of different types of control technology. However, in the present time of intense concern with the environment, these bulletins are being used beyond their intended scope.[3] Consequently, API undertook a major program to update the bulletins. That project is underway at the present time and involves the field testing of some actual tankage.

As a part of the API program, a considerable effort was expended to determine the most effective means of measuring emissions. A method was selected which is known as the density change method. That method was subjected to considerable scrutiny by the API before being adopted.[4]

Density change method. The basic method, common to both API and the present testing, is the determination of emissions by measurement of a change in density of the stored product. When a mixture of various fluids of differing density evaporates, there is a change in the composition as the light ends leave. The effect is an increase in density of the remaining fluid.

This method of measuring emissions is outlined in API Bulletin 2512[5] and has been used for many years. The method involves two basic phases. The change in density of the product in the test tank is repeatedly measured and the change in density over a given time period determined. Secondly, a calibration factor must be generated in bench scale testing which will allow a change in density to be correlated to a percentage of product lost. The second phase is referred to as a determination of the density-evaporation factor, or DE Factor.

Once the DE Factor for the particular gasoline under its storage conditions is known, the calculation of emission levels from the measured change in density is straightforward. The formula is

$$E = 0.62428 \ \frac{(D)(V_i)(P_i)}{(F)(T)}$$

Where:

E = Evaporation loss rate, lb/day
V_i = Initial stock volume, ft^3
T = Test period, days
F = Density-evaporation factor, gm/cc/% wt. loss
P_i = Initial stock density, gm/cc
P_f = Final stock density, gm/cc
$D = (P_f - P_i)$ = Stock density change, gm/cc

At low levels of emission, it is necessary that the measurements be very accurate to obtain meaningful results. Historically, this method has been performed on samples collected with a sampling thief. The density determinations were made using a pycnometer. The major improvement in the current testing results from the recent development of the oscillating tube type densitometer which gives resolution to the sixth decimal place and allows for direct sampling of the tank. The ease with which this instrument can be used permits a greater number of measurements to be made, thereby improving the statistical quality of the data.

TEST PROGRAM

Test tank. The test tank (35 feet in diameter by 10 feet high) was designed according to API 650 Specifications. The test facility (Fig. 5) was located on Neville Island, in the Ohio River, just downstream of Pittsburgh, Pa. The floating roof was 25 inches high at the perimeter and included five water-tight compartments and all of the standard accessories for such a roof. During the test program all the floating roof's penetrations were sealed so that emissions occurred only at the liquid seal.

Seal. A standard liquid seal was installed as the primary seal (Fig. 4). The seal consists of a scuffband of serrated buna vinyl coating on nylon fabric. The scuffband was bolted to the top of the rim plate of the pontoon and to a bolting angle mounted on the rim plate at an elevation within the stored product. A liquid filled envelope or tube was placed inside the scuffband adjacent to the rim plate of the pontoon. Three hundred fifty gallons of kerosine acted as the sealing fluid in the seal tube. The annular space occupied by the seal was nominally 6 inches, which is the normal annular space on floating roof tanks up to approximately 200 feet in diameter.

A standard weathershield was attached to the rim plate of the pontoon above the seal. The weathershield consisted of overlapping sheets of 22 gage Corten steel bolted to the top deck of the pontoon. The bolts used on the weathershield were the same bolts used in supporting the seal scuffband. No special precautions were taken in installing the seal or weathershield. Because of gaps in the overlaps, the weathershield does not qualify as a secondary seal. It was used in this test simply because weathershields are routinely supplied with resilient seals.

Fig. 5—Test facility.

Densitometer. The densitometer was the same model instrument used in the API test program.

In the density evaporation tests, a 3½-gallon sample of gasoline was carefully extracted from the tank, maintained at a reduced temperature, and brought to the laboratory. The gasoline was water displaced from the sphere directly into the sampling ports on the side of the densitometer.

During the density determination at the test tank, insulated lines running from seven points in the tank interior were used to transfer gasoline samples to the densitometer. All lines were cleaned prior to installation and materials used were fluoroplastic tubing with stainless steel valves and fittings. The pump used was an electrically driven, magnetically coupled pump with a sealed fluoroplastic pumping chamber located just beneath the sampling line penetration in the tank shell. Liquid would gravity feed to the pump through a cooling bath which was intended to maintain the temperature of the sample, as it reached the densitometer, 10° F cooler than the bulk storage temperature. Readings from the seven lines were averaged to obtain the density of product at any given time.

Thermocouples. Standard thermocouples and thermocouple recording equipment were used to measure the temperature at significant points in the tank. Fig. 6 shows the location of some thermocouples. In particular, there was a stack of thermocouples at the tank centerline at the top and bottom surfaces of the floating roof and at 5, 15 and 25 inches below the level of the floating roof. In addition, one thermocouple was located at each end of the north/south and east/west diameters of the tank in the seal liquid space bounded by the underside of the seal, tank shell and rim plate of the floating roof. Their elevation was one inch below the seal angle attachment.

Meteorology. Wind speed was measured by an anemometer located on a mast 10 feet above the shell and

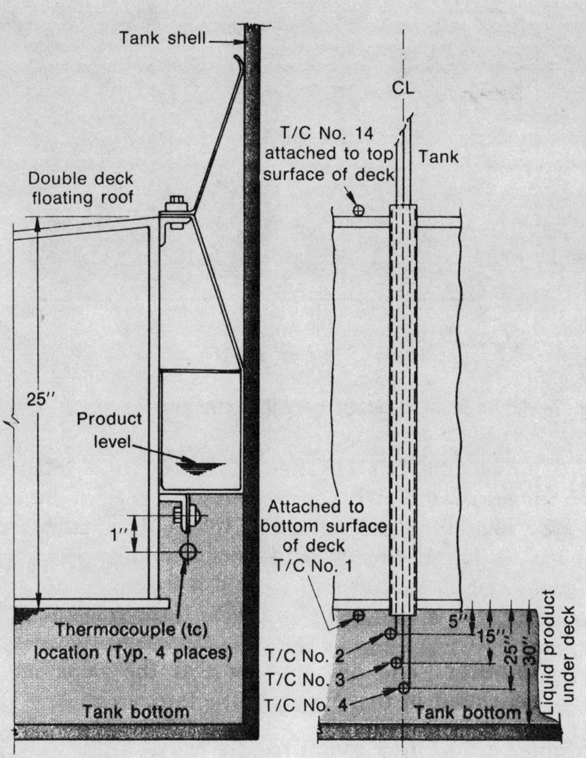

Fig. 6—Location of thermowells in test tank.

bolted to the west side of the tank. Predominate wind in Pittsburgh is from the westerly direction and a continuous recording of wind speed and direction was made using a recorder located in the instrument shed. Barometric pressures were recorded during periods when densitometer readings of various types were being taken.

Gasoline. Eighteen thousand gallons (112,700 pounds) of winter grade gasoline was installed in the tank on April 4, 1978. The quantity of gasoline was very carefully monitored during the tank filling operation so that the total amount was known within an estimated 0.4 percent accuracy. The temperature of the product at that time was about 45° F. On August 29, 1978, the last day of sampling, more than 99 percent of the initial amount remained in the tank.

TABLE 1—Summary of temperature and meteorological data

Test Period	Bulk Liquid Temp. (°F)	Under Roof Temp. (°F)	Under Seal Temp. (°F)	Average Wind Speed, mph	Average Barometric Pressure, psia
6/2—21/78	63.86	68.13	71.08	3.64	14.39
6/22—7/6/78	67.23	70.70	73.18	3.99	14.39
7/7—19/78	68.51	73.64	76.4	3.21	14.39
7/20—8/8/78	72.78	73.28	78.73	3.36	14.41
8/9—29/78	72.79	75.76	78.55	3.97	14.42
6/2—8/29/78	69.28	72.37	75.74	3.65	14.40

TABLE 2—List of days whose peak seal space temperature exceeded 90° F

Date	Peak Temperature, °F	Date	Peak Temperature, °F
July 6	95	July 23	94
7	92	Aug. 2	96
18	92	15	94
19	94	17	93
20	97	18	93
21	95	19	94
22	96	26	92

The gasoline was frequently characterized by its Reid Vapor Pressure (RVP) and the slope of the distillation curve at 10 percent evaporated. The RVP averaged 13.3 psia and slope was approximately 3.1.

On several occasions, vapor samples were removed from the tank through a valved penetration in the roof manhole, and via a tube inserted between the seal scuff-band and the tank shell to approximately the mid-heighth of the seal contact area. The average molecular weight of the vapors at the two locations was substantially the same and averaged 63.3 pounds/pound-mole. The variation in all samplings was from 60 to 63.5 pounds/pound-mole.

Quality control. It was found, prior to beginning the test, that cleaning the densitometer cell was critical in measuring gasoline density. Consequently, the instrument was cleaned by passing xylene and acetone through the cell after each sampling. In addition, the instrument was frequently cleaned with a commercial chromic acid glass cleaning solution. These cleanings helped achieve a standard deviation in the readings of approximately ± 7 ppm.

It was necessary to calibrate the instrument on each day of testing. Typically, the instrument was calibrated in the morning and evening of the day of sampling with ultra pure air and triple boiled and distilled water.

A precision thermistor was used with the densitometer to measure the temperature of the test fluid. Thermistors underwent initial calibration with the digital multimeter used for temperature measurement. It was necessary to know the gasoline temperature in the densitometer's cell to within ± 0.005° C to ensure sufficient accuracy for the test.

It is characteristic of such thermistors to drift with time. Therefore, the test thermistors were frequently checked against a triple point of water cell. The drift in the triple point reading was noted and adjustments were made in the readings of the multimeter to insure operation at the correct temperature.

Density readings were all taken with the cell temperature at 5.000 ± 0.005° C. This was done so that any correction in the thermistor reading when checked against the triple point cell could be more accurately taken into account. Generally, thermistor drift is only an intercept change, not a change in slope of the calibration curve.

Also incorporated was the periodic testing of isooctane as a quality control measure. On one occasion when a significant change in the density of isooctane was measured, it was found that the thermistor being used had failed. A thermistor change was made and the check proved worthwhile. The standard deviation in the 22 replicate density measurements of isooctane was 6.86×10^{-6} grams/cc.

RESULTS AND DATA

The thermocouple system was used to monitor various temperatures in the tank. Certain thermocouple data was then averaged. In particular, the average temperature of the bulk of the liquid during the different test increments was determined and these appear in Table 1. As can be seen, the average bulk temperature was 69.3° F. The thermocouple at the centerline of the tank on the underside of the roof averaged 72.4° F while the temperature underneath the seal averaged 75.7° F. A generally increasing trend in all of the averages was experienced.

Very high temperatures were experienced after mid-July and therefore the maximum daily temperatures occurring in the liquid beneath the seal were examined. A listing of the days when this liquid's temperatures exceeded 90° F and the temperature is shown in Table 2.

If one were to review the thermocouple records for a typical clear summer day, say July 20, 1978, the temperature variation within a particular day could be known. Generally, all thermocouples converged to the same temperature at 7 a.m. That temperature was 72 ± 2° F. As the day progressed, the top of the roof experienced a maximum temperature of 128° F. The four thermocouples under the seal reached 97 ± 2° F and the top of the liquid at the tank centerline was 88° F. The centerline temperatures at 5, 15 and 25 inches below the roof were 76° F, 72° F and 70° F, respectively, when the peak seal space temperatures were achieved. The maximum seal temperatures occurred at 5 p.m. This pattern is typical of that experienced during the test program.

Wind speed and barometer. Average daily wind speed was calculated by averaging the speed of the wind recorded at 3-hour intervals; that is, at midnight, 3 a.m., etc. The direction of the wind was also continuously recorded but is not reported herein. It was, however, anything but steady in direction. The average daily wind speed is shown in Table 1 for the various time periods, and it is seen that the winds were constant and of low velocity over the entire test period from June 2 to Aug. 29. The average was 3.65 mph. The minimum recorded average wind was 1.8 mph, and the maximum was 6.8 mph.

Also the barometric pressure was frequently recorded during the test program and the summaries of the barometric pressures for the different portions of the test program are likewise shown in Table 1. It is seen that the barometer held fairly constant at 14.4 psia over the test period.

Density of gasoline. The gasoline density was determined on June 2 and 21, July 6 and 19, and also on Aug. 8 and 29. The densities occurring on each of these dates is shown on Fig. 7. One notices a distinct change in the slope of the curve on July 19, 1978. The slope of the curves prior to and subsequent to this date were 2.8 ppm/day and 8.4 ppm/day.

Density evaporation factors. A density evaporation factor test was run at 65° F on July 31, and at 72° F on Sept. 8, 1978. These tests were run essentially in accordance with the methods outlined in the API procedures[4] except all-glass flasks were used. The temperatures of 65 and 72° F correspond to the average bulk liquid temperatures for the test period. In the studies of the method sponsored by API, an analysis of the various

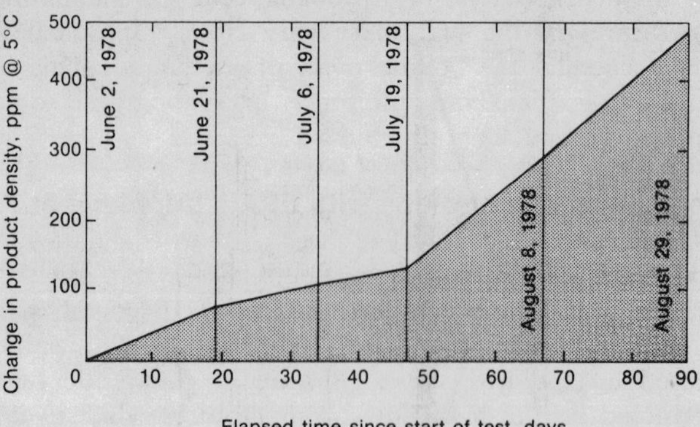

Fig. 7—Change in product density with time of test.

items influencing this DE Factor showed that temperature ranges of ± 20° F changed the DE Factor by only ± 5%. Therefore, a large number of DE Factors were not run. A DE Factor or 1614 × 10^{-6} grams per cc per percent weight loss at 65° F was determined for this gasoline, and 1526 × 10^{-6} grams/cc/% at 72° F. The first figure was used in the data reduction.

ANALYSIS

Calculation of the evaporation loss. Table 3 summarizes the emission data and divides the total test period into a number of increments depending upon the sampling date. The density change in those increments is reported, as is the pounds of gasoline remaining in the tank as of the beginning of the increment. A DE Factor of 1614 ppm was used in calculating the emissions and the total emission over the entire period was 331 pounds which occurred in 88 days. The emission rate is seen to vary from 1.2 to 6.1 pounds/day while averaging 3.76 pounds/day.

TABLE 3—Calculation of liquid seal emissions

Test Periods	Density Change, ppm	DE Factor, ppm	Pounds in Tank	Emission, Pounds	Period, Days	Emissions, Pounds/day
6/2—21/78....	77	1614	112,467	53.65	19	2.82
6/22—7/6/78..	32	1614	112,382	22.3	15	1.49
7/7—19/78....	23	1614	112,322	16.0	13	1.23
7/20—8/8/78..	160	1614	112,297	111.0	20	5.57
8/9—29/78....	183	1614	112,154	127.2	21	6.05
6/2—8/29/78..	475	1614	112,467	331.0	88	3.76

The density varied linearly with time (Fig. 7) in two separate and distinct portions separated by the July 19, 1978, date. In explaining the cause for the abrupt change in slope, it should be recalled that the temperatures being experienced by the gasoline were rising during most of the test period. Subsequent to July 19, a number of days occurred when the measured temperature of the liquid in the space underneath the seal was greater than 90° F. On these same dates, the top of the roof experienced temperatures in excess of 120° F.

The RVP of this product varied from 13 to 14 psia and averaged 13.3 psia. The barometer averaged 14.4 psia. Based on these characteristics and a nomograph in API Bulletin 2513 "Evaporation Loss in the Petroleum Industry—Causes and Control,"[6] the boiling point is estimated in the range of 97 to 102° F. Considering the location of the thermocouples under the seal and the error in such a system, it is probable that the maximum temperatures in the seal space very closely approached or even equaled the boiling point of the liquid. This is thought to explain the difference in evaporation rates occurring around the July 19 date.

Comparison of emissions with EPA's Supplement 7. EPA's Supplement 7 to AP42[7] was used to calculate the emissions expected from such a tank. From this calculation, one would expect an average loss of 15 pounds per day over the period from June 2 to Aug. 29. This value is based on the true vapor pressure of the product as determined by the bulk liquid temperatures which were reported in Table 1. No recognition of elevated rim temperatures is made in this method.

This figure of 15 pounds per day is to be compared with the 3.8 pounds per day measured; 25% of AP42, even though the possibility of boiling in the seal liquid space existed. This confirms what is generally accepted with regard to AP42 and the API Bulletins which preceded it: *These documents overstate emissions from such tankage.*

COMPARISON WITH EPA'S EXTRAPOLATION OF CBI DATA

Attempts were made to compare emissions with the data presented before CARB. For this purpose, a draft copy of "Methods for Extrapolating CBI 20-ft. Ø Test Tank to Full Size Tanks" written by EPA was used.[8] This draft does not evaluate a liquid-filled seal, or any resilient seal mounted in the liquid surface. However, the present test results were compared to data for four seal types or combinations each of which is suitable for use with the method:

- Foam Seal type with no gaps
- Foam Seal/Secondary Seal with no gaps
- Shoe Seal with no gaps
- Shoe Seal|Secondary Seal with no gaps.

Table 4 presents data for the five test periods and the overall summary. Seal combinations used for comparison are shown. Tank emissions are predicted losses expected for a 35-foot diameter tank in pounds per day. Emissions were calculated according to the procedures in the reference and take into account the tank diameter, product true vapor pressure, and wind speed.

Considering the entire test period, the liquid/seal losses amounted to 22 percent of the foam seal's and 46 percent of the shoe seal's predicted losses, when neither of these are equipped with a secondary seal. Obviously, emissions from liquid seals can be reduced

TABLE 4—Comparative estimated loss based on CARB presented data

Test Period	Tank Emissions, Pounds/Day				
	Foam Seal	Shoe Seal	Foam Seal/ Secondary	Shoe Seal/ Secondary	Liquid Seal
6/2—21/78	14.4	6.90	2.71	1.44	2.82
6/22—7/6/78	18.95	8.70	3.03	1.65	1.49
7/7—19/78	13.1	6.67	2.69	1.49	1.23
7/20—8/8/78	17.2	8.26	3.08	1.83	5.57
8/9—29/78	23.32	10.52	3.62	1.97	6.05
6/2—8/29/78	17.0	8.1	3.1	1.65	3.76

using a secondary seal just as the losses are reduced when a secondary seal is used with other types.

In summary, an improved density change method was used to measure emissions from an actual floating roof tank equipped with a liquid-filled seal mounted in the liquid surface. The measurements were made over the summer of 1978 on a winter grade gasoline.

The data collected leads to the following conclusions:

- The density change method is workable and yields data which is consistent and reasonable. At the same time, a considerable learning curve and attention to detail is required.
- Seal pocket temperatures appear to be a significant factor in storage of volatile products. Temperature elevations of 25° F above the bulk liquid temperatures were recorded on a hot summer day.
- Supplement 7 to AP42 overstates emissions occurring from floating roof tanks.
- The liquid type seal reduces emissions below the level attained with both of the other types when used alone and compares favorably with the other types even when a secondary seal is used in conjunction with them.

LITERATURE CITED

1 Jonker, P. E., Porter, W. J. and Scott, C. B.; "Control floating roof tank emissions," *Hydrocarbon Processing*, Vol. 56, No. 5, p. 151, 1977.

2 Good, G. J., "Hydrocarbon emissions from Floating Roof Tanks—The SOHIO-CBI Test Program," presented at the 1977 API Pipeline Conference, March 14-16, 1977.

3 Zabaga, J. G., "Evaporation loss Measurement Activities of the American Petroleum Institute," paper presented at the Air Pollution Control Assoc., New York, April 4, 1977.

4 Anon, "Development of Laboratory Procedures for use in the Field Testing Program to Determine Hydrocarbon Emissions from Floating Roof Tanks," API Committee on Evaporation Loss Floating Roof Tank Subcommittee, May 18, 1978.

5 Anon, API Bulletin 2512, "Tentative Methods of Measuring Evaporation Loss from Petroleum Tanks and Transportation Equipment," July 1957.

6 Anon, API Bulletin 2513, "Evaporation Loss in the Petroleum Industry—Causes and Control," Feb. 1959.

7 Anon, "Supplement No. 7 for Compilation of Air Pollutant Emission Factors," U.S. Environmental Protection Agency, April 1977.

8 Burr, R. and Brothers, K. C., draft copy of "Methods for Extrapolating Chicago Bridge & Iron Co. 20-ft. Test Tank Results to Full Size Tanks," Chemical and Petroleum Branch, Emissions Standards and Engineering Division, U.S. Environmental Protection Agency, April 1978.

How to Prevent Cooling Tower Fog

It's simple. Add heat to the exhaust and raise the dry bulb temperature. Finned-tube exchangers or gas burners are the most promising methods

John C. Campbell, Lilie Hoffmann Cooling Towers, Inc., St. Louis, Mo.

NORMALLY the major portion of the heat removed from circulating cooling tower water results in the mass transfer of a portion of the water to the cooling air. The heat and mass-transfer coefficients are high, and the magnitude of warm, moist air discharged into the surrounding atmosphere is substantial. In cool, humid weather the mixing of the cooling tower exhaust with ambient air often produces a blend containing potentially more water vapor than required for saturation. Condensation of the excess vapor then occurs, and fog is formed in the vicinity of the plume.

One of the most practicable means of preventing cooling tower fog is the addition of heat to the exhaust air, thus increasing its capacity to hold water vapor. The required heat can be added in a number of ways; two of the most promising are *finned-tube exchangers* and *gas burners.*

Specific problems are readily solved with the assistance of a psychrometric chart and other information on the well-charted thermodynamic properties of mixtures of air and water vapor.

Prevention of fog formation. Any plume emitting from a man-made device may be thought of as a form of pollution. In the specific case of water cooling towers, the plume is composed almost exclusively of water droplets, air and water vapor. Thus the true objection is not air pollution in the sense of the addition of poisonous impurities, but rather a matter of *opacity.* The emission of visible plumes near highways and airports can be extremely dangerous.

Psychrometric chart. An interesting and useful property of the psychrometric chart is that it can be used as a "blending" graph; that is, mixtures of two air samples with different enthalpies and moisture contents will be represented on the chart by a straight line connecting the two enthalpy-moisture content points. Condensation of water vapor—and hence fog formation—will occur when the "blend" line representing mixtures of two air samples touches or intersects the saturation curve. That is, hypothetical blends that fall in the area above the saturation curve cannot hold the indicated water vapor, and condensation must occur (see Fig. 5, line M-N).

The problem of preventing water vapor condensation can be thought of as a problem of preventing the "blend" line from intersecting the saturation curve. An obvious answer is to change the properties of one or both components to reposition the blend line as necessary. In the case of hot humid air, such as the cooling tower exhaust (see point N, Fig. 5), discharging into the cool humid atmosphere (point M, Fig. 5), the repositioning effort should be directed toward the *exhaust* stream, since it is

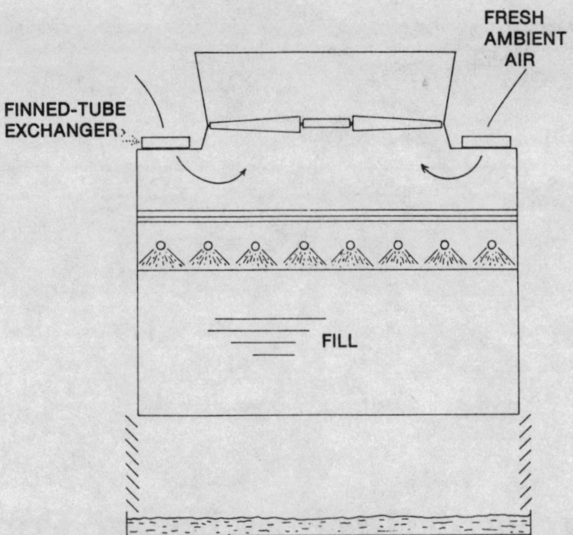

Fig. 1—Prevent cooling tower fog by heating counterflow ambient air with finned-tube exchangers.

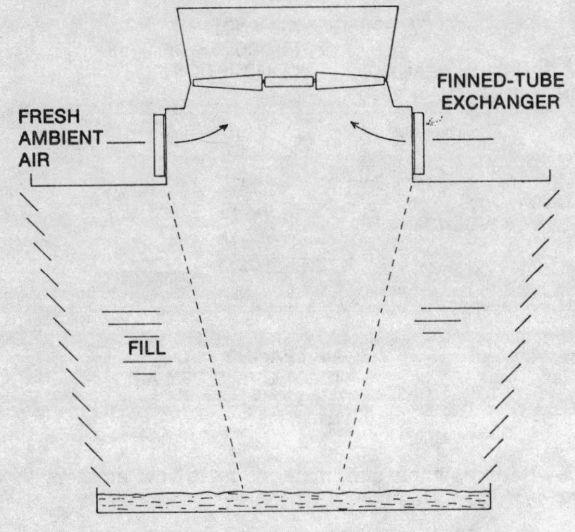

Fig. 2—Prevent cooling tower fog by heating crossflow ambient air with finned-tube exchangers.

not normally practicable to artificially alter the properties of the surrounding atmopshere.

This can be done by decreasing the moisture content at constant dry-bulb temperature, but such a drying operation is generally not feasible for a large cooling tower. The most promising solution is to increase the dry-bulb temperature at essentially constant moisture content. This requires the addition of only a relatively small amount of heat (compared to the cooling tower duty) since only the specific heat of air is involved. From Fig. 5, an increase in the dry-bulb temperature of 11° F (point N to O), requiring only 2.7 Btu per pound of air, is sufficient to move the blend line away from the saturation curve.

PRACTICAL METHODS FOR HEAT ADDITION

Finned-tube heat exchangers. The exchangers may be located in the walls or deck surrounding the cooling tower

MATERIAL BALANCE

Material In

Air leaving fill	lb./min.	lb./hr.	mols/hr.	MW
O_2 = 42,000 (0.232) =	9,744	584,640	18,270	32
N_2 = 42,000 (0.768) =	32,256	1,935,360	69,120	28
H_2O = 42,000 (12.46) (10) =	523	31,399	1,744	18
Fuel gas	7	422	26.4	16
	42,530	2,551,821	89,160	28.62

Notes:
lb.-mols/hr. O_2 reacted = 2 (26.4) = 52.8
lb.-mols/hr. O_2 remaining = 18,270 − 52.8 = 18,217.2
lb.-mols/hr. CO_2 formed = 26.4
lb.-mols/hr. H_2O formed = 2 (26.4) = 52.8

Material Out

	lb./min.	lb./hr.	mols/hr.	MW
O_2	9,716	582,950	18,217	32
N_2	32,256	1,935,360	69,120	28
H_2O	539	32,349	1,797	18
CO_2	19	1,162	26	44
	42,530	2,551,821	89,160	28.62

Notes:
lb. H_2O (v)/lb. dry gas = 32,349/2,519,472 = 0.01284
dew point = 64.09°F
lb.-mols H_2O (v)/lb.-mol dry gas = 1,797/87,363 = 0.02057
Mol fraction CO_2 in dry gas = 26.4/87,363 = 0.000302
The effect of CO_2 on the wet bulb temp. is insignificant.

plenum chamber. Fresh ambient air can then be drawn through the exchangers and mixed with the saturated air emitting from the fill, as indicated schematically in Figs. 1 and 2. Any available fluid with relatively high-level enthalpy, including the inlet water to the cooling tower, can be used as the heating medium. The use of waste steam results in a high temperature-difference driving force and a high over-all heat transfer coefficient. This choice minimizes the required exchanger surface, requiring between 100 and 500 square feet of prime tube surface per million lb./hr. of air flowing through the cooling tower fill.

Alternately, the exchangers may be located in the plenum chamber. For this option no additional fresh ambient air is required, but maintenance will be higher caused by continuous exposure of the heat exchangers to hot humid air.

Gas burners. The required addition of heat can be accomplished by means of gas burners. These may be located in the plenum chamber, in the fan discharge stack (as indicated in Fig. 3) or in a special combustion chamber through which fresh ambient air is drawn into the plenum or stack. Precautions should be taken to minimize the potential danger of this system to the cooling tower and adjacent facilities. A small amount of water vapor will be formed by the combustion process, so somewhat more enthalpy increase will be required by this method.

Example 1: The use of heat exchangers.

Problem: Design fog-prevention equipment to prevent visible exhaust plume from the cooling tower described below, using waste steam as a source of heat. Assume winter conditions are limiting:

Tower description

Number of cells........................ 1

HEAT BALANCE

Heat In — Datum: H_2O(l), gas(v), 32°F

	lb./hr.	C_p	Δt, °F	Btu/hr.
Air at 63.27°F, sat.—O_2	584,640	0.221	31.27	4,040,254
N_2	1,935,360	0.250	31.27	15,129,677
H_2O (v)	31,399	(1089.4)	....	34,206,071
Fuel gas at 32°F—CH_4	422	0.515	0	....
Heat of Combustion	....	....	....	10,000,000
	2,551,821	....	....	63,376,002

Heat Out

	lb./hr.	C_p	Δt, °F	Btu/hr.
CO_2	1,162	0.209	T-32	see below
O_2	582,950	0.221	T-32	
N_2	1,935,360	0.250	T-32	
H_2O(v)	32,349	(x)	....	

Try T = 77°F

Δt = 45°F—CO_2 = (1,162) (0.209) (45) =	10,929
O_2 = (582,950) (0.221) (45) =	5,797,438
N_2 = (1,935,360) (0.250) (45) =	21,772,800
H_2O(v) = (32,349) (1095.4) =	35,435,095
	63,016,262

Try T = 78°F

Δt = 46°F—CO_2 =	11,171
O_2 =	5,926,270
N_2 =	22,256,640
H_2O(v) =	35,448,034
	63,642,115

By interpolation: T = 77.57°F

Cell length, ft.	38
Cell width, ft.	38
Fan diameter, ft.	22
Fan stack height, ft.	14

Thermal design	**Summer**	**Winter**
Water circulation rate, gpm	6,000	6,000
Cooling range, °F	15	15
Wet-bulb temperature, °F	78	30
Dry-bulb temperature, °F	90	31
Air flow, lb./min.	42,000	42,000

Solution: Finned-tube exchangers will be used to transfer heat. Assume the tube sections are located above openings in the roof deck, or in openings in the plenum walls, so that fresh ambient air can be heated in passage to the plenum. Try two tube bundles, with a flow of fresh ambient air of 2,500 lb./min. per bundle (5,000 lb./min. total). Assume the flow of air through the cooling tower fill is 42,000 lb./min. (design) when the fog-prevention louvers are opened.

The properties of the ambient air and the exhaust air emitting from the cooling tower fill are indicated by points M and N, respectively, Fig. 5. In order to prevent the "blend" line from intersecting the saturation line, it can be seen that, if the moisture content is held constant, the dry-bulb temperature of the exhaust air stream must be increased to at least 74° F (point O). Thus the minimum required heat addition is:

$$Q_t = (42{,}000)(60)(31.6 - 28.7) = 7{,}308{,}000 \text{ Btu/hr.}$$

It is recommended that somewhat more than the minimum theoretical amount of heat be added, because of such effects as imperfect mixing of the heated air with

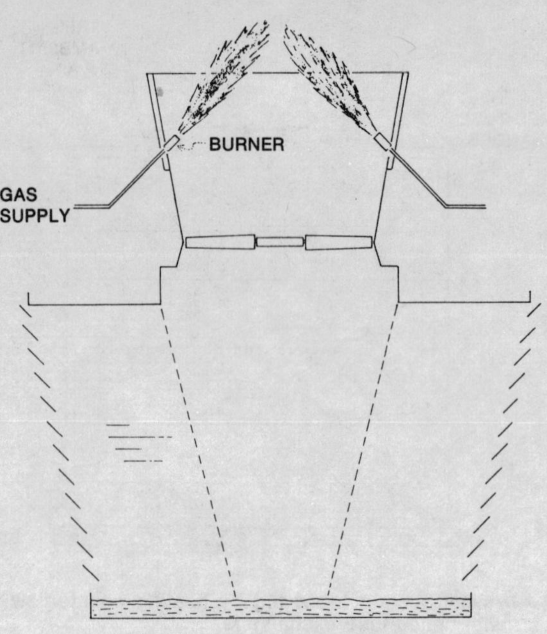

Fig. 3—Gas burners used to heat cooling tower exhaust air.

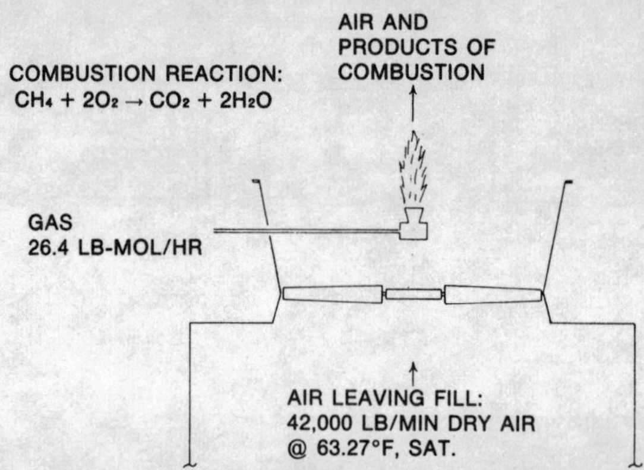

Fig. 4—Heat balance and material balance streams in gas burner system.

the exhaust stream, ambient temperature gradients and variations in the cooling tower heat load. Normally, about 25 percent more than the theoretical minimum is adequate, but unusual conditions and critical fog-prevention requirements may dictate more. In this example, assume a normal increase of 25 percent.

The heat requirement and enthalpy change are determined as follows:

$$Q_a = 7{,}308{,}000\ (1.25) = 9{,}135{,}000 \text{ Btu/hr.}$$
$$\Delta h = 9{,}135{,}000/(60 \times 42{,}000) = 3.625 \text{ Btu/lb.}$$
$$h_2 = 28.7 + 3.625 = 32.325 \text{ Btu/lb.}$$

These conditions establish the location of point P, Fig. 5. The specified heat addition will require the following increase in the dry-bulb temperature of the 5,000 lb./min. fresh ambient air:

$$\Delta t_a = \frac{9{,}135{,}000}{(5{,}000)(60)(0.24)} = 127° \text{ F}$$

Assume waste steam at 230° F is available for the heat-

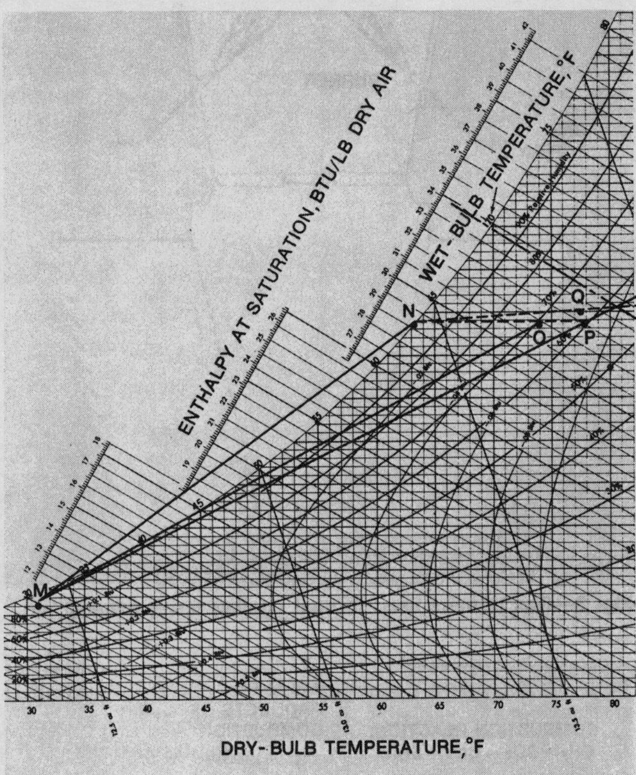

Fig. 5—Psychrometric chart with example problem lines plotted.

ing medium. The effective mean temperature difference will be:

Air	Steam	Δt	
158° F	230° F	72° F	
31° F	220° F	189° F	Effective MTD = 121° F
127° F	10° F	117° F	

The required steam flow rate will be

$$\frac{9{,}135{,}000}{(1{,}157.0 - 188.1)} = 9{,}430 \text{ lb./hr.}$$

Assume an over-all heat transfer coefficient U of 100 can be achieved, based on prime tube external surface. The required total surface for the fog-prevention heaters will then be

$$A = \frac{9{,}135{,}000}{(100)(121)} = 755 \text{ ft.}^2$$

The physical design of the heaters is simply a matter of arranging the required heat transfer surface to achieve the preceding design values, keeping in mind that the air flow resistance must be low enough to permit the use of louvers for air flow control.

Example 2: The use of gas burners.

Problem: Solve the preceding problem using gas burners instead of finned-tube exchangers.

Water vapor formed by the combustion of the natural gas will require somewhat more heat addition for this option. The dewpoint temperature of the exhaust mixture, in this case, is not known, and a convenient solution of the required gas combustion involves a trial calculation to determine the relationship between the exhaust air dewpoint and the amount of heat added for fog prevention.

Assume, as before, a flow of 42,000 lb./min. of dry air through the cooling tower fill. For the trial calculation, assume combustion of sufficient gas in the cooling tower exhaust to add 10,000,000 Btu/hr. to the exhaust stream. For simplicity assume the fuel is pure methane, with LHV = 1,000 Btu/scf.

gas burned = 10,000,000/1,000 = 10,000 scf/hr.
lb.-mols/hr. fuel gas = 10,000/379 = 26.4

The dewpoint and dry-bulb temperatures of the exhaust mixture are determined by material and heat balances. The streams involved are shown in Fig. 4.

The trial calculation thus reveals that the addition of 10,000,000 Btu/hr. to the exhaust air, by means of the combustion of methane, has the following effect:

	No Heat Added N, Fig. 5	10,000,000 Btu/hr. Added Q, Fig. 5
Exit dewpoint temperature, °F	63.27	64.09
Exit dry-bulb temperature, °F	63.27	77.57

These points are plotted on a psychrometric chart, shown in Fig. 5. A straight line passing through them permits straightforward determination of the required enthalpy of the exit air. Using heat addition of 25 percent more than the theoretical minimum, as in the preceding example, the required enthalpy, at the intersection of lines M-P and N-Q is 33.4 Btu/lb. of dry gas. The required heat addition is

$$Q_a = (42{,}000)\ (60)\ (33.4 - 28.7) = 11{,}844{,}000 \text{ Btu/hr.}$$

The required fuel gas will be 11,844,000/1,000 = 11,840 scf/hr.

ACKNOWLEDGMENT

Based on the paper, "The Prevention of Fog from Cooling Towers," originally presented to the Cooling Tower Institute, Houston, Jan. 19-21, 1976.

State of the Art . . . Smokeless Flares

Properly designed mechanical equipment can make any flares smokeless

L. G. Vanderlinde, Imperial Oil Limited, Calgary, Alta., Canada

FLARES ARE MADE smokeless by providing an adequate quantity and distribution of oxygen in the combustion zone. Sufficient primary air is induced in the combustion zone by throttling steam, gas, water or air into the flare gas stream with a flare tip of a design based on various parameters. The costs of the flare system vary with tip type, the hydrocarbon being burned, utilities availability, materials of construction and flare gas rate.

The ideal flare would burn hydrocarbon emissions in a noiseless, odorless and invisible manner. To date no flare ideal system has been developed, but properly designed mechanical equipment can make any flare smokeless.

The key factors affecting design of smokeless flares are:

- Quantity and distribution of oxygen in the combustion zone
- Temperature in the combustion zone
- Type of hydrocarbon being burned.

Other factors to consider when designing a smokeless flare are noise generation, heat radiation, the availability of utilities, and flaring rate and frequency.

Oxygen in the combustion zone—amount and distribution—is the most critical determinant of smoke production. For complete product combustion a stoichiometric quantity of oxygen (or the equivalent quantity of air) is required in the burning zone. The equivalent stoichiometric air requirements vary with the class of combustible (paraffins, olefins, etc.—Table 1) but are essentially constant within a class. For smokeless combustion of a paraffin approximately 20 percent of the stoichiometric quantity of air must be evenly distributed in the primary mixing zone. For smokeless combustion of an olefin, the primary air quantity must be increased to 30 percent of the stoichiometric volume. The remaining air required to complete the combustion process is induced into the flame through aspiration and thermal draft effects. Complete combustion of propane requires 15.7 pounds of air per pound of hydrocarbon (Table 1) with 20 percent or 3.1 pounds per pound of hydrocarbon of the stoichiometric amount in the primary air. This air must be well mixed with the flare gas prior to flame ignition or carbon black and soot will form due to incomplete oxidation taking place.

Combustion zone temperature is the second most important factor in the design of a smokeless flare. Temperature has a direct effect on the amount of smoke formed because it affects the amount of thermal decomposition that takes place. Since decomposition increases with temperature the amount of smoke increases as temperature increases (Table 2). If water or steam is correctly injected into a smoking flare stream, the amount of smoke generally decreases (Figures 1 and 2).

TABLE 1—Stoichiometric air requirements

Gas			Stoich. air
Type		Cu. ft./lb.	Lb./lb. of combustible
Ethane	Paraffin	12.6	16.2
Propane		8.6	15.7
Butane		6.5	15.5
Ethylene	Olefin	13.5	14.8
Propylene		9.0	14.7
Hydrogen sulfide		11.1	6.1

TABLE 2—Effect of temperature on decomposition by heat

Hydrocarbon	Temperature °F	Decomposition %
N-Butane	797	15×10^{-8}
	1202	57
N-Pentane	797	24×10^{-8}
	1112	30

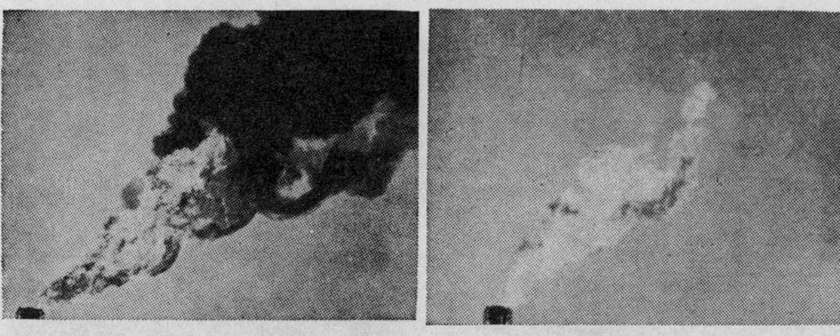

Fig. 1—Flare burning 15,000 lb/hr propane-butane mix before and after adding 0.7 pounds of water per pound of flare gas.

Fig. 2—Flare burning 15,000 lb/hr propane-butane mixture before and after adding 0.23 pounds of steam per pound of hydrocarbon.

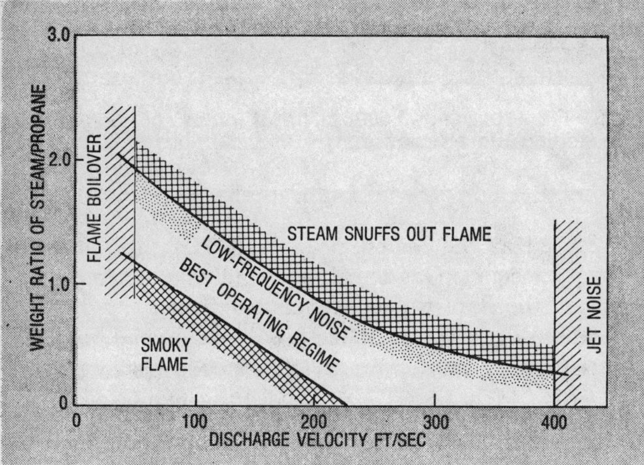

Fig. 3—Performance of propane flare.

Addition of water or steam lowers the combustion zone temperature by dilution, by turbulence which increases the amount of air aspirated and distributed in the flared gas stream, and by creating the endothermic carbon-water reaction. Lowering the combustion zone temperature prolongs the oxidation process and minimizes the permissible amount of hydrocarbon decomposition.

Hydrocarbon type being burned is the last key factor. Tests prove that, at the same temperature, dissociation or decomposition increases as the carbon/hydrogen ratio increases (Table 3). Therefore, unsaturates dissociate easier and make more smoke than saturates. In addition, unsaturates require more primary air; at least 30 percent of the stoichiometric quantity in the primary mixing zone for smokeless combustion to occur. Consequently, a flare designed for unsaturated hydrocarbons requires more air per pound of flare gas than one designed for smokeless operation with saturates. For example, propylene requires 3.4 pounds (30 percent of 14.7) of primary air per pound of gas in the combustion zone compared with only 3.1 (20 percent of 15.7) pounds per pound of propane vapor for smokeless operation. Energy required to mix oxygen or air with a combustible material may be provided by the flare gas stream through pressure reduction and/or thermal draft or by an external source, such as steam injection, power gas assist or a blower fan. The most common technique is air aspiration through pressure reduction of a utility—steam, water or fuel gas—across a nozzle which discharges to the atmosphere. However, if the jet velocity is low very little air is entrained and consequently considerable quantities of smoke are produced. As the gas velocity from a jet is increased, the amount of aspirated air will increase. In a bench test, smokeless combustion of propane was achieved without steam addition by keeping the jet velocity above 220 feet per second (Fig. 3).

A forced air fan blowing the required quantity of primary air directly into the burning gas stream is an example of how smoke can be eliminated with an external energy source (Figs. 4 and 5). This approach requires use of a blower to produce good gas/air mixing characteristics at all flare rates. In addition, an automatic gas/air turndown device is needed to prevent excess air (approximately 400 percent of stoichiometric) from quenching the flame and creating smoke if the flare gas rate is reduced.

Noise. Care must be exercised when designing the exit gas velocity from a flare because increased gas velocity also increases the flare noise.[1] The total noise from combustion systems is composed of two parts: combustion roar and jet noise. Combustion roar, the unique noise spectrum generated by the combustion process, is a linear function of the amount of air mixed with the flare gas. Increasing the air flow increases the combustion roar. The amplitude of combustion roar can be predicted if only one flare is involved and it is in a free environment. If several flares are present, noise predictions must be based on detailed empirical correlations since the interaction between the various noise factors makes quantitative predictions difficult.

The second part of the total noise system is jet noise which is caused by a gas or fluid passing through a restriction. At a constant flow rate, jet noise increases in direct proportion to a pressure drop increase. For this reason, steam or gas aspirator nozzles should be sized to pass the quantity of utility (steam or gas) at the lowest practical pressure drop which is usually not over 100 psig. If a high pressure drop has to be taken, it should be accomplished by taking several small pressure reductions rather than one large pressure drop. When steam is used as the aspirating media, dry steam is preferred over wet steam since the noise level increases

TABLE 3—Effect of hydrocarbon saturation on decomposition by heat

Hydrocarbon	Temperature °F	Decomposition %
Ethane	1067	0.00017
Ethylene	1058	75.6
Propane	1067	0.0026
Propylene	1067	75.0

with wet steam due to the crackling sound of the water molecules exploding in the hot flare stream.

Heat. Heat emissions can be very large through both convection and radiation effects. In most flare systems the convection plume typically rises above the flare and is of little consequence. However, the radiant transfer of heat to the surroundings can be significant.

Radiated heat output is a function of the flame geometry (length, shape and diameter), the type of hydrocarbon being burned and the degree of smoke in the flare. Flame geometry is important because the distance from the center of the flame to the surroundings has an inverse effect on the heat intensity. Assuming the theory of spherical radiation is valid, ball type flames will radiate more heat to the ground than a long cylindrical flame. The type of hydrocarbon being burned has a bearing on the heat level due to the different heating values associated with various hydrocarbons. In the case of smoking or nonsmoking flares, tests indicate that smoking flares radiate approximately 30 percent more heat to the ground than do smokeless flares.

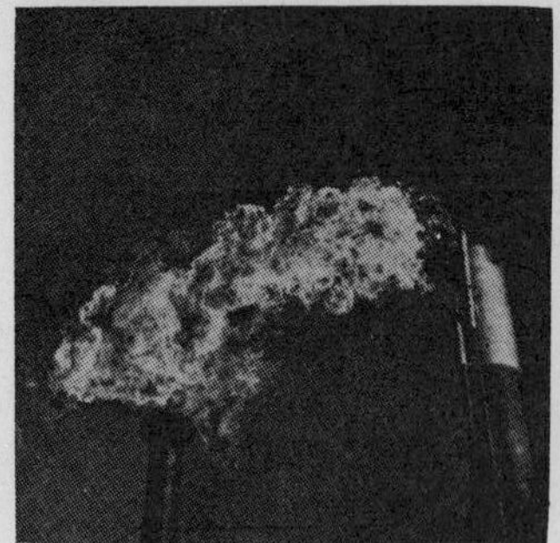

Fig. 4—NAO 4-inch air blower flare burning 4,000 lb/hr propane before and after adding 3 pounds of air per pound of flared gas.

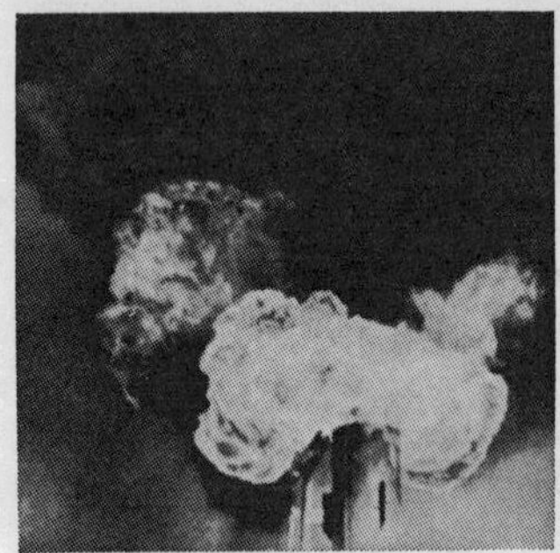

Fig. 5—Flare burning 3,500 lb/hr of propylene before and after starting the air blower.

BURNER CHARACTERISTICS

Flare burners can be categorized as those that reduce smoke by supplying air to the combustion zone and those that reduce smoke by cooling the combustion zone. Regardless of the method selected, the flare burner must be capable of operating over a wide range of flaring rates. To achieve a wide range of turndown the flare must have excellent flameholding ability and mixing characteristics. Flameholding is insured by providing multiple continuous pilots around the combustion tip and by providing a flame stabilization ring on the combustion tip. Efficient mixing of flare gas with air requires a large contact area between the air and gas. This is achieved by using jet action to create high turbulence and/or by dividing the gas flow into small streams through multiple tips or multiple ports on a single tip (Fig. 6). However, the size of the turbulence zone should be minimized in order that smokeless flaring is achieved at the lowest noise level (i.e., at minimum air requirement). Gas burning should take place outside or above the flare tip to extend the tip life.

Many designs are available for smokeless flare burners and the utility consumption (air, water, gas or steam) varies with each design. To achieve the required air/gas mixture, vendors use various types of nozzles on the flare tip. The nozzles may create a circular, swirl, fan, jet or coanda effect. Published test data on the effectiveness of different commercial flare burners is limited, however, a guide is available (Table 4) for selecting the most suit-

Fig. 6—NAO 30-inch NRC flare and 36 inch NFF flare being assembled.

Fig. 7—John Zink Type S flare burning 20,000 lb/hr mixture of ethylene-propylene before and after adding 0.4 pounds of steam per pound of hydrocarbon.

Fig. 9—NAO type NWR flare burning 2,000 lb/hr propylene before and after adding 1.2 pounds of water per pound of flare gas.

Fig. 8—NAO 24-inch NRC flare burning a 60,000 lb/hr mixture of ethylene-propylene before and after adding 0.33 pounds of steam per pound of hydrocarbon.

able type of flare tip. The final utility consumption depends on specific tip design. Figures 7 to 13 illustrate "before and after" effects of several smokeless flares.

Capacity limitations of the various smokeless flare designs are generally set by economic factors rather than flare design parameters. The maximum single flare sizes built to date are shown in Table 5.

Environmental regulations forced the use of smokeless flare systems capable of a very large turndown ratio. To cope with this large variation in smokeless flare rates vendors put multiple flare heads or tips on one stack. A typical tip design might include three steam rings, or a steam ring and a forced air system. In the latter design the forced air unit handles the small flare volumes; the steam unit assists in handling larger volumes. In other cases, staged headers have been used where more burners are ignited as the flare gas rate increases (Fig-

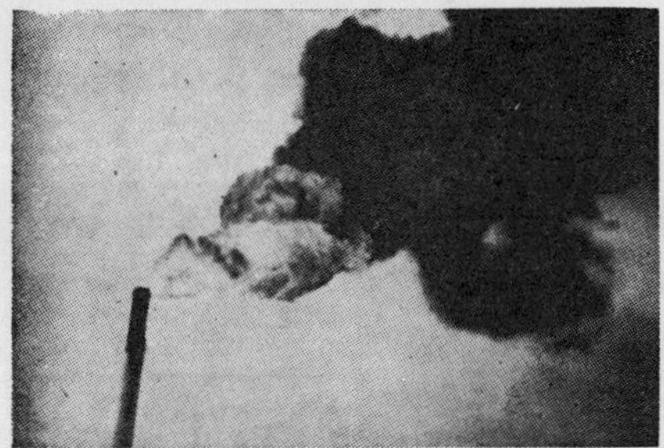

Fig. 10—Flare burning propane and butane with no water, water just turned on, and at a water rate of 20 gpm. Flare rate unknown.

Fig. 11—Ground flare burning 19,000 lb/hr butane with no water and with 14 pounds of water per pound of flared gas. Last picture shows the water spray pattern with no gas being burned.

Fig. 12—Indair flare (left) burning 60 MMcfd of H.P. gas and L.P. flare burning 20 MMcfd (right).

Fig. 13—Indair flare (left) burning 60 MMcfd of H.P. gas and 20 MMcfd of L.P. gas.

Fig. 14—John Zink linear relief gas oxidizer.

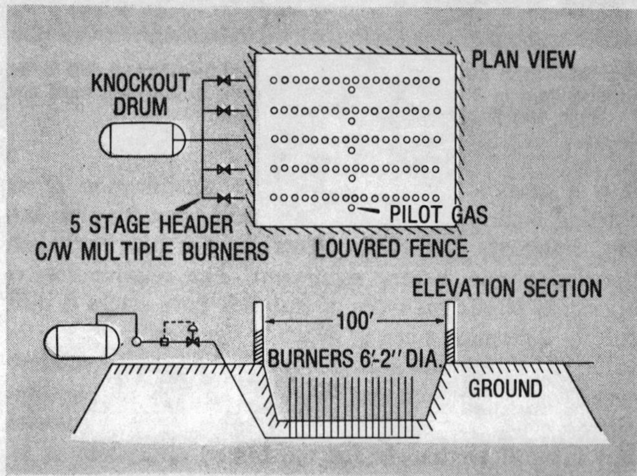

Fig. 15—John Zink linear relief gas oxidizer arrangement.

TABLE 4—Guide to smokeless burner tip selection

Utility	Utility consumption lb./lb. H.C.	Noise level	Use of existing stack
Steam	0.25	Med.	Yes
Water	0.7 or 2.0	Very Low	Yes
Air	3.1	Low	No
Gas	0.7	High	Yes

TABLE 5—Largest smokeless flares

Utility	Hydrocarbon rate #/Hr.
Steam	900,000
Water	100,000
Air	250,000
Gas	105,000

ures 14 and 15). A third system is of particular value where noise, luminosity and smoke formation are critical: an enclosed ground flare or combustor used in conjunction with a stack (Figures 16 and 17). In this case the combustor is designed to handle 80-90 percent of the flare occurrences while both systems are used for the remaining large releases. Economics generally dictate which flare system to use.

COSTS

Capital and operating costs for a given flare burner installation depend upon availability of utilities, whether

Fig. 16—John Zink ZTOF unit capable of burning 25,000 lb/hr. Ringed flare is 350 ft. high, elevated flare is 200 feet high and ZTOF is 100 ft high by 20 feet diameter.

it is a grassroots development or a modification of an existing facility, size of the flare, and frequency of flaring. Since equipment cost generally increases with size, the design rate is very significant. The relative cost of operation of various types of installed flare stacks is difficult to determine since it depends significantly upon the value assigned to the availability of existing utilities. However, relative equipment costs (guyed stack, ignition piping, pilot piping, a specific burner ring and accessories, and integral air blower for the forced air system as required) used in the smokeless flare systems are shown in Table 6. These costs are not stack diameter dependent for all aspiring media except forced air systems.

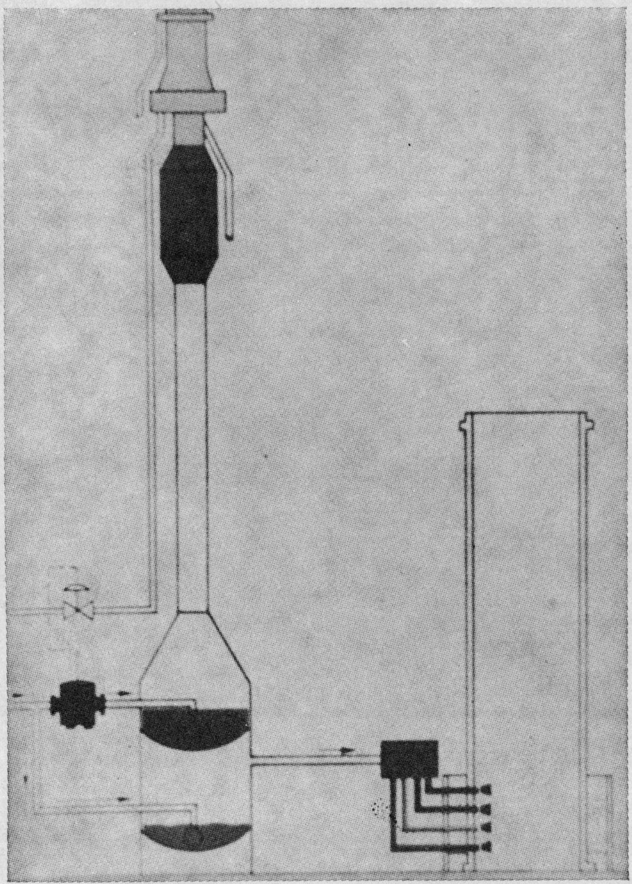

Fig. 17—Schematic showing how elevated flare and ZTOF flare are connected through a seal unit.

TABLE 6—Relative cost of flaring facilities

Type of flare	Equipment costs	
	12″ diameter	24″ diameter
Smoking		
Standard tip	1.00	1.00
Smokeless		
Steam tip	1.25	1.25
Gas tip	1.30	1.30
Water tip	1.20	1.20
Forced air	2.80	3.38

ACKNOWLEDGMENT

Adapted from a paper presented to the Gulf Coast Regional meeting of the Gas Processors Association, Corpus Christi, Texas, February 1974.

LITERATURE CITED

1 "Guide for the Design of Low Noise Level Combustion Systems," American Gas Association, Arlington, Va.

BIBLIOGRAPHY

Cantwell, J. E. and Bryant, R. E., "Failure of Refinery Piping by Liquid Metal Attack and Material Experiences with Smokeless Flares," API Division of Refining, Philadelphia, Pa., May 15-17, 1973.

Flaregas, Trade Literature, 1973.

Giammar, R. D. and Putnam, A. A., "Guide for the Design of Low Noise Level Combustion Systems," A.G.A. Basic Research Project BR-3-5, January, 1971

Gibson, R. D. and Vinson, D. J., "Design and Installation of Smokeless Flare Systems for Gasoline Plants," ASME Publication 72-PET-12

Indair, Trade Literature, 1973, Indair, Birmingham, England

John Zink Company, Test Site Pictures, 1973, John Zink Company, Tulsa, Okla.

John Zink Company, Trade Literature, 1973, John Zink Company, Tulsa, Okla.

Lauderback, W., "Unique Flare System Retards Smoke," *Hydrocarbon Processing*, Vol. 51, No 1, pp. 127-128 January 1972

National Airoil Burner Company, Test Site Pictures, 1973, National Airoil Burner Company, Philadelphia, Pa.

National Airoil Burner Company, Trade Literature, 1973, National Airoil Burner Company, Philadelphia, Pa.

Panitz, T. and Wasan, D. T., "Flow Attachment to Solid Surfaces," *AICHE Journal*, Vol. 18, No. 1, pp. 51-57, January 1972

Reba, I., "Applications of the Coanda Effect," *Scientific American*, Vol. 214, No. 6, pp. 84-92, June 1966

Seebold, J.G., "Curing Flare Systems Noise Provides Major Challenge," *Canadian Petroleum*, Vol. 14, No. 1, p. 53-57, January 1973

Smoke-Ban Manufacturing, Inc., Trade Literature 1973, Smoke-Ban Manufacturing Inc., Pasadena, Texas

Vandaveer, F. E. and Segeler, C. G., "Combustion *Gas Engineers Handbook*, Industrial Press, New York, N.Y. 1966. 2/46-2/100.

Make the Flare Protect the Environment

Design and operation can eliminate environmental problems in flare use

John F. Straitz, III, National AirOil Burner Co., Philadelphia, Pa.

PROPERLY DESIGNED and operated flares protect the environment while being used to safely and economically eliminate gaseous waste streams. Flare use in the hydrocarbon processing industry involves widely variable conditions both for type of installation and individual flare operation. Gas composition depends on the service and on conditions of operation at the time of activation. Temperature depends on the locale and season of the year. Support mediums used to assist flare operation depend on facilities available. Consequently, designs must be made with great care and flare application must be knowledgeably engineered to assure optimum performance.

DESIGN FACTORS

Thermal radiation, liquid carry-over and explosion hazard, resulting from air entry into the stack, are important design factors to be considered[1-10] while operational considerations include factors such as stable and complete combustion, noise, positive piloting, reliable ignition, effective steam or assist gas control, smokeless operation.

Flame stability. A stable flare flame with no possibility of blow-out is of primary concern, since blow-out or

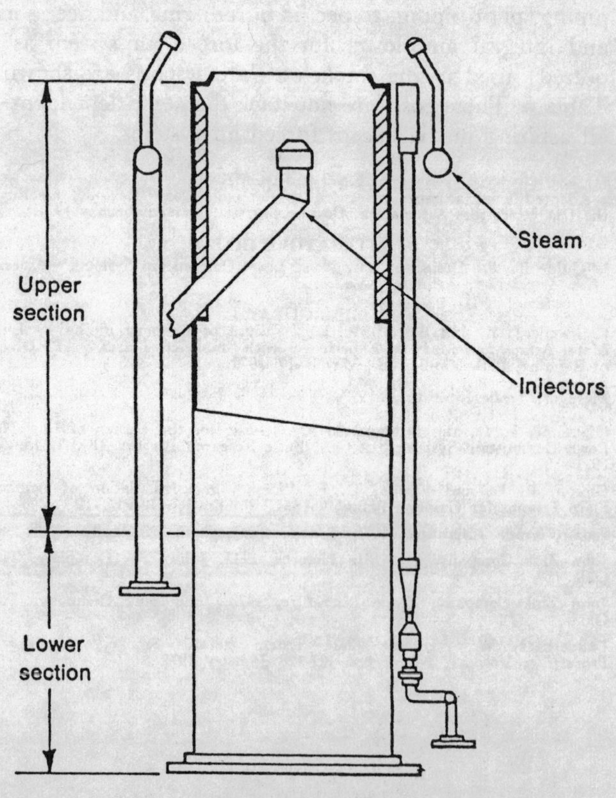

Fig. 1—Steam can be injected into the flare to introduce air to the fuel by use of jets inside the stream and around the periphery.

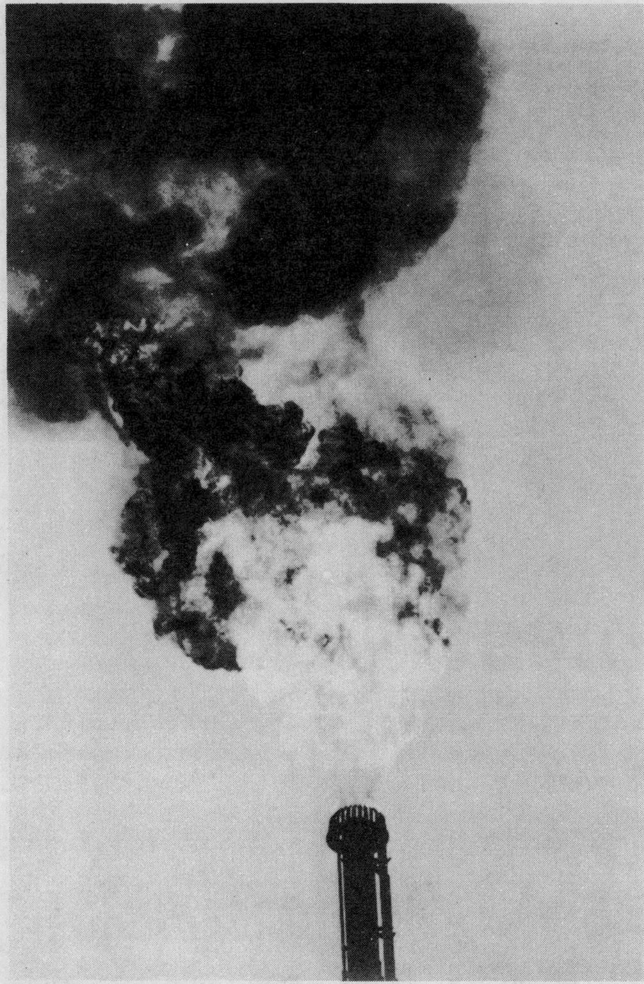

Fig. 2—Low combustion efficiency results when unsaturated hydrocarbons are flared without using some system of enhancing mixing.

flame-out can result in vapor concentrations within the combustible limits at grade, depending upon atmospheric conditions. Thus, a giant "fireball" with devastating resultant consequences could occur. Therefore, safe design practices require adequate flare stack height for dispersion to below flammable limits for protection in the event of flame-out. Also, flaring noxious gases requires sufficient stack height for dispersion to safe exposure levels at grade.

Environmentally acceptable combustion, however, can be realized only with a reliable, stable flame. For a flare built from an open-ended pipe, a maximum exit velocity of 0.2 sonic will prevent flame lift-off. The sonic velocity is based upon waste gas composition and temperature at the flare tip.

An appropriate formula is

$$C = 250\sqrt{\frac{T}{M}} \qquad (1)$$

when C = speed of sound in ft. per sec.
T = gas temperature, °Rankine
and M = gas molecular weight.

The maximum stable exit velocity can be increased by the addition of flame retention devices at the point of gas discharge. A typical device consists of an annular ring with many small piloting holes surrounding a large central orifice. The central orifice is approximately five inches smaller than the flare tip diameter on large size units.

Using these devices, exit velocities of 0.5 sonic can be safely achieved, with reduced possibility of blow-off even at slightly higher velocities. Not only does the flame retention ring insure stable combustion, but it minimizes wind effect, reduces thermal radiation at the base of the flare and helps to increase air entrainment, thereby improving the smokeless operation of the flare.

Complete combustion. An adequate air supply that has been properly mixed with the waste gas in the combustion zone is required for complete burning. For most hydrocarbons, external means of air injection and mixing are required if a high percentage of combustion efficiency is to be attained. Usually, air injection and mixing is accomplished by means of high-velocity steam jets which entrain air via shear forces or directly from a motor-driven air blower. Certain gases inherently produce smokeless operation with complete combustion and without added assist gas. Natural gas, which is mostly methane, is an excellent example.

Tests on flares with flame retention rings, in diameters from 2 to 10 inches, indicated a high combustion efficiency when flaring natural gas. Measurements of unburned hydrocarbons at the tip of the visible flare flame gave a 99+% combustion efficiency or completion rate without steam injection or other assist means. No systematic variation with flare size was observed. The temperature at this location was in the order of 1,800° F and oxygen levels showed approximately 60% excess air. The accuracy of these experiments is difficult to quantify due to spatial variations of the flame, wind effects, time lag and mixing within the sample lines. Use of a continuous chart recorder, video tape recorder and long-time sampling was necessary to achieve confidence in the results.

The effect of steam injection (Fig. 1) on the natural gas flare was also measured. A low rate of steam injection resulted in slight increases in combustion efficiency, flame tip temperature and radiation at grade. As the steam to natural gas ratio was increased, the values of these parameters increased to a maximum level and then began to drop. This drop is indicative of the quenching effect of excessive steam injection. Excessive use of steam also significantly increased the flare noise levels and, if carried to extreme, caused pulsating, unstable combustion. For natural gas, the point of instability occurred at a 1:1 ratio of steam to natural gas on a weight basis.

Measurements for combustion efficiency, temperature, oxygen level, carbon monoxide, NO_x, radiation and noise were also taken for other hydrocarbon streams. As expected, the gases which tended to smoke gave low combustion efficiency readings. Efficiencies in the order of 75% were measured for heavier, unsaturated hydrocarbons. In this case, injection of steam, with its air entrainment and enhanced mixing, improved the efficiency to that of the natural gas test. (Fig. 2 & 3). Steam also minimized the effect of wind on the flame as the steam flow pattern dominates the ambient wind condition.

Fig. 3—Good combustion efficiency can be maintained even when flaring unsaturated hydrocarbons when some system of enhanced mixing is used such as steam injection. (Same as Fig. 2 except that steam has been added.)

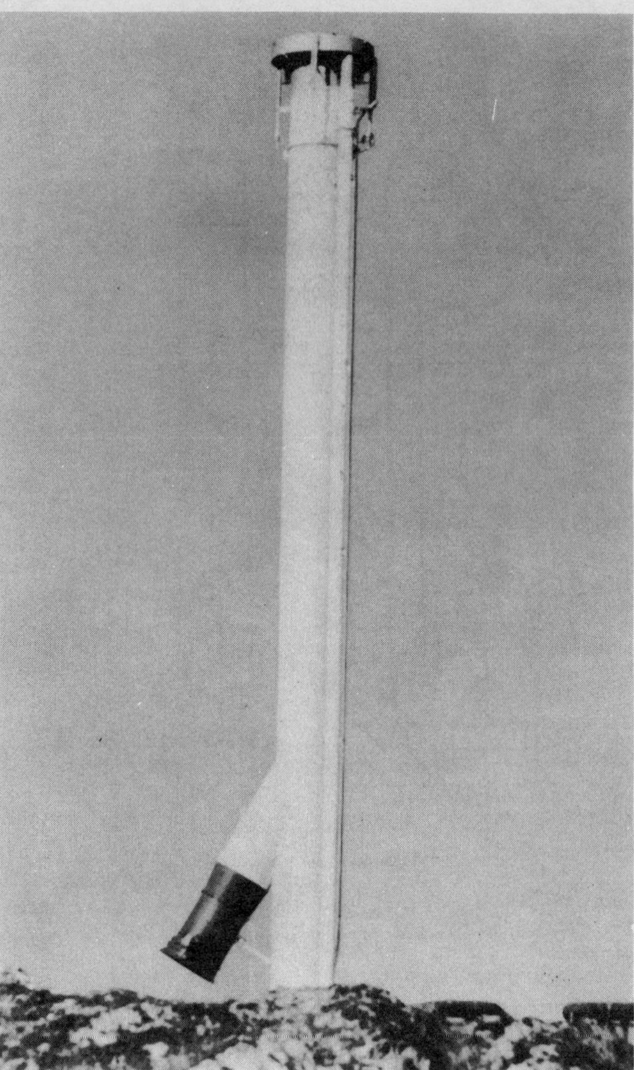

Fig. 4—Flares are available for use with air blowers. This one has a 54-inch air duct feeding air to the 75-foot high flare.

However, excessive steam was always found to be detrimental.

Use and proper control of injection steam is essential to environmental protection. Actual field tests of thermal radiation for varying steam rates confirm the trend and magnitude of the pilot scale test results. Additional laboratory and field testing studies are being conducted. Improved methods and instrumentation for accurate sampling are being reviewed and are under test.

Sampling measurements are quite difficult to carry out even in the controlled environment of a test facility and, at the present time, impossible in the field. Laser or infrared measurements with elaborate electronic computer analysis might make detailed field sampling possible. Extrapolations of the pilot scale data must be used in a combustion and environmental impact model until these new techniques are feasible.

Smokeless operation. Alternate methods of obtaining smokeless operation and complete combustion were also tested. These means are used when steam is not available or its use is uneconomical. In cold climates, steam injection, with its associated freezing problems, creates numerous difficulties. Field production sites and gas processing stations do not have steam boilers. The alternate smokeless methods are: compressed air injection, water spray, high-pressure gas injection, air blower and high turbulence flare tips.

Compressed air injection (Fig. 4) is operationally similar to steam injection. The compressed air entrains additional air and generates turbulent mixing of the air and the waste gas. The cooling effect of compressed air is less than with steam and, therefore, approximately 10% additional mass flow of compressed air is required to achieve the effectiveness of steam. Noise levels are comparable with steam injection but the use of excessive amounts of compressed air is, however, less critical. Compressed air is rarely used because of the high initial and standby compressor costs resulting from the high pressure drop in the header piping and injection system.

Water spray injection into the flame is another alternate. Tests used a fine spray of water in the base of the flare flame. A 4-inch flare was tested, with natural gas, propane and propylene as the flare gases. Water spray ratios to hydrocarbon flow rate ratios were varied from 0 to 5 (weight basis). Excessive water injection, as with excessive steam injection, resulted in an increase in unburnt hydrocarbon emissions. The proper water rate re-

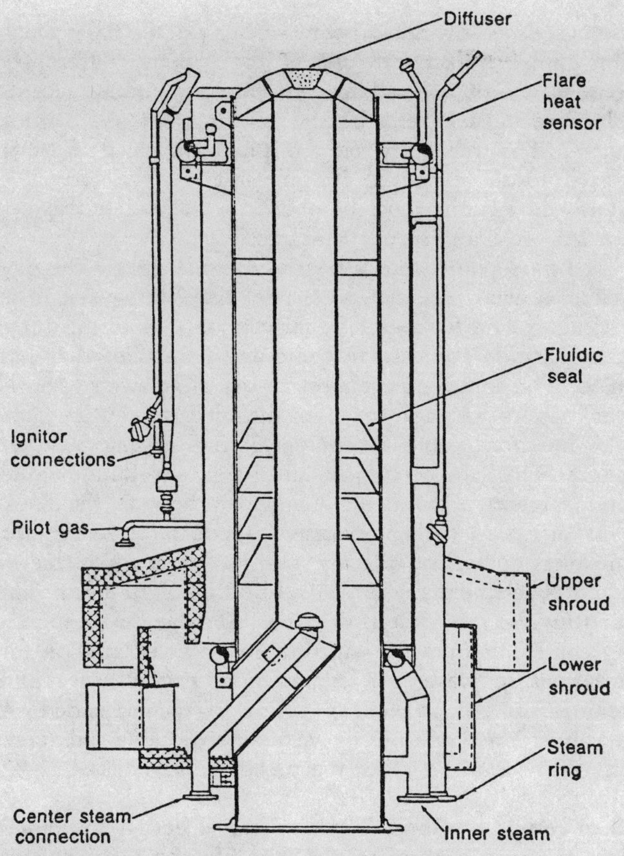

Fig. 5—Special designs are used to minimize the generation of noise.

sulted in smokeless, thorough combustion even of the propylene stream.

A major problem with the water spray was, however, observed. The spray has low momentum and cannot hold the flame in the injection zone, even against mild winds. When the flame moved out of the spray, smoke was produced. A wider spray angle zone might alleviate this problem.

Some water spray flares have been reported using higher water to hydrocarbon ratios.[11] These flares differ in their types of spray as they contain much larger water particles and much of the water passes through the flame without being evaporated and used. Water spray is an effective method of improving combustion efficiency, but it should be applied carefully because it requires critical design and control.

High-pressure gas injection tests were conducted to determine their effectiveness. However, no sampling for unburned hydrocarbons or other plume constituents was conducted. Only qualitative visual observations were made of the flame and its character. These were a definite competition for air between the waste gas and injection gas. The injection gas did generate turbulence and hold the flare flame against the wind, but it also robbed the waste gas flame of air. Unless large quantities of high-pressure injection gas are available at no cost, or with no economic value, then this type of flare is not recommended.

Endothermic or incinerator-type flares are a special case where high-pressure gas injection into the flare flame is not only beneficial but is required to insure combustion of a low Btu waste gas stream. The additional energy provided by the assist gas is necessary to achieve the desired reaction zone temperatures. Any waste gas mixture with a net heating value below 115 Btu/ft.^{3} will require additional energy to bring the over-all mixture of waste and assist gas up to this minimum.[12] Gas streams containing sulfur compounds require that this minimum net heating value be increased for improved environmental dispersion.

Some populated or industrial areas have regulations setting a minimum gas stream heat content. Most areas which permit operation of sulfur or sour gas flares require a pilot thermocouple monitoring system with automatic re-ignition, should a pilot fail. The method of gas injection affects the efficiency of an incinerator flare. Many previous designs injected the assist gas into the base of the flare; therefore, the chemical energy of the assist gas was available for the combustion process. However, the kinetic energy, due to the high gas pressure, was lost. The present practice uses an injection ring at the discharge of the flare which does utilize the kinetic energy. Turbulence and mixing are produced by the injection ring along with reduction of wind effect on the waste gas stream.

Ammonia (incineration) flares. Ammonia applications require special design considerations based upon extensive experience and testing backgrounds. Ammonia has a net heating value of 365 Btu/Ft.^{3}, which is above the previous stated minimums, but assist gas is definitely required for complete combustion. Without assist gas, approximately 8% of the ammonia escapes unburned from the tip of the visible flare flame. If assist gas is added properly, the amount of unburned ammonia issuing from the flare is in the ppm range. Standing within 15 feet of a 4 inch flare at near full capacity, no ammonia odor was detected (the olfactory threshold is 20 ppm). No_x levels were also measured in a low ppm range.

Increased assist gas injection increases NO_x production which could exceed acceptable limits. The amount of assist gas added to the ammonia stream should be automatically controlled to prevent the escape of unburned ammonia or excessive production of NO_x. A flow sensor modulating a control valve on the assist gas line is sufficient if a pure ammonia stream is being flared. If the gas composition is variable, however, then a combination flow/radiant control system should be used to minimize assist gas consumption.

For some ammonia flare applications the possibility of an actual flaring condition is rare. For example, in refrigerated ammonia storage the boil-off is flared when the refrigeration system is down. The use of assist gas for these rare occurrences is not warranted. The cost of the initial equipment and its operation is thereby reduced. When no assist gas is injected, a reduced exit velocity must be used to prevent flame lift-off and blow-out. The flame retention ring alone is not sufficient in this case.

Air blower flares. Smokeless air blower flares produce efficient combustion with low noise levels over wide flow and composition ranges using simple electronic controls.

The basic mechanism of operation is similar to a steam injection flare. The motor driven fan supplies air and turbulent mixing for complete oxidation of the waste gas. The fan supplies primary air for premixing with the waste gas just prior to combustion. Secondary air is drawn in by the natural buoyancy of the flame above the flare tip. The forced, primary air is sufficient to prevent smoke formation and to maintain efficient combustion. The main advantage of the air blower flare is its high energy effectiveness. For a high velocity steam injector about 6-8% of the energy of the steam jet is used for the movement of air. With an air blower, working at only inches of water column static pressure, the energy for moving air is roughly 10 times more efficient. The blower also produces an artificial vertical wind to hold the flare flame in the mixing zone even with strong crosswind conditions.

Air blower flares are being used in Alaska for smokeless operation under all types of weather conditions. The combustion performance of the forced air flare is equivalent to that of the natural gas case. Low noise, less critical air to hydrocarbon ratio and simplified control make the air blower one of the most environmentally compatible flare types.

Multiple high-velocity flares. Complete combustion and smokeless operation can be obtained without steam, gas assist, water spray or air blower by means of multiple high-velocity jet-type burners. These burners are used in elevated, pit and ground flares. Many small flames spread out the waste gas flow to contact the air needed for combustion. The high velocity provides rapid air entrainment. In the previous flare designs discussed, the combustion air was brought to the waste gas; the jet-type burners employ the reverse approach.

For elevated flares, the exit velocity is the key to high combustion efficiency. For pits and enclosed ground flares, the exit velocity is less crucial. The natural draft inherent in a pit or enclosure generates sufficient turbulence to insure smokeless operation and full burning efficiency. Multiple valve staging systems are used for flow turndown. The ground level location of the multiple high-velocity jet-type burners enables easy inspection, sampling and maintenance.

Noise. The flare also represents a significant source of noise. For a given hydrocarbon the noise generated is a function of the flow rate, turbulent mixing and steam injection.[13] Thus flare design must be a trade off between exit velocity and turbulent mixing in the over-all design of an efficient, low-noise flare tip (Fig. 5).

Enclosed ground flares. Enclosed ground flare units provide the closest control on the flare combustion process, with close temperature and air control. The emissions can be easily monitored from an attached platform, since there is no thermal radiation hazard. Sampling an elevated flare is virtually impossible because of inaccessibility due to high thermal radiation near the top of the stack.

Other factors. Numerous control systems have been developed to monitor the various components of flare systems. Especially critical components are the flare pilots. The pilots, in fact, are one of the primary differences between a vent and a flare. An environmentally acceptable flare must have a highly reliable pilot and ignitor system. The reliability of the pilot system is directly related to the potential for an unignited release. The pilots and ignitors must be able to withstand hurricane-strength wind and severe weather.

The flare pilot flame is nearly invisible during the day against a clear blue sky, so a pilot monitoring system is advisable. The flame-out signal can be tied to an automatic ignition panel to re-ignite any flare pilot that has failed. The controls are almost always AC-powered; however, battery backups are often included, to continue the pilot monitoring and automatic ignition during a power failure. This can be critical, since a power failure quite often triggers a maximum emergency flow to the flare.

As discussed earlier, excessive injection of steam, for smokeless operation, or assist gas, for thorough incineration, has detrimental effects. An injection of an insufficient quantity also has obvious ill results. Flexible, fast-responding control equipment, with automatic compensation for variations in waste gas composition, exit velocity and steam/assist gas injection ratios, is recommended. A dual-function system, using velocity and flame radiation, will compensate for all of the preceding variables.

Final considerations. The mechanical design and materials of construction affect the flare life and can influence its environmental impact. A flare with a ruptured steam main or burned out premixing chamber cannot perform as originally specified. The materials of construction in the high-temperature zone should be of high-quality alloy with exacting welding requirements enforced.

Some basic rules for proper flare mechanical design should be observed: *No* moving parts; *No* burning inside the flare tip; and *No* small openings for steam or gas injectors.

LITERATURE CITED

1 Brzustowski, T. A., "Flaring in the Energy Industry," *Prog. Energy Combustion Science*, Vol. 2, pp. 129-141, 1976, Pergamon Press, Great Britain.

2 Brzustowski, T. A., and Sommer, E. C., Jr., "Predicting Radiant Heating from Flares," 38th API Midyear Meeting, May 17, 1973, Philadelphia, Pa., Preprint No. 64.73.

3 Hajek, J. D., and Ludwig, E. E., "How to Design Safe Flare Stacks," *Petro/Chem Engineer*, June 1960, pp. C-31-C-37.

4 Kent, G. R., "Find Radiation Effect of Flares," *Hydrocarbon Processing*, Vol. 47, No. 6, June 1968, pp. 119-130.

5 Kent, G. R., "Practical Design of Flare Stacks," *Hydrocarbon Processing*, 1964, pp. 86-90, Vol. 43.

6 Klooster, H. J., Vogt, G. A., and Braun, G. F., "Optimizing the Design of Relief and Flare Systems," *Chemical Engineering Progress*, Vol. 71, No. 1, January 1975, pp. 39-44.

7 Owen, L. E., and Rothrock, W. C., Jr., "Cantilever Flare Boom Design for Offshore Platform," Offshore Technology Conference, Paper Number OTC 2482, May 3-6, 1976, Houston.

8 Steward, F. R., "Prediction of the Height of Turbulent Diffusion Buoyant Flames," *Combustion Science and Technology 1970*, Vol. 2, pp. 203-212, Belfast, N. Ireland.

9 Straitz, J. F., III; O'Leary, J. A., Brennan, J. E. and Kardan, C. J., "Flare Testing and Safety," AIChE 11th Annual Loss Prevention Symposium, March 20-24, 1977, Houston.

10 Vanderlinde, L. G., "Smokeless Flares," *Hydrocarbon Processing*, October 1974, pp. 99-104.

11 Lauderback, W. H., "Unique Flare System Retards Smoke," *Hydrocarbon Processing*, January 1972, pp. 127-128.

12 Hajek, J. D., and Ludwig, E. E., "Part 2—How to Design Safe Flare Stacks," *Petro/Chem Engineer*, July 1960, pp. C-44 to C-51.

13 Swithenbank, J., "Ecological Aspects of Combustion Devices (With Reference to Hydrocarbon Flaring)," *AIChE Journal*, Vol. 18, No. 3, pp. 553-560, May 1972.

Unique Flare System Retards Smoke

A pollution-fighting smoke suppressant setup uses water spray as the key to its unusual arrangement

William Lauderback, Texas Eastman Co.,
Division of Eastman Kodak Co., Longview, Texas

A NOVEL MEANS of smoke suppression is available for horizontal flares.

At its Longview plant, Texas Eastman has nine horizontal flares varying in size from 8 through 42 inches for which conventional smoke suppression equipment was found to be inadequate. The use of conventional equipment would require conversion from horizontal to vertical flares, a costly and time-consuming process.

This prompted the company to begin experimenting with various concepts of smoke suppression. Eventually, a system that employs finely divided water spray to eliminate smoke from horizontal flares was devised. This system is now installed on the nine existing horizontal flares.

Why smoke suppression? To understand the company's concern with the flare smoke problem, you must first look at the history of the hydrocarbon processing industry (HPI). Almost from the beginning, the flaring of unwanted gases and process upsets has been a problem. As the HPI has grown, so has concern over what to do with smoke from flared gases. In fact, flare lines up to 42 inches in diameter are commonly used to relieve major plant upsets safely.

Industry has always been interested in finding ways to reduce air pollution from large flare lines in its plants. In many cases, these stacks are more than 200 feet high.

Why horizontal flares? In the early 1950s, Texas Eastman constructed a hydrocarbon cracking plant in Longview, but decided not to use the conventional practice of flaring the gas high in the air. Instead, it ran the flare system to a non-hazardous area and discharged the gas at ground level.

There were good reasons for this. One was the necessity to break into the flare system for the addition of new equipment or, more often, to remove pressure relieving devices for testing. When this is done with a high stack, a draft is created which pulls air in through the opening in the lower level. If enough air is pulled in flame propagation starting from the lighted pilot can occur back through the flare system possibly causing an explosion in the knockout drums or operating areas.

When the gas is discharged at ground level, however, the air does not rush back into the plant through the opening in the flare line.

Under non-deluge conditions, the system emits a fine, continuous spray of water mist. At right, intensified flaring is suppressed by a heavy deluge. See the drawing for piping design and sprinkler configurations.

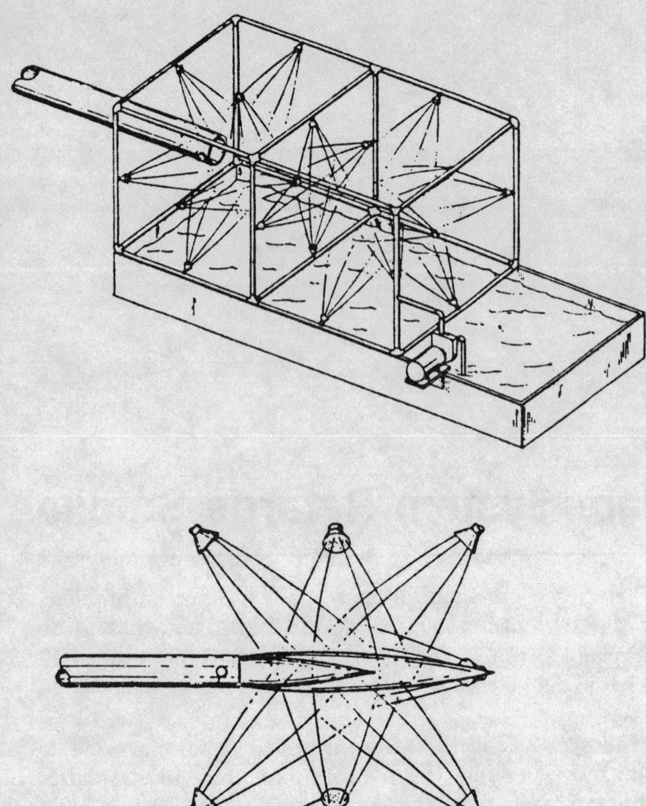

Apparatus is relatively inexpensive to install. Main components are piping, water pump and drive, and spray nozzles. Standard construction materials are suitable, as long as enough water for cooling continuously passes through system.

Injected steam. Steam injection systems, even when backed up by an adequate standby steam supply, have not proven effective on horizontal flares for Texas Eastman. The only other recognized adequate smoke suppression system for horizontal flares requires a costly forced draft air supply system which uses supplementary natural gas firing when flaring even moderately heavy fuels.

Even in vertical flare systems, the amount of steam needed per quantity of input combustible material is high, resulting in a large standby requirement for steam to compensate for sudden, unpredictable discharges.

Suppression equipment. This system consists basically of a number of spray nozzles which produce an extremely fine spray mist designed so that the fire is contained in the water fog. Approximately 0.5 gpm of water is required per cubic foot of the fogged area. When the fire is contained within the fogged area, smoke is eliminated.

During normal pilot operation of the flare, only a small water stream is required to keep the structure, water piping and heads cooled. During heavier flaring, additional water is supplied as necessary to prevent visible smoke.

System's advantages. The Eastman-developed system has the following advantages:

1. Installation cost for most circumstances is considerably less than any other known system. For a system using manual controls, the main items of expense are a water pump and drive, spray nozzles and related piping.

2. Operating costs are low. Nearly any source of available water which can be pumped through spray nozzles may be used, including industrial waste water. The water can be recirculated and make-up water is needed only for evaporation. By using multiple or variable speed pumps, power consumption can be minimized by pumping only enough water at any time to control smoking. In our experience, maximum water rates required have ranged from 1,925 gpm for a 42-inch flare burning 50,000 pounds per hour of a hydrocarbon mixture to 525 gpm for a 12-inch flare burning a lesser amount.

3. The system has been effective at Texas Eastman in eliminating smoke from horizontal flares and complies with the requirements of the Texas Air Control Board. Smoking from heavy flaring from a commercial-size cracking plant is eliminated seconds after the water is directed around the flame from a properly designed set of spray nozzles.

4. Standard materials of construction have been satisfactory, in our brief experience, provided some water flow is maintained through the nozzles at all times for cooling.

5. The system is uncomplicated to operate and maintain. Simple manual controls are adequate, enabling the operator to quickly turn on additional water to eliminate smoke should an upset occur.

6. Liquids in the system drain and burn, eliminating costly knockout drums and related equipment.

A typical flare arrangement is shown in the figure.

Tested locally. This process has been used under operating conditons at Texas Eastman's Longview plant. Variations in operating parameters at other locations or in other HPI operations may affect the operability or safety of the process. Obviously, these aspects should be taken into account before installing the system elsewhere.

How Accurate Are Dispersion Predictions?

Know what you are calculating when you predict the plume dispersion from stacks. All is not as it seems—nor as others would say!

M. R. Beychok, Consultant, Irvine, Calif.

MATHEMATICAL MODELS of stack gas plume dispersion have evolved theoretically since the 1930's and earlier. Stringent environmental regulations have stimulated an immense growth in the use of dispersion calculation models in the last decade. Yet, until recently[1], there has been no single reference text which chronicled the evolution and derivation of the so-called Gaussian dispersion models currently in wide usage. Unfortunately, lacking an in-depth understanding of the many assumptions and constraints involved in the Gaussian models, many users have fallen into the trap of confusing mathematical precision with accuracy. This has lead to a widespread belief that dispersion models predict stack gas plume concentrations within a factor of two or three of the actual concentrations in the real world. This belief is fostered by some EPA computer dispersion programs which print the message:

"Concentration estimates may be expected to be within a factor of three (under some conditions)."

This belief has very little factual foundation. *The consistent prediction of actual plume concentrations within a factor of as high as ten is probably a more realistic assessment for the short time-average models.* Seemingly small inaccuracies in the input data and assumptions can result in changing the plume concentration predictions of the Gaussian dispersion models by factors ranging from 6 to 80.

Most dispersion models in current usage include three steps:

- Calculation of stack gas plume rise after exiting the stack.

- Calculation of dispersion of plume components from effective stack height (source stack height plus plume rise) assuming that dispersion follows a Gaussian distribution.

- Conversion of so-called 'instantaneous' concentrations predicted by the Gaussian model into concentrations reflecting a 1-hour averaging period. The 1-hour averages may then be used in conjunction with hourly meterological data to predict 3-hour, 24-hour and annual average concentrations.

There are literally dozens of methods which have been

TABLE 1—1-Hour ground-level concentrations (ug/m^3) calculated under plume centerline by:

Receptor downwind distance (km)	(1) Base model	Adjusted model	Adjusted model plus wind shift correction	(2) Adjusted model plus wind shift and C_{10}/C_{60} corrections	Over-prediction Ratio, (1)/(2)
2	16.1	1.1	0.5	0.2	80
3	24.3	9.2	4.4	1.7	14
4	20.5	13.8	6.0	2.4	9
5	15.8	13.7	5.8	2.3	7
6	12.0	12.0	4.7	1.9	6
7	9.4	10.1	3.9	1.5	6
8	7.4	8.4	3.1	1.2	6
9	6.0	7.0	2.5	1.0	6
10	5.0	5.9	2.0	0.8	6

(1) BASE MODEL: Briggs' plume rise, Gaussian dispersion, power-law conversion of surface wind velocity (using exponent of 0.15) to obtain wind velocities at stack height (for determining plume rise) and at plume centerline height (for determining dispersion). Concentrations taken to be 1-hour values as per the EPA.

(2) ADJUSTED MODEL: Same as the base model, except that calculated plume rises were increased by 20 percent and Pasquill vertical dispersion coefficients (sigma z) were decreased by 25 percent. Concentrations taken to be 10 minute values.

WIND SHIFT AND C_{10}/C_{60} CORRECTIONS: See text.

proposed for the calculation of stack gas plume rises. However, the EPA has more or less adopted the Briggs' method,[2] causing the method to be widely used. Briggs has modified his equations a number of times and this is his 1972 version:

For unstable or neutral atmospheric turbulence conditions and for buoyancy factors (F) greater than 55 m^4/sec^3, the plume rise (Δh) equations are:

$$\Delta h_{max} = 1.6F^{1/3}(3.5x^*)^{2/3}u^{-1} \qquad (1)$$

at any distance beyond the point of maximum rise which is taken to occur at $3.5x^* = 119F^{0.4}$, and

$$\Delta h = 1.6F^{1/3}\, x^{2/3}u^{-1} \qquad (2)$$

at any distance (x) before the point of maximum rise.

For unstable or neutral atmospheric turbulence conditions and for buoyancy factors (F) less than 55 m^4/sec^3, the plume rise equations are the same as equations (1) and (2) except that the point of maximum rise is taken to occur at $3.5x^* = 49F^{0.625}$.

For stable atmospheric conditions, the plume rise equations are:

$$\Delta h_{max} = 2.4(F/us)^{1/3} \qquad (3)$$

at any distance beyond the point of maximum rise which is taken to occur where $x = 2u/\sqrt{s}$, and

$$\Delta h = 1.6F^{1/3}\, x^{2/3}u^{-1} \qquad (4)$$

at any distance (x) before the point of maximum rise.

In all of the above equations, the variable u is the wind velocity at the top of the stack, the variable s is Briggs' stability parameter and the buoyancy factor F is defined by the sensible heat emission of the stack gas.[1]

Although EPA now uses Briggs' plume rise equations in computer dispersion models, their widely used Turner's Workbook[3] recommends Holland's plume rise equations and the Workbook has yet to be revised. Also, Briggs' many changes in his equations has led to confusion as typified in the 1977 API publication on dispersion models[4] which uses Briggs' 1969 version rather than his 1972 version. There are few who would dispute that calculated plume rises could easily over or underpredict actual plume rises by at least twenty percent. In fact, Briggs' own changes in his equations reflect *at least* that degree of uncertainty.

The Gaussian dispersion equation for predicting ground-level centerline and crosswind concentrations underneath an elevated stack gas plume (in the absence of any inversion layer aloft which would limit the vertical dispersion upward) is:

$$C = \frac{Q}{u\pi\sigma_z\sigma_y} \exp\left(-y^2/2\sigma_y^2\right) \exp\left(-H_e^2/2\sigma_z^2\right) \qquad (5)$$

The variables in equation (5), expressed in a consistent set of dimensional units, are:

C = the predicted ground-level concentration (in ug/m^3) of a given plume component, such as SO_2, at a receptor located x meters downwind and y meters crosswind of the source stack

Q = the mass emission rate of the given plume component (in ug/sec) exiting from the source stack

u = the wind velocity (in m/sec) at the plume centerline height or effective stack height

H_e = the plume centerline height or effective stack height (in m) at the downwind distance of x

y = the crosswind distance (in m) from the plume centerline to the receptor

σ_z and σ_y = the Pasquill vertical and crosswind dispersion coefficients (in m) which are a function of the downwind distance x and of the prevailing atmospheric turbulence conditions

The Pasquill dispersion coefficients are the product of a long history of field experiments and empirical judgments and extrapolations of the data from those experiments. There are few knowledgeable practitioners of the dispersion modeling art who would dispute that the Pasquill coefficients could easily have an inherent uncertainty of plus or minus 25 percent. In fact, Pasquill himself has recently proposed[5] a thorough re-examination of the coefficients and has offered some specific revisions to acknowledge the effect of plume dilution during buoyant rise on the vertical dispersion coefficient and the effect of the turning of the wind direction with height on the crosswind dispersion coefficient.

One of the major questions concerning the Gaussian dispersion equation (5) is that of defining what the calculated concentration C really represents. It is an instantaneous value or does it represent an average concentration over some time period ranging from 3 minutes to 1 hour? Turner's Workbook,[3] which is an EPA publication, says that the Gaussian dispersion equation yields concentrations representing 3 to 15 minute averages. The recent API publication on air dispersion models[4] says that the concentrations probably represent 10 to 30 minute averages. Many, many other publications have said that the concentrations represent averaging times ranging from 5 to 30 minutes with the preponderance of opinion agreeing on a range of 10 to 15 minutes. Despite that body of opinion, most of the EPA's computer dispersion models assume that the Gaussian dispersion equation yields a 1-hour average concentration. *THIS IS A VERY IMPORTANT POINT WHICH MERITS MORE THAN MERE AQUIESCENCE TO AN EPA ASSUMPTION THAT HAS NEVER BEEN RATIONALIZED IN THE TECHNICAL LITERATURE.* One of the commonly accepted conversions of short-term concentrations from one averaging time (t_1) to another averaging time (t_2) involves this equation:

$$C_1/C_2 = (t_2/t_1)^n \quad (6)$$

where n ranges from 0.2 for stable atmospheric conditions to 0.7 for highly turbulent conditions.

Assuming that $n = 0.52$ is representative of moderate-to-strong turbulence, equation (6) indicates that a 1-hour concentration value would be only 40 percent of the equivalent 10-minute value. Therefore, under those conditions of turbulence, the EPA's assumption that the Gaussian models yield 1-hour values (instead of 10-minute values) constitutes a 'built-in' overprediction factor of as much as 2.5 which certainly deserves some explanation.

In addition to the above discussed areas of uncertainty involved with calculated plume rises, the accuracy of the Pasquill dispersion coefficients, and determination of the concentration averaging time resulting from the use of those dispersion coefficients, it is important to understand the host of other assumptions and constraints involved in the derivation of the Gaussian dispersion equation:

- Wind from the source stack to the receptor is constant in velocity and direction. For a wind velocity of 2 m/sec and a travel distance of 10 km, that requires constant conditions for about an hour and a half.

- Atmospheric turbulence conditions are constant and homogeneous throughout the vertical and crosswind regimes from the source stack to the receptor.

- There is no deposition of plume components which reach the ground and such components are totally reflected from the ground back into the plume. There is no absorption of plume components by the ground, bodies of water or vegetation, nor is there any chemical transformation of plume components.

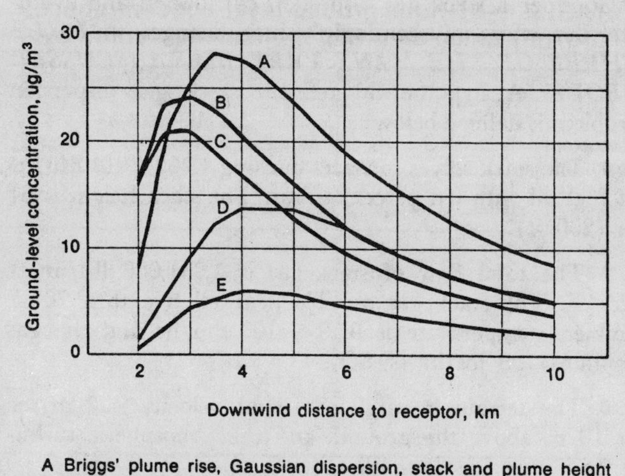

A Briggs' plume rise, Gaussian dispersion, stack and plume height wind velocities the same as the surface wind velocity.
B Briggs' plume rise, Gaussian dispersion, stack and plume height wind velocities obtained by power-law conversion of surface wind.
C Same as B (base model), except that calculated plume rises were increased by 20 percent and Pasquill dispersion coefficients were increased by 25 percent.
D Same as B (base model), except that calculated plume rises were increased by 20 percent and Pasquill dispersion coefficients were decreased by 25 percent.
E Same as D, except that concentrations reflect a 10° shift in wind direction.

Fig. 1—Effect of assumptions on dispersion predictions.

- Dispersion in the downwind direction is negligible relative to the wind transport. Only vertical and crosswind dispersion occurs.

- Diffusion patterns are probabilistic and can be described exactly by Gaussian distribution.

- The plume expands in conical fashion as it travels downwind, whereas the ideal 'coning' plume is only one of many observed plume behaviors.

- Terrain conditions underneath the plume can be accounted for by using one set of dispersion coefficients for rural terrain and another set for urban terrain. The basic Gaussian dispersion models cannot handle complex terrain features such as valleys, mountains or shoreline regimes.

In other words, the basic Gaussian dispersion model represents an ideal steady-state of homogenous meteorological conditions, idealized plume geometry, uniform flat terrain, complete conservation of mass, and exact Gaussian distribution. Such ideal conditions very rarely occur in the real world.

It is equally important to understand that the plume rise equations, which provide the emissions height used in the Gaussian dispersion equation, also assume steady-state homogeneous conditions of wind velocity, wind direction and atmospheric turbulence. Nor do those equations include aerodynamic downwash and wake effects upon the plume.

Consider a stack gas dispersion calculation and evaluate the effect of seemingly minor changes in *ONLY THREE OF THE MANY VARIABLES DISCUSSED ABOVE.* A hypothetical refinery stack gas dispersion problem is defined below:

- The stack serves furnaces burning 1,200 x 10^6 Btu/hr of fuel oil with 0.6 percent sulfur. The stack height is 61 m (200 ft).

- The total flow of stack gas is 1,031,000 lbs/hr at 450° F. The stack gas sensible heat relative to a 70° F ambient temperature is 97.9 x 10^6 Btu/hr and the gas contains 758 lbs/hr of SO_2.

- The terrain is rural, the wind velocity is 2 m/sec at 10 m above the ground, and the atmospheric turbulence is typed as Pasquill Class B.

- The calculated plume buoyancy factor is 251 m^4/sec,[3] and the maximum plume rise is 532 m occurring 1,084 m downwind assuming the wind velocity at the stack top is also 2 m/sec.

This problem is identical with the one illustrated in the recent API publication.[4] The three variable changes to be evaluated are:

1—The calculated plume rises will be increased by 20 percent to reflect a reasonable degree of uncertainty in those calculations.

2—Pasquill's *vertical* dispersion coefficients will be changed by 25 percent to reflect a reasonable degree of uncertainty in their values.

3—A 10° shift in wind direction will be assumed, which means that the downwind receptor concentrations will reflect some crosswind dispersion.

Figure 1 summarizes five sets of Gaussian dispersion calculations to obtain ground-level SO_2 concentrations at receptors ranging from 2 km to 10 km downwind:

A—In this set, the surface wind velocity was used for the plume rise and Gaussian dispersion calculations.

B—In this set, stack top and emissions height velocities were obtained by a power-law conversion of the surface wind velocity (using an exponent of 0.15). Since most of the EPA's dispersion models use a power-law wind profile, this set is defined as our 'base model'.

C—The base model plume rises were increased by 20 percent and the *vertical* dispersion coefficients were increased by 25 percent.

D—Same as C except the vertical dispersion coefficients were decreased by 25 percent.

E—Same as D except a 10° wind shift was assumed (amounting to a crosswind shift of 350 m at 2 km downwind and 1,750 m at 10 km downwind).

Figure 1 illustrates quite graphically how much the base model (calculation set B) overpredicts receptor concentrations compared to the adjusted model (calculation set E).

Table 1 quantifies the results shown in Figure 1 and also includes the effect of the EPA's assumption that the Gaussian models yield 1-hour averages rather than 10-minute averages. When that effect is included, Table 1 shows the cumulative overprediction ratio from our hypothetical example ranges from a factor of 6 to 80 . . . resulting from seemingly minor changes in a few of the many variables and from assuming that the Gaussian models yield 10-minute averages rather than 1-hour averages.

The usefulness of the Gaussian dispersion models should not be denigrated. Such models are the best tools available and they are good tools. However, the models are only tools and do not necessarily reflect the real world. Precision should not be confused with accuracy!

LITERATURE CITED

[1] Beychok, M. R., "Fundamentals of Stack Gas Dispersion," Irvine, California, 1979.

[2] Briggs, G. A., "Plume Rise," USAEC Critical Review Series, NTIS TID-25075, 1969 (subsequently revised in other publications in 1971, 1972 and 1975).

[3] Turner, D. B., "Workbook of Atmospheric Dispersion Estimates," EPA Publication AP-26, revised 1970.

[4] "Gaussian Dispersion Models Applicable to Refinery Emissions," API Publication 952, October 1977.

[5] Pasquill, F., "Atmospheric Dispersion Parameters in Gaussian Plume Modeling, Part II: Possible Requirements for Change in Turner Workbook Values," EPA-600/4-76-030b, June 1976.

Flue Gas Desulfurization Technology

What are the possibilities? This analysis sketches the boundaries in which processes must be designed

Albert B. Welty, Jr.
Esso Research and Engineering Co., Linden, N. J.

FLUE GAS DESULFURIZATION is expected to take a place alongside fuel desulfurization as an important air conservation process for power plants and for the hydrocarbon processing and other heavy fuel-consuming industries. However, selection and design of flue gas desulfurization processes are now confused by uncertain economics, a proliferation of unproven processes and compelling needs. Some confusion arises from the need for a better and orderly understanding of the fundamentals of flue gas desulfurization.

The technology of flue gas desulfurization can be understood better by reviewing the chemical and engineering possibilities for developing such processes. Although no single design best meets the needs of all existing combinations of services, markets and supplies, better evaluations can result from a review of the broad principles involved.

DESIGN CRITERIA

Successful flue gas desulfurization processes must be economical, reliable, simple and designed to fit the operator's physical and business situations. A number of requirements can be generalized for these objectives:

- Installations must be large. Investment, overhead and operating costs preclude economical applications where fuel consumption is small. Most power plant and some HPI and other large industrial installations are worth consideration for flue gas desulfurization.

- Reagent regeneration is desirable. Non-regeneration is practical only where very large amounts of cheap reagent are readily available and where the large quantities of solid waste can readily be disposed of without upsetting the ecology of the surrounding land and water system. Furthermore, a valuable credit can normally be obtained for the usual byproducts, sulfur or sulfuric acid. For example, in a typical power plant burning a high sulfur fuel, this credit could amount to 0.09-0.18 or 0.29-0.58 mils/kwh based on sulfur at 10-20/long ton or acid (100 percent) at $10-20/short ton, respectively.

- Reaction time must be short to avoid unreasonably large reactors. For a 400-mw power plant, a 19,000-cubic-foot reactor is required to provide one-second contact time at 150° C.

- Reactor pressure drop must be small. A drop of only 25 inches of water in a reactor running at 150° C corresponds to an energy loss of 4,600 hp in a 400-mw plant.

- Operation must be relatively trouble-free since flue gas desulfurization is associated with processes requiring long, uninterrupted runs.

- Local or area sulfur limitations must be met while using high-sulfur fuels. Typical flue gas compositions produced from 3-4 percent sulfur fuels are shown in Table 1. Flue gas desulfurization processes should remove about 90 percent of the sulfur oxides, so that the desulfurized flue gas contains only approximately 0.025 percent SO_2, even when burning these very high sulfur fuels. In the desulfurization, no significant amounts of SO_3 or H_2S should be allowed to escape with the flue gas.

- The process preferably should be installed with the minimum disruption of normal plant designs. In power plants this means there is an advantage for placing the reactor in the flue gas after the air preheater or else between the economizer and the air preheater, where normal temperatures are about 150° C and 350° C, respectively.

TABLE 1—Typical flue gas composition

	Volume %	
	Oil-Fired*	Coal-Fired**
SO_2	0.245 } 0.25	0.245 } 0.25
SO_3	0.005	0.005
O_2	1.0	3.0
CO_2	13	15
H_2O	10	7
N_2	76	75

* 4% S fuel. ** 3% S fuel.

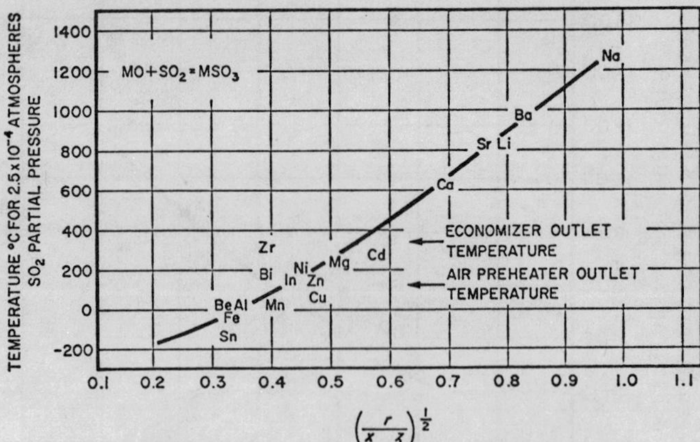

Fig. 1—Equilibrium temperature for 0.025%, vol., concentration of SO_2 for various metal oxide-sulfite systems.

TABLE 2—Oxidation of S(IV) to S(VI)

	Temperature °C	Equilibrium Constant K
$SO_2 + \frac{1}{2} O_2 = SO_3$	1,000	1.5×10^{-1}
	700	2.5
	400	5.4×10^{2}
	100	7.3×10^{8}
$SO_3^{=} + \frac{1}{2} O_2 = SO_4^{=}$	25	10^{43}
$HSO_3^{-} + \frac{1}{2} O_2 = HSO_4^{-}$	25	10^{40}
$CaSO_3 + \frac{1}{2} O_2 = CaSO_4$	700	10^{10}
	300	10^{19}

TABLE 3—Combining tendency of various cations

	$\left(\frac{r}{x\ z}\right)^{1/2}$		$\left(\frac{r}{x\ z}\right)^{1/2}$
Cs	1.4	In(III)	0.43
Rb	1.3	Bi(III)	0.40
K	1.2	Mn(III)	0.39
Na	0.97	Pb(IV)	0.37
Ag	0.96	Ti(IV)	0.36
Ba	0.83	Fe(III)	0.34
Hg(II)	0.81	Ga(III)	0.34
Li	0.79	Al	0.34
Sr	0.76	Be	0.32
Ca	0.69	Sn(IV)	0.32
Cd	0.58	Ge(IV)	0.26
Y(III)	0.52	Sb(V)	0.25
Mg	0.51	Si	0.24
Cu(II)	0.48	As(V)	0.20
Zn	0.47	B	0.18
Ni(II)	0.45	Graphite	0.13
Zr(IV)	0.40		

REACTIONS

The chemical possibilities for flue gas desulfurization are usually based on SO_2-containing gas. However, under special conditions H_2S-containing gas from partially consumed fuel can be treated and the desulfurized gas further burned to recover the remaining heat. Different reactions are involved in these two approaches.

SO_2-containing gas. Several different groups of reactions are possible: conversion to acid, sulfur or other byproducts, dry absorption and wet scrubbing. In all groups two basic conditions must be considered: effect of carbon dioxide on the reaction and oxidation of sulfur(IV) to sulfur(VI).

Fortunately carbon dioxide in the flue gas has little effect on the desulfurization reactions. The equilibrium governing formation of sulfites from sulfur dioxide is generally much more favorable than the equilibrium for the formation of carbonates from carbon dioxide.[1] Practically, there is no significant effect of carbon dioxide on the reaction of sulfur dioxide at most conditions.

By contrast, the equilibrium favoring oxidation of sulfur(IV) to sulfur(VI) by the oxygen remaining in the flue gas is significant at most temperatures (Table 2). Thus it must be assumed that the desulfurization reaction product of all removal processes will contain at least some sul*fate*, not just sul*fite*.

Direct conversion to acid. Most of the sulfur dioxide can be converted to sulfur trioxide at moderate or low temperature (Table 2) and in the presence of an appropriate catalyst. At still lower temperatures the steam in the flue gas converts the sulfur trioxide to sulfuric acid.

Using this technique the flue gas must be cooled enough to avoid unacceptably high sulfuric acid vapor pressure in the flue gas vented to the atmosphere. Assuming that 5 ppm is permissible, the flue gas must be cooled to 110-115° C; the equilibrium concentration of the acid produced is then 74-75 percent.[2] By moving the acid countercurrent to the flue gas, more concentrated acid can be withdrawn; at 275-300° C, 93 percent acid can be made. But this is expensive. Other processes oxidize the sulfur dioxide in aqueous solution, making dilute acid. Corrosion in a large heat exchange system and the water inherently present in the acid are disadvantages of all these processes.

Conversion to sulfur. The Claus reaction is possible by introducing a stoichiometric amount of hydrogen sulfide along with an alumina catalyst into the gas stream cooled to 150° C.[3] The sulfur forms on the alumina surface and may be withdrawn with it for recovery by heating. Care must be taken to avoid escape of any hydrogen sulfide into the effluent flue gas, however.

Dry absorption. A number of reagents, potentially suitable for dry absorption, are more or less effective depending on their thermodynamic properties, the temperature desired to carry out the reaction, the rate at which the

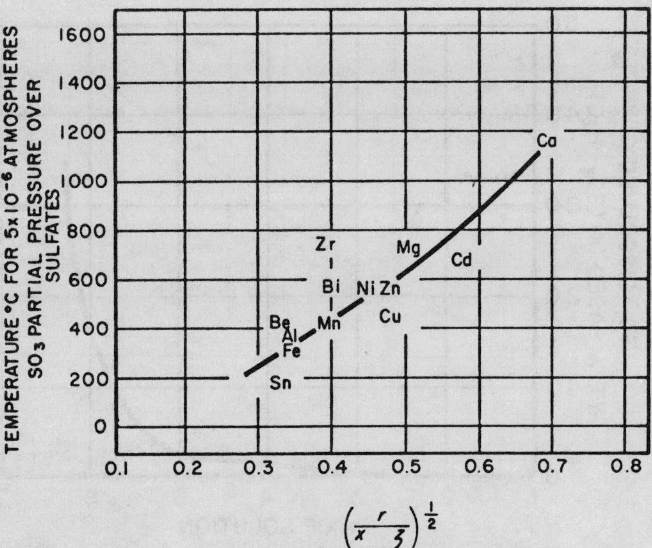

$$\left(\frac{r}{x \cdot z}\right)^{\frac{1}{2}}$$

Fig. 2—Maximum temperature at equilibrium between 5 ppm SO_3 in flue gas and sulfates.

reaction occurs and other considerations. Although these reagents are not normally in the pure state, data for the pure crystalline materials[1,4-6] can be used to evaluate their potential effectiveness. A characterization factor developed for this purpose is:

$$\left(\frac{r}{x \cdot z}\right)^{1/2}$$

where r is the cation radius, x is the Allred-Rochow electronegativity and z is the valence of the cation. The most reduction in the concentration of sulfur dioxide at equilibrium occurs with the most electropositive cations and, for a given electropositive character, for those that are larger and have a lower valence. Values of this parameter (Table 3) suggest that cesium should give the greatest reduction in the sulfur dioxide content, and boron and carbon (graphite) the least, which is in accord with practical experience.

TABLE 4—Sulfate reduction to sulfite

$$MSO_4 + H_2 = MSO_3 + H_2O$$

	Temperature °C	Free Energy Change ΔG_R kcal/mole	Equilibrium[a] Partial Pressure of SO_2 Atmos.
$Sn(SO_4)_2$	180	−6	1×10^5
$Fe_2(SO_4)_3$	360	−12	8×10^4
$Al_2(SO_4)_3$	380	−12	4×10^4
$BeSO_4$	420	−14	3×10^3
$Mn_2(SO_4)_3$	430	−11	3×10^4
$CuSO_4$	460	−10	2×10^4
$NiSO_4$	570	−7	4×10^1
$ZnSO_4$	570	−12	5×10^2
$Bi_2(SO_4)_3$	580	−10	2×10^3
$CdSO_4$	680	−8	1×10^2
$MgSO_4$	983	−8	9×10^3
$CaSO_4$	1170	−1	4×10^1

[a] SO_2 partial pressure over sulfite at equilibrium at the temperature indicated, which is the maximum temperature allowable based on Fig. 2.

Sulfite reaction. When sulfur dioxide is absorbed from the flue gas as a sulfite, the cations that are potentially suitable depend on the temperature, the sulfur dioxide concentration acceptable to leave in the flue gas and how closely equilibrium is approached. The relationship between the characterization factor and the highest temperature that can be used if equilibrium is reached is shown (Fig. 1) for an assumed residual 0.025 percent SO_2 in the flue gas. Only cations with a characterization factor greater than 0.55 could be suitable at the economizer outlet (350° C). These include calcium and more electropositive cations. But at the air preheater outlet temperature (150° C), those cations with a characterization factor greater than only about 0.45 could be suitable; thus the possibilities are extended to include, for example, bismuth, zinc and magnesium.

Active carbons can not be characterized in these "cation" terms. However, they will absorb sulfur dioxide from flue gas at a temperature around 200° C and act as though they had a characterization factor of about 0.48.

Regeneration of the sulfites formed from these reactions is theoretically possible by heating to 100-200° C above the flue gas desulfurization temperature. At these temperatures the equilibrium concentration of sulfur dioxide in the off-gas is calculated to be 1/10 atmosphere, which is sufficient for feeding an acid operation or other recovery process. Unfortunately, however, this is not an adequate procedure since *sulfates* are not regenerated at these temperatures and will accumulate, thus quickly rendering the reagent inactive. These sulfates are formed from the sulfur trioxide in the flue gas, by oxidation of sulfite to sulfate, and also by the disproportionation of sulfite to sulfate plus sulfide, which is prone to occur at the higher temperatures. If these sulfates are to be regenerated, the temperature must be raised several hundred degrees, making the total reheat requirements approach 500-800° C. The very large amount of heat required creates difficult engineering and economic problems.

Sulfate reaction. Since sulfate must be regenerated regardless, there is good reason to oxidize all the sulfur dioxide absorbed from the flue gas "instantaneously" to sulfate. In this way, weaker, less reactive cations can be used to remove the SO_2, and regeneration is made easier.

The equilibrium sulfur trioxide content of the flue gas over the sulfate at the desired reaction temperature will determine which cations may be suitable for this mode of operation (Fig. 2). If, for example, one limits the sulfur trioxide content to 5 ppm of the effluent flue gas and places the reactor after the economizer, cations with characterization factors as low as about 0.35 might be used. These weaker cations require only about 250-400° C heating for regeneration, as compared to the 500-800° C required for the sulfite absorption system. However, even this smaller heat load creates difficult engineering and economic problems.

Other sulfate regeneration systems are possible including use of a reducing gas such as hydrogen. Thermodynamic calculations predict that this system would not be satisfactory since the end product should be sulfide, and formation of sulfide obviously does not regenerate the absorbent. On the other hand, the reduction can occur step-wise to form sulfite first. Then if the temperature is high enough for the sulfite to be unstable and the sulfur dioxide can escape quickly from the system, regeneration would be effected. This is possible from a thermodynamic view as shown by the free energy changes for the reduction of different sulfates and the equilibrium sulfur dioxide partial pressures over the corresponding sulfites (Table 4).

While this approach is possible with cations up through calcium, it is impractical for the more alkaline substances such as sodium. With sodium, a temperature in the range of 1,800° C would be required to generate even one atmosphere sulfur dioxide pressure in equilibrium with sodium sulfite! Even at temperatures much lower than this sulfur dioxide "escape" temperature, any sulfite which tends to form would disproportionate to sulfide and sulfate. Hence in the case of sodium the end-point of reduction is indeed the sulfide. Regeneration of the sodium sulfide so-formed, however, can be effected at moderate temperatures by displacing the sulfide with another anion, e.g., carbonate[8] or aluminate.[9] Of course, this mode of operation consumes more reducing agent than when sulfur dioxide is released from unstable sulfites.

A comparison of free energy changes (Table 5) shows that other reducing agents than hydrogen can be used. From a strictly thermodynamic view, carbon monoxide is preferred at low temperatures and carbon at very high temperatures.

Wet scrubbing. Wet scrubbing systems must be applied after the air preheater since, with the amount of steam usually present in the flue gas, cooling to about 55° C is required to establish a water phase. At this temperature the scrubbed gas usually requires reheating to obtain adequate plume rise from the stack.

Water by itself would only dissolve about 0.02 percent sulfur dioxide at this temperature and at the sulfur dioxide partial pressure in the flue gas. This is impractical. However, sulfur dioxide will react with hydroxyl ions reversibly to form bisulfite ions (HSO_3^-). Thus by maintaining a sufficiently high hydroxyl ion concentration, large amounts of sulfur dioxide can be absorbed.[10]

Maintaining a solution P_H of at least five and preferably above six provides fairly concentrated solutions (Fig. 3). Many alkaline reagents have the capability to maintain this level of P_H, including the hydroxides, sulfites, carbonates, and phosphates of metals such as potassium, sodium, calcium, magnesium, zinc and ammonia. Where some of these reagents are available in low-cost forms such as limestone and dolomite, consideration can be given to non-regenerative processes. However, disposal of the waste product and loss of revenue from the unrecovered sulfur are liabilities. In the long run, regenerative processes are expected to be preferred.

A number of factors must be considered in selecting the appropriate regeneration system: presence of sulfates, solubility of the sulfites and sulfates, conversion reactions and decomposition heat requirements. As with the dry processes, regeneration must account for the sulfates that are always formed, either by direct decomposition or elimination from the system. The simplest regeneration would be removal of sulfur dioxide from the solution by steam stripping. However, a large amount of steam is required[11] so other means need to be considered.

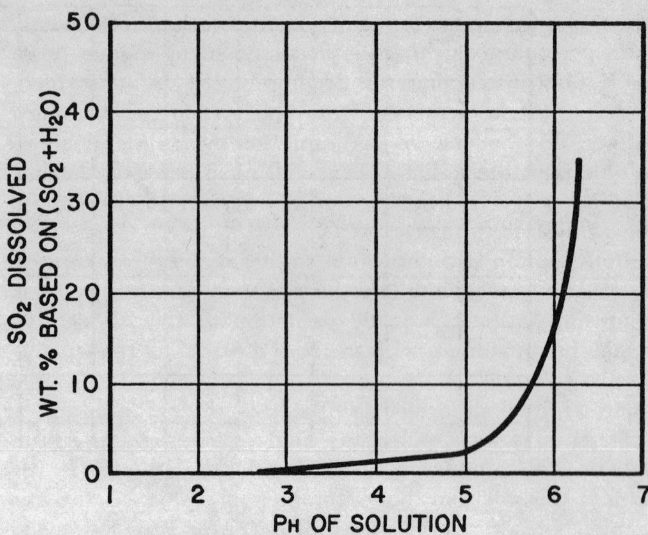

Fig. 3—Amount of SO_2 absorbed in water as HSO_3^-, $SO_3^=$ plus SO_2, at equilibrium at 55°C.

TABLE 5—Relative oxidizing tendency

Temperature °C	Free Energy Change ΔG_R, kcal/mole: $H_2 \rightarrow H_2O$	$CO \rightarrow CO_2$	½($C \rightarrow CO_2$)
300	−51	−55	−47
700	−46	−47	−47
1,100	−41	−39	−47

TABLE 6—Temperature required to generate 0.10 atmosphere sulfur oxide partial pressure at equilibrium

	Temperature °C: Over Sulfite	Over[a] Sulfate
Zinc	240	880
Magnesium	410	970
Calcium	810	1440
Strontium	>1000	>1500
Barium	>1100	>1500

[a] 0.20 atmospheres O_2 partial pressure assumed.

When alkaline earth or zinc scrubbing solutions are used, insoluble sulfites will form; the solubilities at 50° C (mole percent) are: Zn—0.016, Mg—0.15, Ca—0.00065, Sr—0.00036 and Ba—0.00009. The precipitate can be separated and regenerated by heating. Here, as in the dry processes, however, heating must be high enough to decompose the sul*fate* formed, not just the sul*fite* (Table 6). Zinc and magnesium compounds are the most desirable ones because the sulfates can be decomposed at the lowest temperatures.

Potassium and sodium sulfites are very soluble in water and furthermore are not decomposable at practical temperatures. However, the potassium and sodium cations are so electropositive that they can form the pyrosulfite ion ($S_2O_5^=$) from concentrated sulfite solutions and sulfur dioxide at scrubbing conditions, thus desulfurizing the flue gas. Also potassium pyrosulfite is less soluble than the sulfite (6.0 vs. 12.2 mole percent at 50° C) and so can be precipitated readily from solution. The potassium pyrosulfite can then be decomposed by modest heating.[12] This process does not decompose the sulfate formed, and it must be removed from the system as potassium sulfate. In the case of sodium, the pyrosulfite is more soluble than the sulfite (7.1 vs. 5.0 mole percent) so that it is necessary to heat the entire solution to decompose the pyrosulfite formed.

Both sodium and potassium sulfites may be regenerated by other means including reduction with an agent such as potassium formate,[13] or by reduction to the sulfide followed by treatment with carbon dioxide to release hydrogen sulfide.

An entirely different approach is the addition of a stronger acid than sulfurous acid to release the sulfur dioxide and produce a strong acid salt. In special instances the salt may have commercial value, as for example, potassium phosphate for fertilizer use. However, in most cases the strong acid salt will need to be regenerated. In the case of the sulfate this can be accomplished by electrodialysis. Electrodialysis separates dilute sodium sulfate solution into dilute sodium hydroxide (for use in the scrubber to form sodium bisulfite) and dilute sulfuric acid (to react with the sodium bisulfite to form sodium sulfate).[14] The sulfate formed from sulfur trioxide in the flue gas or by oxidation in the system is decomposed in the dialyser also and appears as "excess" dilute acid.

Another system which has been proposed[15] is to use citric acid solution for scrubbing. Citric acid forms a complex with bisulfite ion ($HSO_3 \cdot H_3Cit^{=}$) thus removing sulfur dioxide from the flue gas. Hydrogen sulfide subsequently decomposes the complex in regeneration, with precipitation of sulfur.

Organic solvents. Use of organic solvents is an alternate to wet scrubbing systems. However, with this approach, solvent loss is a major problem. The large volumes of gases to be scrubbed accentuate vaporization losses and create both cost and air pollution problems. For example, about 5 tons of solvent would be lost per day in a system serving a 400-mw power plant if the partial pressure of the solvent were only 0.01 mmHg. Any entrainment would add to this. One solution to this problem would be to put the solvent on an absorptive solid support. Alternatively, the "solvent" could be used in the form of a solid membrane with selective diffusion of sulfur dioxide to the other side. The necessary pressure differential would be maintained by applying a vacuum or using a sweep gas.

With organic solvents in whatever form, sulfate formation again raises a serious problem to which no practical solution has yet appeared. Regeneration of these sulfates must be accomplished in order to have an economical process. Heating is not an adequate solution since the redox reactions that occur gradually destroy the solvent.

H_2S-containing gas. Where it is practical to carry out an initial combustion with less than stoichiometric air, sulfur will appear as hydrogen sulfide rather than as sulfur dioxide. Following removal of the hydrogen sulfide, combustion can be completed, producing a desulfurized flue gas. However, to ensure good combustion in the second stage and to make good use of the heat subsequently produced, cooling after the initial step must not be carried too far. Calcium oxide is a suitable reagent for this reaction.[16, 17, 18] At 830° C, for example, the equilibrium hydrogen sulfide concentration over a mixture of calcium oxide and calcium sulfide would be only 0.004 percent by volume which corresponds to about 99 percent removal. Magnesium oxide would not remove the hydrogen sulfide at this temperature because magnesium sulfide is too unstable in the presence of the steam contained in the combustion gas. Sodium carbonate could, at equilibrium, remove about 85 percent of the hydrogen sulfide.

Calcium sulfide can be regenerated by partial oxidation with air, with continuous removal of sulfur dioxide to minimize sulfate formation. At 830° C, for example, the sulfur dioxide partial pressure over calcium sulfite is 0.15 atmospheres at equilibrium; at 1,000° C it is 4.1 atmospheres.

ACKNOWLEDGMENT

Adapted from a paper presented at the Technical Symposium of the Southern California Section of the American Institute of Chemical Engineers, April 1971.

LITERATURE CITED

1 Erdos, E., "Thermodynamic Properties of Sulfites. I. Standard Heats of Formation," p. 1428-1437, "Equilibria in the Systems SO_2 —CO_2 —$M^{II}O$," p. 2152-2167, "Thermodynamic Properties of Sulfites II. Absolute Entropies, Heat Capacities and Dissociation Pressures," p. 2273-2283, *Collection Czechoslov. Chem Commun.*, Vol. 27, 1962.

2 Gmitro, J. I. and Vermeulen, T., "Vapor-Liquid Equilibria for Aqueous Sulfuric Acid," University of California, 1963.

3 French Patent No. 1,428,721, Peter Soence & Sons Ltd.

4 Parsons, T. et al, "Applicability of Metal Oxides to the Development of New Processes for removing SO_2 from Flue Gases," work carried out under National Air Pollution Control Administration Contract PH 86-68-68, Tracor, Austin, Texas, 1969.

5 Stern, K. H. and Weise, E. L., "High Temperature Properties and Decomposition of Inorganic Salts, Part 1. Sulfates," National Bureau of Standards report NSRDS-NBS 7, 1966.

6 Wicks, C. E. and Block, F. E., "Thermodynamic Properties of 65 Elements—Their Oxides, Halides, Carbides and Nitrides," *Bureau of Mines Bulletin* 605, 1963.

7 "JANAF Thermochemical Data," Dow Chemical Co., Midland, Mich.

8 Oldenkamp, R. D. and McKenzie, D. E., "The Molten Carbonate Process for Control of Sulfur Oxide Emissions," preprint APCA Annual Meeting, 1968.

9 Bienstock, D. et al, "Removal of Sulfur Oxides from Flue Gas with Alkalized Alumina at Elevated Temperatures," *Journal of Engineering Power, Transmissions. ASME,* Series A 86, No. 3, 353-360, 1964.

10 Johnstone, H. F., "Recovery of Sulfur Dioxide from Waste Gases," *Industrial and Engineering Chemistry,* Vol. 27, 587-93, 1935.

11 Johnstone, H. F., et al, "Recovery of Sulfur Dioxide from Waste Gases," *Industrial and Engineering Chemistry,* Vol. 30, 101-109, 1938.

12 "Wellman-Lord SO_2 Recovery Process," paper presented at APCA meeting, June 1969, Wellman-Lord Inc., Lakeland, Fla

13 Yavorsky, P. M. et al "Potassium Formate Process for Removing SO_2 from Stack Gas," ACS Meeting, Houston, Feb. 22-27, 1970.

14 Humphries, J. J. and Mac Rae, W. A., "The Stone and Webster/Ionics Sulfur Dioxide Removal and Recovery System," American Power Conference, April 23, 1970.

15 "Citric acid used in SO_2 recovery," C&EN, June 14, 1971.

16 Moss, G. "The Desulfurizing Combustion of Fuel Oil in Fluidized Beds of Lime Particles," First International Conference on Fluidized Bed Combustion, Hueston Woods State Park, Oxford, Ohio, Nov. 18-22, 1968.

17 Skopp, A. P., "Use of Fluidized Beds of Limestone Based Materials for Desulfurizing Flue Gas," First International Conference on Fluidized Bed Combustion, Hueston Woods State Park, Oxford, Ohio, Nov. 18-22, 1968.

18 Zielke, C. W. et al, "Sulfur Removal During Combustion of Solid Fuels in a Fluidized Bed of Dolomite," *Journal of Air Pollution Control Association,* Vol. 20, No. 3, 164-9, March 1970.

FGD—A Viable Alternative

Flue-gas desulfurization (FGD) is likely to become the prime control technology for the increased usage of high-sulfur coal while meeting current SO_2 emission standards

Aziz A. Siddiqi and **John W. Tenini,**
Atlantic Richfield Co., Houston

WE HAVE LEARNED the dangers of oil embargo and dependence on foreign sources of fuels. Today's international oil situation dictates that we minimize dependence on foreign oil as an energy source and we must rely upon our own resources and our own technology.

We are in the fortunate position of having within our national borders 60 percent of the world's coal. Some 219 billion tons of it are considered technologically and economically available.[1] U.S. coal reserves have been estimated to be larger in Btu content than the oil reserves of the Middle East and should be enough to last us for about 300 years at the rate we are using today. But coal's participation in the energy market has been in a steady decline for nearly 60 years because of the abundant availability and relatively low cost of oil and natural gas. We are now at the point where we must again focus on coal as a major energy source.

America's national energy policy also places a great importance in the usage of coal as a major alternate energy source. However, coal can pose an environmental problem. Many supplies of coal are high in sulfur content and on combustion, yield unacceptable levels of sulfur oxides emissions.

Use of low-sulfur coal can avoid such environmental problems, but supplies of low-sulfur coal are less abundant and would not be able to meet our future energy demands. The availability of low-sulfur coal depends on the expeditious development of our domestic resources and of the transportation network required to get the fuel to the market. In 1980, coal demand for utility boilers alone is projected to be about 620 million tons per year.[2] However, low-sulfur coal production in 1980 could supply less than 44% of the 620 million tons demand. The potential production estimates for low-sulfur coal are summarized in Table 1.[2]

The combustion of coal in conventional utility and industrial boilers will undoubtedly play an important role in meeting future U.S. energy needs. In order to utilize the vast reserves of high-sulfur coal, we will have to take measures to make it environmentally acceptable.

The urgent need to become a coal-fired economy, while simultaneously safeguarding quality of environment at reasonable cost, presents a complex energy/environment/economic problem.

SULFUR DIOXIDE STANDARDS

The Environmental Protection Agency (EPA) has established a primary standard for sulfur dioxide (SO_2) of 0.03 ppm as an annual arithmetic mean, and 0.14 ppm as a maximum 24-hour concentration not to be exceeded more than once a year. A secondary standard for SO_2 also limits maximum three-hour exposure to 0.5 ppm, again not to be exceeded more than once a year. In addition, a performance standard for new or modified coal-fired steam generators limits sulfur dioxide emissions to 1.2 lb./million Btu of heat input as a maximum two-hour average.

SULFUR DIOXIDE EMISSION CONTROL TECHNOLOGIES

Principal technologies that will be available to control sulfur dioxide emissions from the combustion of high-sulfur coal through 1980 include:

- Physical and chemical coal cleaning
- Coal gasification
- Coal liquefaction
- Fluidized-bed combustion
- Flue-gas desulfurization (FGD).

Coal-cleaning techniques are designed to remove dif-

TABLE 1—Potential low-sulfur* coal production in 1980[2]

Region	Millions of Tons
Appalachian	38
Midwestern	...
Montana/Idaho/Wyoming	194
Rocky Mountain	36
Pacific	...
Total	268

* Coal that would conform to New Source Performance Standards which limit emissions to 1.2 pounds of SO_2 per 10^6 Btu.

ferent types of sulfur compounds. Sulfur occurs in coal in two forms—as organic sulfur chemically bound to the hydrocarbon constituents, and as inorganic pyritic and sulfate sulfur in the minerals associated with the coal bed. Physical coal cleaning removes up to 80% of pyritic sulfur and a significant portion of ash. Chemical cleaning might remove all forms of sulfur. Present indications are that physical coal-cleaning might be competitive with flue-gas desulfurization before 1980.

Coal gasification, coal liquefaction and fluidized-bed combustion processes are not expected to make a significant contribution until 1985. Therefore, the increased use of high-sulfur coal may have to be accompanied by stack gas cleaning, at least until clean coal or coal-derived fuels become available at competitive prices.

Based on comparison of these sulfur oxide control technologies and also considering near-term needs (through 1980), FGD seems the only viable alternative, other than the burning of scarce clean fuels, that is compatible with current emission standards.

FLUE-GAS DESULFURIZATION PROCESSES

Flue-gas desulfurization (FGD) is an end of process pollution abatement technique primarily utilized for removing SO_2 from utility and industrial boiler combustion gases. This technique has also been successfully applied to Claus sulfur recovery unit tail gas and sulfuric acid plant tail gas streams.[3] Flue-gas desulfurization processes contact the stack gas stream with a sorbent which absorbs and/or reacts with the sulfur dioxide, except for catalytic oxidation which converts SO_2 to SO_3 for subsequent recovery of sulfuric acid.

Flue gas desulfurization processes are usually categorized as **non-regenerable** or **regenerable** depending on the fate of the reactive component in the sorbent. In non-regenerable processes the reactive component in the sorbent chemically combines with sulfur dioxide to form a sludge which consists of fly ash, water, calcium sulfate and calcium sulfite. There is no regeneration step and the SO_2 remains in the sludge for eventual disposal.

In regenerable flue-gas desulfurization processes, the sorbent reacts with, absorbs or adsorbs SO_2 and is subsequently regenerated and recycled. Sulfur dioxide from the flue gas stream is recovered and converted into marketable byproducts such as elemental sulfur, sulfuric acid or concentrated sulfur dioxide gas.

Non-regenerable processes. Non-regenerable flue gas desulfurization processes which have been commercially tested include:

- Lime scrubbing
- Limestone scrubbing
- Double alkali soluble salt scrubbing
- Dilute sulfuric acid scrubbing
- Limestone injection.

In lime and limestone scrubbing processes, SO_2 reacts with the absorbent slurry to form calicum sulfite and calcium sulfate, which are released from the process as waste byproducts. Flue gas first passes through an electrostatic precipitator or venturi scrubber for particulate removal and then into a multi-stage absorber where it contacts the lime or limestone slurry. Scrubbed gas passes through a demister and is reheated prior to discharge to the atmosphere to provide plume buoyancy and prevent condensation. Sulfur dioxide removal is normally 85-90%. Sludge disposal in the United States generally consists of ponding, landfill or chemical fixation and landfill. In Japan, an oxidation step is included to yield dry gypsum as a byproduct.[3] A schematic flow diagram of a typical lime/limestone scrubbing system is presented in Fig. 1.

In double alkali processes, a soluble salt solution such as NaOH captures SO_2 as sodium sulfite/sulfate. This sulfite/sulfate solution is then regenerated by reacting with lime or limestone to form calcium sulfite/sulfate sludge for waste disposal. Regenerated salt solution is recycled to the absorber, but this process is classified non-regenerable since the sulfur dioxide is not recovered. Scrubbed gas passes through a demister and reheater prior to release to the atmosphere. Sulfur dioxide removal normally exceeds 90%.[4] A schematic flow diagram appears as Fig. 2.

Sulfur dioxide is also removed from flue gases by contacting in an absorber/oxidizer with dilute sulfuric acid as in the Chiyoda Thoroughbred 101 Process.[5] Dilute sulfuric acid is continuously removed from the absorber, neutralized with limestone and centrifuged to remove gypsum. Sulfur dioxide removal ranges 90-95%.

In another system, limestone injected in a boiler firebox reacts with sulfur dioxide in flue gas formed and is then scrubbed out with water. This system has not been very successful, although commercial installations are still operating.[3] Sulfur dioxide removal efficiency ranges 25 to 35%.[6]

Regenerable processes. Regenerable flue gas desulfurization processes which are currently available and tested on commercial scale include:

- Magnesium oxide slurry scrubbing
- Wellman-Lord sodium sulfite scrubbing
- Catalytic oxidation.

In the magnesium oxide process, magnesium oxide is utilized to absorb sulfur dioxide from flue gas, after particulate removal, in a wet scrubber as magnesium sulfite. This aqueous slurry is centrifuged and $MgSO_3/MgSO_4/MgO$ solids are recovered and calcined at elevated temperatures, with coke, to regenerate the absorbent back to magnesium oxide. An SO_2 rich gas stream is recovered which can be used to produce elemental sulfur or sulfuric acid. Scrubbed flue gas then passes through a demister and reheater to provide buoyancy and prevent condensation. Sulfur removal normally exceeds 90%. A schematic flow diagram is shown as Fig. 3.[7]

The Wellman-Lord Sulfite Scrubbing Process is based on absorption of sulfur dioxide by sodium sulfite to form sodium bisulfite, which can be thermally regenerated. Flue gas enters the process from an electrostatic precipitator and passes through a venturi or tray prescrubber where additional ash and chlorides are removed. Cooled, humidified gas from the prescrubber is contacted with sodium sulfite and sodium carbonate in the absorber to form sodium bisulfite, which is regenerated, and some sodium sulfate, which is purged from the system as solids following centrifuging. Regeneration of sodium sulfite is accomplished thermally in evaporators forming an SO_2/water vapor mixture. Sulfur dioxide is concentrated in this stream to 90% by water removal via condensation and stripping and further processed to produce elemental sulfur or sulfuric acid. Scrubbed flue gas is reheated prior to discharge to the atmosphere to provide plume buoyancy and prevent condensation. Sulfur dioxide removal in excess of 95% can be realized. A schematic flow diagram is shown in Fig. 4.[7]

Catalytic oxidation of sulfur dioxide to sulfur trioxide, followed by absorption to produce concentrated sulfuric acid byproduct, is the basis for Cat-Ox processes. For

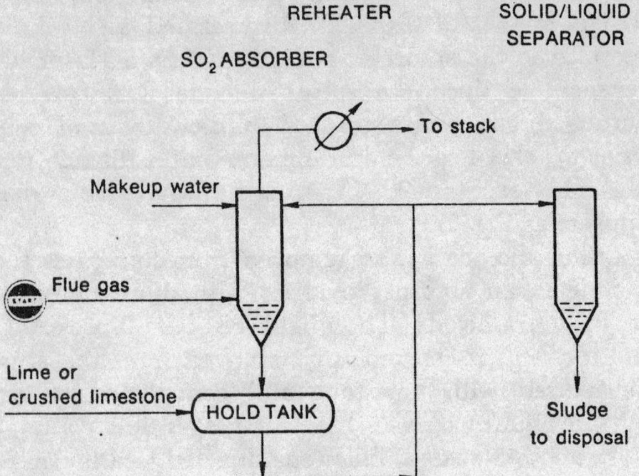

Fig. 1—Lime/limestone flue-gas scrubbing schematic.

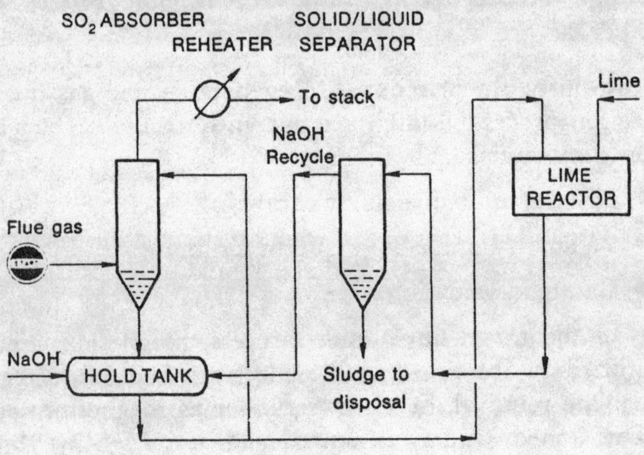

Fig. 2—Double alkali flue-gas scrubbing process schematic.

these processes, flue gas first enters high efficiency electrostatic precipitators to reduce particulate loading to 0.005 grains per standard cubic foot, or less, to reduce catalyst deactivation. Flue gas at 850° F to 900° F (reheaters normally required on retrofit units) enters the converter, containing a vanadium pentoxide catalyst, which oxidizes sulfur dioxide to sulfur trioxide. Flue gas is then cooled and sulfur trioxide is absorbed by a circulating stream of sulfuric acid in the absorption towers. A sulfuric acid concentration of 78% is maintained in the circulating stream by temperature control and byproduct sulfuric acid is continuously removed from the system. Scrubbed flue gas from the absorbers passes through high efficiency demisters to remove entrained sulfuric acid prior to discharge to the atmosphere. Sulfur dioxide removal efficiency ranges from 85% to 97%.[7] A schematic diagram is shown in Fig. 5.

Processes under development. Several other regenerable flue-gas desulfurization processes are also in various stages of development and testing. These include:

- Aqueous sodium carbonate scrubbing
- Activated carbon adsorption
- Copper oxide dry scrubbing
- Dry char adsorption
- Ammonia scrubbing-liquid Claus regeneration
- Citrate or phosphate scrubbing
- Ammonia scrubbing-ammonium sulfate byproduct
- Sodium sulfate/caustic scrubbing
- Potassium thiosulfate scrubbing
- Glyoxylic acid scrubbing.

The Aqueous Carbonate Process, developed by the Atomics International Division of Rockwell International Corp., utilizes an aqueous sodium carbonate solution to scrub sulfur dioxide from flue gas in a spray dryer. Dry sodium sulfite and sodium sulfate removed from the spray dryer are reduced to Na_2S in a molten salt pool at a temperature of 1,700° F to 1,900° F in the presence of coke or coal and combustion air. Reduced melt flows continu-

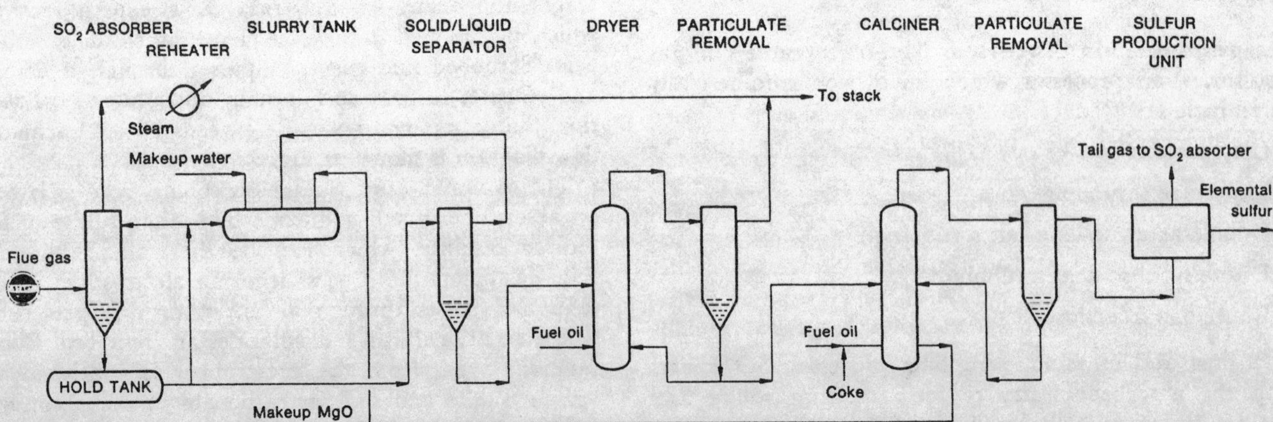

Fig. 3—Magnesia slurry absorption flue-gas scrubbing process schematic.

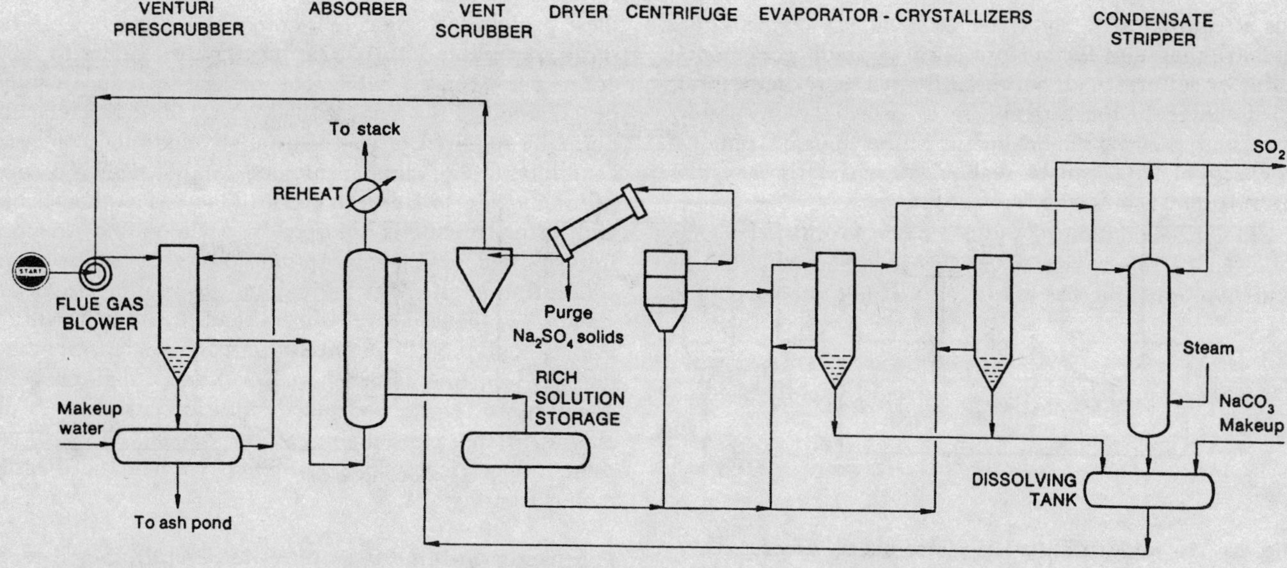

Fig. 4—Wellman-Lord sulfite scrubbing process schematic.

ously to the quench/dissolver where it is dispersed by steam jets and dissolved in solution (green liquor) near the boiling point. Fly ash and unreacted coke are removed by vacuum filtration and the green liquor is contacted with CO_2-rich reducer off gas in carbonation towers. These towers produce a concentrated sodium carbonate solution for recycle to the spray dryer absorber and an H_2S-rich Claus plant feed gas for elemental sulfur recovery. Sulfur dioxide removal efficiency of 95% has been demonstrated in pilot unit tests.[7]

Activated carbon processes have been developed by Reinluft, Bergbau-Forschung, Lurgi, Hitachi and Westvaco.[7] The Westvaco process is unique among these since the flue gas is contacted with activated carbon in a multistage, fluidized bed unit. Flue gas passing up through the unit maintains the fluidized state. Carbon catalyzes the oxidation of SO_2 to SO_3 in the fluidized bed and absorbs the sulfuric acid formed by injection of cooling spray water. Acid rich carbon is regenerated by reducing the acid with hydrogen in a three step process. In the first step, sulfuric acid is reduced to sulfur and sulfur dioxide in a fluidized bed contactor, with H_2S in the fluidizing gas stream, at a temperature of 300° F. The carbon is then preheated to 750° F with hydrogen rich regeneration gas and is introduced into the H_2S generator/sulfur stripper, which converts a portion of the sulfur to H_2S for sulfuric acid reduction. Regenerated carbon is recycled to the flue gas scrubber and sulfur is condensed from the sulfur stripper/H_2S generator off gas. Sulfur dioxide removal efficiencies as high as 99% can be achieved through proper design.[7]

The Shell Flue Gas Desulfurization Process is a dry copper oxide adsorption system. Parallel reactors utilize a copper oxide on alumina acceptor for sulfur dioxide removal. Oxygen in the flue gas entering the acceptor initially oxidizes copper to copper oxide which reacts with sulfur dioxide to form copper sulfate. When the copper is all converted to copper sulfate and SO_2 breakthrough occurs, flow is directed to the parallel reactor. Copper sulfate is regenerated with hydrogen to produce sulfur dioxide which can be recovered as elemental sulfur, sulfuric acid or concentrated SO_2 gas. Sulfur removal efficiency for this process is 90%.[7,8]

Another flue-gas scrubbing process which utilizes char as adsorbent in the Bergbau-Forschung Process which is licensed by Foster Wheeler Energy Corp. In this process, sulfur dioxide, oxygen and water vapor present in the flue gas are absorbed within the char pores where they react to form sulfuric acid. Char then enters the regenerator where it is heated to 1,200° F with hot sand in an inert atmosphere. Sulfuric acid is reduced to sulfur dioxide and char is recycled to the flue gas scrubber. The concentrated SO_2 gas stream then enters the Foster Wheeler RESOX process where it is reduced to elemental byproduct sulfur in a vessel containing crushed coal. Sulfur dioxide removal efficiencies up to 99% have been realized in pilot test units.[7,9]

The Catalytic/IFP flue gas desulfurization process is a combination of Catalytic's Ammonia Scrubbing Process with a regeneration system developed by Institut Francais du Petrole (IFP). Sulfur dioxide in flue gas is contacted with aqueous ammonium sulfite and bisulfite solution to form an ammonium sulfate brine solution. Brine solution passes through an evaporator, a sulfate reducer, reheater and SO_3 reactor where it is reduced to NH_3, H_2O, SO_2 and SO_3. This mixture is then converted to hydrogen sulfide in the H_2S generation reactor and fed to the IFP Liquid Claus System, where byproduct elemental sulfur is recovered. Regenerated ammonia is recycled to the flue gas scrubber. This process can be designed for up to 99% sulfur dioxide removal efficiency.[7]

Citrate/phosphate flue gas desulfurization processes utilize aqueous sodium citrate or sodium phosphate for absorption of sulfur dioxide. These processes are currently under development by the U.S. Bureau of Mines, Arthur G. McKee and Co., Peabody Engineering, Pfizer, Inc., and Stauffer Chemical Co. The Stauffer Process is available through Chemical Construction Corp. (CHEMICO). After contact with flue gas in the ab-

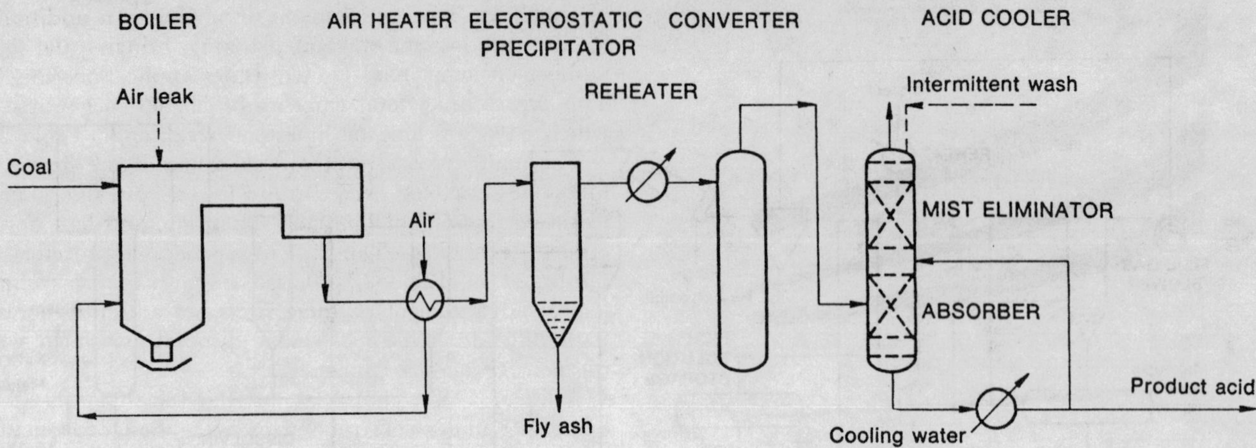

Fig. 5—The reheat Cat-Ox flue-gas scrubbing process.

sorber, the SO_2-rich citrate or phosphate solution enters a series of regeneration reactors. An H_2S-rich stream is bubbled countercurrent through the absorbing solution to reduce the sulfur dioxide to elemental sulfur. Sulfur slurry from the last regenerator reactor enters the sulfur flotation unit where sulfur and citrate or phosphate liquor are separated. Citrate or phosphate solution is recycled to the flue gas scrubber and sulfur slurry is continuously drawn through a sulfur melter to a decanter for final separation. Byproduct sulfur is withdrawn for sale with a portion recycled for H_2S production. This process can be designed for 99% sulfur dioxide removal from flue gas. Pilot unit tests have demonstrated removal efficiencies of 95% to 99%.[7]

Another flue gas desulfurization process which utilizes an aqueous ammonium sulfite and bisulfite absorption solution to remove sulfur dioxide from flue gas is under development by the Tennessee Valley Authority and Environmental Protection Agency. Sulfur dioxide in flue gas reacts with the ammonium sulfite/bisulfite to form an ammonium sulfate liquor solution. Part of this solution is recycled to the flue gas scrubber and the rest enters an acidulator where it is mixed with ammonium bisulfate to release sulfur dioxide gas. Remaining SO_2 in the liquor is removed in the sulfur dioxide stripper and combines with the acidulator stream for further processing to byproduct sulfuric acid or elemental sulfur. Liquor from the stripper enters the evaporator/crystallizer where water is evaporated and a slurry of ammonium sulfate crystals is produced. Ammonium sulfate crystals are recovered by centrifuging, with liquor recycled to the evaporator/crystallizer. Ammonium sulfate crystals are recycled to an electrical thermal decomposer to form ammonia for the flue gas scrubber and ammonium bisulfate for the acidulator. Excess ammonium sulfate crystals are sold as byproducts. This process can be designed for sulfur dioxide removal efficiency as high as 99%.[7]

The Ionics Electrolytic Regeneration Process utilizes a sodium sulfate/caustic scrubbing solution to remove SO_2 from flue gas. Absorbent is regenerated with their SULFOMAT Process to produce sulfuric acid and caustic. Flue gas entering the absorber is contacted with a mixture of sodium sulfate and caustic and sulfur dioxide is removed as sodium bisulfites. Flue gas is reheated prior to release and the liquor from the absorber is acidulated with a sulfuric acid/sodium bisulfate solution to form sulfur dioxide. This stream then enters a stripper where SO_2 is released from conversion to byproduct elemental sulfur or sulfuric acid. Liquor from the stripper contains residual sulfur dioxide and is fed to the SULFOMAT cells for regeneration. These cells produce the sulfuric acid/sodium bisulfate solution utilized for acidulation and the sodium sulfate/caustic solution recycled to the flue gas scrubber. This process is in a very rudimentary stage of development and good pilot unit data are not available.[7]

Another process is Conoco Coal Development Co.'s Consul Process. A concentrated, aqueous solution of potassium thiosulfate is utilized to remove sulfur dioxide from flue gas in packed bed absorber. Absorption liquor is regenerated and recycled to the flue gas scrubber. This process is currently being pilot tested.[10]

Another emerging process is that of Nobel Hoechst Chimie S.A. (France), Spring Chemicals, Ltd. (Toronto). A highly selective organic solvent (glyoxylic acid) is utilized to absorb sulfur dioxide from flue gas at a temperature of 95°F. Absorbing solution is then regenerated and produces a 95% sulfur dioxide gas stream and glyoxylic acid for recycle to the flue gas scrubber.[10]

COMMERCIAL APPLICATIONS

There are currently 37 flue gas desulfurization units

TABLE 2—Status of flue gas desulfurization units in the U.S.[3, 7, 10, 11, 12]

Status	No. of Units	Total MW	Percent Lime or Limestone (by MW)
Operational	37	7,441	82
Under construction	31	13,309	86
Planned:			
Contract signed	20	9,981	98
Letter of intent	2	365	52
Requesting/evaluating bids	4	2,327	14
Considering	37	16,726	26
Total	131	50,149	64

Note: Utility and industrial boilers only—does not include sulfur or sulfuric acid plants.

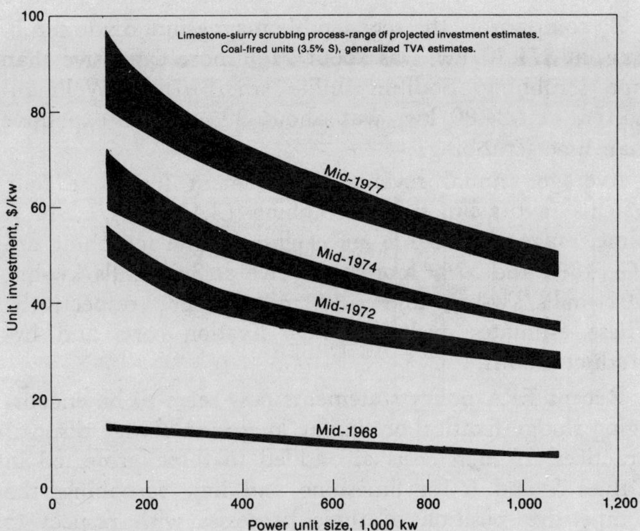

Fig. 6—Changes in investment costs for flue-gas scrubbing processes versus size and time.[14]

operating in the United States on utility and industrial boilers, representing a total capacity of 7,441 megawatts. Of these, the most popular process is lime or limestone scrubbing which accounts for 82% (by MW) of operating capacity.

There are also 31 units in some phase of construction representing a total capacity of 13,309 megawatts, 86% of which are lime or limestone scrubbing. An additional 63 units are in some stage of planning, bringing the total number of commercial United States applications to 131. This represents a total capacity of 50,149 megawatts of which 64% are lime or limestone scrubbing (Table 2).

The main reasons for the dominance of the lime and limestone scrubbing processes are lowest cost, due mainly to the availability of lime and limestone, and best development technology. The shift to regenerable processes, as illustrated in Table 2, is based on anticipated technological advancement of these processes and the increasingly difficult problem of sludge disposal associated with the non-regenerable processes.

A list of U.S. commercial installations that are either operating, under construction or being designed, including process type and supplier, is shown in Table 3.

In Japan, there were 333 operational flue gas desulfurization units on commercial installations at the end of 1976.[13] A listing of these plant installations, including type of scrubbing and byproducts, is shown in Table 4. The most utilized processes are lime scrubbing, limestone scrubbing, and the double alkali process, with waste sludges oxidized to byproduct gypsum.

ECONOMICS OF FLUE-GAS DESULFURIZATION PROCESSES

Economics can be a critical factor in the wide-spread use of flue-gas desulfurization technology. From the earliest cost estimates of the late 1960s to now, the estimated and actual costs of FGD systems have been increasing to higher levels. Some perspective can be gained from Fig. 6 which shows the capital cost of the generalized limestone slurry scrubbing process as estimated by Tennessee Valley Authority (TVA) over the past

TABLE 3—Major flue gas desulfurization installations in the United States[3, 5, 7, 12, 13]

Process	Supplier	Number of Plants	Total Capacity, MW
Lime scrubbing	Chemico	5	3,005
Lime scrubbing	UOP	3	1,330
Lime scrubbing	SCE	1	160
Lime scrubbing	American Air Filter	3	667
Lime scrubbing	Combustion Engineering	2	248
Lime scrubbing	Combustion Equipment Associates	3	1,170
Lime scrubbing	Research Cottrell	1	20
Limestone scrubbing	Peabody	3	630
Limestone scrubbing	Research Cottrell	8	3,942
Limestone scrubbing	Riley Stoker Environeering	3	460
Limestone scrubbing	Babcock & Wilcox	3	1,127
Limestone scrubbing	Combustion Engineering	4	2,760
Limestone scrubbing	Zurn Air Systems	1	37
Limestone scrubbing	UOP	2	250
Limestone scrubbing	TVA	1	550
Lime/limestone scrubbing	UOP	1	10
Lime/limestone scrubbing	Chemico	1	10
Double alkali	ADL Combustion Equip. Assoc.	1	20
Double alkali	General Motors/KOCH	1	32
Activated carbon	Foster-Wheeler	1	20
Sulfuric acid scrubbing	Chiyoda	1	23
Sulfuric acid scrubbing	Stone & Webster	1	70
Limestone injection	Combustion Engineering	4	765
Sodium carbonate scrubbing	Combustion Equipment Associates	4	500
Wellman Lord/Allied Chemical	Davy Powergas	3	830
Wellman Lord/Allied Chemical	Davy Powergas	3	sulfuric acid plants
Magnesium oxide scrubbing	United Engineers	4	846
Magnesium oxide scrubbing	Chemico	2	250
Soda ash	Combustion Equipment Associates	1	250
Wellman Lord	Davy Powergas	2	1,000
Wellman Lord	Davy Powergas	2	Claus units
Catalytic Converter	Enviro Chem Systems, Inc.	1	110
Magnesium Promoted Lime scrubbing	Kellogg-Weir	1	170

TABLE 4—Major flue gas resulfurization installations in Japan[13]

Process Supplier	Absorbent	Byproduct	Number of Units
Mitsubishi (MHI)	Lime, limestone	Gypsum	31
Chiyoda	Sulfuric acid	Gypsum	15
Babcock-Hitachi	Limestone	Gypsum	9
Wellman-MKK	Sodium sulfite	Sulfuric acid	10
Fuji Kausi-Sumitomo	Lime, limestone	Gypsum	7
Oji	Sodium hydroxide	Sodium sulfite	49
Kureha-Kawasaki	Sodium sulfite-limestone	Gypsum	3
IHI-TCA	Sodium hydroxide	Sodium sulfite	26
Mitsui-Chemico	Limestone	Gypsum	5
Tsukishima-Bahco	Sodium hydroxide	Sodium sulfite	20
Kurabo	Sodium hydroxide	Sodium sulfite	70
Showa Denko-Ebara	Sodium sulfite-limestone	Gypsum	16
Chemico-IHI	Limestone	Gypsum	2
Kureha	Sodium hydroxide	Sodium sulfite	8
Wellman-SCEC	Sodium sulfite	Sulfuric acid	6
Nippon Steel	Slag	Gypsum	2
Kobe Steel	Lime	Gypsum	3
Nippon Kokan	Ammonia	Ammonium sulfate	1
Tsukishima	Sodium sulfite-lime	Gypsum	4
MKK	Sodium hydroxide	Sodium sulfate	17
IHI-TCA	Lime, limestone	Gypsum	3
Hitachi Ltd	Carbon	Sulfuric acid, gypsum	2
Kurabo	Ammonium sulfate-lime	Gypsum	5
Kawasaki	Lime, limestone, magnesium oxide	Gypsum	3
Showa Denko	Sodium hydroxide	Sodium sulfite	2
Chuba-MKK	Limestone	Gypsum	2
Nippon Kokan	Lime	Gypsum	2
Nippon Kokan	Ammonium sulfite-lime	Gypsum	1
Sumitomo H.I.	Carbon	Sulfuric acid	1
Shell	Copper oxide	Sulfur	1
Onahama-Tsukishima	Magnesium oxide	Sulfuric acid	1
Mitsui Mining	Magnesium oxide	Sulfuric acid	1
Chemico-Mitsui	Calcium hydroxide	Calcium sulfite	1
Dowa		Gypsum	4
Chemico-Mitsui	Magnesium oxide	Sulfur	1
Total			333

decade.[14,15] For a 500-megawatt application, cost estimates have escalated from about $18/kw in 1968 to about $72/kw in 1977 (Fig. 6). This 300% increase cannot be entirely attributed to inflation. Some increase is due to improvement in the knowledge of process requirements. Nevertheless, these figures do emphasize the difficulty of accurately projecting control costs over a period of a few years.

TVA has updated generalized, comparative estimates of capital cost and annual revenue requirements for various FGD processes (Tables 5 and 6).[14,15] Installations were designed to remove 90% of sulfur dioxide from a 500-megawatt new boiler fired with 3.5% sulfur coal. A lime system was estimated to cost $61.10/kw and a limestone system was about 12% more expensive at $68.40/kw.

TABLE 5—Capital requirements[14]

Case	Life, years	Investment, $/kw* Limestone	Lime	Magnesia	Sodium sulfite
Coal-fired power unit, 90% SO_2 removal: on-site solids disposal**					
200 megawatts, new, 3.5% S	30	88.4	79.9	95.6	112.0
500 megawatts, new, 3.5% S	30	68.4	61.1	71.7	84.8
1,000 megawatts, new, 3.5% S	30	51.4	44.9	53.0	64.2

* Total capital investment. Midwest plant location represents project beginning mid-1975, ending mid-1978. Average cost basis for scaling, mid-1977. Minimum in process storage; only pumps are spared. Investment requirements for disposal of fly ash excluded. Construction labor shortages with accompanying overtime pay incentive not considered. These investment costs depend heavily on project definition. Working capital has been included.

** Sludge-disposal pond with clay liner.

TABLE 6—Annual revenue requirements[14]

Case	Life, years	Mills/kw-hr.* Limestone	Lime	Magnesia	Sodium
Coal-fired power unit, 90% SO_2 removal; on-site solids disposal					
200 megawatts, new, 3.5% S	30	4.20	4.54	5.03	6.60
500 megawatts, new, 3.5% S	30	3.41 (4.01)†	3.65 (4.15)†	4.02 (3.23)‡	5.37 (4.99)§
1,000 megawatts, new, 3.5% S	30	2.74	2.94	3.26	4.46

* Cost basis 1978. Power unit on-stream time 7,000 hr./year. Midwest plant location, 1978 revenue requirements. Investment and revenue requirements for disposal of fly ash excluded. These revenue requirements reflect capital investments shown in Table 4 (updated); byproduct credit and sludge fixation costs excluded.

† Assumes sludge fixation service fee by contract treating sludge in the utility's pond at $10/ton of 100% solids.

‡ Includes byproduct credit for 100% H_2SO_4 at $25/ton.

§ Includes byproduct credit for elemental sulfur at $40.18/ton.

By comparison, the regenerable magnesium oxide scrubbing, at $71.70/kw, was about 17% more expensive than lime scrubbing. Sodium sulfite scrubbing (Wellman-Lord), at $84.80/kw, was about 39% more expensive than lime scrubbing.

Average annual revenue requirement for these four systems favors limestone scrubbing (3.41 mills/kw-hr.). Lime, magnesium oxide and sodium sulfite scrubbing are 7%, 18% and 57% more expensive at 3.65 mills/kw-hr., 4.02 mills/kw-hr., and 5.37 mills/kw-hr., respectively. These estimates exclude sludge fixation costs and byproduct credits.

Recent EPA policy statements now seem to be encouraging sludge fixation or similar long term sludge disposal practices. If such costs are added to those projected in Tables 5 and 6 for limestone and lime scrubbing, the competitive position of these processes with respect to regenerative processes could change significantly.

ACKNOWLEDGMENTS

The authors gratefully acknowledge, the support and interest of Mr. W. D. Haney, Jr., and Mr. A. A. Muse, Jr., and help of Dr. Wade H. Ponder (EPA) in providing reference material. Opinions expressed in this paper are strictly those of the authors and do not necessarily reflect the official views of Atlantic Richfield Co.

LITERATURE CITED

1 Scollon, T. Reed, "An Assessment of Coal Resources," *Chem. Eng. Prog.*, June 1977, Vol. 73, No. 6, pp. 25-30.
2 Report to Congress on Control of Sulfur Oxides, EPA 45/1-75-001, February 1975, pp. 1.
3 Beychok, M. R., "Coping with SO_2," *Chem Eng.*, Oct. 21, 1974, pp. 79-85.
4 Gall, R. L., and Piasecki, E. J., "The Double Alkali Wet Scrubbing System," *Chem. Eng. Prog.*, May 1975, Vol. 71, No. 5, pp. 72-76.
5 Tamaki, A., "The Thoroughbred 101 Desulfurization Process," *Chem. Eng. Prog.*, May 1975, Vol. 71, No. 5, pp. 55-58.
6 Maurin, P. G., and Jonakin, J., "Removing Sulfur Oxides from Stacks," *Chem. Eng.*, April 27, 1970.
7 *Evaluations of Regenerable Flue Gas Desulfurization Procedures*, Electric Power Research Institute Report, EPRI FP-272, Project 535-1, January 1977, Vol. II.
8 Pohlenz, J. B., "The Shell Flue Gas Desulfurization Process," Paper presented at the Third Annual International Conference on Coal Gasification and Liquefication, Pittsburgh, Pa., Aug. 3-5, 1976.
9 Bischoff, W. F., and Habib, Y., "The FW-BF Dry Adsorption System," *Chem. Eng. Prog.*, May 1975, Vol. 71, No. 5, pp. 59-60.
10 Greene, R., "Utilities Scrubout SO_x," *Chem. Eng.*, May 23, 1977, pp. 101-103.
11 Devitt, T. W., Isaacs, G. A. and Laseke, B. A., "Status of Flue Gas Desulfurization Systems in the United States," Proceedings: Symposium on Flue Gas Desulfurization, New Orleans, La., March 1976, pp. 13-51.
12 Edwards, W. M., and Huang, P., "The Kellogg-Weir Air Quality Control System," *Chem. Eng. Prog.*, August 1977, Vol. 73, No. 8, pp. 64-65.
13 Ando, J., "Status of Flue Gas Desulfurization and Simultaneous Removal of SO_2 and NO_x in Japan," Proceedings: Symposium on Flue Gas Desulfurization, New Orleans, La., March 1976, Vol. I, pp. 53-78.
14 Ponder, W. H., and McGlamery, G. G., "SO_2 Control Methods Compared," *Oil & Gas Journal*, Dec. 13, 1976, pp. 60-68.
15 McGlamery, G. G., Faucett, H. L., Torstrick, R. L., and Hensen, L. J., "Flue Gas Desulfurization Economics," Proceedings: Symposium on Flue Gas Desulfurization, New Orleans, La., March 1976, Vol. I, pp. 79-99.

Control NO_x Emissions from Fixed Fireboxes

Current processes need improvements; more technology needed; combustion modification offers best ultimate solution

Aziz A. Siddiqi, John W. Tenini, Larry D. Killion,
Atlantic Richfield Co., Houston

NO_x CONTROL IS UNLIKELY to be economically satisfying if accomplished after firebox combustion. Such techniques almost certainly result in excessive cost for commercial operations. On the other hand, control of NO_x formation through combustion modifications has been successful for both new and existing industrial steam boilers and appears to be relatively easier and economical. This combustion modification technology is currently being studied aggressively in the United States and Japan.

Considerable data have been reported on percent reductions in NO_x emissions for steam boilers and process heaters through combustion modifications. These data show wide ranges from 10 to 70 percent due to initial operating mode of each particular boiler (Table 1), therefore, they serve only as a guide to reducing NO_x emissions. Actual reductions achieved will vary widely, for example, the maximum reported value of 70 percent reduction for a large, natural gas fired industrial boiler at high excess air and peak flame temperatures.

A better guide for gaging the magnitude of emissions from a given boiler or process heater is to compare NO_x emissions in pounds per million Btu with Environmental Protection Agency (EPA) New Source Standards (Table 2) or typical NO_x emission data presented (Table 3).

CHEMISTRY

Nitrogen oxides, NO_x, primarily nitric oxide (NO) and nitrogen dioxide (NO_2) are formed during combustion of fossil fuels with air. They are considered together because they are readily interconvertible in the atmosphere.

Nitrogen dioxide is a pollutant in itself because of its toxic properties and also a precursor of other pollutants. In the presence of certain hydrocarbons and sunlight, it is involved in the production of photochemical smog. The role of nitrogen dioxide as a precursor of photochemical oxidants (ozone, peroxyacyl nitrate and related compounds) can be shown by the following photolytic cycle:

$$2\,NO + O_2 \rightarrow 2\,NO_2 \quad (1)$$

$$NO_2 \xrightarrow{h\mu} NO_2^* \rightarrow NO + O \quad (2)$$

$$O + O_2 \rightarrow O_3 \quad (3)$$

$$O_3 + NO \rightarrow NO_2 + O_2 \quad (4)$$

$$O_3 + HC + NO_2 \rightarrow \text{Photochemical Smog Products} \quad (5)$$

During natural gas combustion, NO_x forms via a thermal fixation mechanism. During heavy oil and coal combustion, NO_x is formed via this fixation mechanism and through conversion of fuel bound nitrogen to NO. The

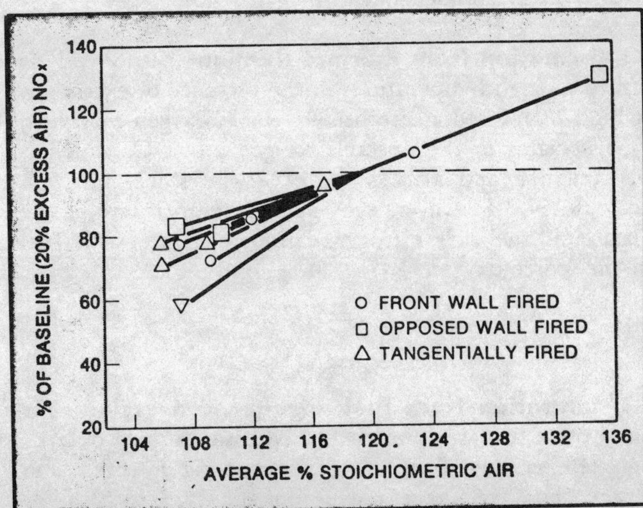

Fig. 1—Effect of excess air on NO_x emissions under normal operation. (Coal fired boilers).[9]

TABLE 1—Effectiveness of NO_X control techniques for industrial boilers[5-10]

Type of Fuel	Typical % Reduction in NO_x Emissions			
	Low Excess Air	Staged Combustion	Flue Gas Recirculation	Staged Combustion and Low Excess Air
Gas	10 to 38	15 to 35	10 to 70	16 to 68
Oil	10 to 19	10 to 22	10 to 23	15 to 45
Coal	17 to 27	13 to 54	13 to 17	20 to 55

NOTE: Larger percentage reductions are normally associated with oil boilers operating under high NO_x formation conditions.

TABLE 2—Standards of performance for new stationary sources[15]

	Fuel Burning Emissions, lbs/MM BTU		
	Gas	Oil	Coal
New Sources	0.2	0.3	0.7

TABLE 3—Typical NO_X emissions levels for industrial boilers[5-10]

Fuel Type	NO_x EMISSION LEVEL (PPM IN STACK GAS)			
	Without Controls		With Controls	
	Range	Average	Range	Average
Natural Gas	50 to 1000	200	50 to 350	110
Oil	65 to 650	280	150 to 365	210
Coal	164 to 1500	475	200 to 700	370

fixation mechanism depends strongly on both local gas temperature and local stoichiometry; therefore, temperature reduction techniques such as flue gas recirculation or water injection and reductions in oxygen availability through low excess air firing or staged combustion are effective in practical NO_x control. Conversion of fuel nitrogen seems to depend on the amount of nitrogen present in the fuel and oxygen availability.

Both mechanisms result primarily in NO because residence time in most stationary combustion processes is too short for oxidation of NO to NO_2. NO, however, does oxidize in the atmosphere to NO_2, which is the primary participant in photochemical smog reactions.

NO_x formation from thermal fixation. Nitric oxide is formed at high temperatures in the presence of excess air. At high temperature, the usually stable oxygen molecule, O_2 dissociates to the unstable oxygen atom, O, which is very reactive and attacks the otherwise stable nitrogen molecule, N_2.[1] However, even at high temperature, if the mixture is fuel-rich nitrogen cannot compete with fuel for the scarce oxygen.

$$N_2 + O \rightleftarrows NO + N \qquad (6)$$

$$N + O_2 \rightleftarrows NO + O \qquad (7)$$

NO_x formation from fuel nitrogen conversion. For many years it was assumed that NO was formed only by high temperature fixation of atmospheric nitrogen and oxygen. However, recent experimental studies[2,3,4] show that conversion of chemically held nitrogen fuel may be of equal importance in formation of NO_x during coal and fuel oil combustion. The relative importance of this oxidation process, which has a lower activation energy than N_2 fixation, varies with the nature of the flame process and with nitrogen content of the fuel.[3,4]

CONTROL TECHNIQUES

As mentioned, NO_x emissions can be effectively reduced by

- Reducing maximum peak temperature and residence time in combustion zone
- Decreasing oxygen concentration in combustion zone
- Limiting fuel nitrogen content.

To accomplish this, several techniques have been employed on existing industrial steam boilers and reported. These techniques include flue gas recirculation, staged combustion, reduced excess oxygen level, reduction in air preheat temperature, coolant injections such as water or steam and fuel switching.

These techniques have been incorporated in the design of new steam boilers in order to meet EPA New Stationary Source Performance Standards for nitrogen oxides emissions. Other design changes for burners, liquid and solid fuel handling systems, burner arrangement and firebox construction have also been utilized. In addition, several techniques for stack gas cleanup have been proposed and tested for both steam boilers and certain chemical processes.

Little data have been published on NO_x emission control for petroleum or chemical processing heaters due to the low level of emissions when compared to industrial steam boilers. Techniques that are employed include low excess air firing, flue gas recirculation and staged combustion. Discussion of each control technique is based mainly on industrial steam boiler data. While these techniques are applicable to process heaters, corresponding reductions in NO_x emissions may not be as large.

Flue gas recirculation. Recirculation of essentially inert products of combustion (flue gas) decreases NO_x emissions by lowering peak temperature and excess oxygen level in the primary combustion zone. This technique is accomplished by routing flue gas to the primary combustion zone at 15 to 20 percent of total combustion air. Recirculation rates as high as 35 percent have been reported with no flame stability problems,[5] but are normally limited below this level due to fan capacity.

Reductions in NO_x emissions of 10 percent (oil fired) to 70 percent (large natural gas boilers) have been reported.[5,6,7] Since flue gas recirculation primarily affects formation of thermal NO_x, emission reductions are much lower (approximately 6 to 10 percent) when this technique is added to staged combustion or low excess air firing. Boiler efficiency and particulate emission levels are essentially unchanged, but vary with individual boilers.[5,6,7]

Staged combustion. Staged or off stoichiometric combustion decreases NO_x emissions by lowering peak flame temperature, reducing residence time and creating fuel rich conditions in the primary combustion zone. This limits formation of both contributions from thermal NO_x and fuel NO_x since complete combustion is not achieved in the primary combustion zone. Additional air is introduced at a different location where complete combustion can be achieved at a lower temperature.

Staged combustion is accomplished on existing industrial

boilers by selectively removing burners from combustion service and using them as air registers and/or by adjustment to existing air registers. New boilers normally have over fire or "NO Ports" for staged addition of combustion air. For existing oil and coal fired boilers removal of the upper burner rows from combustion service normally gives the maximum reduction in NO_x emissions.[5, 8]

Reductions in NO_x emissions of 9 to 54 percent for burners out of service, 6 to 22 percent for air register changes and 10 to 47 percent for over fire air ports have been reported.[5-10] Variation in these data illustrates the importance of individual boiler design on actual emissions reduction which can be achieved. Boiler efficiency data show a variation of $\pm$ 3 percent with all types of staged combustion. Particulate emission levels are normally constant, but did increase by 5 to 50 percent in some cases.[5]

Reduced excess oxygen level. Maintaining a reduced excess O_2 level in industrial steam boilers or process heaters decreases both thermal and fuel NO_x formation. This technique is accomplished by operating test runs to determine the minimum excess air requirement for a given boiler or heater. This is normally limited by stack gas particulate emissions or CO levels. Reductions in NO_x emissions of 17 to 38 percent have been reported by reducing excess air to the minimum.[7, 8]

One should also realize the great interdependence of excess oxygen level and other control techniques. An example is staged combustion where a portion of the NO_x emissions reduction is from insufficient oxygen in the primary combustion zone. A meaningful method of illustrating these data is to plot average percent stoichiometric air through the firebox versus percent of baseline NO_x at 20 percent excess air (Fig. 1).[9]

A combination of staged combustion and minimum ex-

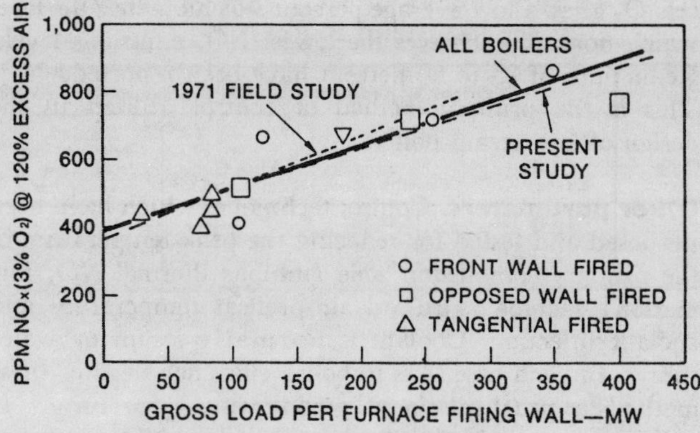

Fig. 2—Uncontrolled NO_x emissions versus gross load per furnace wall (Coal fired boilers).[9]

TABLE 4—Conversion of fuel nitrogen to NO_x[6, 9, 11]

Fuel Type	Nitrogen Content, Wt. %	Conversion to NO_x, %
#2 Oil.........	0.006-0.02	100*
#5 Oil.........	0.1	56-60
#5 Oil.........	0.13	70
#5 Oil.........	0.2	41
#5 Oil.........	0.28	44
#6 Oil.........	0.27	52
#6 Oil.........	0.44	43-51
#6 Oil.........	1.5	45

* Fuel nitrogen content too low to determine realistic values. Near 100%.

TABLE 5—NO_x abatement techniques currently in use in Japan[14]

Process	Description	Advantages	Disadvantages
Reduction by Sodium Sulfite (CCIC-JECCO Process)	Sodium sulfite reduces NO_x to N_2.	NO_x reduced from 3000 to 50 ppm. Sodium sulfite usually available from desulfurization processes.	Large flow rates and excess O_2 hinders reduction.
Sodium Scrubbing (Fujikasui Process)	ClO_2 oxidizes NO to NO_2. NO_2 scrubbed with $NaClO_2$.	Up to 90% NO_x removal. Low cost, easy operation.	SO_2 hinders RXN — must be prescrubbed. Wastewater treatment required for NaCl & $NaNO_3$ from scrubber.
Alkali Permanganate (MON Process)	NO_x & SO_2 absorbed and oxidized with alkali permanganate to form alkali nitrate and sulfate. Manganese dioxide precipitate is reduced for reuse.	Nitrate is used for fertilizer. Pilot plant shows 90% removal of NO_x. No waste material.	Alkali KOH and permanganate production is expensive.
Sodium-Potassium Permanganate (Nissan Process)	70-80% of NO_x removed in scrubber with NaOH. Off gas is oxidized with permanganate and NO_2 absorbed with NaOH. Two scrubber process.	High NO_x removal and flexible operation. NO_x off gas reduced to less than 100 ppm.	Na_2NO_3 waste treatment is required.
Alkali Scrubbing (Shinko Process)	NO_x is scrubbed with NaOH for 50% removal. Off gas sent to packed tower containing granular alumina sprayed with NaOH and $NaClO_2$-dried. 95% NO_x is removed with products of $NaNO_3$, NaCl & gaseous ClO_2.	High removal of NO_x (K95%).	$NaClO_2$ is expensive.
Sodium Scrubbing (Sun Mec SV Process)	High NO_x concentration gas, such as from pickling plants, is first scrubbed with NaOH. Off gas of 1000-3000 ppm NO_x is oxidized with a catalyst, steam sprayed to produce HNO_3 which is subsequently washed with NaOH.	Easy operation and maintenance. NaOH prescrubbed NO_x is reduced to 1000-3000 ppm and further reduced to 200 ppm.	Process *not* suited for flue gas treatment. Wastewater containing $NaNO_3$.
Alkali Scrubbing (Kyowa Kako Process)	Waste gases from pickling plants are first scrubbed with NaOH, off gas is oxidized with hydrogen peroxide and then washed with alkali hydrogen sulfide or alkali sulfide.	High NO_x removal ratio. 4000 ppm NO_x reduced to 50 ppm.	Difficult wastewater treatment and sludge disposal.
Sodium Scrubbing (Ube Process)	Nitric acid plant process gas NO_2/NO ratio is adjusted to 1.0. The gas is scrubbed with NaOH to form $NaNO_2$ and $NaNO_3$.	Simple process, low investment and operating costs. NO_x reduced to 200 ppm.	Produces large quantities of $NaNO_2$ and $NaNO_3$.

Other Alkali Scrubbing Processes

Mitsubishi Heavy Industries
Pickling plant gases are first scrubbed with water to recover nitric acid, off gas is scrubbed with NaOH. NO_x reduced from 700-2100 ppm to 280-490 ppm.

Hishinaka Industries Process
NO is oxidized to NO_2 in an activated carbon layer and then scrubbed with NaOH to form $NaNO_2$ and $NaNO_3$.

cess O_2 gives a low average percent stoichiometric air level which normally achieves the lowest NO_x emissions levels. Reductions of 25 to 60 percent have been reported.[5,7,9,10] This is the primary method of control utilized in the design of new steam boilers.

Other parameters. Control techniques which have been proposed and tested for reducing the peak temperature in the primary combustion zone (inhibits thermal NO_x formation) include reduced air preheat temperature and coolant injection. Coolant is normally steam or water sprays. In each case, loss in boiler efficiency negates these methods as practical control techniques.

Firing rate or load will normally affect NO_x emissions (Fig. 2).[5,9] NO_x increase with load is due to corresponding increase in maximum firebox temperature. In some cases, however, NO_x emissions are unchanged or, for coal firing may increase with decreased load. This is mainly due to operational problems in controlling excess O_2 level.

For liquid fuel burning, viscosity, atomization method and atomization pressure have affected NO_x emissions by ± 6 to 10 percent.

Tilting burners and new burner designs can affect NO_x emissions if swirl or mixing patterns are altered. This affects peak flame temperature and residence time which limits formation of thermal NO_x.

Fuel switching. Emissions of NO_x from steam boilers and process industry heaters are greatly dependent on fuel type and nitrogen content. In properly operated low NO_x boilers emission levels are largest with coal firing, then oil, and natural gas (Table 2—summary of EPA New Stationary Source Standards of Performance for fossil fuel fired steam generators).

For liquid fuels, nitrogen present is converted to NO_x based on concentration. Conversions in the range of 100 percent for low N_2 content No. 2 Fuel Oil to 45 percent for high N_2 content No. 6 Fuel Oil have been observed.[6,9,11] Typical NO_x conversion data for different types of liquid fuels are presented in Table 4.[6,9,11]

Stack gas cleanup. Removal of NO_x, SO_x and particulates should be a cumulative objective in NO_x abatement. Abatement problems center around gas/liquid contacting, equipment size, pressure drop, flow rates, temperature and corrosion. Non-selective catalytic reduction of NO_x with noble metals, CO, H_2 and CH_4 shows good promise in nitric acid tail gas emissions. Supportive copper oxide also is applicable. Selective NO_x reduction by ammonia injection has been successful in test chambers but not field tested.[12] Problems with SO_2 and particulate poisoning have been encountered.

Other flue gas treatment methods include differences based on physical properties such as molecular size, magnetic capabilities and condensation potential.[13] Dry absorbents that are used to absorb NO_2 include silica gel, alumina, molecular sieves, char and ion exchange resins.[13] Disadvantages include low capacity absorptivity. Some success has been found in absorbing equimolar concentrations of NO and NO_2 in aqueous alkaline solutions or sulfuric acid.[13] A tabulation of leading NO_x abatement techniques, currently in use in Japan, is presented in Table 5.[14]

ACKNOWLEDGMENTS

The authors gratefully acknowledge support and interest of W. D. Haney, Jr. and help of Dr. William Bartok (Exxon), Bob Hall (E.P.A.) and Don Bartz (KVB Inc) in providing reference material. L. D. Killion is currently working with J. E. Sirrine Co.

LITERATURE CITED

[1] Zeldovich, Y., *Acta Physiochim*, URSS, 21, 577 (1946).
[2] Bartok, W., *et al*, "Systems Study of Nitrogen Oxide Control Methods for Stationary Sources," ESSO Research and Engineering Company, Final Report No. GR-2-NOS-69, Contract No. PH22-68-55 (PB 192789), November, 1969.
[3] Turner, D. W., Andrews, R. L. and Sieginund, C. W., "Influence of Combustion Modification and Fuel Nitrogen Content on Nitrogen Oxides Emissions from Fuel Oil Combustion," ESSO Research and Engineering Company Report, Linden, N.J., 1971.
[4] Bartok, W., *et al*, "Basic Kinetic Studies and Modelling of Nitrogen Oxides Formation in Combustion Processes," 70th National Meeting AIChE, Atlantic City, August, 1971.
[5] Thompson, R. E., McElrov, and Carr, R. C., "Effectiveness of Gas Recirculation and Staged Combustion in Reducing NO_x on a 560 MW Coal-Fired Boiler," presented at EPRI NO_x Control Technology Seminar, San Francisco, California, February 5 and 6, 1976.
[6] Cato, G. A., Muzio, L. J., and Hall, R. E., "Influence of Combustion Modifications on Pollutant Emissions from Industrial Boilers," presented at the Stationary Source Combustion Symposium, Atlanta, Georgia, September 24-26, 1976.
[7] Bartok, W., *et al*, "Reduction of Nitrogen Oxide Emissions from Electric Utility Boilers by Modified Combustion Operation," presented at American Flame Days," September 6-7, 1972.
[8] Crawford, A. R., *et al*, "The Effect of Combustion Modification on Pollutants and Equipment Performance of Power Generation Equipment," presented at the Stationary Source Combustion Symposium, Atlanta, Georgia, September 24-26, 1975.
[9] Crawford, A. R., *et al*, "Field Testing—Application of Combustion Modifications to Control NO_x Emissions from Utility Boilers," Environmental Protection Technology Series, Publication No. E.P.A.-650/2-74-066, June, 1974.
[10] Bartok, W., Crawford, A. R., and Piegari, G. J., "Systematic Investigation of Nitrogen Oxide Emissions and Combustion Control Methods for Power Plant Boilers," AIChE Symposium Series, Vol. 68, No. 126.
[11] Cato, G. A., Muzio, L. J., and Shore, D. E., "Field Testing—Application of Combustion Modifications to Control Pollutant Emissions from Industrial Boilers—Phase II," Environmental Protection Technology Series, Publication No. E.P.A.-600/2-76-086a, April, 1976.
[12] Muzio, L. J., Arand, J. K., and Teixeira, D. P., "Gas Phase Decomposition of Nitric Oxide in Combustion Products," presented at the EPRI NO_x Control Technology Seminar, San Francisco, California, February 5-6, 1976.
[13] Bartok, William, "Methods of Nitrogen Oxide Emission Control for Large Stationary Combustion Sources," U.S.-Japan Joint Symposium on NO_x Control, Tokyo, Japan, June 28-29, 1974.
[14] "Nitrogen Oxide Abatement Technology in Japan," EPA Technology Series, EPA-R2-73-284, June 1973.
[15] Federal Register, Vol. 36, No. 247, Dec. 23, 1971, Subpart D—for Fossil Fuel Fired Steam Generators.

Reduce No_x in Stack Gases

K. Sridhar Iya, The Pennsylvania State University, University Park, Pa.

NITROGEN OXIDES POLLUTION is a major concern for the HPI. Approximately 60 percent of all nitrogen oxide emissions in this country come from burning fossil fuels in stationary heaters and furnaces like those used in the HPI.

Since nitrogen oxides cause eye and nose irritation and contribute to photochemical reactions forming smog, stringent regulations are being imposed in regard to emission control. Therefore, designers and equipment operators should seriously consider methods to cut emissions.

Kinetics of formation. The principal chemical reactions responsible for forming oxides of nitrogen are shown in Table 1. Chemical species such as O, H, OH and N are formed as reaction intermediates of the fuel-oxygen reaction. Reactions 1 and 3 and their reverse reactions (2 and 4) are known as the Zeldovich mechanism which controls the formation and disappearance of NO at flame temperatures of the order of 3,000° F and above.

In many combustion processes residence times are too short for oxidation of NO to NO_2. Thus, emissions are predominantly NO.

A rigorous theoretical estimate of the concentration levels of NO depends on the solution of equations of flow for the particular configuration together with the conservation equation for each of the chemical species present in the reaction mechanism. This approach is far too complex and generally involves a number of simplifying assumptions. Alternatively, the available residence time is assumed to be sufficient for combustion to reach equilibrium, and the equilibrium concentrations of NO_x are calculated using the technique of minimizing free energy.

A simple expression for the rate of production of NO considers only Reactions 1-4 in Table 1. The net rate of production of NO is given by the relation:

$$\frac{a}{dt}(NO) = k_1(N_2)(O) - k_2(NO)(N) + k_3(O_2)(N) - k_4(NO)(O) \quad (1)$$

where k_1-k_4 are rate constants of reactions 1-4.
Assuming steady state concentration for N atoms,

$$(N) = \frac{k_1(N_2)(O) + k_4(NO)(O)}{k_2(NO) + k_3(O_2)} \quad (2)$$

Therefore,

$$\frac{d}{dt}(NO) = \frac{2(O)[k_1 k_3 (N_2)(O_2) - k_2 k_4 (NO)^2]}{k_2(NO) + k_3(O_2)} \quad (3)$$

Simplifying,

$$\frac{d}{dt}(NO) = \frac{2k_1(N_2)(O)\left[1 - \frac{k_2k_4}{k_1k_3}\frac{(NO)^2}{(N_2)(O_2)}\right]}{1 + \frac{k_2(NO)}{k_3(O_2)}} \quad (4)$$

At time $t = 0$. NO concentration is zero. Initially, that is, for small time intervals, Equation 4 simplifies to

$$\frac{a}{dt}(NO) = 2k_1(N_2)(O) \quad (5)$$

The concentration of O atoms is frequently approximated by assuming that it equals the equilibrium concentration in the hot products. In many low temperature flames, however, the O atom will exceed the equilibrium concentration.

The validity of Equations 4 or 5 is established by comparing the predicted rate with the experimentally measured rate. Where predicted rates do not agree with experimental measurements for a particular flame, additional chemical reactions must be taken into account.

Table 2 shows the calculated equilibrium concentrations of NO_x (NO and NO_2) for four different fuels. Temperatures shown correspond to adiabatic flame temperatures. Fuel-air ratios are adjusted for similar adiabatic flame temperatures with the maximum difference less than 200° F. The NO_x concentration follows the same trend as temperature.

This flame temperature effect is further illustrated

TABLE 1—Reactions for forming nitrogen oxides

Reaction	Rate constant (cm-mole-sec units)
1. $N_2 + O \rightarrow NO + N$	$1.36 \times 10^{14} \exp(-75.4/RT)$
2. $NO + N \rightarrow N_2 + O$	$3.10 \times 10^{13} \exp(-0.33/RT)$
3. $O_2 + N \rightarrow NO + O$	$6.43 \times 10^{9}\, T \exp(-6.25/RT)$
4. $NO + O \rightarrow O_2 + N$	$1.55 \times 10^{9}\, T \exp(-38.64/RT)$
5. $NO + O + M \rightarrow NO_2 + M$	$1.05 \times 10^{15} \exp(-1.87)/RT)$
6. $NO_2 + O \rightarrow NO + O_2$	2.0×10^{13}
7. $NO_2 + H \rightarrow NO + OH$	3.0×10^{14}
8. $NO + OH \rightarrow NO_2 + H$	$3.0 \times 10^{12} \exp(-29.35/RT)$

TABLE 2—NO_x equilibrium concentrations for different fuels

Type of flame	Inlet gas composition % volume					Equilibrium combustion		
	CH_4	C_2H_2	C_2H_4	H_2	Air	T_{ad}* °F	NO ppm	NO_2 ppm
CH_4/Air......	4.5	...	...	...	95.5	2039	520	2.8
C_2H_2/Air.....	...	3.0	...	...	97.0	2149	804	3.8
C_2H_4/Air.....	...	...	3.0	...	97.0	2208	880	3.5
H_2/Air.......	...	...	...	14.0	86.0	2075	594	3.1

*Adiabatic temperature.

TABLE 3—Flame temperature effects on NO_x emission

% C_2H_2	% Air	T_{ad} °F	NO ppm	NO_2 ppm
2.0	98.0	1522	82.6	1.98
3.0	97.0	2149	804	3.80
4.5	95.5	3009	4228	5.01

(Table 3) showing that NO_x equilibrium concentration increases sharply with flame temperature.

NO_x emission factors. Important factors known to influence NO_x emission in a combustion system are:

- Type and composition of fuel
- Rates of heat release and heat loss
- Amount of excess combustion air
- Mixing and distribution of fuel/air in combustion chamber

Fuel affects NO_x formation because the type of fuel determines flame temperature and rate of radiative heat transfer and both increase in the order gas- oil- coal such that in general, NO_x formation in small to intermediate installations increases in the order gas- oil- coal.

Fuel composition controls NO_x emission for a given fuel-air ratio since raising the heat of combustion increases flame temperature. Nitrogen content in the fuel is also important because part of the nitrogen is converted to NO. Generally, the role of fuel nitrogen is dominant at low temperatures and small at high temperatures.

The rates of heat release and heat loss determine the temperature in the combustion chamber. Since the kinetics of NO_x formation is very sensitive to peak temperatures, high heat release rates lead to increased NO_x emission, and high heat removal rates lead to a decrease in the NO_x formation.

The amount of excess air used for combustion can affect the availability of oxygen for reactions leading to the formation of NO_x. Also, the combustion zone temperature is influenced by excess air. Low excess air firing favors reduced NO_x emission. However, very low levels of excess air also lead to unburned fuel, smoke and CO emission. Thus, permissible levels of excess air for any given combustion process should be determined with caution. Oil burners, particularly of the conventional atomizing type, generally require greater amounts of excess air than gas burners and thus may be expected to have higher NO_x emission.

Distribution of fuel and air and their mixing with the combustion gases affect NO_x emission. Internal recirculation of combustion gases leads to dilution of the primary flame zone resulting in reduced peak temperature and decreased NO_x formation. Tangential firing also favors decreased NO_x emission.

Reducing NO_x emission. Important techniques for reducing NO_x emission are flue gas recirculation, low excess air firing, water or steam injection, two-stage combustion and combinations of these.

Recirculation of flue gas to the combustion zone lowers both the peak flame temperature and the oxygen concentration, thus reducing NO_x formation. Low pressure created in the upstream side of a burner may be effectively utilized to set up recirculation of the hot flue gases.

Low excess air firing effectively reduces NO_x emission provided problems involving unburned fuel, smoke and CO emission can be overcome. Oil burners can be fired with low excess air if the fuel oil is vaporized before ignition, and if combustion is essentially carried out in the gas phase. Such burners compete well with gas burners in reducing NO_x emission. Also, they provide high heat release and clear, short flames. Vaporization heat for the fuel oil can be provided by recirculating combustion products. The combination of flue gas recirculation and low excess air firing is an ideal method of reducing NO_x emission.

Injection of water or steam into the combustion zone tends to dilute gases and lower flame temperatures, thus reducing NO_x emission. In smokeless flaring and incineration of some waste gases, steam is injected into the flame zone to reduce soot formation. In such cases, NO_x emission is also reduced.

Two-stage combustion is another important technique for NO_x reduction. In this approach, all fuel is fired to the first stage along with substoichiometric quantities of primary air. NO_x formation is limited by the unavailability of oxygen. Complete fuel burn-out is achieved by injecting secondary air in the second stage. This reduces the flame temperature and NO_x formation in this stage. This method is effective and very promising in reducing NO_x emission in different applications.

Catalytic treatment of flue gas. Catalytic treatment of flue gas to reduce NO_x emissions should be a useful alternative. Two treatment processes are catalytic decomposition of NO and catalytic reduction of NO.

Decomposition of NO to its elements is thermodynamically favorable following the reaction:

$$2\,NO \longrightarrow N_2 + O_2. \qquad (6)$$

Because of the tremendous incentive much effort has been spent to find a suitable NO decomposition catalyst that works at reasonable temperatures. However, no practical NO decomposition catalyst has yet been found.

Catalytic reduction of NO is accomplished at a temperature of about 1,000° F by mixing a reducing agent such as CH_4, CO or H_2 with the gas. The mixture is then passed through a NO reduction catalyst chamber. Noble metal catalysts such as platinum, palladium or rhodium are recommended. The chemical reactions expected to occur in the chamber are:

$$CH_4 + H_2O \longrightarrow CO + 3\,H_2 \qquad (7)$$

$$2\,NO + 2\,CO \longrightarrow N_2 + 2\,CO_2 \qquad (8)$$

$$2\,NO + 2\,H_2 \longrightarrow N_2 + 2\,H_2O \qquad (9)$$

The reaction chamber effluent contains mostly N_2, CO_2 and H_2O, with the NO concentration reduced to acceptable limits.

Catalyst life is limited by the temperature to which the catalyst is exposed and the oxidizing tendency of the flue gas. Honeycomb matrix catalyst structure extends the catalyst life to a suitable length. This technique combines advantages of low pressure drop, good distribution of gas to catalyst surface, thermal stability and high volumetric efficiency.

REFERENCES

Bartok, W.; Crawford, A. R., and Skopp, A., *Chem. Eng. Progress*, Vol. 67, No. 2, p. 64 (1971).
Faith, W. L., *Chem. Eng. Progress*, Vol. 52, No 8, p. 342 (1956).
Fenimore, C. P., and Jones, G. W., *J. Phys. Chem.*, Vol. 61, p. 654 (1957).
Fristrom, R. M., and Westenberg, A. A., Flame Structure, McGraw Hill, New York (1965).

Thermal DeNO$_x$: How it Works

Exxon's Thermal DeNOx process is based on a simple concept, but engineering considerations are critical. Here are results from commercial units

R. K. Lyon, Exxon Research and Engineering Co., Linden, N. J.

THIS PROCESS for control of NO_x emissions from stationary sources such as power plant boilers and industrial furnaces is based[1,2] on a new homogeneous gas phase reaction: the selective reduction of NO by NH_3 in the presence of O_2. Test results are given from the application of the process to gas and oil fired industrial and utility boilers and furnaces and to a coal fired pilot plant unit. The chemistry of the reactions, process side effects and prospects of the process's commercial application are also given.

A simple concept. The Thermal DeNO$_x$ reaction provides the basis for a conceptual simple process of NO_x control, the sole step of that process being the mixing of NH_3 into the flue gas. Engineering considerations, i.e. the manner in which that mixing is done, are critical.

While ideal laboratory conditions might show that 90+% NO reductions are possible in principle, the actual reductions which may be achieved in practice are typically 45 to 60% for the process in its present state of

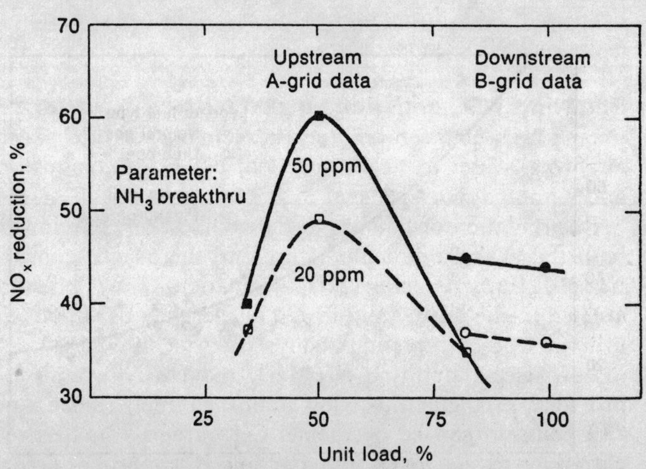

Fig. 1 — Typical data for a 156 megawatt utility boiler.

engineering development.

The amount of NH_3 to be mixed into the flue gas is relatively small, the mass ratio being of the order of 10^{-4}. It is difficult to provide such a small amount of gas with enough momentum to mix completely with a large gas flow and accordingly the NH_3 is first diluted with a carrier gas such as steam or air and then the carrier gas-ammonia mixture is injected into the flue gas.

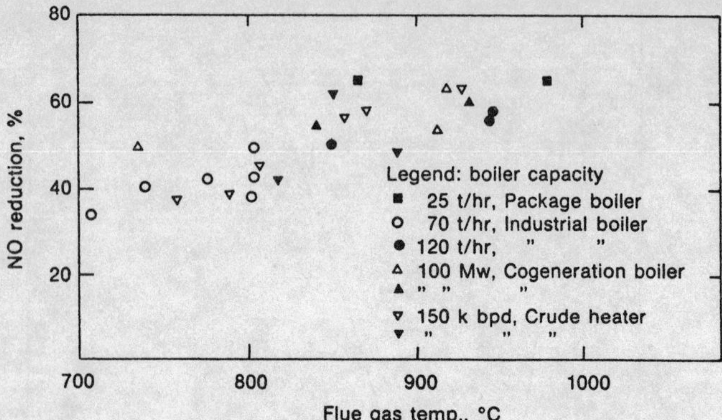

Fig. 2 — Commercial data showing effect of temperature.

This mixing is done under conditions such that NH_3 has a very finite lifetime and hence the mixing must be rapid. This necessitates the use of a multipoint injection grid.

If it were desired to achieve the maximum possible NO reduction, then the optimum design of this multipoint injection grid should take into account and compensate for a number of factors. The flow of the flue gas passing through any cross section of a furnace or boiler will in general be non-uniform. This means that more NO_X flows through some portions of the cross section than others and the amounts of NH_3 injected must be correspondingly adjusted.

Flue gas temperature variations are important also. Not only must the injection grid and the associated control system accomodate temperature changes caused by normal load and operating variation, but the grid and control system must also allow for fluctuations of temperature across the injection zone which occur because of non-uniformities in flow and heat transfer.

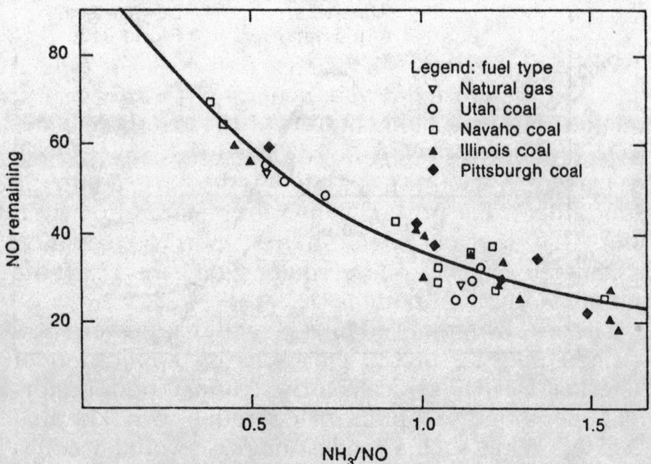

Fig. 3 — Process is equally effective for various fuels.

Details of how an injection grid may be designed to meet the above requirements are beyond the scope of this paper. It is, however, to be acknowledged that the extent to which these requirements can be met varies from unit to unit with the result that the effectiveness of the Thermal $DeNO_X$ process also varies.

Preliminary estimates of applicability and NO_X reducing capability can be provided by reviewing the original equipment design specification. More accurate estimates can only be provided if flue gas temperatures and local conditions are measured. Individual process designs are then custom fitted on the basis of such data.

Performance in commercial units. Thermal $DeNO_X$ has been demonstrated in twelve industrial and utility boilers and furnaces, eleven in Japan and one in the U.S. Fig. 1 shows typical data from one of the Japanese installations, a 156 megawatt utility boiler. Several points are to be noted. NO_X reduction is a function of both load and grid location, these being the factors which dictate how close the flue gas is to optimum temperature when it is injected with NH_3. Secondly as the amount of NH_3 injected is increased, the NO_X reduction improves but the amount of NH_3 remaining unreacted also increases.

A correlation for process performance in a number of units is shown in Fig. 2 and further illustrates the effect of flue gas temperature on NO_X reduction. The data in this figure are up to date to mid '78, while more recent test results from Japan have not been added to this correlation.

Application to coal firing. While all the commercial applications have, to date, been on gas and oil fired units, KVB has, under joint Exxon/EPRI funding, carried out a study using a 3 M Btu/hr modified package boiler, of the application of the process to coal firing.[3] The results of this study as they relate to side effects have been incorporated into the following discussion. The effectiveness with which the process reduces NO was examined for four different coals and for gas firing and, as is illustrated by Fig. 3, the process was found to be equally effective for all fuels.

In this pilot plant study it was also found that the temperature of optimum NO reduction was apparently 50°C higher for Illinois coal than for the other fuels. Later laboratory work,[4] however, indicated that the temperature for optimum NO reduction was the same for all fuels and it is believed that the earlier results represent an artifact of temperature measurement.

Using the prediction procedures developed for gas and oil fired units, Exxon Research and Engineering under an EPA contract has carried out an engineering assessment of the application of Thermal $DeNO_X$ to coal fired utility boilers of eight different designs.[5] The numerical average for NO reduction predicted for these eight boilers at 100, 75 and 50% load, without use of H_2 but with use of two injection grids, was 54%. Differences among the boilers were minor, the range averaged over load being 49 to 60% $DeNO_X$. The absolute best and absolute worst for any boiler, any load were 63 and 45% $DeNO_X$.

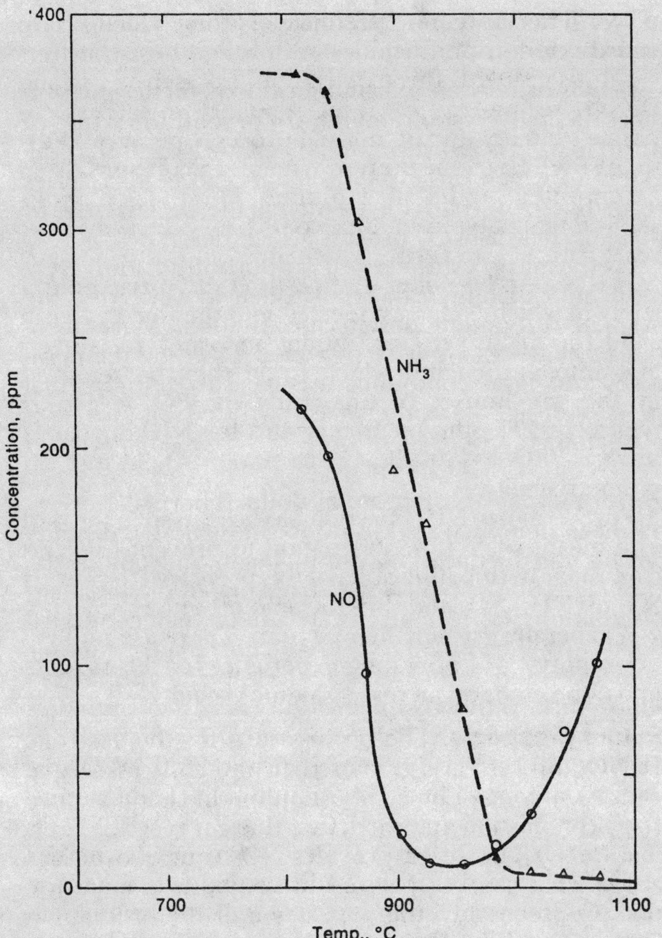

Fig. 4 — Nearly complete NO reduction and NH_3 consumption can be achieved. Lab data for 2% O_2, NH_3/NO = 1.7 and time is 0.2 seconds.

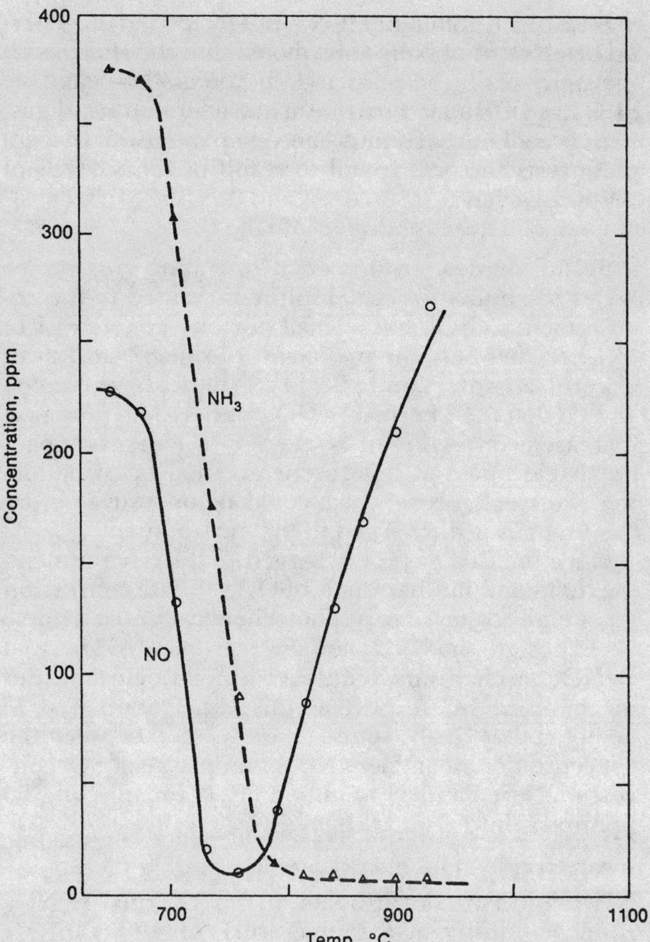

Fig. 5 — Addition of hydrogen decreases the optimum temperature. These data same as for Fig. 4 except hydrogen addition is same as NH_3.

Chemistry of the thermal DeNO$_X$ reaction process. In the Thermal DeNO$_X$ process NO is reduced via a complex free radical chain reaction.[1,2] The overall net reaction is

$$NO + NH_3 + 1/4\ O_2 \longrightarrow N_2 + 3/2\ H_2O \quad (1)$$

Concurrently with this NO reduction reaction, however, ammonia oxidizes to form NO via the overall reaction

$$NH_3 + 5/4\ O_2 \longrightarrow NO + 3/2\ H_2O \quad (2)$$

Since the reduction and oxidation reactions are in competition and are both highly temperature sensitive, there is a narrow range of conditions in which the balance between oxidation and reduction is favorable and in which both nearly complete NO reduction and NH_3 consumption can be achieved. Laboratory data illustrating this are shown in Fig. 4.

As illustrated by Fig. 5, the addition of hydrogen decreases the optimum temperature for NO reduction but for a given H_2/NH_3 ratio the temperature range in which efficient NO reduction may be achieved is not increased. The amount of H_2 required to shift the optimum reaction temperature increases exponentially with the size of the temperature shift. Thus the practical lower limit at which efficient NO reduction may be achieved is roughly 700°C.

Side effects. Studies have been done to determine what, if any, adverse effects the Thermal DeNO$_X$ process may have, both on the operation of the boiler and on the emission of secondary pollutants.[6] As is demonstrated below, the presently available evidence indicates that the process has no unmanageable side effects.

N_2O. While N_2 is the major product from the Thermal DeNO$_X$ reaction, N_2O is also formed in yield of 1 to 2% of the NO reduced. Since N_2O is generally accepted as harmless and, harmless or not, is an abundant natural material with an ambient concentration worldwide of 0.3 ppm, minor N_2O emissions would not appear to be a matter of concern.

CO. The Thermal DeNO$_X$ reaction does not reduce CO_2 and therefore does not generate CO. However, it can inhibit the oxidation of CO to CO_2, so if there is any CO left unburned when the NO is reduced, that CO will be present in the flue gas when it is discharged into the atmosphere. For normally operating gas and oil fired units, this is not a problem. In these units CO oxidation is complete long before the combustion gases reach the ammonia injection point.

The tests in a coal fired pilot plant did, however, show a slight increase in CO emissions on application of Thermal DeNO$_X$, typically a rise from 50 ppm in normal operation to 80 ppm for operation with Thermal DeNO$_X$. While such a modest increase would seem to indicate no significant problem exists, only full scale testing can resolve the matter.

HCN. The pollutant HCN is not formed in the Thermal $DeNO_X$ unless hydrocarbons are co-injected with the ammonia. In some special situations 50/50 mixtures of H_2 and CH_4 are available at considerably lower cost than pure hydrogen and hence may be used. In pilot plant tests this was found to result in HCN levels of 10 ppm or less.

Sulfur oxides. In a normal operating furnace or boiler the bulk of the fuel sulfur is emitted to the environment as SO_2, but a small portion, generally 1 to 5%, becomes SO_3, the precursor of sulfuric acid mist. Careful laboratory and pilot plant studies have demonstrated that the Thermal $DeNO_X$ process does not cause increased conversion of SO_2 to SO_3 by either homogeneous gas phase or heterogeneous catalytic oxidation, the two mechanisms which could be operative before the flue gas is discharged to the environment.

Once the flue gas is discharged to the environment, the dominant mechanism for SO_2 to sulfate conversion is generally believed to be photochemical, and it is not to be expected that NH_3 left over from the Thermal $DeNO_X$ mechanism would have any influence on this mechanism. In support of this expectation it is to be noted that Healy[7] found no correlation between the concentration of ambient NH_3 and the rate of SO_2 conversion. Friedlander[8] found no effect for injecting 50 ppm NH_3 into a smog chamber.

NH_4HSO_4. The reaction $NH_3 + H_2O + SO_3 \rightarrow NH_4HS_4$ occurs downstream of the Thermal $DeNO_X$ zone beginning at a temperature of 400-500°F. It is experimentally difficult to separately measure the amounts of NH_4HSO_4 particulates and gaseous $NH_3 + SO_3$ which are present in boiler flue gas. From our data, however, it would appear that the total quantity of sulfate particulates emitted from the process would not be affected by NH_3 injection. This assumes that SO_3 emissions are counted as sulfate particulates in the form of sulfuric acid mist and neglects the slight molecular weight differences between sulfuric acid mist and NH_4HSO_4.

Fouling. While the formation of NH_4HSO_4 within the boiler represents a potential problem of air preheater fouling, operating experience with low sulfur oil has been that fouling occurs only in some boilers. Long term tests conducted in two oil-fired boilers which did experience fouling revealed that these ammonium sulfate deposits, being highly water soluble, could be readily removed by water washing the air preheaters, while the units remained onstream. The waste water from the washing process was neutralized with a caustic solution using industrial-type waste water treatment facilities.

Plume visibility. The process does not increase the particulate content of the plume but merely changes the form of the particulates which are there. Since both the sulfuric acid mist and the ammonium sulfates are formed by condensation from the vapor phase, they should have similar size distribution and light scattering power. Therefore, no difference in plume visibility is to be expected.

Consistent with this expectation, no user of Exxon's $DeNO_X$ technology has ever reported an increase in plume visibility during normal process operation. Further ER&E has done thermodynamic calculations using a computer model of a plume. These calculations showed that at typical plume conditions NH_3 would not react with SO_2 to form particulates.

It is also to be noted that Davis et al.[9] carried out a study in which NH_3 was injected into the flue gas from a 50 ton/hr stoker-fired boiler in order to convert SO_X into particulates which could then be removed by the bag house. In this study the SO_2 level was typically 1500 ppm or greater and for NH_3 injections of $NH_3/SO_2 < 0.25$, there was no change in plume visibility.

Furthermore, NH_3 injection has been used in the past as a means of SO_3 neutralization to prevent backend corrosion. In this application it was preferred to convert SO_3 to $(NH_4)_2SO_4$ and thus the amount of NH_3 injected was generally greater than 20 ppm. There are numerous reports[9-19] of operating experience for this process and none mention increased plume visibility.

Future prospects. The process costs for the Thermal $DeNO_X$ process are greater than the costs associated with NO_X control by combustion modification via two stage combustion and moderate flue gas recycle. Thermal $DeNO_X$, however, provides NO_X control over and above what can be achieved by combustion modification. Consequently, the application of the process has been and will be limited to these situations in which greater NO_X control is required than can be achieved by combustion modification alone. The general application of the process is thus probably limited to those geographical regions, such as Japan and Southern California, in which the regulatory authorities decide that their air quality problem demands emission control beyond what can be achieved by combustion modification. Since the general trend is toward more stringent regulations, the number of such regions may increase.

(Original presentation was before a joint meeting of the ACS and CSJ in Honolulu, April 2-5, 1979.)

LITERATURE CITED

1. Lyon, R. K., *Inter. J. Chem. Kin.* Vol. 3, 1976, pp. 315-318. Also see Lyon, R. K., U.S. Patent 3,900,554.
2. Lyon, R. K., 17th International Combustion Symposium, Leeds, England, August 1978.
3. Muzio, L. J., et al., KVB Report No. 15500-717B, also see Muzio, L. J., et al., 17th International Combustion Symposium, Leeds, England, August 1978.
4. Lyon, R. K., et al., Third International Symposium on SO_2 and Other Gaseous Emissions, April 1979, University of Salford, and *I. Chem. E.* Symposium Series No. 57 in press.
5. Varga, G., Jr., et al., Applicability of the Thermal $DeNO_x$ Process to Coal Fired Utility Boilers, EPA Report 60017-79-079.
6. Lyon, R. K. and Longwell, J. P., EPRI NO_x Workshop, San Francisco, February 1976.
7. Healy, T. V., *Atomospheric Environment*, Vol. 8, 1979, pp. 81-83.
8. Friedlander, S., Discussion following his paper, "Relationship Between Emissions and Air Quality", Conference on Coal Combustion Technology and Emission Control, California Institute of Technology, February 5-7, 1979.
9. Davis, W. T., et al., Paper 77-26.3, 70th APCA Annual Meeting.
10. Rendle, L. K., *World Petrol. Congr. Proc.*, 5th, NY 1959, 7, pp. 155-78.
11. Wilkinson, T. J., and Clarke, D. G., *J. Inst. of Fuels*, Vol. 32, 1959; pp. 61-72.
13. Murray, G. F. J., *Schweiz. Arch. angew. Wiss. a. Tech.* Vol. 23, 1957, pp. 280-292.
14. Gundry, J. T. S., et al, *J. Inst. of Fuels*, Vol. 37, 1964, pp. 178-186.
15. Rendle, L. K., and Wilsden, R. D., *J. Inst. of Fuels*, Vol. 29, 1956, pp. 372-380.
16. Kato, R., and Paris, B. E., Amer. Soc. Mech. Engrs., Paper No. 60-WA-255, 1960.
17. Gvozdetskii, L. A., et al., *Elektr. St.* Vol. 37, No. 5, 1966, pp. 23-27.
18. Nakazawa, O., et al., *Tech. Rev. Mitsubishi Heavy Ind.*, Vol. 3, No. 3, pp. 185-191.
19. Mosse, G., *Rev. Gen. Thermique*, Vol. 5, No. 53, 1966, pp. 445-453.
20. Gvozdetskii, L. A., and Bonvech, V. E., *Elekt. Stantsii*, Vol. 37, 1966, pp. 17-22.

Tests Quantify Emissions from Ship Loadings

Hydrocarbon losses are less than normally estimated

Aziz A. Siddiqi, Larry D. Killion, John W. Tenini and **James T. Adams, Jr.,** Atlantic Richfield Co., Houston

HYDROCARBON EMISSIONS from loading gasoline into compartments on ships are less than normally estimated. Furthermore, prior cleaning of compartments significantly reduces these emissions.

In an effort to assess the impact of proposed regulations[1,2] on loading operations of ships and barges, accurate estimates for hydrocarbon composition of gaseous emissions from cargo compartments are required. Literature data are almost nonexistent, therefore, monitoring tests were conducted during loading operations.

TEST RESULTS

Three ships were monitored for hydrocarbon emissions during gasoline loading. Sampling was carried out in February, April and November to provide representative data for year-round conditions. Samples were collected at intervals during the entire loading operation to give a representative curve.

Loading conditions and previous cargo history of the compartments are given in Table 1. Results of a typical hydrocarbon analysis of these samples are shown in Table 2. Hydrocarbon concentration of the vapors leaving the cargo compartment as a function of percent cargo volume loaded has been plotted in Figs. 1-3.

Two sources of hydrocarbon emissions are generated from cargo compartments: residual hydrocarbon vapor attributed to the previous cargo and vapor generated during the loading.

Vapors left from the previous cargo can be reduced by compartment pre-treatment by "butterworth cleaning," ballasting and ventilating the compartment. Such cleaning processes are normally carried out enroute to harbor in order to expedite dock time.

The effect of prior compartment cleaning is clearly evidenced (Figs. 1 and 2). Hydrocarbon content of vapors leaving the compartment is insignificant with respect to 90 percent of the cargo loaded. Only the last 10 percent of the cargo shows any significant hydrocarbon emissions to the atmosphere. In comparison, compartments which were not cleaned prior to loading showed somewhat higher hydrocarbon emissions throughout the entire loading operation (Fig. 3).

Previously reported emission estimates from ships and barges have primarily been based on data reported by the American Petroleum Institute[3,4] and the Air Pollution Control District County of Los Angeles.[5]

API Bulletin #2514[3] Vessel Loading Loss Estimates included eight tests on loading gasoline into ships and barges. The loading-loss test methods were described as "stock-property change (vapor-pressure change and density change), vapor analysis via the air-balance technique and direct measurement of loss via repeated transfer." The bulletin emphasized that the loss correlations for marine vessels are not supported by enough data to be accepted as final. Recognizing that the test methods and data employ wide confidence limits, additional data and refined experimental techniques were needed.

Another API Bulletin #4080,[4] presented the same emission estimates as Bulletin #2514 for estimating vapor

TABLE 1—Hydrocarbon emission tests for ships (gasoline)

Test no.	Cargo loaded	Month	Previous cargo	Cargo compartment pretreatment	Ullage height, ft.	Ambient temp., °F	Cargo temp., °F	RVP	TVP, psia	Average fill rate, bbls./hr.	Cargo load, barrels	Load time, hours	Average percent hydrocarbon displaced
1	Gasoline	November 1974	Furnace oil	Ballasted	50	75	68	11.0	6.6	3,360	11,200	3.3	2.13
2	Gasoline	February 1975	Gasoline	Strip dry	50	45—70	70	13.5	9.0	482	7,573	15.7	5.73
3	Gasoline	February 1975	Furnace oil	Flood bottom Strip dry	50	45—70	70	13.5	9.0	406	7,272	17.9	....
4	Gasoline	April 1975	Gasoline	Ballasted	45	84	87	9.7	8.0	2,883	12,974	4.5	2.06

Geographical location —Houston
Type loading —Bottom loading, submerged fill pipe

Test method —Gas chromatography
Type of tank filled from—Floating roof

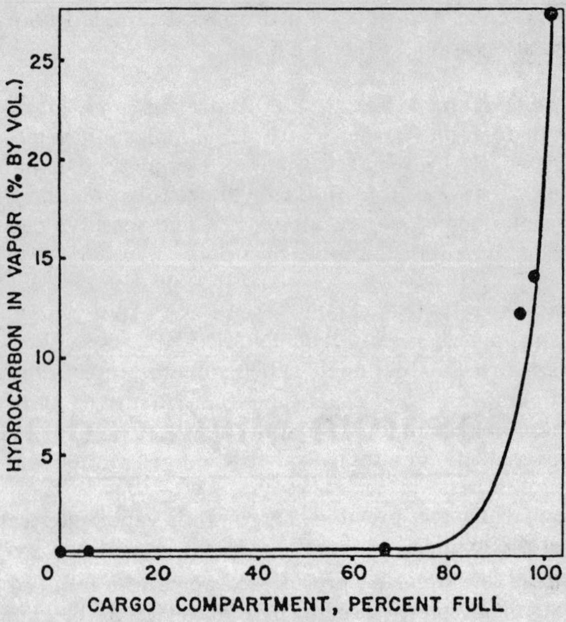

Fig. 1—Hydrocarbon emission as a function of gasoline volume-loaded, ballasted compartment. Previous cargo: furnace oil (November 1974).

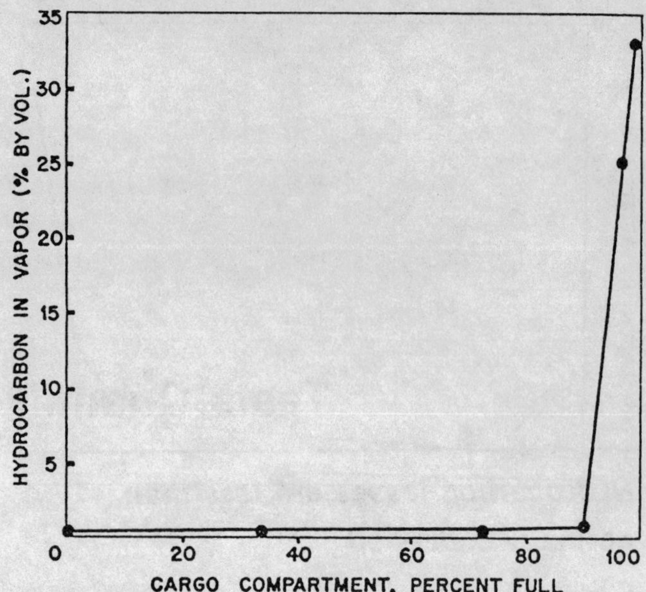

Fig. 2—Hydrocarbon emission as a function of gasoline volume-loaded, ballasted compartment. Previous cargo: gasoline (April 1975).

losses from tankers and barges. The correlation equation was:

Loading loss % by volume = 0.008 x TVP (true vapor pressure).

TABLE 2—Hydrocarbon composition of vapor displaced from cargo compartment during gasoline loading (27% full compartment)

Hydrocarbon	Ppm (by vol.)
Methane	185.76
Ethane & ethylene	3.37
Propane	25.38
Propylene	....
Isobutane	17.41
n-butane	120.96
1&i-butene	....
t-2-butene	1.65
c-2-butene	1.35
Isopentane	77.33
n-pentane	47.52
1-pentene	neg.
t-2-pentene	1.67
c-2-pentene	2.65
2,2 Dimethylbutane	1.98
2,3 Dimethylbutane	16.92
3-methylpentane	11.16
n-hexane	10.08
C_6+ hydrocarbons	37.75
Total HC_s =	562.94 (0.06 vol. % in vapor space)
Total non methane HC_s =	377.18

TABLE 3—Comparison of hydrocarbon emission loss estimates (gasoline loaded into ships)

Method	Hydrocarbon emission loss # /1,000 gals.	Hydrocarbon emission loss Pounds	Average percent hydrocarbon displaced
API	3.94	2,147	16.46
Los Angeles	13.01	7,087	54.34
ARCO-Houston Refinery	0.49	269 (pre-cleaned)	2.06
	1.22	665 (uncleaned)*	5.73

Basis: Gasoline loaded—9.7 RVP (8 psia)
87°F gasoline —84°F ambient
12,974 bbl. load
71 molecular weight, 60° API
* Uncleaned compartment calculation:

$$269 \text{ (precleaned tank)} \times \frac{5.73\% \text{ (uncleaned)}}{2.06\% \text{ (cleaned)}} \times \frac{8 \text{ psia}}{9 \text{ psia}} \text{ (Vapor pressure correction)}$$

The Los Angeles Report[8] presented a correlation which was applicable to tank trucks, railroad cars and marine vessels. The equation and its assumptions used are as follows:

$$W = \frac{PVw}{14.7}$$

where:

W = Weight of hydrocarbon vapors vented, lbs.
P = Partial pressure of hydrocarbon vapors in the air vapor mixture (considered to be true vapor pressure of liquid in vessel before filling), psia
V = Volume of product loaded, cu. ft.
w = Specific weight of hydrocarbon vapors, lbs./cu. ft.

Several key assumptions were involved

- Volume of gases vented equals volume of liquid loaded
- Displaced gases are saturated with hydrocarbon vapors
- Hydrocarbon-air mixture conforms with Dalton's Law of Partial Pressures
- Hydrocarbon vapors vented occupy 6 cu. ft./lb.
- Early average temperature of the product remaining in the vessel before filling is 75° F.
- Products loading under vapor recovery was accomplished without any hydrocarbon emission to the atmosphere.

Because of the large volumes involved in ship and barge cargo compartments, assumptions of hydrocarbon saturation in the vapor space result in an exorbitant emission. This assumption may accurately predict emissions from the turbulent loading of tank trucks or railroad cars but unrealistically over estimates emissions from ships.

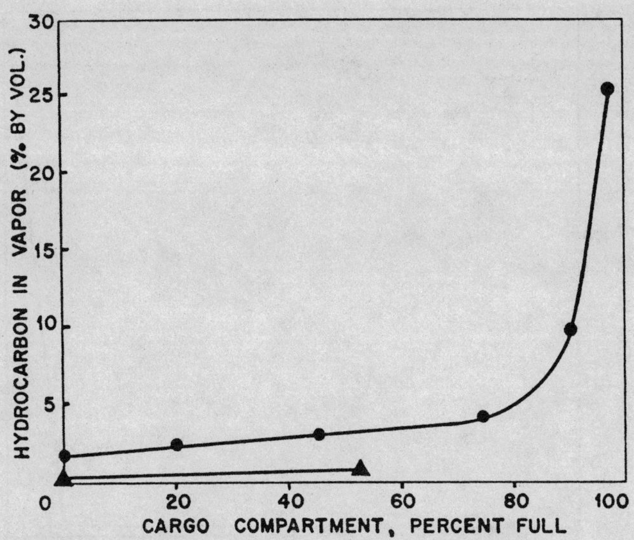

Fig. 3—Hydrocarbon emission as a function of gasoline volume-loaded. Strip-dried compartments. Previous cargo: gasoline and furnace oil (February 1975).

Loading time has a significant effect on the hydrocarbon emissions (Table 1). In tests 1 and 4, higher filling rates were employed and hence reduced filling time which in turn resulted in lower hydrocarbon emissions. Table 3 lists a comparison of the hydrocarbon emissions loss estimates. It is apparent that the emission estimates based on our tests give significantly lower values than the previously reported estimates.[3,4,5]

EXPERIMENTAL CONDITIONS

Sampling. Samples of the vapor displaced from the cargo compartments during gasoline loading were collected in evacuated glass bombs which were purged with dry nitrogen prior to evacuation. Bombs were analyzed immediately to avoid any sample deterioration. A gas-tight syringe was used to inject the hydrocarbon sample into a gas chromatograph.

Instrumentation. A Varian 1700 model research grade gas chromatograph equipped with flame ionization detector was used for hydrocarbon analyses (response of flame ionization detection for C_1-C_6 hydrocarbons is linear with the number of carbon atoms).[6,7] The sensitivity of the instrument was of the order of 5 ppb.

Column. The analytical column[8] used was 15 ft. x 1/4-inch OD polypropylene glycol, (UCON LB 550X, 20% w/w on chromosorb P, 60-80 mesh). The column temperature was maintained at 50° C for about 35 minutes until n-hexane eluted and then programed to 150° C to get C_6^+ hydrocarbons. The total analysis time per sample was about 90 minutes.

Calibration. The gas chromatograph was calibrated before and after each set of analyses using an ethylene standard that had been double checked against a methane standard and a propane standard. Results agreed within ±5%.

Caution. These data should not be extrapolated to estimate emission losses from compounds other than gasoline. However, hydrocarbon compounds with lower vapor pressures generally have a correspondingly lower emission loss.

ACKNOWLEDGMENTS

The authors wish to express their gratitude to Mr. W. J. Racine and Mr. W. D. Haney, Jr., for support and interest during this work. Opinions expressed in this paper are strictly those of the authors and do not necessarily reflect the official views of Atlantic Richfield Co.

LITERATURE CITED

1 *Federal Register*, 38 (213), 30645 (Nov 6, 1973).
2 Proposed Amendments to the Texas Hydrocarbon/Photomechanical Oxidant Strategy (Preliminary) Environmental Protection Agency, July 25, 1975.
3 API Bulletin #2514, "Evaporation Loss From Tank Cars, Tank Trucks, and Marine Vessels, American Petroleum Institute, Washington, D.C., (November 1959).
4 API Bulletin #4080, "Recommended Procedures for Estimating Evaporation and Handling Losses of Volatile Petroleum Products in Marketing Operations (1971).
5 "Emissions to the Atmosphere from Petroleum Refineries in Los Angeles County," Report #9, Joint District, Federal and State Project for Evaluation of Refinery Emissions, 1958, APCD County of Los Angeles.
6 McEwen, D. J., *Analy. Chem.*, 31, 1047 (1966).
7 Dimitriades, B., Seizinger, D. E., *Environ. Sci. & Technol.*, 5, 223 (1971).
8 Siddiqi, A. A., Worley, Jr., F. L., "Hydrocarbons in Houston's Atmosphere," Proceedings of Conference on Ambient Air Quality Measurements, Air Pollution Control Association, Austin, Texas, March 10-11, 1975.

Part V.

Odor Control

Identify Refinery Odors

Focal point of the HPI's concern should be causative compounds, methods of analysis, reporting and sources—in a setting of sensible measurement practices

J. Charlton, R. Sarteur and **J. M. Sharkey,** Stichting CONCAWE, The Hague, The Netherlands

GREAT STRIDES HAVE been made by the hydrocarbon processing industry in the identification and measurement of plant odors. From the various research efforts and practices, four key points emerge:

1. Refineries generally do not have a continuous odor problem. Most complaints are due to intermittent faults during plant operation.

2. Sulfur compounds, particularly H_2S and mercaptans, are the main source of odor from refineries. Unsaturated hydrocarbons and aromatics can give odor problems but less frequently.

3. Sulfur compounds can be determined down to 1 ppb using preconcentration techniques followed by GLC with a flame photometric detector. Hydrocarbons can be determined down to 1 ppm directly by GLC and down to 1 ppb if preconcentration is used.

4. Odor panels should be used where an instrumental method is not available or to assist in maintaining good public relations in the refinery area.

Nuisance. Refiners have, for the most part, whipped the toxic-odor problem. However, because some compounds can be detected by the nose at very low concentrations, e.g., 1 part in 10^9 or 10^{10} parts there still remains some nuisance. The assessment of the degree of odor nuisance is difficult. Individual reaction to odor is highly subjective. Threshold concentrations at which odors can be detected vary from person to person by as much as a factor of 100.

The relationship between odor intensity and concentration is rarely linear. Characteristics of odors may change at some critical concentration. Meteorological conditions, particularly wind direction and strength have a profound effect on the degree of odor nuisance. For these reasons it is not possible to set definite conditions for an acceptable odor level which is universally applicable. There are few statutory levels set by governments, unlike the wide variety of toxic limits.

The only Western European country that has specific restrictions for odorous compounds is Germany, where emission values for H_2S around new installations have been introduced. However, in most countries, the authorities can, and do, apply sanctions on companies through laws governing general nuisance. Public opinion brings pressure to bear. It is essential therefore that companies have satisfactory means of identifying and measuring odors to pinpoint sources of emission.

CONCAWE's report. In 1963 the Oil Companies' International Study Group for "Conservation of Clean Air and Water in Western Europe" (CONCAWE) came into being and now has 23 participants representing over 70 percent of the refining capacity in Western Europe. The group's basic function is to examine and to promote the use of means for preventing air, water, soil and noise pollution attributable to oil industry activity and the use of oil products.

In 1969, CONCAWE produced a report entitled "An Investigation into the Causes of Refinery Smells" (Doc. No. 3875) which summarized the results of questionnaires sent to 80 refineries to determine the frequency and causes of reported smells. Since the 1969 report, analytical techniques have been devised to identify and measure the concentration of odorous compounds. This article collects the knowledge and expertise available both from the literature and member companies in this field.

Dealing with complaints. There are several different systems for receiving and dealing with odor complaints. Probably the most sophisticated is the one operated by the Rijnmond Authority in Holland. This Authority covers the city of Rotterdam and the major industrial area to the west which includes several refineries as well as numerous other industries. The public complains to the Authority on a special, well publicized telephone number. Each complaint is logged by a central controller who then contacts one of the patrolling inspectors by radio. These inspectors investigate the complaint to validate it and also attempt to identify possible sources. In the last three years the Authority received about 15,000 calls/annum of which approximately 90 percent were about odors.

Several other authorities in industrial areas have similar, but less comprehensive, systems. Isolated refineries tend to receive and deal with odor complaints directly. Certain refineries use the local newspaper to inform residents in the vicinity of the refinery when there is likely to be an odorous emission. This can be done during a planned shutdown when degassing and washing operations can cause odorous emissions even though every effort is made to minimize them. Plant malfunction or maloperation can also generate odors. These tend to be difficult to identify and quantify. However, recently developed analytical techniques, in particular chromatographic methods, have provided the tools for identifying and measuring most refinery odors. Some of these techniques will be discussed later.

Odor-causing compounds. The nose is an extremely sensitive detector for odors. It can be matched only by the most sensitive instrumental techniques. Many smells are identified by comparison with pure compounds diluted by odor-free air until they are near the odor threshold limit. However, with the introduction of gas chromatography coupled to mass spectrometry, it has been possible to confirm the identification of many of the compounds giving rise to odors.

The main offenders have been shown to be sulfur compounds such as H_2S, mercaptans and disulfides which are known to be present in crude oil fractions and have been detected in refinery atmospheres.[1, 2] Because of their offensive smell and low olfactory levels, these compounds give rise to the majority of complaints. Any olfactory level quoted in the literature must be treated with caution. Typical detectable limits for the most important sulfur compunds are:

	ppm v/v
Hydrogen sulfide, H_2S	0.001-0.014
Methyl mercaptan, CH_3SH	0.001-0.0085
Ethyl mercaptan, C_2H_5SH	0.001-0.0026
Dimethyl sulfide, $(CH_3)_2S$	0.002-0.0052

Hydrocarbons can cause offensive odors. But the olfactory level of detection is generally about 1,000 times higher than for sulfur compounds. Most complaints are due to unsaturated C_5/C_6 and aromatic compounds. These emanate from plants where these compounds are concentrated. Occasional maloperation results in a detectable leak (Table 1).

Complaints have been reported recently about odors from a few biological treatment plants for purifying refinery effluents. The complainants often quote a "sewer"-type smell from dimethyl sulfide, ethyl and amyl mercaptans.[2] This problem can generally be overcome by adequate steam-stripping of the feed to the biotreater.

Methods of analysis. Three different approaches have been applied to refinery odor problems: instrumental measurements, olfactometry (dilution techniques) and odor panels. Where possible, a precise instrumental method is preferable to other methods, because the instrument measures quantitatively and reproducibly, has a wide linear concentration range and can usually be arranged to eliminate interference from other odors. However, if the compounds have not been identified, or an instrumental method cannot be devised, then another method must be used: the olfactometer for quantitative measurement, and odor panels to carry out "on the spot" investigations into public complaints.

A major problem with all methods of analysis is the transient nature of many odor problems. Very often after a complaint, the instrument or odor panelist cannot detect the smell even if the measurement is attempted only a short time after the initial complaint.

Instrumental. Gas-liquid chromatography (GLC) is the most universal method used for the determination of air pollutants because of:

- The high separation efficiency of the chromatographic columns. Capillary columns are available which will separate most of the components present in gasoline.
- The sensitivity and specificity of the detectors.[3]

A flame ionization detector (FID) will detect carbon compounds at the ppm level without preconcentration. The flames photometric detector (FPD) is specific for sulfur compounds, being 20,000 times more sensitive to sulfur than to carbon. However, preconcentration by 10^3 or 10^4 is still necessary to accurately quantify sulfur com-

TABLE 1—Olfactory and threshold limit values

	Odor threshold limits ppm vol./vol.	Threshold** limit values ppm vol./vol.
n-Pentane	....	500
n-Octane	....	400
Benzene	5	25
Toluene	2	100
p-Xylene	0.5	100
Cyclopentadiene	....	75
Butadiene	....	1000
Hydrogen sulfide	0.004*	10
Dimethyl sulfide	0.002*	....
Carbon disulfide	0.21	20
Methyl mercaptan	0.0008*	0.5
Ethyl mercaptan	0.0003*	0.5
Methylene chloride	214	250
Methyl ethyl ketone	10	200
Ammonia	50	25
Phenol	0.05	5

* Wilby F. V., **J. A. P. C.**, February 1969.
** Threshold limit values refer to airborne concentrations of substances and represent conditions under which it is believed that nearly all workers may be repeatedly exposed day after day without adverse effect. In other words a time weighted average for a 8-hour day, 40-hour work week.

pounds at the ppb range required for analysis of refinery odors.[4, 5, 6]

A recent invention is a microwave-plasma detector, which can simultaneously detect Cl, S. N or O and C. The selectivity and sensitivity are not as good as, for example, those of the FPD, but will probably be adequate for many pollution problems. A unique attribute of this detector is that oxygenated compounds can be detected in the presence of hydrocarbons at the ppm level.[7]

Two useful methods. Two analytical methods are now described in more detail because of their general applicability in refinery odor analysis.

1. The determination of hydrocarbons in air at the ppm level. This method determines the C_5-C_{12} hydrocarbons in air down to 0.1 ppm by separation on two 10-meter x 2 percent OV101 columns on 80-100 chromosorb, temperature programed from 0° C-150° C with dual flame ionization detectors.

One hundred mls of sample are drawn through a stainless steel trap, packed with 2 percent OV101 on chromosorb which is cooled to −70° C with solid CO_2/acetone. The trap, still maintained at −70° C, is then fitted to the inlet of the chromatograph, and carrier gas allowed to flow through the trap to remove residual air. The trap is then heated to 80° C with hot water and the hydrocarbons are desorbed on to the GLC column. As soon as this desorption commences, the column is heated from 0° C to 150° C at 3° C per minute and the peaks, which are eluted in approximately the order of their boiling points, are integrated. The actual values are then calculated by comparison with calibration mixtures containing 10 ppm v/v of hydrocarbons in air.

2. The determination of sulfur compounds in air. This method will determine H_2S, SO_2, CH_3SH, CS_2 and $(CH_3)_2S$ down to 1 ppb v/v by trapping these components from a kown volume of air on a Porasil D trap at −80° C, desorbing at 40-100° C directly on to a polyphenyl ether/phosphoric acid column at 40° C and detecting the eluted components with a flame photometric detector. The U tube in the trap must be made of glass, with stop valves in passivated INOX steel or made completely in PTFE because of the reactivity of sulfur compounds.

Samples can be taken in the field using a trap and a battery operated pump and then analyzed in the laboratory. The reproducibility of trapping is between 5 and 10 percent at the ppb range if the samples are analyzed immediately. It is recommended that samples are analyzed within three hours of taking to avoid oxidation problems. The concentration of each component is determined by measuring the peak area and reading off from a previously prepared calibration graph the equivalent in μ gram. The calibration is carried out using permeation tubes to produce gases of known concentration. Though flame photometric detectors are much more selective to sulfur compounds than to carbon compounds, it is possible to get a quenching effect when a large quantity of hydrocarbon is eluted with a sulfur compound. This has not been a problem, however, with the polyphenyl ether/phosphoric acid column, as the sulfur compounds are eluted well before the hydrocarbons.

Olfactometry. Olfactometry, or dilution techniques, involves quantitative dilution of odorous materials with odor-free air until the olfactory threshold limit is reached for a given observer. The strength of the odorous source can then be defined in terms of odor units, i.e., the number of dilutions necessary to reduce the odor to the threshold limit.[8] This technique is useful for:

- Determining how far above the detectable limit a given odor is, and thus gaging the magnitude of the control problem;
- Determining the relative efficiency of the various steps which may be taken to effect control;
- Using pure substances as odorous materials, determining and identifying smells and olfactory levels.

Various commercial instruments are available for this type of work.[9]

Odor panels. Mention has already been made of the use of odor panels when instrumental methods are not available, but consideration should be given to the direct value of odor panels in the field of public relations. Most complaints from the general public are detected by the human nose and the public readily accept the odor panel approach to investigating complaints. But panels can only assist in the qualitative identification of odors and possibly in locating the source of an odor. A typical scheme for selecting and training odor panels is outlined above.

The panelists were required to have the following qualities:

- Sensitivity—able to detect low concentrations of odorants;
- Reliability—able to reproduce consistently accurate results;
- Honesty—always say exactly what he (or she) perceives;
- Good odor memory—able to recall different refinery odors and differentiate between similar odors.

Screening tests were carried out on volunteers from the

staff of a refinery to determine the people who had the above qualities. The first screening test consisted of asking the volunteers what odors they thought they could recognize from a list of 40 chemicals, fragrances and flavors. They were then tested on 15 of these and 5 other odors. This test indicated whether people really could identify a substance from its odor and also how many odors they thought they knew.

In the second test, dilute aqueous solutions of a chemical, including unknown "blanks," were used to assess the sensitivity of the volunteers who were asked to indicate whether they could detect an odor. This test was repeated at weekly intervals to determine honesty and sensitivity. The tests were also carried out in the vapor phase.

The final screening test involved finding the "odd man out" of a group of two identical samples of the same substance and one of a contaminated sample of the same substance, which ranged from markedly different to very similar. A series of these triangular tests was carried out to determine whether people could hold the memory of an odor and compare it with another, albeit very similar.

Once the panel had been selected on the results of these tests, the members were allowed to familiarize themselves with the typical refinery odors during short training sessions. Routine tours were made around the refinery. These helped the panel to experience the effct of dilution of an odor under ambient conditions and to determine which were persistent. Panelists operated in pairs, but a compound was only deemed to have been identified if both recognized it independently.

More details on the selection of odor panels, and also the chromatographic methods for measuring hydrocarbons and sulfur compounds, will be included in the CONCAWE report entitled: *The Identification and Measurement of Refinery Odors.*

Reporting of results. Sarteur has devised a novel way of reporting results which highlights the compounds causing odor problems by producing an odor spectrum. The odor intensity of each component is determined from the formula:

$$\log \text{odor intensity} = 10\,k \log \frac{\text{Measured concentration}}{\text{Olfactory threshold level}}$$

k = Slope of response curve for component.

Odor presentations can be similar to those for acoustic measurements. But remember that odor measurements have not yet reached the precision of acoustics. The olfactory threshold levels quoted in the literature show considerable variation. It is not possible to add up intensities to give an over-all intensity, as odors can either intensify or suppress one another. However, the odor spectrum is a useful method of presenting results to illustrate which classes of compound are odor sources.

TABLE 2—Sources of odors

Type of smell	Source	Odorous compounds
Bad eggs	Crude storage Distillation of gases Sulfur removal Flare stacks (extinguished)	H_2S + trace of disulfides
Sewer smell	Effluent water Biological treatment plants	Dimethyl sulfide, ethyl and amyl mercaptans
Burnt oil	Catalytic cracking unit	Unsaturated hydrocarbons
Gasoline	Product storage API separators	Hydrocarbons
Aromatics (benzene)	Aromatic plants	Benzene, toluene

Refinery odor sources. Refineries are not normally considered "bad smell" producing units and do not give off a continuous obnoxious odor. Most problems are due to leakages or other maloperations which are intermittent in nature and which may usually be prevented by good housekeeping and maintenance.

Odor suppression. The measurement techniques described here have helped refineries to pinpoint odor sources and in some cases determine the odor strength. These types of data have given refinery management the semi-quantitative information required for deciding on where odor suppression equipment should be installed and also a measure of the efficiency of the equipment when installed.

(Originally presented at API Division of Refining Mid-year Meeting, Chicago, May 12-15, 1975.)

LITERATURE CITED

[1] Contribution à l' estimation quantitative d' une odeur par analyse spectrale. R. Sarteur and J. C. Fayard. *Pollution Atmospherique* October-December 1974 pp 398-404.

[2] A Method of Analysis of Trace Odour Components for Odor Pollution Control—*Matsumura Sekiyu Gakkai-shi*, Vol 15, No 7, pp 596-600 (1972) (in Japanese)

[3] Gas Chromatography Detectors—C Harold Hartman *Analytical Chemistry*, Vol 43, No 2 February 1971—113A.

[4] Gas Chromatographic Determination of Sulphur Compounds in North Sea Natural Gases by a Flame-Photometric Detector—K A Goode, *Journal of the Institute of Petroleum*—Volume 56 No 547—January 1970.

[5] Modern Aspects of Air Pollution Monitoring—R K Stevens and A E O'Keeffe, *Analytical Chemistry*, Vol 42 No 2 February 1970-143A.

[6] Automated Gas Chromatographic Analysis of Sulphur Pollutants—R E Pecsar and C H Hartmann. *Journal of Chromatographic Science*—Vol 11, P 492-502 September 1973.

[7] A Quantative Tunable Element—Selective Detector for Gas Chromatography —W R McLean, D L Stanton and G E Penketh. *Analyst*, June 1973 Vol 98 pp 432-442.

[8] Comparison of Techniques for Organoleptic Odour-Intensity Assessment—D Cormack, T A Dorling and B W J Lynch. *Chemistry and Industry*—2nd November 1974. pp 857-861.

[9] Measuring Industrial Odours—A Draunicks—*Chemical Engineering*/Deskbook Issue October 21, 1974. pp 91-95.

Refinery Odor Control

Check your approach against these recommendations

H. J. Klooster, G. A. Vogt and **D. G. Bernhart,** Fluor Engineers & Constructors, Los Angeles

Odor control is best accomplished by early planning which results in economical and totally compatible designs. This is true of modern refineries which include latest, most efficient processing units and many ancillary systems for environmental protection. Many techniques of proven technology are used to control malodors to varying degrees. Although many controversies exist over air and water pollution concentration limits, values are readily quantified. However, if the ultimate goal is to regulate odors through strict quantitative guides, then analytical techniques used to detect them must be reliable and conclusive. Actual shortcomings and limitations of many methods are noted at great length.[1,2,3] Inside the United States, the Environmental Protection Agency has classified odors as noncriteria pollutants[4] and has not set regulations

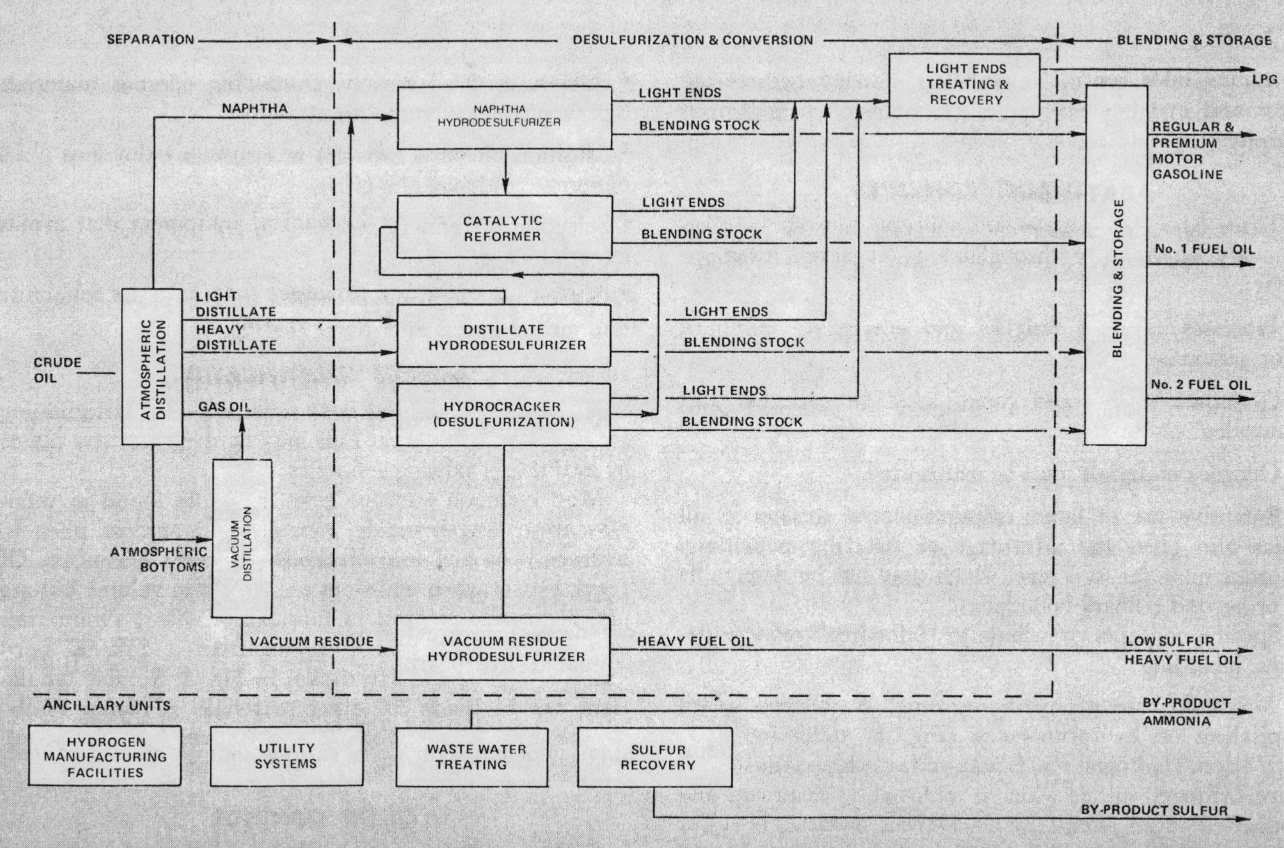

Fig. 1—Maximum use of hydroprocessing minimizes odor.

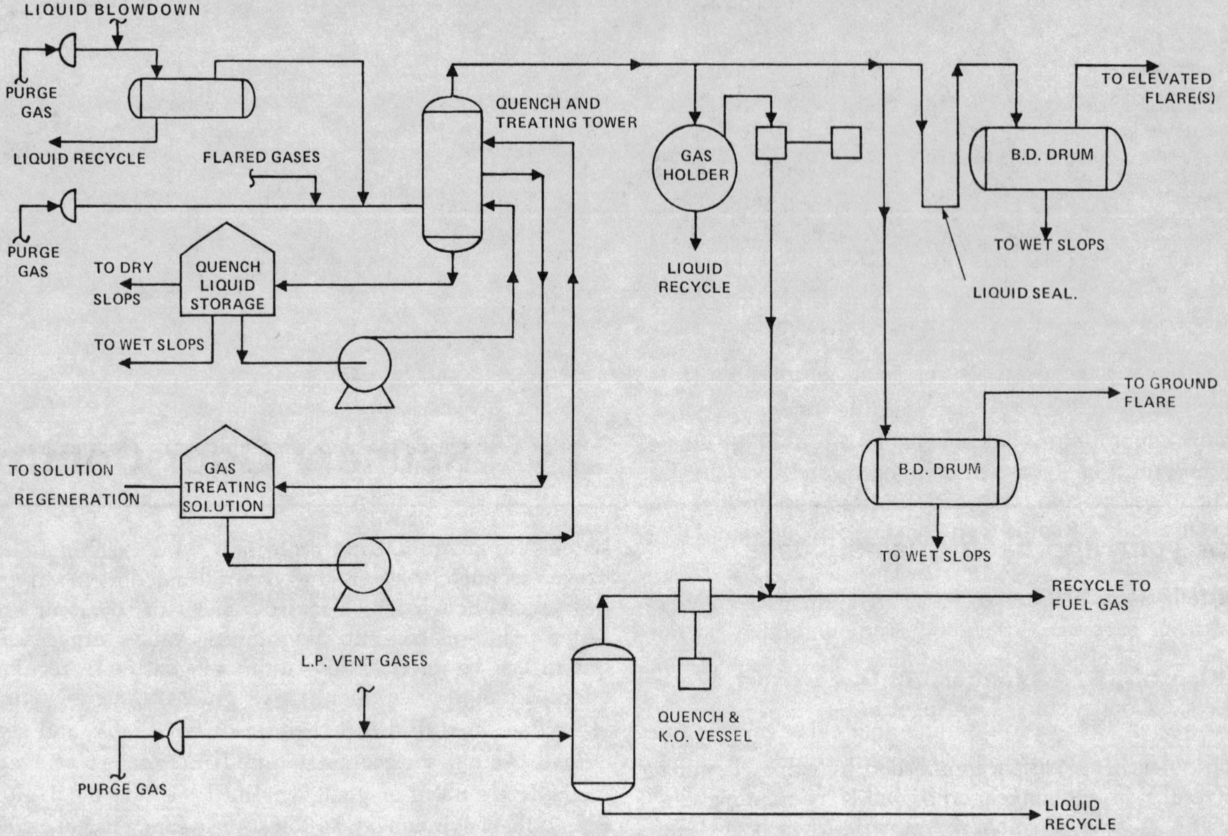

Fig. 2—Modern flare system for minimizing liquid and gas losses.

governing odor control. Nevertheless, design engineering firms and operating companies are obliged to minimize emissions.

ABATEMENT CONCEPTS

Three basic odor abatement concepts must be applied at every opportunity throughout plant design development:.

- Processes must be designed and selected to minimize odor generation.
- Odorous vapors and fumes must be collected and controlled.
- Odorous materials must be neutralized.

Extensive use of latest emission-control devices in all areas also gives the advantage of reducing potentially odorous material to a level which may not be detectable at or beyond refinery boundaries.

Specific concepts contribute to "odor-free" refinery designs, including:

- Maximum use of hydroprocessing. A refinery which capitalizes on hydroprocessing (Fig. 1) minimizes odor formation. Hydrogen reacts with sulfur compounds to produce hydrogen sulfide which is removed by treatment and converted to an essentially nonodorous marketable sulfur product. Other odorous materials such as phenols and asphaltenes are also minimized by hydroprocessing.
- Enclosure of all systems containing odorous materials to prevent release to the atmosphere.
- Addition of extra capacity in emission reductions units to increase over-all reliability.
- Selection of refinery mechanical equipment that avoids potential leaks.
- Choice of operating processes that improve reliability to minimize upsets and avoid venting.

SOURCE IDENTIFICATION

Recognizing potential odor sources aids in determining best control techniques. This may be done for new plants by evaluating existing refineries.

Most common odorous gases generally found in refineries are hydrogen sulfide, mercaptans, ammonia, phenols, hydrocarbons and miscellaneous treatment chemicals. Of these, hydrocarbon emissions are largest in volume but are not necessarily greatest in nuisance. Table 1 summarizes estimated hydrocarbon emissions from a 250,000 bpcd refinery such as the one shown in Fig. 1. Similar tabulations can be made for other materials. A review of the amount of emission from various sources aids in concentrating design efforts in areas of greatest concern.

ODOR CONTROL

Control techniques and other mitigating factors must be used to prevent known emission sources from becoming

TABLE 1—Hydrocarbon Emissions—Total

	lb/day
Valves	2,490
Flanges	500
Boilers and heaters	2,092
Blowdown system	1,250
Compressor seals	1,250
Cooling tower	432
Process drains	275
Pump seals	4,250
Vacuum jets	(negligible)
Sampling, misc.	2,500
Relief valves	(negligible)*
Tanks	4,609
TOTAL	19,648

* All relief valves are connected to the vapor recovery or blowdown system.

a nuisance. Specific examples of such controls illustrate typical situations encountered in a refinery.

- Steam eductor and barometric condenser effluents (vacuum unit, Fig. 1) are discharged to a closed sour gas collection system and treated with amine to remove hydrogen sulfide which is then converted to elemental sulfur in the sulfur unit.
- Phenolic odors are prevented (Fig. 1) by extensive use of hydrogen processing. Phenolic compounds that do not react with hydrogen leave the refinery in products or in effluent recycle water streams.
- Sour water streams containing hydrogen sulfide and ammonia (from various units including crude, vacuum, naphtha and distillate hydrotereaters vacuum residuum hydrodesulfurizer, tail gas treating and hydrocracker) are collected in closed systems and treated in a sour water treatment unit. Sufficient totally enclosed storage capacity is provided to allow storage of sour water during unit downtime for maintenance.
- Spare sulfur recovery capacity is provided to ensure the reliability needed for continuous hydrogen sulfide conversion. Three parallel sulfur recovery and tail gas treating units are supplied (Fig. 1), each of which is rated at 50 percent of total design load. This concept, of spare capacity, not typical of past refinery designs, is almost mandatory today to avoid emergency venting of gases containing hydrogen sulfide.
- API oil/water separators are equipped with floating roof covers to minimize escape of hydrogen odors from exposed liquid surfaces. Drainage systems which collect oily water upstream of the API separator are closed and equipped with liquid seals to prevent hydrocarbons from escaping through drainage boxes.
- All relief valves on hydrocarbon systems discharge to a closed flare system. Recovered hydrocarbons are preferentially recycled to storage or to refinery fuel. Excess hydrocarbons above recovery system capacity are flared. Optimization of relief and flare systems (Fig. 2) can actually increase refinery profits.[5] Pump vents are also connected to the flare system to contain and retain hydrocarbons.
- Centrifugal compressors are used in sour gas service because they tend to lower initial losses from seals and these seals may be purged with nitrogen to a vapor recovery system to prevent leaks to the atmosphere.
- Mechanical seals are used wherever possible on pumps to reduce hydrocarbon leakage to an absolute minimum. Packing is less desirable because of higher leakage rates.

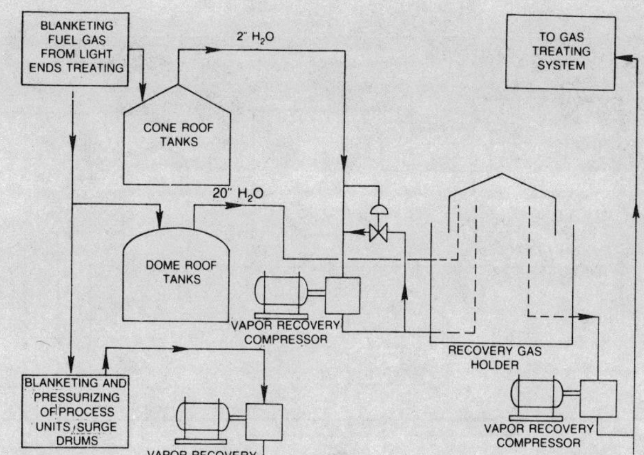

Fig. 3—Gas blanketing and vapor recovery are required for odor control.

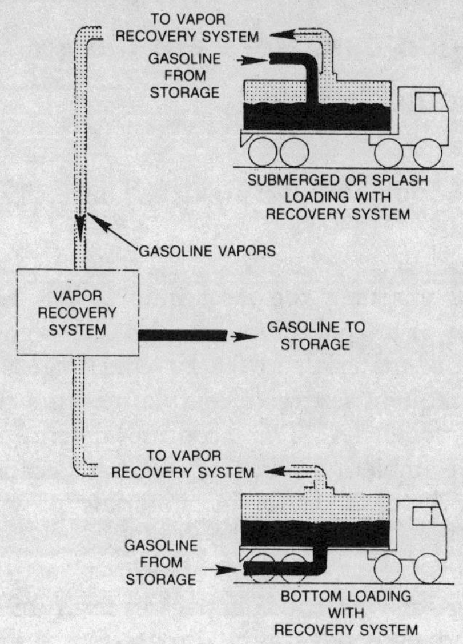

Fig. 4—Loading facilities must be equipped with vapor recovery facilities.

- All petroleum storage tanks are equipped with floating roofs, vapor recovery systems, or gas blanketing (Fig. 3).
- Loading facilities for gasoline, LPG and ammonia into tank trucks and cars are equipped with vapor recovery systems (Fig. 4).

In addition to these specifics, general refinery housekeeping, preventive maintenance and periodic environmental system audits are extremely important for over-all refinery odor control.

LITERATURE CITED

1 G. Leonardos, "A Critical Review of Regulations for the Control of Odors," J.A.P.C.A. **24**: 456 (1974)

2 Air Pollution Control Association, *Specialty Conference on: The State-of-the-Art of Odor Control Technology*, Air Pollution Control Association, Pittsburgh, Pennsylvania, 1974.

3 T. M. Hellman and F. H. Small, "Characterization of the Odor Properties of 101 Petrochemicals Using Sensory Methods," J.A.P.C.A. **24**: 979 (1974)

4 John O. Copeland, "Federal Perspectives in Odor Pollution Control," paper 29 presented at The New York Academy of Sciences; Conference on Odors: Evaluation, Utilization and Control. October 1973.

5 H. J. Klooster, G. A. Vogt & G. F. Braun, "Optimizing the Design of Relief and Flare Systems," Chem. Eng. Prog. **71**: 39 (1975)

Can You Measure Odor?

Industrial odors are difficult to measure, although the public is quick to notice them. A panel of trained observers can be used: chemical methods are of limited use. Here are the current methods

Hal B. H. Cooper, Jr., Texas A&M University, College Station, Texas

ODORS ARE PERHAPS the most immediately noticeable manifestations of air pollution in community atmospheres. The presence of materials in the air which cause odor responses is a frequent source of complaints from the public to regulatory agencies. The accurate determination of odor levels in ambient air is very difficult because of the variations in types of different materials present, especially when these are in combination.

Methods for direct evaluation of odors provide a means for evaluating their potential impact in terms of intensity, threshold, and objectionability. Progressive syringe dilutions using a panel of trained observers in a remote laboratory location provides one useful method for determining odor levels in ambient air and source gas streams. Dynamic dilution systems and the Scentometer may also prove useful for odor evaluation under some conditions. Chemical methods are of limited usefulness in making direct odor measurements, particularly in ambient air.

There are a number of problems involved in making determinations of odor levels in ambient air or flue gas streams. Physical problems include losses of materials by adsorption on container and tubing walls, losses by condensation in the presence of water, and reaction of odorous gases during storage or transport. Problems in organoleptic response of observers include meteorological variations in humidity, temperature, wind speed and direction. Also affecting olfactory response of observers include physical condition, smoking, prior history of exposure, and olfactory fatigue after prolonged exposure to odorous constituents.

Regulatory agencies. Several air pollution control agencies in the United States have regulations regarding either discharges of odorous constituents to the atmosphere or ambient odor levels in the atmosphere.[1] A number of agencies, including the Bay Area Air Pollution Control District in California, employ a general nuisance regulation regarding adverse odor levels in the atmosphere. Knox County, Tennessee,[2] employs regulations which limit the level of objectionable odors beyond the property line of a source which may result in either damage or interfere with enjoyment of life or property. The regulation specifies an odor level violation when 30 percent or greater of a panel of 20 or more persons observe the odor at a dilution of 1 to 8 with odor free air or more.

The Mid-Willamette Valley Air Pollution Authority in Oregon employs ambient odor regulations based on public nuisances. Direct evaluation of ambient odor levels is also made by means of progressive dilutions by an inspector using a device known as a "Scentometer."

The State of California has a specific regulation regarding maximum ambient concentrations of hydrogen sulfide as an odorous constituent of 30 parts per billion by volume. Other ambient odor regulations include community response programs, direct odor level evaluations by an inspector, and direct measurement of specific odorous compounds.

Major odorous compounds of interest in air pollution studies include compounds containing sulfur, nitrogen and chlorine, plus numerous organic materials. Oxidized organic materials such as styrene and butadiene, as well as halogen-containing chlorine and hydrogen chloride, tend to have irritating odors. Reduced sulfur compounds such as hydrogen sulfide and methyl mercaptan, and amine compounds from protein decay tend to have offensive nauseous odors.

CLASSIFICATION

Description. Odors can be described in terms of their character, intensity, pervasiveness and acceptability.[4] The character of an odor depends on the type or types of chemical constituents present, where the action of an odorous component in the pure form may be different from its activity in a mixture. The intensity of an odor is a function of concentration, where the odor intensity level tends to increase with the logarithm of concentration, according to the Weber-Fechner law.[5] Odor levels can be rated on arbitrary scales on a semiquantitative basis in terms of their relative intensities in varying degrees of being weak or strong by a panel of observers. The panel is asked to make subjective evaluations of odor intensities (see Table 1).

Two additional characteristics regarding odors are per-

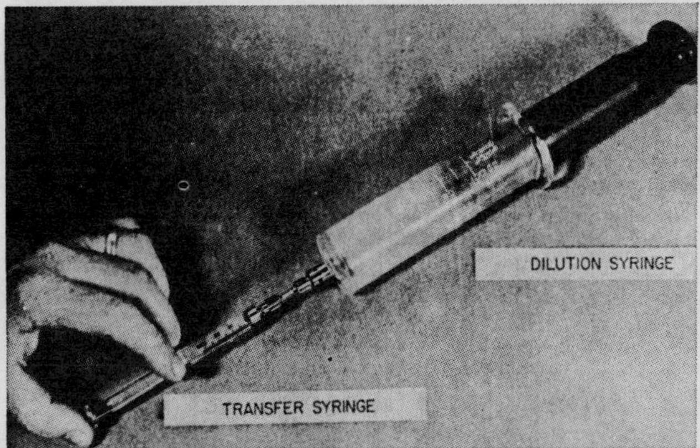

Fig. 1—ASTM syringe dilution method. Photo shows transfer of odor sample to dilution syringe.[11]

vasiveness and acceptability.[4] The pervasiveness of an odor describes the degree of dilution required to reduce the odor level of a pollutant to below its threshold level. It can be described as the ratio between a given odorant concentration and its threshold level. Pervasiveness describes the degree of dilution required by ventilation to alleviate a potential odor problem caused by a particular contaminant. Odor acceptability is a subjective evaluation of the character of an odor in terms of being pleasant, neutral or unpleasant. The relative acceptability of an odor depends on its chemical form and the specific response of the individual involved.

Thresholds. Considerable work has been done involving odor threshold levels for specific compounds. [6,7,8,9] There is often considerable variation between values reported in the literature, depending on the manner in which studies are conducted and the relative sophistication of the odor panel. A summary of typical odor threshold levels for specific compounds is listed in Table 2.

TABLE 1—Odor intensity evaluation scales

Number	Odor intensity level	
	System 1[4]	System 2
0	No odor	No odor
1	Detectable	Detectable
2	Definite	Faint
3	Strong	Noticeable
4	Overpowering	Strong
5	....	Overpowering

TABLE 2—Odor threshold levels in ambient air[6]

Compound	Threshold ppm by vol.	Compound	Threshold ppm by vol.
Acrolein	0.21	Hydrogen sulfide	0.004
Methyl amine	0.021	Methyl mercaptan	0.002
Dimethyl amine	0.047	Dimethyl sulfide	0.004
Trimethyl amine	0.00021	Dimethyl disulfide	0.006
Ammonia	46.8	Paracresol	0.001
Butyric acid	0.001	Phenol	0.047
Chlorine	0.314	Phosgene	1.00
Diphenyl sulfide	0.0047		

An additional factor which must be considered is that odorous gases or vapors by themselves may exhibit different threshold response characteristics than odorous gases which may become absorbed to the surfaces of particles.[10]

Measurement. Odor levels can be measured in both source gases and in ambient air, where the concentrations in flue gases are often on the order of 1,000 to 10,000 times as high as in ambient air. The three general approaches to odor level determinations are as follows: (1) direct organoleptic subjective determination of odor threshold levels; (2) nonselective chemical determination as indicators of odor levels; (3) selective chemical determination of individual odorous constituents. Considerations in odor level evaluation include both collection and analyses of samples.[11]

DILUTION METHODS

Dilution techniques. The four major techniques for determination of odor threshold levels and odor intensities by dilution with air are as follows: (1) mixing of odor-containing air and odor-free air in a container, bringing the nose in contact with the container, and inhaling the mixture; (2) mixing of odor-containing air and odor-free air in a container, and injecting the gas mixture into the nose (syringe); (3) mixing of odor-containing and odor-free air in a duct, which is then supplied to an observer's nose through a face mask or enclosed hood (dynamic dilution); (4) mixing of odor-free and odor-containing air in varying dilutions by the act of inhalation (Scentometer).[5]

Odor observations. Olfactory measurements of odor levels from ambient air or flue gas can be made using either an odor survey or an odor panel.[5] The odor survey employs a representative group from the over-all population without regard to training and sensitivity.[12] It has the advantage of being representative, but is not necessarily precise, accurate or sensitive. The odor panel consists of a group of trained observers who have been specifically selected because of their sensitivity and accuracy in being able to perceive odor levels.[13] The odor panel tends to be more specific and accurate than the odor survey, but must be calibrated against comparable results obtained for the general population for similar odor level exposures.

Members of an odor panel are selected in terms of their sensitivity, accuracy, reproducibility and speed in evaluating odor threshold and intensity levels. Prince and Ince[14] observed variations in responses to given concentrations of plus or minus 40 percent between given days for a single observer, and plus or minus 20 percent for a panel of observers. Olfactory response tends to undergo fatigue after prolonged exposures with resultant insensitivity to changes in odor levels. Panelists should not undergo continuous testing for more than 15 minutes at a time, with rest periods of at least 30 minutes between exposures. In addition, exposure to initial high odor levels tends to deaden response to lower odor levels. Exposures should be from low to high odor levels if at all possible.

Sample collection. Samples of odorous gas can be piped directly from the ambient air or source to a panel, but this practice may be either impossible or may involve exposure of panelists to excessive levels of odorous gases prior to testing. The alternative technique involves the collection of given volumes of sample from the source or ambient air, followed by transport to the laboratory for evaluation by an odor panel. Samples may be collected in plastic bags (air) or in heated evacuated bottles (source), where the method avoids the problems of prior panel exposure. Potential odorous gas losses on container walls by condensation or other means must be minimized by heating or flushing, or be compensated for in testing, particularly for odor determinations of flue gas stream samples.

Methodology. Organoleptic methods for odor level determination include the following: (1) static progressive dilutions by means of syringes; (2) dynamic dilution of odorous gases by means of odor-free air; (3) vaporization of odorous compounds in a continuous flow system; (4) dilution of odorous air with odor-free air by means of respiration.

Syringe dilution. The dilution of odorous gases from sources and ambient air by means of syringes provides a reliable and inexpensive method for determining odor threshold levels. The method employed by the American Society for Testing and Materials[15] requires collection of an odorous gas sample in a 100-milliliter (cc) glass syringe, followed by sealing to prevent leakage and transport to an odor-free laboratory for evaluation. Small volumes of 1, 2, 5 or 10 milliliters of odorus gas sample can then be removed from the large syringe by means of small glass syringes with attached needles, as shown in Fig. 1.

The aliquots of odorous gas are injected into additional 100-cc glass syringes to make up a series of progressive dilutions, which can then be evaluated by an odor panel. The odor strength is directly proportional to the degree of dilution required to reach the threshold level. For low odor levels as found in ambient air the method may be simplified by taking variable odorous air volumes in 100-cc syringes. Following sealing, the syringes are taken to the laboratory and diluted with odor-free air and the odor threshold levels then determined.

It is necessary to clean syringes thoroughly following use to avoid contamination and to be sure syringes are dry to avoid odorant condensation.

Dynamic dilution. Several devices are available for obtaining odor levels by dynamic dilution with odor-free air. An Osmo device developed by Gex and Snyder[16] uses dynamic dilution of odorous gas with activated carbon-purified odor-free air in a cylindrical flow proportioning chamber. Flow proportioning is adjusted to alter the degree of dilution by movement of an axially located piston along a cylinder with 528 equally sized holes along the outer cylinder. The particular odorous gas stream is fed to a face mask following mixing for evaluation by an odor panel, as shown in Fig. 2.

Nader[17] developed a portable dynamic dilution system for determining odor threshold levels from either flue gases or ambient air. Odorous air samples or isolated odorous compounds are collected and drawn into the dilution system and mixed with purified air at a flow rate of 1 to 2 cfm, and then fed to an enclosed hood. The subject is kept in the enclosed hood and isolated from his surroundings during odor response tests to minimize outside interferences. Tubing and wall connections are constructed of either Teflon or glass to minimize potential absorption problems. The system is illustrated in Fig. 3. A diffusion cell is added containing a known odorant vapor as means of calibration.

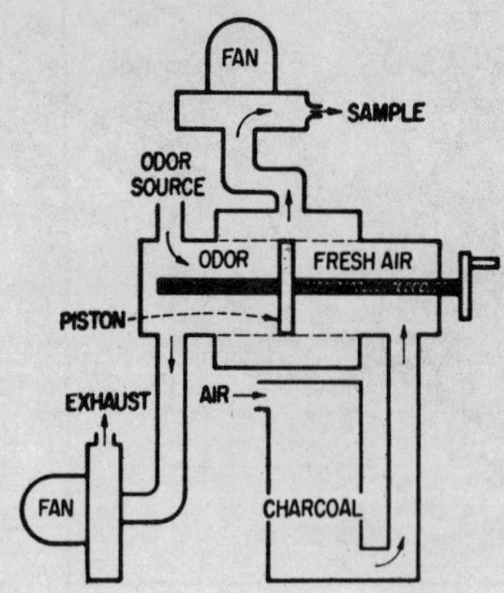

Fig. 2—Osmometer dynamic dilution system. Sample is delivered to face mask for evaluation at various dilutions.[11]

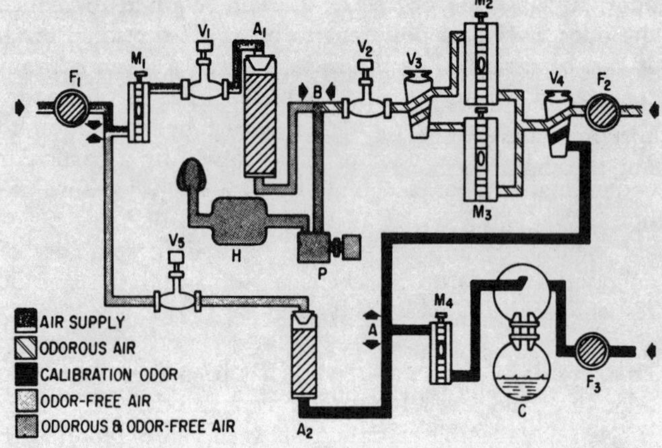

Fig. 2—Nader dynamic dilution apparatus. F_1, F_2, F_3 glass-wool filters; M_1 30 lpm plastic rotameter; M_2 1.5 ml/min, M_3 15 to 425 ml/min, and M_4 9 to 900 ml/min glass rotameters; V_1, V_2, V_5 brass needle valves; V_3, V_4, glass two-way valves; A_1 2.25-in. OD x 15-in. length, 0.125-in. wall, and A_2 1-in. OD x 12-in. in length, 0.125-in. wall, plastic activated-carbon absorber columns; P, carbon-vane oil-less air pump, approx. 25 lpm free air flow; 110-v. AC motor for lab or 6.3-v. DC for field use; H, Teflon hood with glass viewing window; C, odorant diffusion cell, approx. 4 mm dia. and 10 cm capillary.[11]

Hemeon[18] developed a system similar to that of Nader, except that much larger air flow rates of 30 to 70 cfm were involved. The system was not portable and samples were brought to the laboratory in flexible plastic bags. Dilution of odorous gas samples was obtained in one step with air for ambient air, and two dilution steps in series for flue gases.

The Institute of Hygiene of the Swedish National Institute of Public Health[19, 20] employs collection of odorous gas samples from either source gas streams or ambient air in large, flexible plastic bags. The bags are then returned to the laboratory and odorous gas is drawn by a diaphragm pump through one of a series of calibrated capillary tube flow meters to achieve a low constant flow rate. The odorous gas is then mixed with purified odor-free air in a dynamic mixing chamber, and passed to an exposure hood for evaluation by an odor panel, as shown in Fig. 4.

Vaporization dilution. Two methods employ collection and isolation of odorous gas constituents from ambient air prior to evaluation of odor levels by a panel. Odorous gas components can be selectively removed from air sam-

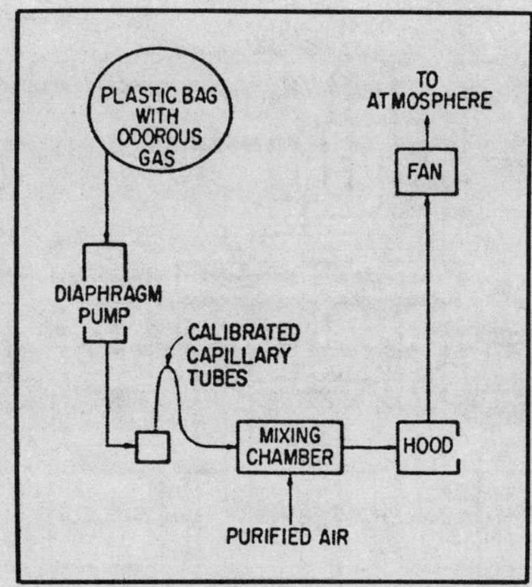

Fig. 4—Swedish dynamic dilution system. Plastic bag sample is obtained at plant, diluted and evaluated at laboratory.

ples by either absorption onto activated carbon, or condensation in cold traps using dry ice or liquid nitrogen.[4] The method provides for isolation and concentration of odorous constituents from ambient air or flue gas streams, but water condensation, retentivity losses, and interfering substances may limit the ability to determine potential odor threshold levels.

Fair and Wells[21, 22] developed an osmoscope which is primarily useful for measuring odor levels in water, but can also be used for odorous compounds collected by condensation. Purified air is passed over the liquid material to vaporize the odorous constituents and then diluted with additional purified air in a concentric tube arrangement. The degree of dilution of odorous air in the inner tube is controlled by addition of purified air in the outer tube at the face mask by the observer, where the respective flow rates of air streams indicate the relative degree of dilution.

Katz and Allison[23] developed an odorometer which could be used to determine concentrations of odorous constituents in parts per million by volume. Purified air at a known flow rate is passed through a U-tube containing an odorous material of known volume, where the rate of evaporation is held constant at a constant temperature. The odorous material of known concentration in an air mixture is then passed through a face mask or hood for assessment by an odor panel.

Scentometer dilution. Direct measurement of the odor level in the atmosphere is made by an inspector employing a progressive dilution device known as a "Scentometer." The system employs drawing in variable amounts of odorous gas or air through an orifice, followed by dilution in a stream of air prepurified by passage through a bed of activated carbon.[24] The relative amount of odorous air entering in relation to the prepurified air is governed by a series of four orifices of varying diameters.[25] Motive power is supplied by the normal breathing of the observer through the device.

The normal procedure in operation is to start by breathing prepurified air only for one minute, then switching to the smallest orifice (greatest dilution) and breathing for approximately 5 minutes. If no odor is perceived the next larger orifice is placed in line. The dilution is progressively reduced by changing the orifice size until an odor is just barely perceptible, where 0, 2, 7, 31 or 170 dilution volumes with odor-free air can be sequentially obtained with the device, as listed in Table 3. The

TABLE 3—Effect of orifice diameter on Scentometer dilution and odor intensity classification[24, 25]

Scentometer number	Orifice diameter (inches)	Number of dilutions to threshold	Odor intensity
0...........	Closed	0	None
1...........	½	2	Barely perceptible
2...........	¼	7	Distinct, persistent
3...........	⅓	31	Strong, persistent
4...........	1/16	170	Intolerable

odor instantly shows a marked increase with the number of dilutions required to bring it to the odor threshold level.

To date, the Scentometer has been used only for qualitative evaluations of ambient odor levels. No convictions for violations have yet been obtained with the device by the Mid-Willamette Valley Air Pollution Authority in Oregon. The device is also useful for qualitative evaluation of the effectiveness for removing malodorous constituents in source flue gas streams following an odor control device.[26] Problems with the device include variability in sensitivity between observers, variability in response of an observer, history of prior exposure to the odor, wind speed, temperature, relative humidity, whether the observer smokes, and previous exposure history.[22] A drawback of the device is that the observer may receive a full dosage of odorous air before he begins to use the Scentometer which may tend to deaden his sense of smell because of olfactory fatigue.

CHEMICAL METHODS

Both nonselective and selective chemical methods can be used to characterize odor levels in the atmosphere and determine the presence of specific odorous compounds. Nonselective methods such as absorption and condensation for collection of odorous components are simple and inexpensive, but do not provide much information regarding odor levels unless detailed chemical analyses are made. Chemical analyses for specific odorous components can provide an indirect indicator of odor levels, but often require sophisticated analytical procedures and equipment.

Nonselective methods. The major methods for nonselective chemical evaluation to provide an indicator of odor levels are as follows: collection by absorption or condensation and gravimetric weighing,[27] and oxidimetric chemical conversion of reduced malodorous constituents by consumption of an oxidizing agent such as potassium permanganate.[11] Odorous organic vapors can be adsorbed onto activated carbon, moisture content determined and an increase in weight obtained for the organic materials. The method is nonspecific and not accurate because interfering materials can also be adsorbed onto the carbon.

Oxidimetric conversion of an oxidizing agent can also be used as an indicator of odorous gas levels where air is drawn through impingers containing an oxidizing solution.[28] The degree of consumption of oxidizing solution is proportional to the total amount of reduced materials present, but does not differentiate between the individual concentrations or types of materials present. It does not provide data on odorous oxidizing agents. The above methods are rather crude, and provide only general indicators of odorous gas levels.

Selective methods. The concentrations of specific odorous constituents in the atmosphere can be determined by particular chemical analyses. One method used is to employ continuous monitoring or wet chemical analyses for specific chemical constituents. A number of methods are available for continuous monitoring of odorous reduced sulfur compounds in the atmosphere, including coulometric titration, flame photometry, electrochemical membrane cell conversion and ultraviolet spectrophotometry.[29] These methods may or may not differentiate between specific compounds and provide an indicator of odor levels due to sulfur compounds.

Gas chromatography may be used to identify a wide variety of odorous compounds depending on the separation column and detection system employed. Samples must first be collected in heated containers, plastic bags, syringes or continuous sampling lines. Losses of materials may occur on the walls or the sample container or in the chromatographic column. Detectors which may be employed include flame ionization, flame photometry, electron capture and thermal conductivity.

ACKNOWLEDGEMENT

Originally presented as "Ambient and Source Odor Measurements" to the Air Pollution Control Conference, University of Tennessee, March 29, 1973.

LITERATURE CITED

1 Rossano, A.T., and Cooper, H.B.H., "Regulatory Aspects of Odor Control," Special Report, University of Washington, Department of Civil Engineering, Seattle, Wash., October 1970.

2 "Regulation of Odors," Section 21.0, Knox County Air Pollution Regulations, Knoxville, Tenn.

3 Sullivan, R.J., "Preliminary Air Pollution Survey of Odorous Compounds—A Literature Review," Contract No. PH 22-68-25, Report No. APTD 66-42, National Air Pollution Control Administration, Raleigh, N.C., October 1969.

4 Clayton, G.D. ed., Air Pollution Manual. Part I—Evaluation, "Odors," ch. 11, American Industrial Hygiene Association, Detroit, Mich., 1960.

5 Byrd, J.F., and Phelps, A.H., "Odor and Its Measurement," ch. 23 in Stern, A.C., ed., Air Pollution, vol. II, 2nd ed., Academic Press, Inc., New York, 1968.

6 Air Pollution Abatement Manual, "Physiological Effects," ch. 5, Manufacturing Chemists Association, Washington, D.C., 1951.

7 Leonardos, G., Kendall, D., and Barnard, N., "Odor Threshold Determinations of 53 Chemicals," *Journal of the Air Pollution Control Association, 19*, (2), 91-95, February 1969.

8 Benforado, D.M., Rotella, W.J., and Horton, D.L., "Development of an Odor Panel for Evaluation of Odor Control Equipment," *Journal of the Air Pollution Control Association 19*, (2), 101-105, February 1969.

9 Wilby, F.V., "Variation in Recognition Odor Threshold of a Panel," *Journal of the Air Pollution Control Association, 19*, (2), 96-100, February 1969.

10 Cooper, H.B.H., and Rossano, A.T., "Particulate Matter and Odor Control," Special Report, University of Washington, Department of Civil Engineering, Seattle, Wash., January 1971.

11 Jacobs, M.B., The Chemical Analysis of Air Pollutants, "Odor Analysis," ch. 16, Interscience Publishers, Inc., New York, 1960.

12 Rossano, A.T., ed., "The Air over Louisville," Special Air Pollution Study, Louisville and Jefferson County, Kentucky, U.S. Public Health Service, Robert A. Taft Sanitary Engineering Center, Cincinnati, Ohio, 1957.

13 Gruber, C. W., "Odor Control from an Official's Viewpoint," Symposium on Odor, Special Technical Publication No. 164, Amercan Society for Testing and Materials, Philadelphia, Pa., 1954.

14 Prince, R.G.H., and Ince, J.G., "The Measurement of Intensity of Odor," *Journal of Applied Chemistry, 8*, (5), 314-321, May 1958.

15 "Standard Method for Measurement of Odors in Atmospheres (Dilution Method)," ASTM Standard D 139-57, pp. 185-188, American Society for Testing and Materials, Philadelphia, Pa., 1957.

16 Gex, V.E., and Snyder, J.P., "New Device, Wider Concept Helps to Measure Odors Quantitatively," *Chemical Engineering, 59*, (12), 200-201, 372, 374, December 1952.

17 Nader, J.S., "An Odor Evaluation Apparatus for Field and Laboratory Use," *American Industry Hygiene Association Journal, 19*, (1), 1-7, February 1958.

18 Hemeon, W.C.L., "Technique for quantitative Measurement of Odor Emissions," *Journal of the Air Pollution Control Association, 18*, (3), 166-170, March 1968.

19 Cederlof, R., Edfors, M.L., Friberg, L., and Lindvall, T., "On the Determination of Odor Thresholds in Air Pollution Control — An Experimental Field Study on Flue Gases from Sulfate Cellulose Plants," *Journal of the Air Pollution Control Association, 16*, (2), 92-94, February 1966.

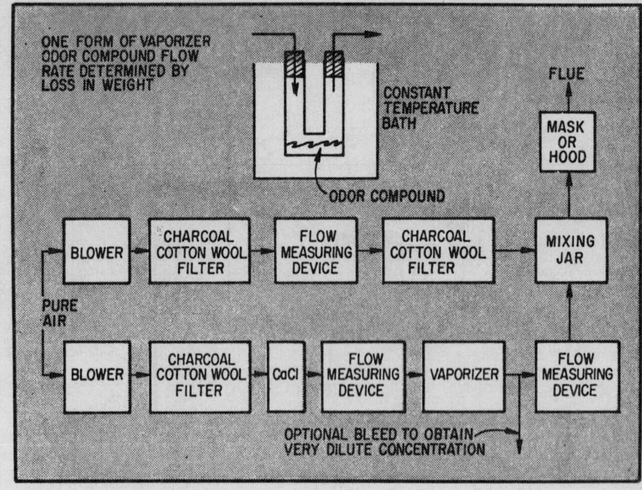

Fig. 5—Katz-Allison odorometer. Relatively pure air is rendered odor free by charcoal and filtered. Air for vaporizing is dehumidified and passed over compound in vaporizer tube. Measured flow of odorous and pure air are mixed and evaluated by panelists.[11]

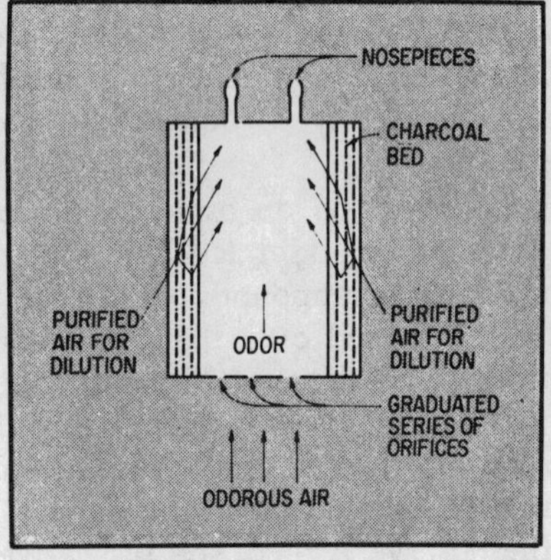

Fig. 6—Scentometer odor dilution device. Odorous air passes through graduated orifices and is mixed with air from the same source, which is purified by passing through charcoal beds. Dilution rates are fixed by orifice selection.[11]

20 Cederlof, R. Edfors, M.L., Friberg, L., and Lindvall, T., "Determination of Odor Thresholds for Flue Gases from a Swedish Sulfate Cellulose Plant," *Tappi, 48*, (7), 405-411, July 1965.

21 Fair, G.M., and Wells, W.F., "The Air Dilution Method of Odor Determination in Water Analysis," *Journal of the American Water Works Association, 26*, (11), 1670-1677, November 1934.

22 Mateson, J.F., "Olfactory: Its Techniques and Apparatus," *Journal of the Air Pollution Control Association, 5*, (3), 167-170, November 1955.

23 McCord, C.P., and Witheridge, W.N., Odors Physiology and Control, pp. 45-56, McGraw-Hill Book Co., New York, 1949.

24 Gruber, C.W., Jutze, G.A., and Huey, N.A., "Odor Measurement Techniques," Proceedings of the 52nd Annual Meeting of the Air Pollution Control Association, Paper No. 59-68, June 1959.

25 Huey, N.A., Broering, L.C., Jutze, G.A., and Bruber, C.W., "Objective Odor Pollution Control Investigations," *Journal of the Air Pollution Control Association, 10*, 441-446, December 1960.

26 Rowe, N.R., "Odor Control with Activated Charcoal," *Journal of the Air Pollution Control Association, 13*, (4), 150-153, April 1963.

27 Turk, A., "Odorous Atmospheric Gases and Vapors: Properties, Collection and Analysis," *Annals of the New York Academy of Sciences, 58*, (3), 194-198, March 1954.

28 Kuehner, R.L., "The Validity of Practical Odor Measurement Methods," *Annals of the New York Academy of Science, 58*, (2), 175-186, February 1954.

29 Cooper, H.B.H., and Rossano, A.T. Source Testing for Air Pollution Control Environmental Science Services Corp., New York, 1971.

Part VI.

Noise Control

Assess and Control HPI Plant Noises

Once you have a plan, then the job of surveying, finding noise areas, training employes and making noise-reduction recommendations can be done by people who have little technical expertise, but lots of common sense

G. Robinson
Gulf Oil Refining Ltd.,
Milford Haven, South Wales

PROTECTION OF EMPLOYES' hearing can be achieved by teamwork of acoustical experts, practical technologists, cooperative manufacturers, medical officers and a committed management.

In 1972 the U.K. Department of Employment published a *Code of Practice for Reducing the Exposure of Employed Persons to Noise.*[1] Prior to this, the less stringent U.S. Walsh-Healey Public Contracts Act limits had been used. Because of the introduction of the *Code of Practice* and because there had been much success in reducing noise within the plant, it was decided to initiate a new hearing conservation program. All measures taken for reducing noise at source also had a favorable effect on the over-all noise level in the refinery and hence on the parallel aim of reducing impact on the surrounding environment. Following is a description of action taken.

Plan of campaign. A plan of campaign was formulated into a logical sequence of events.

(1) Objectives to be defined by management
(2) Formation of an action committee
(3) Noise survey
(4) Noise reductions at source
(5) Definition of noise limits
(6) Marking areas and machines
(7) Education
(8) Protection
(9) Analysis
(10) Review.

Objectives. As in all jobs, precise definition of objectives is important. It is management's responsibility to set objectives and ensure that they are achieved. Objectives of the hearing conservation program were:

1. Define and mark areas where personnel must take preventative action.

2. By education convince all employes exposed to harmful noise of the need for hearing protection.

3. Reduce noise where possible.

Although of high priority, (3) was a longer term problem.

Action committee. To achieve the objectives in a relatively short time, the action committee had to be carefully chosen. Not only is practical acoustical knowledge required, but also acceptance of hearing conservation by operating and engineering personnel had to be considered. The following personnel were chosen as the action committee.

- Environmental pollution control supervisor
- Safety supervisor
- Operations supervisor
- Operator
- Maintenance supervisor

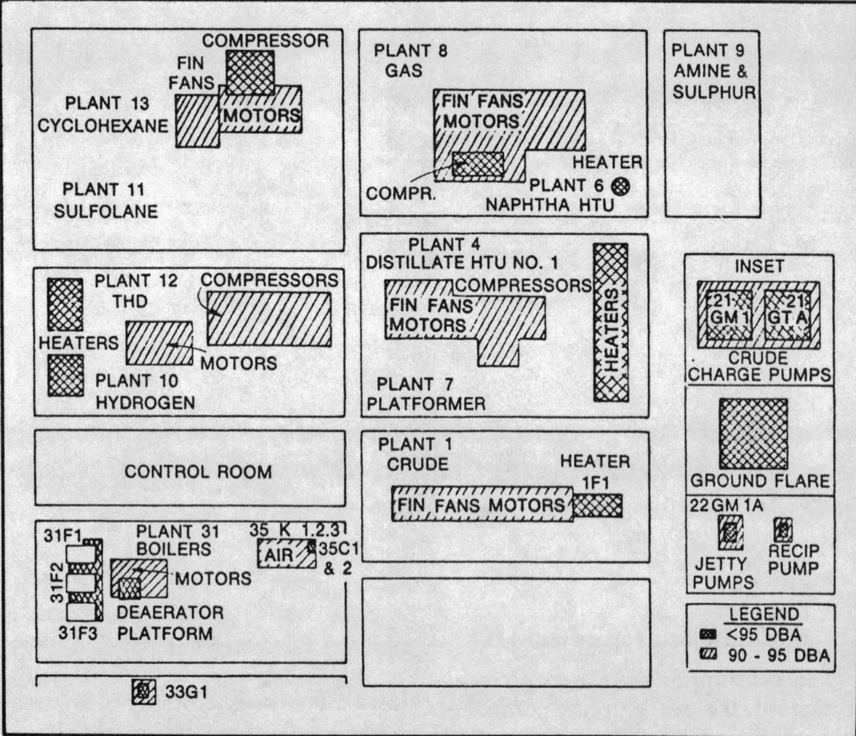

Fig. 1—A 1969 noise survey included analysis of these plant areas and items of equipment.

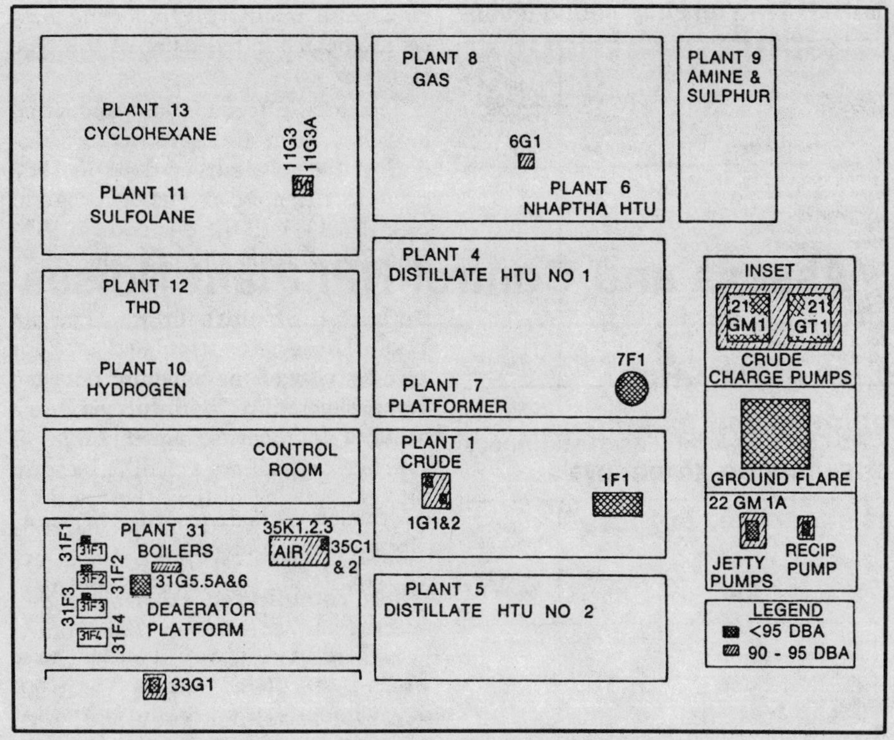

Fig. 2—A 1974 noise survey included these areas of the plant and items of equipment.

Fig. 3—Burners of the low-noise type have been installed.

- Fitter
- Engineer
- Medical officer (part-time).

This constitution gave a good cross-section of the work force pertinent to both the accomplishment and the acceptance of the hearing conservation program. The work force at the refinery is not unionized but at other locations trade union representation might be advantageous.

Noise survey. The timing of an area survey is important. It should occur when the maximum amount of noise is emitted. It is of little use doing a survey to find hazardous areas if some plants are shut down or at low thruput. Account should also be taken of the fact that maximum plant noise may occur during start-up, shutdown or while special operations are being carried out. It could be misleading if surveys were only conducted with plant at maximum thruput. In summary each plant has its own noise personality. This should be allowed for when performing the survey.

Steam leaks which are often present even in a well-maintained plant can prove to be a problem when carrying out the survey. As far as possible all atypical noises should be cured before the survey is conducted.

The survey should not be confined to the ground level. It should take into account all platforms and places where an operator is likely to work. It should also include facilities not in the plant area such as jetties, flare areas, blending areas, pump stations, railway sidings and workshops. Both mobile and stationary machines used on the site should be included.

Weather; people; instruments. Extreme weather conditions should be avoided. To obtain realistic sound level readings, choose a fine, fairly calm day. Even if a plant is totally shut down it can create some noise due to the effect of wind.

The use of non-intrinsically safe equipment, such as sound level meters, necessitates authorization for their use. This requires gas tests for explosive atmospheres over the whole survey area before starting.

Two persons should do the sound level survey. One to use the sound level meter and one to put down the results on a plant plot plan. Personnel must be trained in noise measurement techniques.

The instrument should ideally be a precision grade instrument which is typically accurate to ± 1 dBA. It should be calibrated, fitted with a wind shield and set to "slow" response. There is an exception to setting the sound level meter to "slow" response. This is fully defined in the *Code of Practice* under Overriding Limits.

Which measurements? Initially only the sound level in dBA should be measured. Enough readings should be taken to define the hazardous areas accurately. When doing the survey, account should be taken of the *Code* *of Practice* so that areas with a sound level of 90 dBA and above can be accurately defined.

In areas with a sound level above 90 dBA do an octave band analysis. Thus, any high value pure tone components can be measured. This will give information on the attenuation required for ear protection.

Workshop noise is variable. It depends on the type of work, the material being worked upon and the machines/tools being used at the time. Each type of workshop function should be examined. Grinders, impact extractors, air-operated riveters and concrete impingers all have high sound levels.

Take account of not only the operator doing the job, but also personnel in the vicinity which might be affected.

Mobile equipment such as cranes, tractors, compressors, locomotives and welding sets should also be covered by the survey.

Survey results. Typical results of surveys at one refinery are shown in Figs. 1 and 2, and Tables 1 and 2. More detailed surveys showed that the main sound power level sources in the plant areas, in order of importance, were: air coolers (fin fans); heaters; motors; compressors.

Noise reduction at the source was instigated not only to reduce the exposure of employes working close to the equipment, but also to reduce the general level of noise from the refinery plant as a whole. We felt it could have an impact on the community at large.

As well as continuous noise sources, intermittent sounds such as steam venting and steam leaks contributed to the problem.

The first major noise source to be reduced was heaters. The reason for not starting on the air coolers was that the technology was not available. Heater noise was reduced by the installation of low noise burners (see Fig. 3). The burners were fitted to 10 heaters. A total of 180 burners were changed. This led to a reduction of 13-22 dBA under the heaters with an average reduction of 16 dBA. Ear protection is now required under only two of the furnaces.

A larger problem. Induced draft air coolers proved more of a problem. However, in 1971, certain technological advances in air cooler design gave a significant decrease in sound level with no decrease in duty. Four-bladed fan was replaced with a double-decked six-bladed fan. This allowed a tip speed reduction from 12,000 to 7,000 ft. min.$^{-1}$ with the same cooling effect. After a trial period with prototype equipment, 90 fans were changed to the new design. All those areas where air cooler noise had been the main contributor to over-all levels higher than 90 dBA were eliminated.

TABLE 1—Workshop sound levels

Machine/area	Typical sound level (dBA)	Distance required to reduce noise to 90 dBA (ft.)
General workshop noise	78	—
Welding bay with three welding units in operation	84	—
Welding bay grinder	90	—
Pedestal grinder	92	5
Mobile grinder	91	5
Angle disc grinder	101	12
Impact extractor	111	28
Blasting machine	82	—
High-speed lathe	83	—
Concrete impinger	115	20
Air-operated riveter	115	20

TABLE 2—Mobile and stationary engine sound levels

Engine	Sound level (dBA)	Distance required to reduce noise to 90 dBA (ft.)
Small crane	90 (in cab)	—
Large crane	90 (in cab)	40
Carry lift	96 (in cab)	8
Tractor	92 (in cab)	6
Welding set	93	12
Compressor	90	—
Loco	98 (in cab)	—
Fork lift	88	—

Electric motor noise—the next high noise source—was reduced by the use of undirectional fans and silencers.

Compressor noise, the last major noise source, was reduced by the use of enclosures.

Use of lagging. Small noise sources such as pipe and vessel noise were reduced by the use of lagging. The lagging consisted of 1½ inch of mineral wool followed by lead sheet (1 lb.ft.^{2}) covered with the normal aluminum sheet (26 SWG). Silencers were fitted to steam lances.

The intermittent noises were easy to reduce. Steam vents were resited and silencers fitted.

Steam leaks surveys were instigated, followed by rapid repairs, to reduce noise from this source. Also the education of personnel and the operation of a noise monitoring/reporting procedure helped in reduction of this type of noise.

The total effect of noise reduction measures taken are illustrated by comparing the noise survey done in 1969 prior to the noise reduction program (see Fig. 1) with the one done in 1974 (see Fig. 2) after the program.

Definition of noise limits. The old Walsh-Healey Act stipulated that for each halving of the exposure time the sound level may be increased by 5 dBA. The more stringent *Code of Practice* only allows a 3 dBA increase for each halving of the exposure time. The limits prescribed by the code should be taken as the maximum acceptable levels, not as desirable levels. The limits are:

1. The maximum level of noise which is acceptable over an 8 hr. work day, without ear protection, is 90 dBA, with an increase of 3 dBA for each halving of the daily exposure time, e.g. 4 hr. at 93 dBA, 2 hr. at 96 dBA.

2. The unprotected ear must not be exposed to a sound pressure level, measured with an instrument set to the 'fast' response, exceeding 135 dB (linear) or in the case of an impulse noise, an instantaneous sound exceeding 150 dB (linear).

3. No part of the body must be exposed to a sound pressure level measured with an instrument set to 'fast' response, exceeding 150 dB (linear).

Based on these limits the following rules were implemented:

1. *Areas with a Sound Level Greater than 95 dBA.* Hearing protection must be worn at all times.

2. *Areas with a Sound Level 90-95* dBA. Hearing protection must be worn if the total exposure is greater than 2 hr. per shift.

Similar regulations were issued to maintenance personnel which were:

1. *Machines Emitting a Sound Level Greater than 95 dBA.* Hearing protection must be worn at all times.

2. *Machines Emitting a Sound Level 90-95 dBA.* Hearing protection must be worn if the total exposure is greater than 2 hr. in 8 hr.

Similar regulations were issued also when using mobile or stationary engines.

Marking. Based upon the defined noise limits the appropriate signs were erected and the areas marked (see Fig. 4).

The areas with a sound level greater than 95 dBA were described as Red Areas. The signs had a warning symbol as described an the *Code of Practice,* were edged with red. They carried the wording "Hearing Protection Must be Worn." The size of the signs was 30 by 15 in. The boundaries were defined by painting the ground and stanchions with a red line on a yellow background.

The areas with a sound level 90-95 dBA were described as Green Areas. The signs were edged with green and carried the wording "Hearing Protection Must be Worn if Exposure Exceeds 2 Hours in 8 Hours." The boundaries were defined by painting a green line on a yellow background.

Machines, mobile and stationary engines were similarly marked.

Fig. 4—Signs have been installed calling for hearing protection requirements of the given machine or area.

Training. A training program was devised to educate and convince all employes of the need to use hearing protection. As well as issuing a booklet *Hearing Conservation—Noise Limits and Regulations,* the following questions were posed and answered during the program.

1. Why is hearing conservation necessary? This was illustrated with two films titled *Listen While You Can* and *Medical Aspects and Hearing Conservation.*[2]

2. Where and when is hearing protection required? This was answered using the previously described defined noise limits, the results of the noise survey and showing how and where the areas and machines had been marked.

3. How is hearing conservation monitored? This was illustrated by the use of the films referred to in (1) above.

During training, instruction was given in the use of a sound level meter made available for anyone to use. The use of ear muffs was also described.

The total time for training was 1½ hr. for each 12 sessions which included ¼ hr. for discussion. As suggested in the *Code of Practice,* a register of all personnel undergoing training was kept. Contract personnel were included in the training program.

Protection. At the initial stages of the program, it was decided that the only form of ear protection would be ear muffs and that these would be issued on a personal basis.

To encourage the use of ear muffs, various types with the necessary attenuation were issued to potential users for them to assess their acceptability. It is important to involve operations and maintenance personnel in the choice of hearing protection. This helps in the acceptance of this added protection device in their ever increasing armory.

Analysis. To measure the success of the hearing conservation program, it is necessary to monitor individuals' hearing. This can help to detect at an early stage, anyone whose hearing has been affected, possibly due to exceptional susceptibility to noise or failure to use ear protection.

To establish a base line level of hearing, an audiogram is an essential part of the pre-employment medical examination. Subsequent audiograms can then be compared with the original performance. This is carried out as a regular routine with operations and maintenance people known to be exposed to noise.

Review. As part of the daily environmental pollution control routine, a daily subjective evaluation of noise is done within the plants. If any high sound level conditions are found, the cause is investigated and, if possible, corrected.

To help in the control of steam leaks which in low noise plant can contribute significantly to the noise, there is a regular steam leak survey. It is normally carried out once a month and before shutdowns so that repairs which cannot be done without shutdown are remedied at that time.

I have described results achieved by one particular company in one particular industry. The methods of tackling the problems are of wide application. I hope some of the ideas will be useful to others.

LITERATURE CITED

[1] *Code of Practice for Reducing the Exposure of Employed Persons to Noise,* Department of Employment, HM Stationery Office, 1972.

[2] *Dangerous Noise, Parts 1 and 2,* written and produced by G. Fergusson, Royal Naval Film, distributed by Stewart Films.

Ways to Reduce Plant Noises

Selecting and sizing equipment silencers is made easier with this guide to noise control

B. G. Golden
Burgess-Manning Co., Dallas, Texas

MOST OF THE NOISE from hydrocarbon processing plants comes from vents and blowdowns to atmosphere, reciprocating and turbine engines, blowers and compressors, pressure reduction valves, piping systems and other related sources. Increased process pressures, higher speeds, multiplicity of sources, and economy of layout and construction, all lead to added noise.

Area ordinances and controls do not permit unabated plant noise.

A quiet operation does not merely happen. It is planned. Noise control and noise abatement should be included in the initial planning, not as an afterthought, once a problem has developed. To insure adequate noise control, the unsilenced noise levels expected from the various sources throughout a project plant must be established, preferably by actual measurement. When measured data from other or similar installations is not available, then the noise levels should be calculated or estimated. Secondly, the area criteria must be established and, as above, by actual measurements where possible. Once the ambient or background is recorded the criteria or allowable levels can be fixed. Thirdly, the parameters of silencer selection and sizing must be based upon sound practices of both theory and experience. Most commercially available silencers come in three grades of silencing—commercial, standard and residential.

The reactive type silencer is generally restricted to low frequency applications and depends upon a volume change to reflect sound energy back to the source. When out of phase with the incoming wave, reduction of noise occurs. If in phase, a resonant increase in noise may result. Multi-chamber reactive silencers are available in straight-through tube arrangements for low-loss applications and in labyrinth like, tortuous path, tube configurations for added performance where pressure losses are not critical.

The dissipative type silencer is essentially a high frequency attenuator and depends upon sound absorbing material to absorb or dissipate the sound energy. Dissipative silencers are usually in the form of straight runs of internally acoustically treated piping, annular or parallel baffles of felt, rockwool, fiberglass, and the like.

The combination reactive/dissipative silencer is functionally a reactive silencer with sound absorptive material to provide added high frequency noise control. The perforated or ported tube, when applied to the reactive design acts as a dissipative element in reducing or eliminating troublesome pass-bands inherent to the basic reactive design.

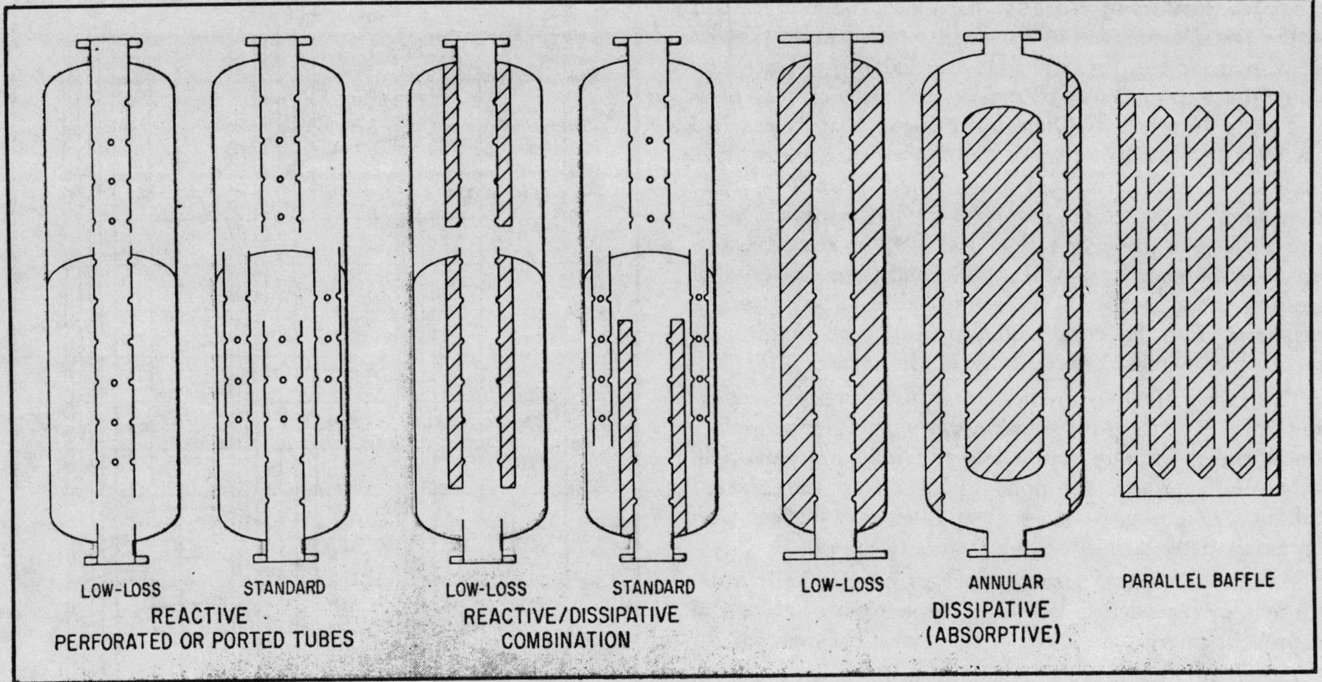

Fig. 1—Typical configuration for various silencer types.

The selection of silencer type is a function of both the frequency content of the noise and the specific application requirements. However, the basic type is generally set by the fundamental or peak frequency of the noise as shown in Table 1.

Silencer performance or attenuation is a function of the input amplitude of the unsilenced noise, silencer length and flow. The greater the input, the more the attenuation. Generalized performance curves without regard to input and velocity are at best an approximation—not a guarantee.

A silencer, regardless of application and service, is rated not only on its effectiveness, but on its durability and life expectancy, for no matter how good the performance, if the unit fails structurally, costs repairs and many times outright replacement may be required.

VENTS AND VALVES

Vents and Blowdowns to Atmosphere. Vent and blowdown noise increases with the valve "jet" velocity, the density of the flowing gas and the valve throat diameter. Vent noise is normally continuous in duration whereas blowdown noise is intermittent and of short duration. Most accepted noise standards permit intermittent levels of 5 to 15 dbs above those set for continuous service. Vent noise is relatively broad-band peaking at the so-called "Strouhal" frequency, which is a function of the valve throat diameter and the jet exit velocity. The selection of silencer type is based upon the peak frequency parameters established for the various silencer types.

In high pressure applications an impact or flow diffuser is normally provided in the inlet section of the silencer to prevent beaming and abrupt expansion of the gas being vented. Design limitations are shown in Table 2.

The sizing of a vent or blowdown silencer is based almost entirely upon the maximum flow rate as designated in Table 3. When pressure drop across the silencer is a factor the velocity must be reduced accordingly. Blowdown flow rates quite often are questionable. It is especially difficult to predict the flow rate in instances where the duration of blowdown is 30 seconds or less.

TABLE 1—Maximum Performance Range for Silencers

Silencer type	Peak freq., Hz
Reactive, Multi-chamber	Up to 150
Reactive/Dissipative Combination	150-1,000
Dissipative, Absorptive	Above 1,000

TABLE 2—Upstream Pressure Limitations for Silencers

	Upstream pressure, psig		
Silencer Typ	I	II	III
Reactive	75	150	Above 75
Reactive/Dissipative	100	300	Above 100
Dissipative	50	75	Above 50

Column designation:
I Normal maximum
II Maximum where at least 3 to 5-inlet nozzle diameters of straight run of piping is installed immediately upstream of silencer.
III With inlet diffuser and when close-coupled to valve.

TABLE 3—For Sizing Vent or Blowdown Silencers

	Maximum sizing velocities, ft./min.	
Silencer Type	Intermittent	Continuous
Reactive	25,000	20,000
Reactive/Dissipative	20,000	15,000
Dissipative	15,000	10,000

Pressure Reducing Valves. Regulator noise is caused by the rapid expansion of gas as in high pressure vent or blowdown service. Pressure reducing valve noise increases with flow and pressure reduction until the critical pressure ratio is reached. Once the critical ratio is reached the noise level will increase only with an increase in flow. Sonic or acoustic velocities occur when the critical ratio is reached or exceeded. For most gases the critical pressure ratio is slightly less than 2:1. Sonic shock waves create turbulence and peak noise conditions. A series of pressure reductions, each below critical ratio, will produce less noise. But the costs would probably be prohibitive, due to the increased complexity of the system.

The most effective method to control pressure reduction noise is at the source inside the pipe. Pressure reduction silencers directly coupled to the reducing valve will substantially reduce the noise at its source and prevent radiation of piping noise. A dissipative type silencer with extremely dense acoustical material is required.

Maximum silencer sizing velocities of 10,000 ft./min. are seldom exceeded to prevent regeneration of noise and to prevent migration or loss of the acoustical material.

The predominant noise that requires the most attention is from the low pressure piping immediately downstream of the valve. The silencer is normally of code design based upon the upstream pressure. The reducing valve may require acoustical treatment. Only rarely is an upstream silencer required.

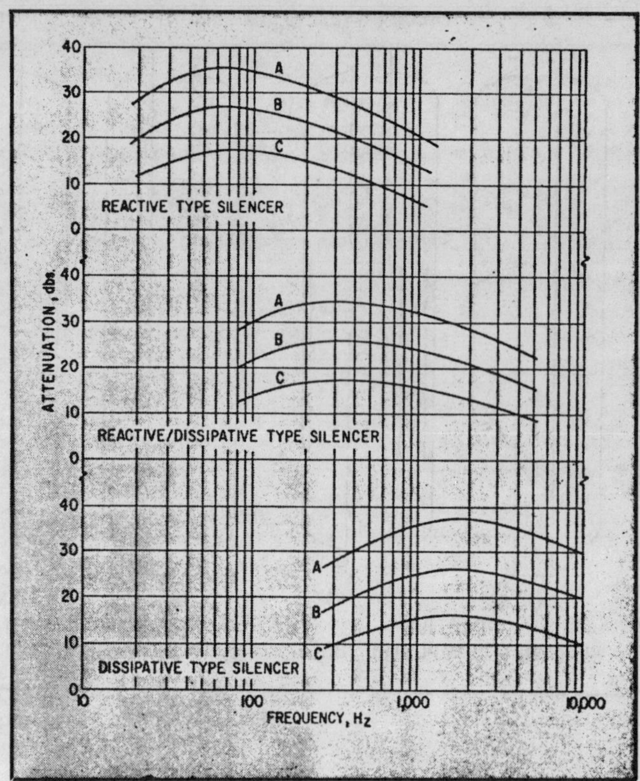

Fig. 2—Generalized performance curves for: A—Residential, B—Standard and C—Commercial.

DRIVERS, BLOWERS AND PUMPS

Engines. Engine noise is a function of horsepower, speed and to some extent fuel type. Engine noise increases primarily with horsepower. The over-all noise level increases by about 3 dbs each time the engine horsepower is doubled. The predominant sources that make up engine noise are intake, exhaust and mechanical noise.

Engine exhaust noise is the predominant source . . . generally peaking at the fundamental engine firing frequency of the engine. Intake noise is usually 8-10 dbs less than the over-all exhaust levels. Turbocharged engines develop a high frequency component at one or more discrete frequencies. Even so, the turbocharge engine does not normally produce as much noise as the non-turbocharged engine. The difference in over-all sound levels is about 6 dbs for equivalent size engines. The exhaust driven turbocharger removes some of the sound energy from the gas resulting in less noise. The length of the intake and exhaust piping may become a factor but for the most part is ignored.

Engines are usually installed within a building or enclosure. The intake (larger engines) and exhaust (all engines) are ducted to the outside. Engine and turbocharger casing noise within the building may become excessive at the higher operating speeds. Operator or workman hearing protection may be required when the levels exceed 95 or 100 dbs. Portions of the intake and exhaust piping and the turbocharger housing may require acoustical treatment to prevent radiation of high frequency noise. The interior walls of the building may also require some form of treatment.

Most engine silencers are made of a plurality of chambers or compartments. Commencing with the first chamber, the succeeding chambers become progressively smaller. When the tubes are sparsely perforated or ported, the first chamber is most effective between 20 and 150 Hz (cycles per second), the second betwen 150 and 850 Hz, and the third (when applied) is most effective between 850 and 2400 Hz. The scavenging means generally dictates the type silencer as well as the number of chambers that are needed to silence the engine. This applies to both the intake and exhaust. Types of silencers for engines are shown in Table 4.

Engine silencer sizing is based upon the volume and flow requirements of the engine. Four-cycle (natural aspirated) and certain two-cycle engines require added volume for optimum engine operation and for low frequency noise attenuation. The silencer volume should be determined by the silencer manufacturer. Velocity limitations apply to all engines dependent upon allowable pressure drop and to some extent upon the degree of silencing that is required. Typical limitations are shown in Table 5.

Gas Turbines. The primary sources of gas turbine noise are, intake, exhaust and radiated noise from the casing. Turbine noise is a function of horsepower or mass flow and speed of rotation. Intake noise is inherently high pitched. These discrete components in the higher frequencies require at minimum dissipative or absorptive silenc-

TABLE 4—Engine Silencers

Silencer type	Engine type
For engine intakes	
Reactive	4-cycle, naturally aspirated 2-cycle, rotary-positive blower
Reactive/Dissipative	2-cycle, turbocharged
Dissipative	2 and 4-cycle, centrifugal blower 4-cycle, turbocharged
For engine exhausts	
Reactive, standard	4-cycle, naturally aspirated 2-cycle, rotary-positive blower 2-cycle, centrifugal blower 4-cycle, turbocharged, Δ P permitting
Reactive, low loss	2 and 4-cycle, turbocharged
Reactive/Dissipative	2 and 4-cycle, turbocharged selected use

TABLE 5—Typical Silencer Intake Velocities

Engine type	Average silencer intake velocity, ft./min. Intake	Exhaust
4-cycle, naturally aspirated	3,000-5,000	4,000-10,000
2-cycle, except turbocharged	3,000-5,000	4,000-10,000
2 and 4-cycle, turbocharged	4,000-5,000	up to 12,500

TABLE 6—Silencer Selection for Blowers and Compressors

Blower and compressor type	Silencer type Intake	Exhaust
Lobe, PLV below 2,700 ft./min.	Reactive or Dissipative (depending on size and application)	Reactive
Lobe, PLV above 2,700 ft./min.	Reactive/Dissipative	Reactive/Dissipative
Vane	Reactive or Dissipative	Reactive
Screw or Axial, Low speed	Reactive	Reactive or Reactive/Dissipative
Screw or Axial, High speed	Reactive/External Lagging or Dissipative	Reactive/External lagging (extra heavy construction)

ing in the form of treated ducts, lined elbows and the like. Exhaust noise is the predominant source and most noticeable in the lower frequencies due to the combustion process and the increased gas flow. Exhaust silencers are usually in the form of parallel or annular baffles. The depth and density of the acoustical material making up the baffles as well as the spacing between the baffles depends upon the amplitude and frequency content of the noise. The length of the baffles varies with the degree of silencing that is required. The turbine housing may require acoustical treatment typically in the form of an acoustical enclosure to present radiation of casing noise.

Turbine intake silencer velocities vary from 4,000 to 10,000 ft./min. Exhaust velocities range from 7,500 to 15,000 ft./min., depending upon the allowable pressure drop, degree of silencing required and space limitations.

Rotary-Positive Blowers and Compressors. Rotary-positive blower noise increases with flow, compression ratio and speed. In many instances, depending upon the type of blower, the noise that is produced is a function of the peripherical velocity of the gearing. The pitch line velocity (PLV) is numerically equal to the product of the gear circumference and the blower rpm.

Rotary-positive blowers and compressor noise may be either broad-band or of discrete frequencies. Discrete frequencies occur at the fundamental and multiples of the fundamental. Casing noise usually occurs in sharp peaks at one or more discrete frequencies.

The noise is the most predominant in the discharge piping at the source, followed by the intake, casing and driving sources. In other than closed systems, where air is drawn from within the blower and compressor room, intake noise may become predominant.

Silencer sizing is based upon pressure drop which is primarily a function of the flow and density of the gas. For most reactive designs, the maximum velocity is 5,500 ft./min.

Rotary-positive blowers and compressor silencers are selected on the basis of type, size and speed as shown in Table 6.

Centrifugal Blowers and Compressors. Centrifugal blowers and compressors produce relatively broad-band high frequency noise which is also a function of flow, compressor ratio and speed. Normally only intake silencing of the dissipative type is required. Sizing velocities are controlled by the allowable pressure drop but seldom exceed 5,000 ft. min.

Vacuum Pumps. Vacuum pumps are ordinarily of two types: (a) reciprocating and (b) water sealed rotary types.

The reciprocating types requires only discharge silencing of the reactive type. The sizing is based upon volume as in the case of certain internal combustion engine silencers and limiting velocities. The start-up velocity should not exceed 7,000 ft./min. The maximum velocity for operating conditions based on the discharge rate generally does not exceed 3,000 ft./min.

Water sealed rotary pumps normally do not require intake silencing but may require a separator in the intake system for separation of liquids in the vacuum line ahead of the pump. The discharge being to atmosphere requires either a separator/silencer or a reactive silencer with an over-sized drain where complete separation is not required. The sizing velocity is again 7,000 ft./min. at startup and 5,500 ft./min. for operating conditions.

Valves Can Be Quiet

Understanding reasons for control valve noise is the starting place for designing for quiet operation. Research involving over 1 million data points shows the way

Ernest E. Allen, Fisher Controls Co., Marshalltown, Iowa

CONTROL VALVE NOISE can be designed out of a valve for a given application. Recent technology makes its possible to accurately predict the level and characteristic of noise radiated to the atmosphere and offers a parametric understanding of the noise generators. Use of this technology requires both a quantitative knowledge of the noise characteristic of the valve and an understanding of sound transmission loss through the sound boundaries that contain the flow stream.

NOISE SOURCES

The major sources of control valve noise are mechanical vibration of components, hydrodynamic noise, and aerodynamic noise.

Mechanical noise. Vibration of valve components is a result of random pressure fluctuations within the valve body and/or fluid impingement upon the movable or flexible parts. Noise that is a byproduct of the vibration of valve components is usually of secondary concern and may even be beneficial since it warns that conditions exist which could produce valve failure. Mechanical vibration has for the most part been eliminated by improved valve design and is generally considered a structural problem rather than a noise problem.

Hydrodynamic noise. The major source of hydrodynamic noise (noise resulting from liquid flow) is cavitation which is caused by implosion of vapor bubbles formed in the cavitation process. Cavitation occurs in valves controlling liquids when the service conditions are such that static pressure downstream of the valve is greater than the vapor pressure and at some point within the valve the local static pressure, either because of high velocity and/or intense turbulence, is less than or equal to the liquid vapor pressure.

Fig. 1 depicts the pressure profile of a cavitating flow stream as a function of distance along the stream. Vapor bubbles are formed in the region of minimum static pressure and subsequently are collapsed or imploded as they pass downstream into the pressure recovery region. Noise produced by cavitation has a broad frequency range which is often described as a rattling sound similar to that which would be anticipated if gravel were in the fluid stream.

Test results and field experience indicate that noise levels from non-cavitating liquid applications are quite low and generally would not be considered a noise problem. *Fig. 2* depicts the typical characteristic of hydrodynamic noise as a function of the ratio of differential pressure across the valve (ΔP) to the static pressure at the inlet (P_1 – psia) minus the vapor pressure (P_V – psia).

Cavitation may produce severe damage to the solid boundary surfaces that confine the cavitating fluid. Generally speaking, noise produced by cavitation is of secondary concern.

Aerodynamic noise. The major source of valve noise is aerodynamic noise generated as a by-product of a turbulent gas stream, or noise produced without the inter-

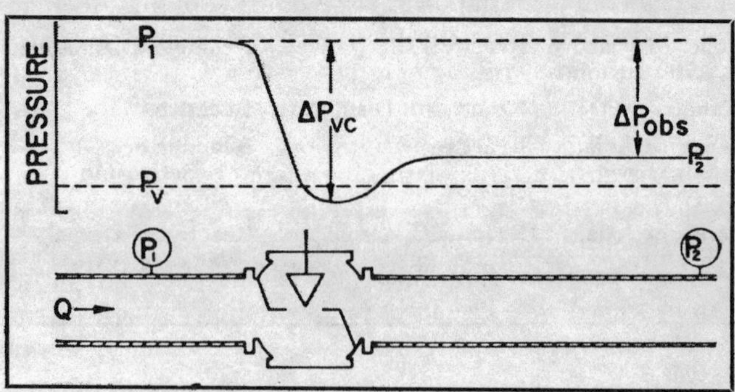

Fig. 1—Static pressure along a stream line cavitation flow.

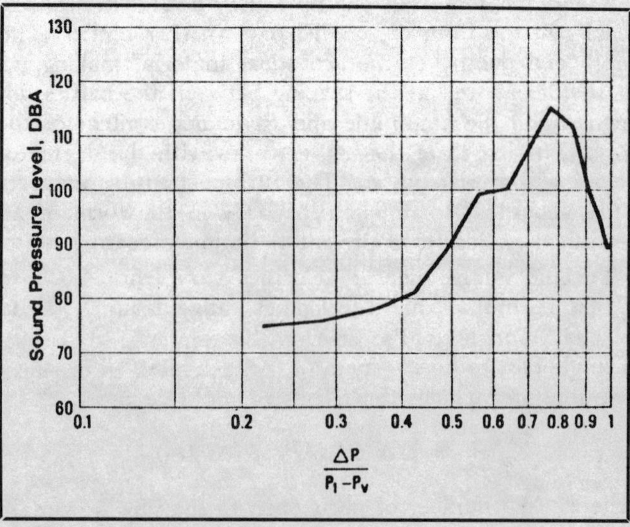

Fig. 2—Typical liquid noise characteristic.

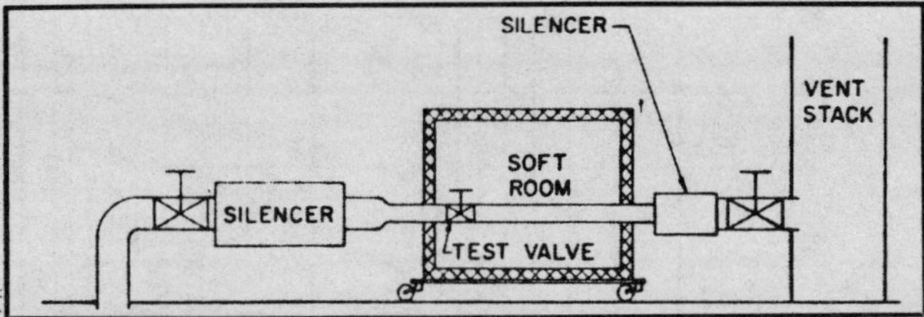

Fig. 3—Noise test facility for control valves.

action of the fluid with vibrating boundaries or other external energy sources.[1]

Aerodynamic noise is a result of the Reynolds stresses or shear forces created in the flow stream as a result of deceleration, expansion or impingement. The principal area of generation is the recovery region immediately downstream of the vena contracta where the flow field is characterized by intense turbulence and mixing, and is of chaotic quality with phase being completely random and discontinuous. The concept of spectral density or amplitude as a function of frequency for the fluctuating physical quantities (pressure, density, and velocity) is without meaning.

NOISE PREDICTION

The noise characteristic and/or potential of each element in the system should be known in order to design a quiet system. Thus the need for an accurate technique for predicting valve noise is self-evident.

The greatest limitation in development of a mathematical model that maps the noise characteristic of a control valve to the known flow parameters is lack of either quantitative or qualitative knowledge concerning the aerodynamics of a bounded flow stream. The noise generators are random functions of space and time, thus only statistical properties (such as autocorrelation functions) of these quantities can be measured or predicted.

Many problems that defy classical analytical techniques can be solved with elegantly simple models fashioned from dimensional reasoning. Valve noise prediction is in this category.

Aerodynamic dimensional analysis. A comprehensive dimensional analysis of aerodynamic valve noise[2] hypothesizes that acoustic power generated by compressible flow thru a control valve obeys the following relationship:

$$W \propto C_g^2 \, (\Delta P)^2 f\,(\Delta P/P_1) \qquad (1)$$

where: W = acoustic power

C_g = gas sizing coefficient (directly proportional to area of restriction)[3]

ΔP = pressure differential across valve

P_1 = absolute inlet pressure

The rationale used to develop this relationship assumes a single port or noise generator. An extension to the basic relationship includes the effects of multiple ports.

$$W \propto 1/N \, C_g^{\bar{2}} \, (\Delta P)^2 f\,(\Delta P/P_1) \qquad (2)$$

where: N = number of ports open to the flow stream

Proportionalities obtained in dimensional formulae are frequently inexact and must be verified by experimental findings before they can be regarded as acceptable.

Testing program. Test work to verify Equation 2 used a controlled sound environment (mobile soft room) to provide isolation of the test valve and/or piping from other noise sources.

Fig. 3 shows a schematic of the facilities. The mobile soft room (large rectangular-shaped structure, cross sectioned) parts longitudinally to facilitate piping changes and movement from one test line to another. The room is sealed around four thru twelve-inch piping and provides 50 decibels (dB) attenuation in ambient sound pressure level (SPL) from outside to inside the room in the frequency range of interest.

The flow facilities provide metered flow up to 10,700 gpm of water and 12,500 scfh of air. Maximum flow capabilities for tests of short duration are 16,000 gpm of water or 18,000 scfh of air. Absorption-type inline silencers are located upstream and downstream of the test section for isolation from fluid-borne noise generated at other points in the system.

Sound measurement is made with a Real Time Audio Spectrum Analyzer that provides a CRT display and digital printout of ⅓-octave band analysis of the audio spectrum for test runs with time duration as short as 28 milliseconds.

Noise levels are measured for a large variety of valve styles in sizes one inch through twelve inches as a function of valve travel, $\Delta P/P_1$ ratio, and adjacent piping configuration.

Test results. Only a small sampling from over 1,000,000 accumulated data points is presented to exhibit correlation of test data with Equation 2. These serve to demonstrate the basis of a graphical technique for prediction of aerodynamic valve noise.

Fig. 4 shows the change in *SPL* outside the pipe as a function of the ΔP across the valve. This curve represents the best fit of data for all valve styles and sizes tested. The slope of the experimental curve is 17.7 dB per decade as compared to a slope of 20 dB per decade predicted by Equation 2.

Fig. 5 shows a plot of change in *SPL* (ΔSPL) as a funtion of C_g for all valves that modulate flow by increasing or decreasing the area of a given number of openings or restrictions exposed to the flow stream. The C_g curve is independent of valve size. The best fit of experimental data exhibits a slope of 23 dB per decade as compared to a slope of 20 dB per decade predicted by Equation 2.

Figs. 6, 7 and 8 show the noise characteristic as a function of $\Delta P/P_1$ and valve geometry. These test curves provide a very useful basis for comparing the noise characteristic of different valve styles.

The relationships (Figs. 4-8) provide the basis for a very expeditious technique for prediction of valve noise radiated to the atmosphere. The base SPL is taken from the ΔP curve with additive corrections then made for C_g and for valve geometry as a function of $\Delta P/P_1$ ratio. The

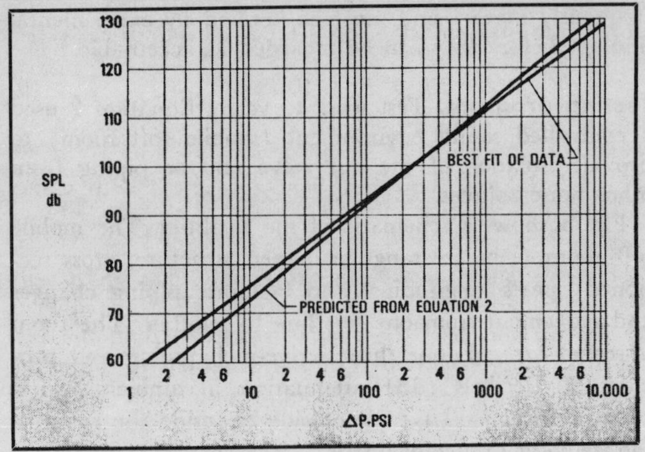

Fig. 4—Effect of pressure drop on external sound pressure

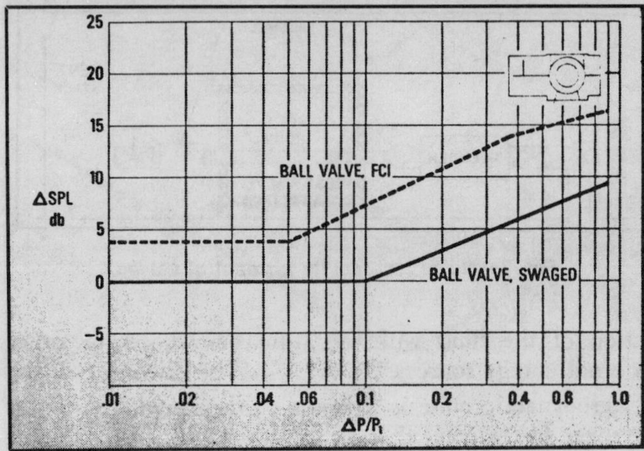

Fig. 6—Correction to base SPL as a function of $\Delta P/P_1$ ratio and valve style, ball valves.

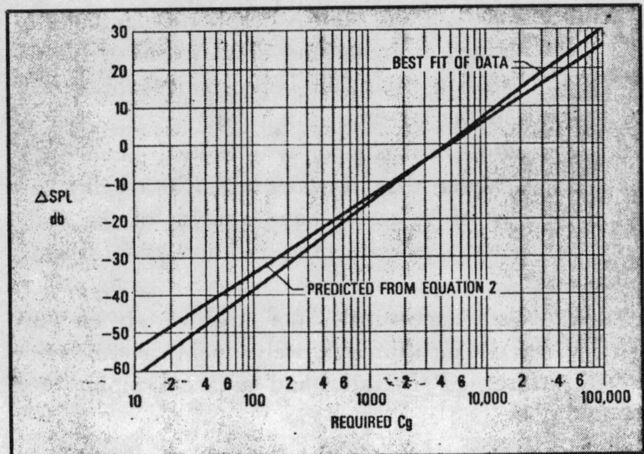

Fig. 5—Effect of gas sizing coefficient on external sound pressure level.

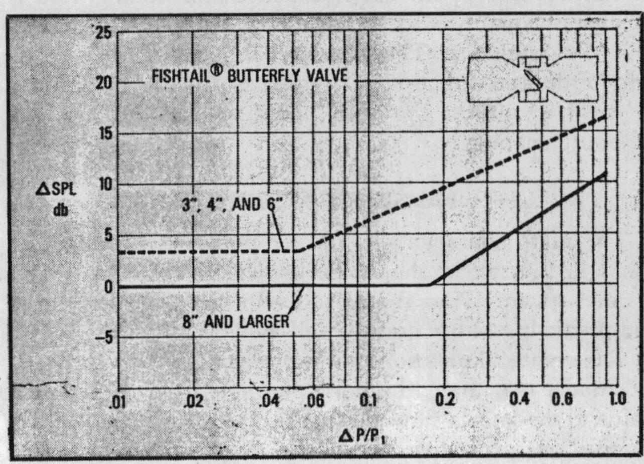

Fig. 7—Correction to base SPL as a function of $\Delta P/P_1$ ratio and valve style, butterfly valves.

result based on standard weight pipe gives the contribution to the ambient SPL at a distance of 48 inches downstream from the valve and 29 inches from the pipe surface. Since most noise measurements of valves are made on a complete installation, quantitative information on the acoustic behavior of the valve itself would be of questionable value.

Hydrodynamic dimensional analysis. The following relationship exists for hydrodynamic valve noise.[4]

$$W \propto C_v^2 f\left(\Delta P, \frac{\Delta P}{P_1 - P_v}\right) f\left(\frac{\Delta P}{P_1 - P_v}\right) \qquad (3)$$

where: C_v = Liquid sizing coefficient
P_V = Vapor pressure

An expeditious graphical solution is utilized to predict hydrodynamic noise.

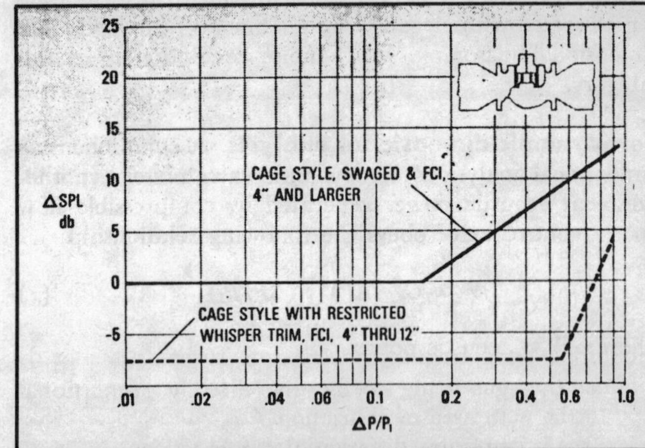

Fig. 8—Correction to base SPL as a function of $\Delta P/P_1$ ratio and valve style, cage style valves.

NOISE TRANSMISSION

The primary concern with valve noise is contribution to the overall ambient noise level. In closed systems (not vented to atmosphere) any noise produced in the process becomes airborne only by vibration of the solid boundaries that contain the flow stream. Thus, an understanding of the relative transmission loss as a function of the physical properties of the solid boundaries is imperative to accurate noise prediction.

The transmission loss characteristic for cylindrical piping is shown in Fig. 9. Transmission loss is stiffness con-

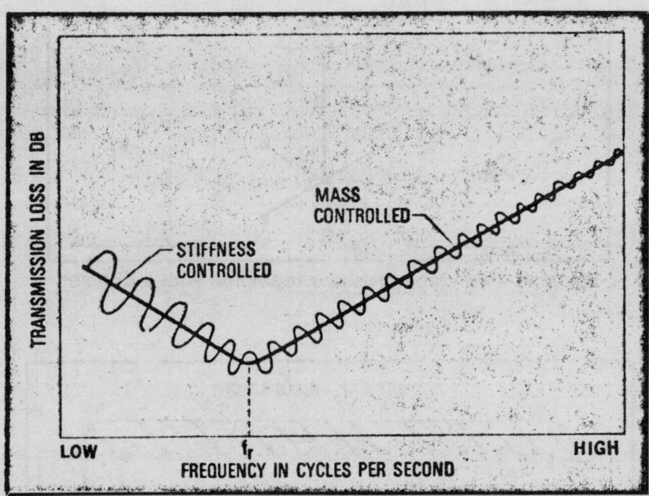

Fig. 9—Transmission loss is controlled by stiffness and mass at frequencies below and above resonance, respectively.

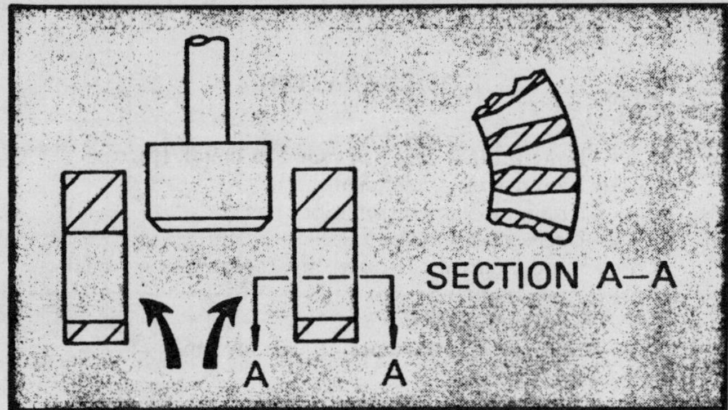

Fig. 10—A valve trim designed for noise attenuation.

trolled at frequencies below the ring frequency and mass controlled at higher frequencies. Ring frequency (f_r) can be calculated using the following equation:

$$f_r = \frac{C_L}{\pi D_p} \qquad (4)$$

where: D_p = Nominal diameter of pipe
C_L = Longitudinal speed of sound in metal

The problem of predicting pipe transmission loss has not yet been solved. The multiplicity of response modes and broad band frequency excitation force (noise field) enhance the complexity of calculating the response of the pipe to the noise within. The theory of vibration of cylindrical shells can be used to calculate the various resonant frequencies. At high frequencies the resonances are grouped sufficiently close together to approach a condition of continuous response (Fig. 9). Even with the enormous computational capacity of computers, the large number of resonant frequencies and modes that exist impose a practical limitation on the utility of a theoretical approach to calculation of transmission loss.

Table I presents a tabulation of empirically determined values for the change in pipe transmission loss, relative to standard weight, as a function of pipe schedule.

A theory recently advanced suggests the following relationship for transmission loss (T.L.) of cylindrical pipe.[5]

$$TL \propto K + 10 \, Log \, t^4 \qquad (5)$$

where: t = Pipe wall thickness

This relationship shows excellent agreement with values published in Table 1.

The dissipation of acoustic energy into heat by viscosity and heat conduction is a slow process; i.e., at a frequency of 4K Hz only half the energy is dissipated in the first 5000 ft. of propagation thru the atmosphere. Thus the noise generated within a closed transmission system is frequently propagated for long distances in the fluid stream with little attenuation. Consequently, any change in piping schedule at relative large distances from the valve may effect a change in the ambient noise level.

DESIGN OF QUIET VALVES

The parameters that determine the level of noise generated by compressible flow thru a normal control valve for a given application are: valve geometry, number of ports or restrictions exposed to the flow stream, total C_g, differential pressure across the valve, and ratio of differential pressure to absolute inlet pressure.

The noise characteristic of a control valve is dependent on valve geometry (Figs. 6, 7, and 8). Fig. 10 shows a cross-sectional view of a cage style trim designed specifically to provide a favorable velocity distribution in the recovery region and hence minimize the turbulence level or noise. This approach can provide substantial noise reduction (15-20 dB) with little or no decrease in total flow capacity.

From Equation 2, the acoustic power of a single flow restriction increases as a function of C_g^2. Changing the area by a factor of 2 results in a corresponding 6 dB change of power level, whereas, the power level is changed only 3 dB when the number of equal noise sources is changed by a factor of two. Thus noise reduction to be derived from utilization of many small restrictions rather than a single or few large restrictions is self-evident.

The noise characteristic or noise potential increases as a function of ΔP^2 *and* $\Delta P/P_1$ (Figs. 4, 6, 7 and 8). Thus for high pressure ratio applications ($\Delta P/P_1 > 0.7$) an appreciable reduction in noise can be effected by staging the pressure loss thru a series of restrictions to produce the total pressure head loss required. This approach to quiet valves can easily be incorporated into cage style trim fabricated from stacks of discs machined to stage the total pressure drop thru a series of circumferential restrictions.

For control valve applications operating at high pressure ratios ($\Delta P/P_1 \geqq 0.8$), splitting the total pressure drop between the control valve and a fixed restriction (diffuser) downstream of the valve can be very effective in minimizing the noise. In order to optimize the effectiveness of a diffuser, it must be designed (special shape and sizing) for each given installation so that the noise levels generated by the valve and diffuser are equal. Fig. 11 depicts a typical valve-plus-diffuser installation.

Design of quiet valves for liquid application resolves itself to designs to eliminate cavitation. Service conditions that will produce cavitation can readily be calculated.[6]

The use of staged or series reductions provide a very viable solution to cavitation and hence hydrodynamic noise.

PATH TREATMENT

A second approach to noise control is path treatment. Sound is transmitted via longitudinal waves thru the elastic medium or media that separate source and receiver. The speed and efficiency of sound transmission is dependent on the properties of the medium thru which it is propagated. Path treatment consists of increasing the impedance of the transmission path so as to reduce the acoustic energy that is communicated to the receiver.

Dissipation of acoustic energy by use of acoustical absorbent materials is one of the most effective methods of path treatment. Whenever possible the acoustical material should be located in the flow stream either at or immediately downstream of the noise source. This approach to abatement of aerodynamic noise in gas transmission systems is accommodated by inline silencers (Fig. 12) that effectively dissipate noise within the fluid stream and attenuate the noise level transmitted to the solid boundaries. Where high mass flow rates and/or high pressure ratios across the valve exist, inline silencers are often the most realistic and economical approach to noise control. Use of absorption-type inline silencers can provide almost any degree of attenuation desired. However, economic considerations generally limit the insertion loss to approximately 30 dB.

Noise that cannot be eliminated within the boundaries of the flow stream must be eliminated by external treatment which suggests the use of heavy walled piping, acoustical insulation of the exposed solid boundaries of the fluid stream, use of insulated boxes, buildings, etc. to isolate the noise source.

Benefits to be derived from use of heavy-walled pipe are shown in Table 1.

TABLE 1—Change in transmission loss as a function of pipe wall thickness

Nominal Pipe Size, Inches	SCHEDULE NUMBER							
	30	40	80	120	160	STD	XS	XXS
2	...	0	+ 6	....	+12	0	+6	+16
4	...	0	+ 7	+10	+13	0	+7	+16
6	...	0	+ 8	+12	+15	0	+8	+18
8	−3	0	+ 9	+14	+18	0	+9	+16
10	−3	0	+ 9	+14	+19	0	+6	....
12	−4	0	+10	+16	+20	0	+5	....

Acoustical insulation of the exposed solid boundaries of the fluid stream is an effective means of noise abatement for localized areas. Test results indicate that ambient noise levels can be attenuated as much as 10 dB per inch of insulation thickness.

Path treatment such as heavy wall pipe or external acoustical insulation can be a very economical and effective technique for localized noise abatement; however, because noise is propagated for long distances via the fluid stream, the effectiveness of the heavy wall pipe or external insulation terminates where the treatment is terminated.

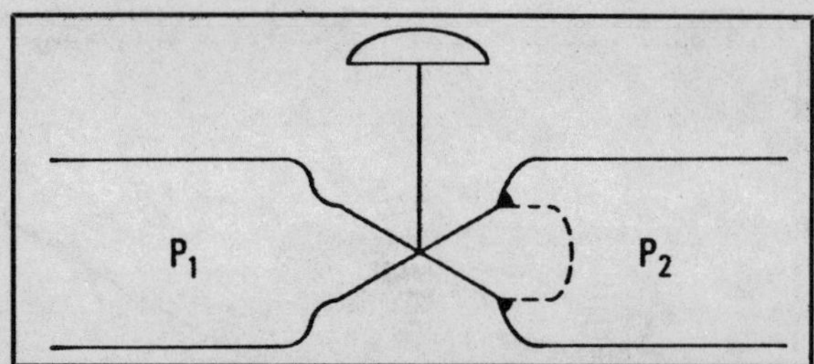

Fig. 11—Two stage pressure reduction with a diffuser.

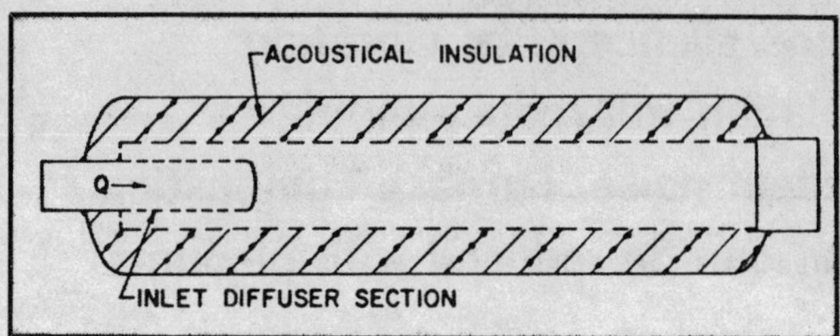

Fig. 12—Typical inline silencer.

NOMENCLATURE

C_g — Gas sizing coefficient
C_L — Longitudinal speed of sound in metal, fps
C_V — Liquid sizing coefficient
D_P — Nominal diameter of pipe, feet
dBA — Decibels, Scale A
FCI — Fluid Control Institute, line-to-valve ratio equal one
f — Ring frequency, cps
K — Equation constant
N — Number of ports open to flow stream
P — Pressure, psia
ΔP — Pressure differential across valve, psi
P_V — Vapor pressure, psia
Q — Volumetric flow rate
SPL — Sound pressure level
t — Pipe wall thickness, inches
TL — Transmission loss
W — Acoustic power, dB

Subscripts:

1 — upstream
2 — downstream

LITERATURE CITED

[1] M. J. Lighthill, "On Sound Generated Aerodynamically," Department of Mathematics, The University of Manchester, England.
[2] C. B. Schuder, "Coping with Control Valve Noise," CHEMICAL ENGINEERING, October 19, 1970.
[3] J. F. Buresh and C. B. Schuder, "Development of a Universal Gas Sizing Equation for Control Valves," TM-15, Fisher Controls Company.
[4] R. E. Nugent, "Prediction of Noise Generated by Incompressible Flow Through Control Valves," 18th Annual Southeastern Regional ISA Conference and Exhibit, April, 1972.
[5] P. S. Malloy, "Control Valve Noise," 1972 Annual ISA Conference (72-536), New York.
[6] G. F. Stiles, "Development of a Valve Sizing Relationship for Flashing and Cavitating Flow," ISA Final Control Elements Symposium, Wilmington, Del., 1970.

BIBLIOGRAPHY

C. M. Harris, "Handbook of Noise Control," McGraw-Hill Book Co., Inc., 1957.
R. W. B. Stephens & A. E. Bates, "Acoustics and Vibrational Physics," (London) Edward Arnold (Publishers) Ltd., 1966.
E. J. Richards & D. J. Mean, "Noise and Acoustic Fatigue in Aeronautics," John Wiley & Sons Ltd.

How to Reduce Control Valve and Furnace Combustion Noise

Quiet control valve designs, inline silencers, or vibration isolation will reduce piping system noise. Furnace combustion roar can be reduced with plenums

James G. Seebold,
Standard Oil Co. of California, San Francisco

PIPING SYSTEM NOISE comes mostly from control valves and high-speed machines. Flow noise from milder discontinuities—bends, tees, swages and the like—is negligible by comparison. When valve noise dominates, smoothness in valve manifolds and piping systems is not very important.

Serious problems can usually be predicted by any of several published schemes for standard valves in gas service. Standard designs are almost always too noisy in choked service at substantial flow rates. No reliable methods are yet available for liquid valves; although if selected to avoid cavitation and erosion, they usually do not produce excessive noise.

A solution is the selection of special, quiet, valve designs that change the internal physics of pressure reduction, or piping system treatment to prevent noise propagation. Appropriate treatment depends on the propagation path — gas, pipewall, or both. When the energy is gas borne, an inline silencer works well. Pipe-borne energy might be damped by vibration isolation. Lagging is generally ineffective, except locally, unless long lengths of pipe are treated.

Piping flow noise. Gas (or steam) flow in process plant piping is virtually always in the low subsonic range. The radiated sound power is proportional to the square of the pressure drop, while the characteristic frequency is proportional to the square root of the pressure drop.[1] Therefore, a control valve in subsonic flow, with any substantial pressure drop at all, is bound to be much noisier than such comparatively mild discontinuities as bends, tees, swages, and the like. Also, we would expect a considerable disparity in the characteristic frequencies, the control valve's being much higher and even more so, for choked control valves.

It is sometimes suggested that *smoothness* is important in piping system layout, particularly in valve manifolds and the like. In the light of the foregoing discussion, it is hard to see how smoothness, or lack of it, could have any important effect on noise at all, except in systems which are already relatively quiet.

On the other hand, a rule of thumb can be developed that should prevent noise problems, if any, caused by the lack of smoothness. It is a generally accepted design practice to limit the velocity at valve outlet flanges to about Mach 0.3 to make sure no noise problems develop. If we interpret this as a limit on dynamic pressure change, we may impose the same limit on static pressure change at a piping system discontinuity. The equivalent limit on pressure drop is about 5 percent of line pressure, which is quite liberal. Alternatively, the fitting loss should not exceed about 10 velocity heads, while losses for standard fittings (entrances, exits, elbows, tees, and reducers) rarely exceed about one velocity head.

As far as flow itself is concerned, noise generation is related to the dynamic pressure, which determines the turbulence intensity and hence the fluctuating pressure levels in the turbulent boundary layer. A thumb rule is that as long as the velocity (in fps) does not exceed about 100 times the square root of the specific volume (in cubic feet per pound) for gases and two-phase flow, and 30 fps

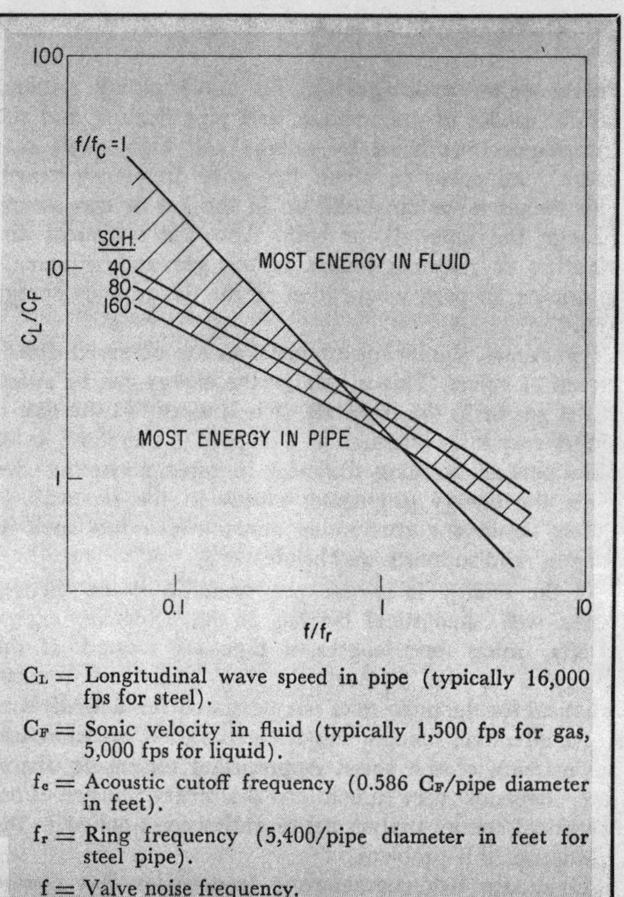

C_L = Longitudinal wave speed in pipe (typically 16,000 fps for steel).

C_F = Sonic velocity in fluid (typically 1,500 fps for gas, 5,000 fps for liquid).

f_c = Acoustic cutoff frequency (0.586 C_F/pipe diameter in feet).

f_r = Ring frequency (5,400/pipe diameter in feet for steel pipe).

f = Valve noise frequency.

Fig. 1—Energy distribution by statistical energy analysis.

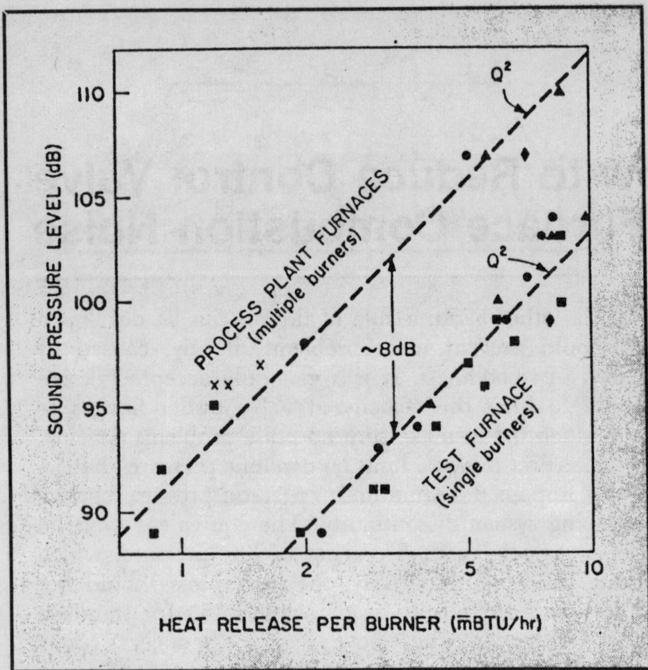

Fig. 2—Natural draft furnaces—peak octave band noise levels at 3 feet from the burners.

for liquids, no noise problems should result. Standard piping practice rarely exceeds these velocities.

Therefore, it appears that we need not concern ourselves too much with flow noise in most piping systems of practical engineering interest in process plants. The real culprits are control valves (and high speed machines).

Valve noise propagation. In many piping systems, natural modes of gas response and pipe flexure, and the disturbance introduced by valves (and high speed machines), all occur in about the same frequency range. This means noise can build up in the gas or can escape through the pipewall, or both. Also, the statistical distribution of response modes in the gas and pipewall determine, in part, where most of the disturbance energy goes.

Sometimes, significant attenuations are observed downstream of valves. This is because the energy can be either in the gas or in the pipewall. If it is mostly in the gas, it suffers very little attenuation. Frequently, therefore, valve noise persists for long distances in pipings systems. But when the energy propagates mostly in the pipewall, it suffers significant attenuation across constraints such as flanges, rigid supports, and bends.

If the energy is in the gas, then an inline silencer works well. Acoustical lagging is not effective, except locally, unless long lengths of pipe are treated. If the energy is in the pipewall, lagging is somewhat more practical for the noise does not persist so far downstream. More effective, though, might be vibration isolation just downstream of the valve. Appropriate treatment, therefore, depends very much on the energy distribution. Statistical energy analysis has recently been applied in the solution of this problem.

There are two characteristic frequencies that govern the propagation of energy. One is the ring frequency (the frequency of the extensional *breathing* mode of vibration in which the pipe cross section remains circular). The energy distribution is different above and below the ring frequency.

The acoustic cutoff frequency is the other characteristic frequency that governs the propagation of energy. The cutoff frequency is that frequency below which the higher-order acoustic modes will not propagate and only plane waves are present. These waves have no radial or angular dependence as do the higher modes. They simply echo up and down as in an organ pipe.

Fig. 1 provides a ready comparison with typical process plant conditions. In the upper right part of Figure 1, above all of the diagonal lines shown, multimodal behavior obtains ($f/f_c>1$), and the number of response modes in the fluid exceeds that in the pipewall. Assuming strong coupling, this means most of the disturbance energy should propagate in the fluid for systems and disturbances characterized by points that lie well into the upper right part of Fig. 1.

In the lower left part of Fig. 1, below the cutofl frequency, the picture is not as clear[2]. There are questions about the propagation of the higher-order acoustic modes. Also, the role of coupling strength between the various acoustical and mechanical response modes remains to be clarified, as does the role of the manner of introducing the disturbance, whether mechanical or acoustical. For valves, however, it seems clear that the excitation is almost purely acoustical. Thus, for valves, considerable progress has been made toward understanding noise propagation in piping systems so optimum path treatments can be devised.

Experimental and analytical studies seem to confirm that in general, for valves in gas service, most of the energy will remain in the gas, and inline silencers are most effective; whereas for valves in liquid service, it appears that in general most of the energy is transferred to the pipe, suggesting that vibration isolation should be more appropriate. And too, special valves are available that don't make noise in the first place.

Furnace combustion noise. In natural draft furnaces in the 100 to 500 million Btu/hr. heat-release class, the inherent combustion roar of the turbulent-diffusion flames is a serious problem. There appear to be no proven burners available today, acceptable to operators, that will eliminate the combustion roar. Its escape from the firebox has to be prevented by using acoustically treated, air-intake plenums in natural-draft furnaces.

Certainly, there are many other elements of the combustion system that may, at times, complicate the furnace-noise problem. However, they seem relatively unimportant compared with the inherent combustion roar that is always present, simply because eliminating the complicating influences will not eliminate the inherent combustion roar. Yet, preventing the escape of combustion roar can generally prevent the escape of noise from the other elements as well, whenever they may be important. On this point, most investigators seem to be in general agreement, as least for the large furnaces used in outdoor petroleum processing complexes.

Flame control. When combustion is rough, with consequent poor flame control, additional noise over and above the inherent combustion roar of a properly tuned industrial burner usually results. Generally, these conditions are easily recognized by operating personnel. A burner change that improves the flame pattern is also likely to reduce the noise more nearly to expected levels.

Predicting noise. There is no known anaytical method for accurately predicting the complete noise spectrum from a single burner, let alone from multiple burner installations. However, it is safe to say that virtually all burners firing above a heat-release rate of about 2 million Btu/hr. in a natural-draft furnace represent a noise abatement problem, and it is possible to have problems at even lower rates. This will not be news to anyone familiar with process plants. The current understanding does permit appropriate steps to be taken at the design stage to control combustion noise.

Noise data. Noise data obtained on a full-scale test furnace, and on untreated process plant furnaces, is summarized in Fig. 2. Plotted are the peak sound pressure levels, at three feet from the burners, which occurred very broadly in the octave bands centered at 125 hz through 500 hz, the combustion roar frequencies. These flames, produced by standard commercial burners, ranged over 1 to 10 million Btu/hr. heat release. These are the flames of industrial importance. The heat release squared trend of noise generation is evident. The test furnace data are for single burners. Industrial furnaces use the same burners producing the same size flames, but in multiple installations. The data of Fig. 2 suggest that a difference of approximately 8 db in the combustion roar frequencies is reasonable to account for multiple burners, although a range of about 5 to 10 db is evident.

Field data is sometimes difficult to interpret, often containing apparently unaccountable anomalies. Sometimes the secondary air registers account for supposed anomalies, although other factors can be important, too. At the same combustion rate requiring the same quantity of air, a high-draft furnace may be less noisy than one with lower draft. This is because the higher draft requires a smaller secondary air register opening. An aperture that is small compared to the wavelength transmits sound in proportion to its area, so each time the secondary air register is closed half way, the combustion roar heard outside the firebox should be reduced about 3 db. In spite of these and other differences in field installations, noise proportional to the square of the heat release is apparent in Fig. 2.

Fuel pressure at burners. We sometimes hear that lower fuel-gas pressure will somehow reduce combustion noise. Reducing the gas pressure on an operating furnace does indeed reduce combustion noise, because the heat release is reduced. Turning it off will reduce the noise still further.

Even at constant heat release, however, changes in fuel-gas pressure can have substantial effect on the noise produced by primary air mixing in inspirating burners. This high-frequency jet mixing noise characteristically peaks above 2,000 hz and has nothing at all to do with combustion roar. It would be there whether or not combustion were taking place.

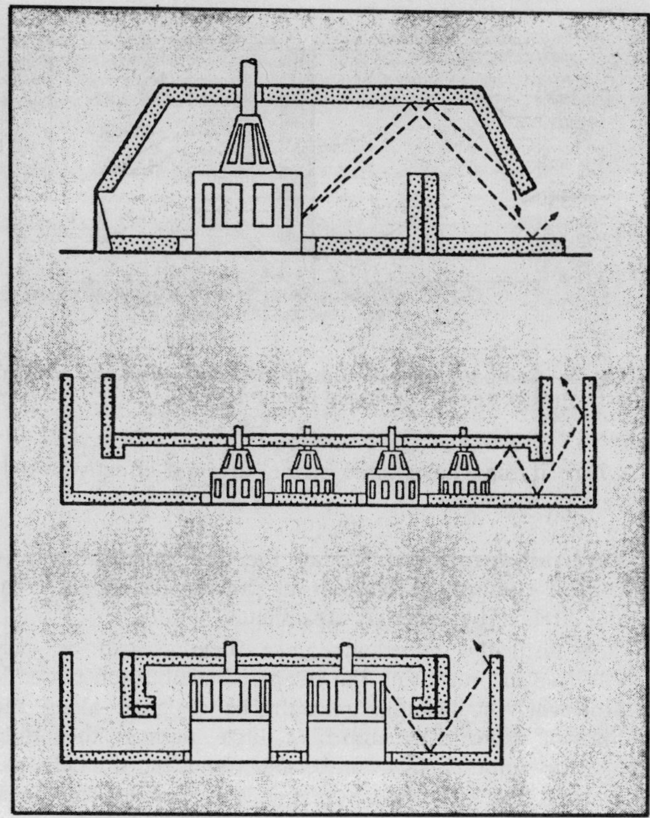

Fig. 3—Practical plenum designs.

The noise of primary air inspiration, when inspirating burners are used, can usually be adequately controlled with primary air mutes.[3] Uncontrolled, this high-frequency jet noise may exceed even the inherent, lower frequency combustion noise.[4] In such cases, if the primary air mixing can be improved, gas pressure can be reduced at constant mass flow, while the heat release remains the same. A marked reduction in the high-frequency mixing noise may result but combustion roar remains little affected. Generally, the noise of secondary air mixing at the burner spuds in the absence of combustion is negligible.[5-7] At constant heat release the fuel pressure has to be reduced to ridiculous extremes before any significant reduction in combustion noise is achieved.[8]

Number of burners. More, smaller burners are sometimes suggested as a means of reducing combustion noise at the source. Since individual burner noise varies approximately as the square of the heat release, only about three decibles reduction in total combustion noise would be expected each time the number of burners is doubled. Thus, this idea appears to have little merit.

As a matter of fact, rather than more smaller burners, the furnace designer probably should consider fewer larger ones. The smaller number of air registers ought to be easier to enclose.

Firebox size. It is also sometimes suggested that increasing the size of the firebox will somehow reduce combustion noise. This idea seems to have little merit since firebox cost increases significantly with size.

Simple reverberation calculations show that increased firebox size should have no important effect on reverberant noise buildup in large furnaces. On small furnaces,

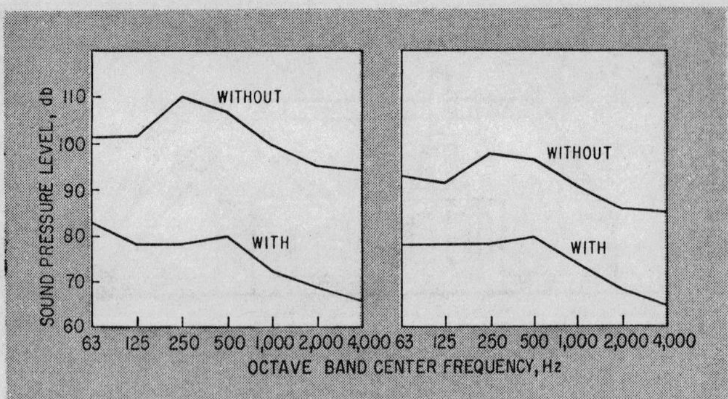

Fig. 4—Comparable natural draft furnaces—with and without acoustically treated air intake plenums.

where it may have some effect, the cost effectiveness is very poor.

External treatment. Today, the only practical way to control combustion noise is to prevent its escape from the firebox by external treatment.

Acoustically treated air-intake plenums can be very effective in preventing the escape of combustion noise.[9,10] However, they cannot generally be designed along the lines of acoustically absorbent ducts, because they then become impractically cumbersome and inordinately expensive. The practical alternative is to provide a tortuous, multi-turned path for the air to come in and the noise to get out.

Flow resistance can sometimes be a problem in this kind of plenum design, however. If this becomes critical in a given design, the conclusion seems obvious. On the one hand, an acoustically treated air-intake plenum has to be used to achieve satisfactory noise control. On the other hand, the plenum may restrict the natural draft too much. When it does, use of an air preheat or forced draft system is effective; the acoustic plenum is practically built-in already, and air is no longer a problem.

An alternative to the natural-draft plenum is the individually enclosed burner. These enclosed burners are usually somewhat less effective because of volume and absorbent thickness limitations.

Plenum design. Some practical plenum designs are shown schematically in Fig. 3. It is very important to line the entire plenum, including the furnace wall, with good quality acoustic material having high absorption coefficients in the relatively low frequencies of combustion roar (at least 0.80 at 250 hz).

Sound rays from the burner air intakes and the plenum interior must suffer several reflections off absorbent surfaces before escaping. For greatest effectiveness, acoustic material density should be 6 pounds per cubic foot. The thickness should exceed 4 inches, and the plenum volume should exceed 15 cubic feet.

Plenums should be fabricated of relatively heavy plate (about 10 BWG), undercoated with a sound-deadening material, and constructed so as to completely enclose the burner registers. The acoustic lining can be retained with an expanded metal screen. If the furnace is oil fired, provision has to be made for inspection and drainage of oil leaks. Acoustic lining should be omitted where drips collect.

All the combustion air, both primary and secondary, must be taken from the plenum chamber. There must be no line-of-sight from the burner or the plenum interior to the outside. Lined baffles are used to provide at least two reflections of escaping sound on absorbent surfaces. Hand holes, penetrations for external air register control, oil gun, etc., must be well sealed.

The dramatic effect of employing this kind of external treatment is illustrated in Fig. 4. Here we see noise level measurements taken three feet in front of the firing wall of similar natural-draft heaters doing exactly the same job in two different process plants. In each case, one of the furnaces has an air intake plenum and the other one does not. The difference is obvious.

While this comparison is more qualitative than quantitative, it makes clear the advantage, indeed necessity, of using properly designed air intake plenums for suppression of combustion noise in modern natural draft process plant furnaces.

This article is an abstract of two papers presented by the author. The papers were: "Valve Noise and Piping System Design," First Symposium on Flow—Its Measurement and Control in Science and Industry, May 9-14, 1971 (by ISA, ASME, American Institute of Physics, and the Bureau of Standards.); and "Combustion Noise and Its Control in Process Plant Furnaces," ASME Petroleum Mechanical Engineering Conf., Sept. 19-23, 1971, Houston.

LITERATURE CITED

1 Gordon, C. G., "Noise Generation by Fluid Flow Through Pipes," Paper No. 69-WA/GT-9, ASME Winter Annual Meeting, Nov. 16-20, 1969.
2 Sawley, R. J. and White, P. H., "Energy Transmission in piping Systems and Its Relation to Noise Control," Trans. ASME *Journal of Engineering for Industry*, to be published in 1972 in expanded form (available as ASME preprint No 70-WA/Pet-3).
3 Judd S. H., "Engineering Control of Furnace Noise," *American Industrial Hygiene Association Journal*, Vol. 30, No. 1, January-February 1969, pp. 35-40, Figs. 6 (a), 7(a).
4 Reed, R. D., "Environmental Noise Levels Versus Chemical Engineering Progress," John Zink Co., Tulsa, Okla., p. 6.
5 Putnam, A. A., "Flame Noise from the Combustion Zone formed by two Axially Impinging Fuel Gas Jets." *University of Sheffield Fuel Society Journal*, Vol. 19, 1968, pp. 18-19.
6 Giammar, R. D., and Putnam, A. A., "Noise Generated by Free and Impinging Fluid Jets," A Phase Report on AGA Basic Research Project BR-3-5, Battelle Memorial Institute, May 1969.
7 Bragg, S. L., "Combustion Noise," *Journal of the Institute of Fuel*, January 1963, p. 15.
8 Bitterlich, G. M., "Some Findings on Burner Noise and Its Suppression," Symposium on Burners and Noise, 35th Mid-Year Meeting of the API Division of Refining, Houston, May 13, 1970, Fig. 5.
9 Bitterlich, G. M., *op. cit.*, Fig. 12, 15, and 17.
10 Judd, S. H, *op. cit.*, Fig. 5 (a)
11 Bitterlich, G. M. *op. cit.*, Fig. 18.

Use Lead to Block Noise

Application procedures are simple

B. R. Schmidt, Federated Metals Corp., Somerville, N.J.

Noise control problems cannot be solved as quickly and easily as we might like. There are no simple problem/solution classifications to substitute for an understanding and application of the basic science involved—acoustics—and solutions that are developed must not only be effective but economical and flexible.

An attempt is made here to explain and illustrate in general terms how engineers take advantage of lead sheet sound barriers to create noise control solutions that meet these criteria. Those who lack a working knowledge of acoustical principles must rely on the expertise of noise control consultants and material suppliers to help them use sheet lead sound barriers alone or in combination with other materials to design and install systems suited for their unique needs.

Why lead? Lead sheet serves as the ideal sound *barrier* (attenuator) because it is both dense and limp. Other acoustical materials in its sound transmission class can be applied in greater thicknesses in an attempt to achieve the same transmission loss, but there is a point at which they gain stiffness that negates the effectiveness of their density (Table 1).

Porous acoustical materials *absorb* sound, preventing reflection/reverberation, but allowing transmission at only slightly reduced intensity. Thus, they are often used together with lead sheet in order to achieve sound attenuation.

Other factors have contributed significantly to the widespread acceptance of lead sheet as an effective noise control tool. Since its limpness allows easy cutting, folding, forming and crimping, lead can be applied with the same flexibility as absorbers that are far less dense. This, coupled with the fact that it can be readily adhered or fastened to other materials, facilitates the development of effective noise control systems that do not cause costly disruption of plant operations. The system can be designed so that it is easily integrated into a plant both physically and operationally and machinery or equipment can often continue to function during installation. In addition, lead sheet is unaffected by coolants, cutting oils, drawing compounds and other common industrial materials.

Lead sheet, generally produced in thicknesses of $\frac{1}{64}$ and $\frac{1}{32}$ in. weighing 1 lb. and 2 lb. per square foot, respectively, is supplied in rolls 3 to 4 feet wide. Some manufacturers can produce lower weights per square foot for special applications.

APPLICATION

Common sources of noise in refineries and processing plants include pipes and ducts carrying high velocity process gas or high pressure steam and fuel; gears, gearboxes and drivetrains; intakes, exhausts, blowers and compressors, and miscellaneous sources such as transformers. Any tied-in equipment often becomes a source in itself through vibration, which can radiate elsewhere and com-

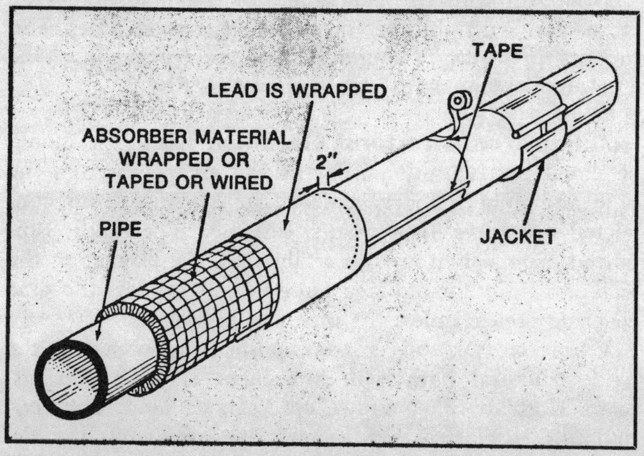

Fig. 1—Installation of lead on pipes.

TABLE 1

Approximate thickness of single, simple partitions to meet permissible transmission loss requirement.

Material	Thickness required in inches		
	Moderately noisy office	Quiet office	Very quiet office
Fir plywood	3.67	6.67	13.33
Sand plaster	0.20	4.45	8.9
Glass	0.13	4.70	9.3
Dense concrete	0.14	4.85	12.2
Aluminum	0.13	5.7	11.3
Steel	0.045	0.20	7.0
Lead	0.030	0.135	0.54

The stepped line indicates the break between those materials which are limp enough to use their weight efficiently (below the line) and those which are stiff enough to lose the effectiveness of their weight (above the line).

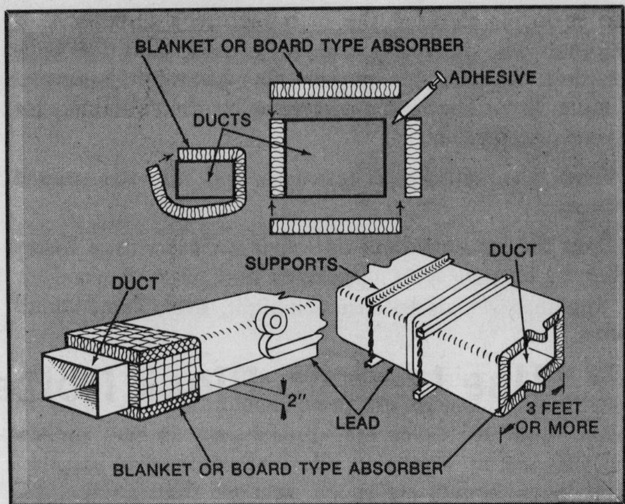

Fig. 2—Installation of lead on ducts.

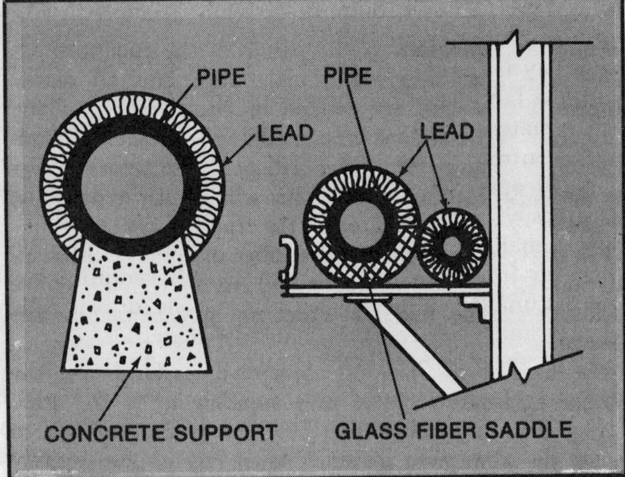

Fig. 3—Use of lead to isolate vibration from structures.

plicate the problem. The following procedures are intended to serve as general guidelines for the use of lead sheet as a tool to control noise from these sources.

Although specific solutions can be developed only on a case-by-case basis, two examples are cited.

Pipe. Noisy pipes are quieted by covering them first with a layer of sound-absorbing material, then a wrapping of sheet lead, and finally an aesthetic covering such as aluminum or plastic.

The absorber layer is usually glass fiber or mineral wool, though soft open cell foam is also used. Excellent results are obtained with absorber densities ranging from three to 18 pounds per cubic foot (pcf). The choice of absorber is governed by:

- Surface temperature of the pipe. Where no absorber is available for the expected temperature, thermal lagging such as calcium silicate insulation must first be used to drop the temperature to the absorber rating.

- Size of the pipe or duct. Low density absorbers are crushed by the covering weight on large pipes. Low density materials, therefore, should not be used on ducts larger than 18 inches or pipes with diameters larger than 12 inches.

As shown in Fig. 1, the absorber is wrapped around the pipe and temporarily fastened by means of tape, wire or banding. Absorber coverage of the surface need not be perfect so long as 90 percent or more of the surface is covered and no gap exceeds an inch or so. For irregular surfaces, such as branches or valve bodies, it is helpful to shred scrap absorber and stuff any inescapable gaps. Lead is now wrapped over the absorber and dressed to a smooth, snug, puncture-free fit.

Lead sheets are normally supplied in 4-foot x 25-foot rolls. Pipes are best covered using the 4-foot dimension along the run of the pipe, then cutting the lead sheet to match the pipe's circumference plus overlap. Where the diameter of the absorber-wrapped pipe is 14 inches, the lead sheet may be cut in 8-foot lengths and applied lengthwise along the run of the pipe.

Best results are attained with a two-inch overlap of the lead sheet at all seams. The overlap on longitudinal seams may vary somewhat because of elbows, support structures, etc. Laps of less than one inch should be avoided. For top performance, seems should be sealed by taping.

Other methods of sealing (e.g., mastic material, clamps) may also be used where the job requires them. Irregular surfaces may be "pieced out" using lead trimmings without loss of acoustical performance so long as a good seal is maintained. Unlike the absorber, the lead covering should be *gap free.*

Any desired outer casing or cosmetic covering may now be applied. Care should always be taken that the lead is not punctured. The outer casing should be light enough so that the absorber is not crushed. Finally, the lead sheet should be isolated from any stiff, heavy outer casing by a resilient material acting as a vibration break and applied in the same manner as the absorber.

Ducts. The same general method for installing lead on pipe is also used for ducts, or any other situation where a close fitting enclosure is indicated, such as reducing valves. In the case of rectangular ducts, it is often more convenient to use an absorber of the board type rather than a blanket. Glass and mineral wool batts and boards are suitable absorbers. Fiber building board materials do not give good acoustical results in close-fitting enclosures.

Where a board absorber is used, duct sealing tape or spot application of adhesive is a convenient way of anchoring the absorber while applying the lead sheets. Clips or studs which penetrate the absorber and the lead can be used *if all penetrations are sealed* in the completed job.

Where the width of rectangular ducts exceeds 30 inches (three feet with absorber in place), it is good practice to support the lead on the underside to prevent sagging. One quick method of doing this is shown in Fig. 2.

Isolation from structural vibration. Briefly, it should be noted that vibrational energy can escape through the pipe and duct supports and radiate as noise from connected structures. In the worst case, i.e., pipe rigidly supported from a light metal curtain wall, the leakage to the wall and resultant radiated noise usually renders the pipe covering useless unless the pipe can be isolated.

Vibration isolation is accomplished by breaking the rigid structural path with a resilient element. Springs, foam, cloth or other straps, etc., can be used. For continuously supported, lightweight pipe and duct, the absorber layer of the covering can provide the isolation if its density is carefully selected. A more general solution,

Fig. 4—Sound level around this blower was reduced by more than eight decibels by adding the lead sheathing.

Fig. 5—Noise levels permitted by OSHA were realized after adding lead barriers.

highly desirable for point support, is the use of 80 pcf glass fiber saddles which are available in standard pipe sizes. These short half-cylinders are substituted for the absorber at support points and may be nested where it is necessary to match the absorber layer thickness. Examples are shown in Fig. 3.

Machinery enclosures. Enclosures are required to reduce noise from compressors, generators, pumps and other machinery. The interior walls of the enclosure should be covered with a fibrous or foamed absorber material to help absorb the sound, while the panels should contain a barrier material (sheet lead) to block out the noise. Such an enclosure, which is easy and inexpensive to construct, can reduce noise by 25-30 dB in the more harmful, higher frequencies.

Panel construction. Plywood, hardboard, sheet metal or gypsum board are suggested materials for the panel facing. The panel should be laminated with sheet lead as follows:

- Lay lead flat on a working surface at least as large as the panel to be laminated. Smooth out wrinkles by rolling or dressing with a small board.
- Apply either neoprene or urethane cement to cover lead surface, following the manufacturer's directions. If neoprene base (contact) cement is used, coat the back surface of the panel facing sheet also. Allow 10-15 minutes or more, according to the directions on the container, for solvent evaporation.
- When the adhesive is tack-free, join the two coated surfaces.
- Press the lead firmly to the other surface with a heavy roller, by tamping with a hammer and piece of wood, or by applying weights to ensure a tight, even bond at all points.

Two-part urethane adhesives are suggested where strength requirements are more demanding, especially in doors. These adhesives are applied only to one surface and they set up slowly to allow adjustment in register. Urethane adhesives are much stronger than contact cements and more resistant to moisture and temperature cycling. Clamp finished laminates together or apply sufficient weights and allow to cure overnight.

Panel installation. If the panels of the enclosure are to be fabricated over studs, make sure that all plates, runners or channels are bedded in elastomeric caulking compound. Affix sealant tape, caulking or gaskets to both faces of the framework at the critical points before applying the leaded surface skins. This will greatly reduce the telegraphing of sound through the framing system.

Fill the wall cavity with glass fiber or mineral wool insulation. Tightly seal all joints and seams. Seal the entire enclosure at the floor in order to prevent any noise leakage.

We suggest the use of absorptive material on the interior enclosure surfaces as a supplement to the wall. Cavity walls or panels, which have one perforated skin to expose the absorptive material, must rely almost entirely on the other or outer leaded skin of the panel to provide the transmission loss. If high performance is required, use a double skin and cavity design with absorptive material both in the cavity and on the panel facing the noise source.

Sound rated doors. Doors often are a weak link in an acoustically insulated enclosure. However, you can readily make high performance doors—1¾ or 2 inches in thickness—from facings of ⅛, 3/16 or ¼ inch plywood or hardboard.

- Laminate lead to the inside surface of both panels following the directions for panel construction outlined above.
- Join one panel to the door frame, using urethane cement.
- Fill the cavity with low density absorptive material.
- Cement second face in place and clamp or weight for overnight cure.
- Apply gaskets on three sides of the door and a seal at the bottom. If not gasketed, the crack around even a well-fitted door could reduce its performance by 10 to 20 dB.

Supplemental walls. You can improve walls by as much as 10-12 dB by installing leaded supplements directly to them.

- Attach furring strips to the existing wall.

• Fill the cavities between the furring strips with absorptive material.

• Apply layer of caulking or butyl tape to the furring strips before the leaded facing laminate is fastened in place.

An alternative construction method is to substitute fiber board or insulating board for both the furring and absorptive material. In this case the facing laminate is a sandwich of insulating board, lead and outer (finish) sheet. Apply it directly to the existing wall with adhesive or butyl building tape. This construction is not quite as effective acoustically as the first supplement described, but it's convenient, saves space and in some situations, it is adequate.

The major difference between the two methods are at the low frequencies. When these are the source of the noise problem—such as transformer hum—use the furred supplement with very deep furring.

It is vital that doors and other parts of the enclosure have transmission loss comparable to that of the walls. If you use hung ceilings, be sure to extend the wall system through the plenum to the floor above by installing lead barriers above them. This will prevent room-to-room sound transmission. Remember, poor performance in any area of the enclosure can seriously impair over-all performance.

EXAMPLES

Refining equipment. Gearboxes, blowers, steam chests and some sections of the pipe network were the primary sources of noise. Corrective action called for shrouding noisy equipment in thick layers of glass fiber and thin layers of sheet lead, and then jacketing the noise shroud to heavy-gage aluminum.

In order to modify the gearboxes, anchor pegs were bonded to gearbox casings to support the 1.5 to 2-in. thick fiberglass layers. Bonding was chosen over welding since the latter might have created a spark hazard in an area where large volumes of flammable materials are processed.

Fiberglass was then covered with sheet lead. Generally, 2 psf (0.0312-in. thick) of the sheet lead was used although some 1 psf (0.0156-in. thick) sheet lead was also utilized on other parts of the equipment.

Sound attenuation on valve handles and gages was accomplished by adding silicone rubber and aluminum sliding doors. Silicone rubber added to the valve handles and sliding aluminum doors covering them eliminated potential sources of noise. Similar doors combined with clear plastic were installed around gages.

Enclosures for the blowers (Fig. 4) were of different construction from gearboxes (Fig. 5) and steam chest housings (Fig. 6). Blowers were first layered with glass fiber and sheet lead, then encased in sound-insulated outer panels of aluminum. Movable panels permit easy access to the equipment when required.

Fig. 6—Steam chest noises are muffled by lead panel enclosures.

Olefins plant piping. High velocity process gas and high pressure steam and fuel gas caused considerable noise in 5,000 feet of welded carbon and low alloy steel pipe ranging in diameter from 8 to 36 inches.

Pipe and control valves were wrapped in 1½ in., 7½ lb. fiberglass absorber, which in turn was covered with 1 pound lead sheet. Joints for the latter were overlapped 2 inches and sealed with duct tape. An aluminum jacket was added to protect the malleable lead. Lead requirements amounted to 42,000 square feet.

Flare Noise: Causes and Cures

Most flare noise problems can be cured today except intense combustion noise. Grade-level flares are 10 decibels quieter than elevated flares

James G. Seebold, Standard Oil Co. of California, San Francisco

THE ROAR of the combustion process is the most serious noise problem in an elevated flare. Potential aggravating noise sources include steam injection, moisture condensation shock, seal drum sloshing, low-flow instability, and combustion-driven oscillations. Causes and cures are considered, together with sound level estimates, and a brief comparison with grade-level flares.

Combustion roar. Because it is inherent, the roar of steady combustion is the one source that really cannot be effectively dealt with in an elevated flare. At moderate release rates, this source may not, by itself, be objectionable once disturbances related to unsteady combustion and the steam injection systems are eliminated. If the roar remains objectionable, the moderate (and most frequent) releases can be burned in a grade-level flare, saving the elevated flare for outright emergencies.

Intensely turbulent flames are inherently noisy. While a buoyant flame, nearly laminar and noiseless like a candle, would ordinarily occur at the flare tip, a high-velocity steam-air mixture is usually injected into the combustion zone. This greatly increases the turbulence intensity, and produces the combustion roar characteristic of a turbulent diffusion flame.

By considering a small volume element in a turbulent burning zone, Bragg[1] arrived at an expression for the sound energy radiated by the flame. Bragg's expression suggests the noise should be proportional to the second power of the heat release, which we can consider to be proportional to the hydrocarbon mass flow rate (assuming all is burned). Others have found this to be a reasonably accurate description.[2-6] As long as the proportionally correct amount of steam is injected, this seems to hold quite well for elevated flares, as well, as shown in Fig. 1.

Bragg also found that, at constant heat release (hydrocarbon mass flow), noise should be proportional to the second power of the mixing velocity. This is where the artificially-induced turbulence of steam-air injection comes in. Apart from the higher-frequency noise the steam jets produce by themselves, steam-air injection raises the intensity of the lower-frequency combustion roar in accordance with the increased mixing velocity, which is related to the mass flow rate of steam per pound of hydrocarbon burned. (The usual rates are in the order of about ½ pound-steam per pound-hydrocarbon.) The typical change in combustion roar with steam flow, at constant hydrocarbon flow, is shown in Fig. 2.

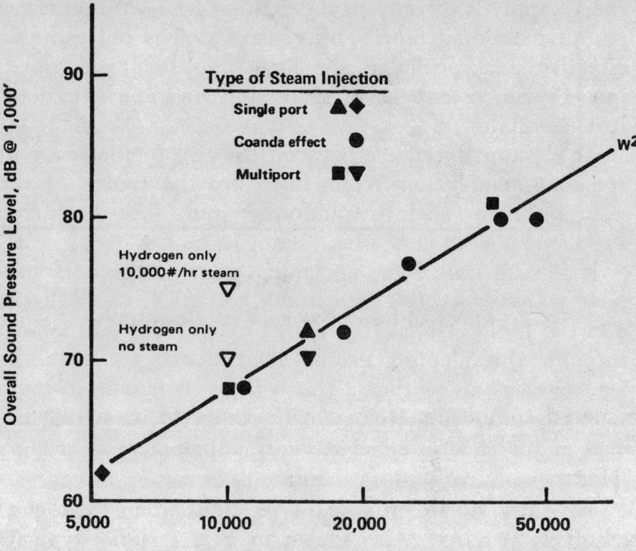

Fig. 1—Elevated flare noise vs. hydrocarbon flow rate.

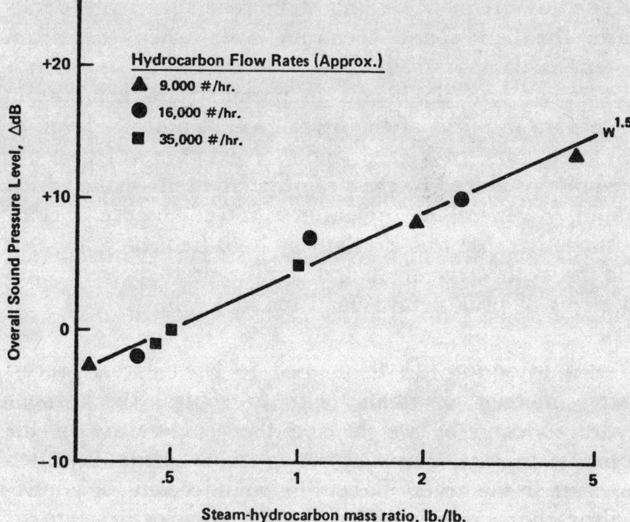

Fig. 2—Effect of steam injection on combustion noise.

Intuition suggests that the role of the gas should primarily be expressed by the air-gas mass ratio theoretically required for complete combustion, and the calorific value. Bragg's expression suggests that combustion noise should be proportional to the second power of the quotient of the air-gas ratio and the calorific value. Hydrogen has the capacity of making large changes in the air-gas mass ratio and calorific value of flare gas mixtures. When large amounts of hydrogen are sent to flare stacks, substantially

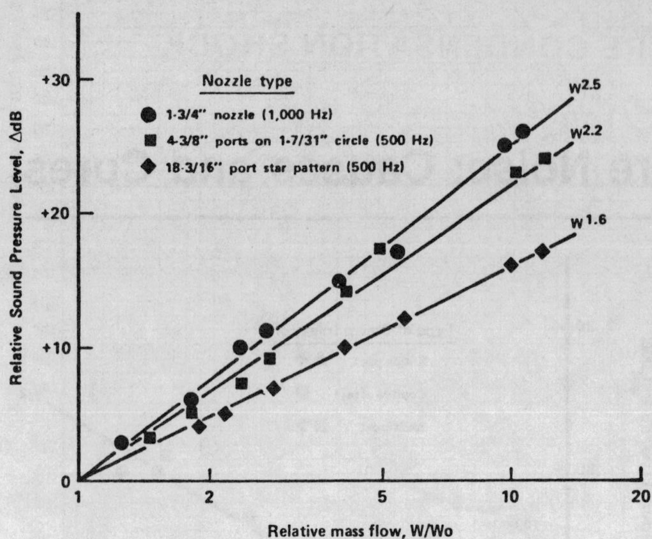

Fig. 3—Effect of mass flow rate on steam noise.

increased combustion roar can be expected as shown in Fig. 1.

Hydrogen burns without luminosity or smoke, of course. Consequently, no steam is really required when hydrogen exclusively is flared. Also shown in Fig. 1 is the typical effect of cutting the steam at a constant hydrogen flow rate. This illustrates the effect of the artifically-induced turbulence of steam-air injection.

The low-frequency combustion roar invariably dominates the flare sound spectrum, even when the steam system is comparatively noisy. Consequently, it is both convenient and significant to present flare combustion noise data in terms of the overall sound pressure level, as has been done in Figs. 1 and 2. Next, we will consider the noise produced by the steam system itself—noise which would, in general, be present even in the absence of combustion. Because this is relatively high-frequency noise, it will be both convenient and significant to present data in terms of peak octave band sound levels.

Steam injection. In flares used to burn excess process gases, injection of steam helps to control the carbon-hydrogen ratio by weight and the temperature in the combustion zone, thus suppressing smoke. Usual injection rates are in the order of about ½ pound-steam per pound-hydrocarbon. Also, injection of the high-velocity steam-air mixture into the combustion zone greatly increases the turbulence intensity, improving combustion and reducing the size of the flame. All of these are very desirable effects. Unfortunately, as we already pointed out, steam injection increases the combustion roar as well.

Additionally, the steam system has its own (higher-frequency) noise that is present whether or not combustion is taking place. In the simple, singleport steam injector, the major source of noise is the highly sheared mixing zone just downstream of the nozzle. The noise of elevated flares can be reduced to some extent by using multiport steam nozzles.[7] Most commercially-available flares now include some form of multiport nozzle for steam injection.

In most flare systems, steam is supplied to the injectors at 100 to 150 psig. The nozzles are therefore highly choked (pressure ratio far exceeds the critical value at which the steam flow becomes sonic at the throat). Under these conditions, the noise production is approximately proportional to the second power of the steam mass flow rate as shown in Fig. 3.

Because of this approximate second-power relationship, N ports passing the same total flow as one port would be expected to produce only about $1/N$ as much sound power. Actually, if the holes are placed so that the jets coalesce into a single larger jet before much mixing takes place with the surrounding air, somewhat greater than $1/N$ sound reduction can be obtained due to the increased mixing perimeter at the point of maximum shear. (This can be seen, for example, in the test results in Reference 8). This shortens and broadens the mixing zone, further reducing the radiated sound power.

The quicker the steam becomes thoroughly mixed with the inspirated air, the less the noise produced by the steam system itself. However, it is worth remembering that there is no accompanying reduction in combustion roar.

Moisture condensation shock. The noise-producing potential of moisture condensation shocks appears to have had little study, but they are another possible source of noise in steam injection systems. This phenomenon may account for the gravel-like noise (similar to cavitation) that has occasionally been observed.

Moisture condensation shocks first received attention in connection with the flow through the nozzles of steam turbines.[9] Sometimes the supply pressure, supply temperature, and pressure ratio are such that at some point in the nozzle process the saturation condition is reached, and a wet mixture of saturated vapor and saturated liquid might be expected upon further expansion of the stream. Experiments show, however, that the precipitation of moisture is delayed beyond the point where saturation is reached. The phenomenon is called supersaturation, and, during the period when the vapor pressure is greater than the saturation pressure corresponding to the temperature, a state of metastable equilibrium exists. Precipitation probably is connected with the appearance of nuclei, perhaps comprising foreign particles or accidentally formed groups of steam molecules. On the Mollier diagram, the line of precipitation states, called the Wilson line, has been found by experiment to lie about 60 Btu/lb. below the line of saturation states.

The precipitation of moisture is quite sudden, as the state of the fluid jumps quickly towards a state of stable equilibrium. The droplets formed during condensation apparently are exceedingly small, although little data are available concerning the size of droplets initially formed, and virtually no data are available concerning the rate of growth of the droplets. The detailed mechanisms are not clear; it may be that steady flow is impossible and an oscillatory (hence highly acoustically efficient) flow occurs.

A typical steam-injector nozzle process is shown on the Mollier diagram of Fig. 4. The process proceeds isentropically to a pressure equal to about 55 percent of the supply pressure, at which point the velocity is sonic. Thereafter, the process proceeds discontinuously across the normal shock, seeking equilibrium at atmospheric pressure. The steam is at the critical pressure at the nozzle exit and, when it suddenly enters a region of lower pressure, it expands explosively. The steam particles are accelerated outwards from the jet axis, and because of their inertia overshoot the equilibrium position, producing a rarefaction in the core of the jet, which then sucks them inward again, thus producing the familiar pattern of the choked jet.

We can see from Fig. 4 that the line of precipitation states (Wilson line) is typically reached when the supply stream moisture content is as little as one or two percent, and moisture condensation shocks can be expected to form. While the noise-producing potential of these shocks has had very little study, it would seem prudent to preclude their formation by making sure dry or somewhat superheated steam is supplied for flare injection.

Seal drum sloshing. The first of several sources of unsteady combustion to be discussed is seal drum sloshing. Sloshing of the water level in a seal drum can lead to periodic surges in the gas supply. For typical seal drums, the period of the fundamental sloshing motion is of the order of one second. This is typical of combustion pulsations observed in elevated flares.

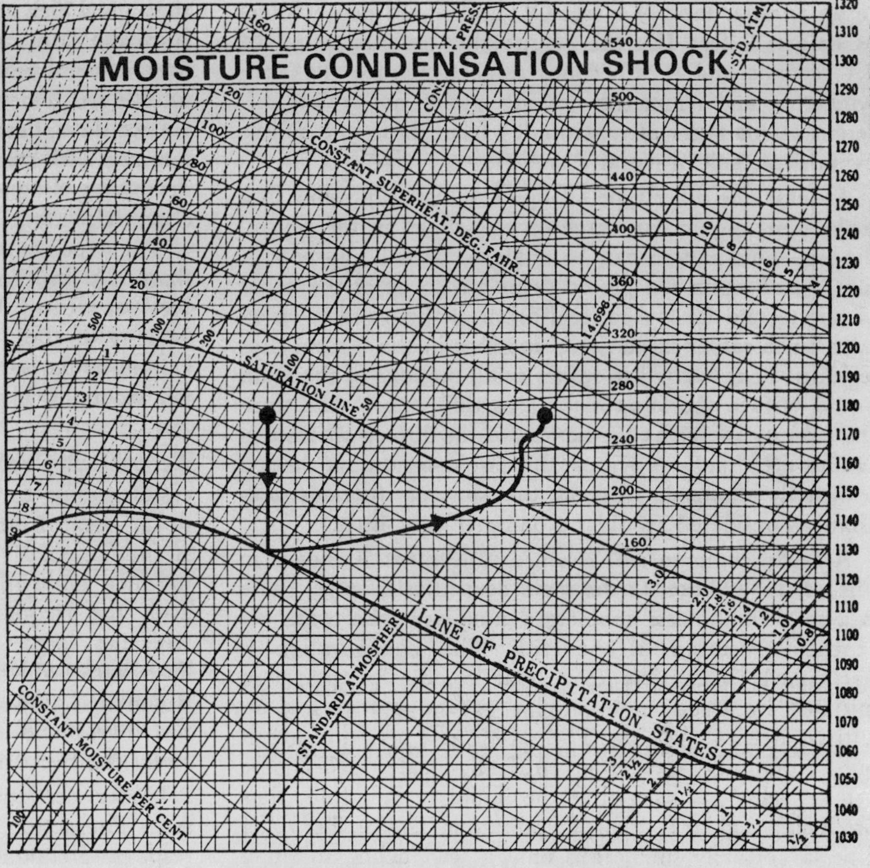

Fig. 4—Typical steam-injector nozzle process.

The displacement of the free surface from the equilibrium liquid level in a vertical cylindrical seal drum satisfies the Wave Equation in cylindrical coordinates.[10] Application of the boundary condition of no radial motion at the wall gives the frequency equation. The solutions come out in terms of Bessel functions of the first kind. Solution in tanks of other configuration is similar.

The period of the fundamental antisymmetrical sloshing mode shown in Fig. 5a is (0.301) $D/H^{1/2}$, where D is the drum diameter and H is the liquid depth, both in feet. For a drum six feet in diameter and liquid five feet deep, the period is 0.81 seconds. The period of the fundamental symmetrical sloshing mode shown in Fig. 5b is (0.145) $D/H^{1/2}$; for the tank used in the previous example, the period would be 0.39 seconds.

Strictly speaking, the seal drum is an annular tank because of the central relief pipe dip-leg. The solution for the annular tank comes out in terms of Bessel functions of both the first and second kind because the surface is discontinuous at the center. However, sloshing periods differ little from those in simple cylindrical tanks as long as the radius-ratio is not too close to one. For example,

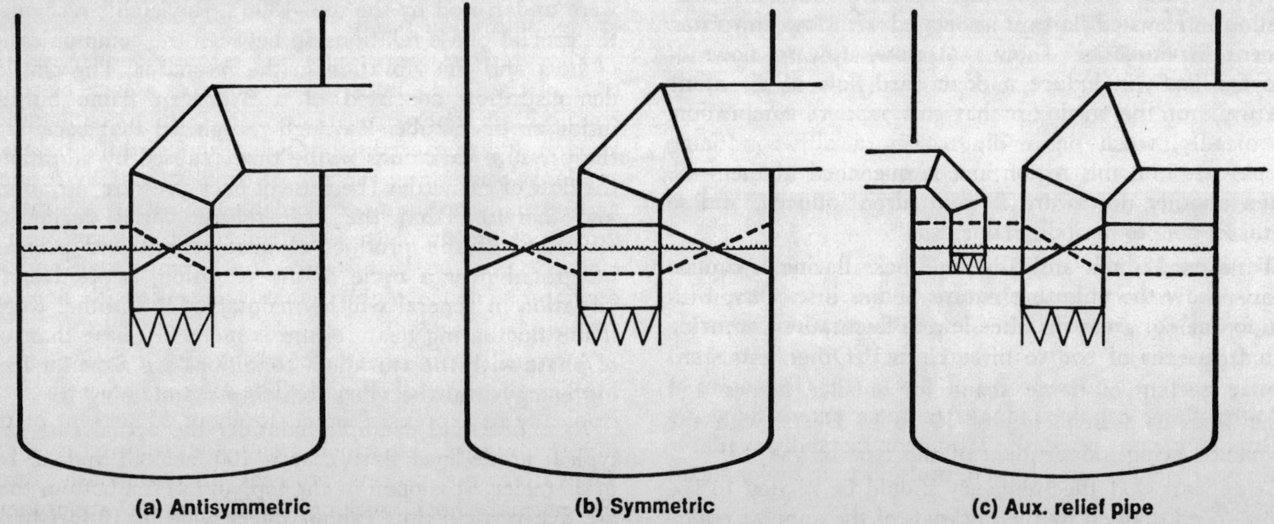

(a) Antisymmetric (b) Symmetric (c) Aux. relief pipe

Fig. 5—Seal drum sloshing modes.

in an annular tank of 6 feet OD and liquid depth 5 feet as before, 2 feet ID; the comparable periods would be 1.1 seconds and 0.23 seconds respectively for the anti-symmetric and symmetric sloshing modes, not greatly different (same order of magnitude) from those in a simple cylindrical tank of equal diameter.

At least one other type of liquid motion is possible in a seal drum. Liquid can surge up and down in the central dip-leg. This motion can be analyzed as a liquid-pendulum,[11] and again the period turns out to be of the order of one second for typical seal drum configurations. However, since rather large liquid motions in the central dip-leg must be converted to rather small motions of the main liquid surface (because of the area difference), and since the velocity is high at the bottom of the dip-leg where serrations are usually provided, this liquid-pendulum motion is heavily damped, whereas the sloshing motions previously considered would be damped very little.

A fairly common seal drum arrangement is shown in Fig. 5c. In addition to the main central relief pipe, there is a smaller and less submerged auxiliary relief pipe eccentrically located in the tank. This configuration is particularly susceptible to regular combustion pulsation initiated by a sudden main-pipe gas surge that sets up antisymmetrical sloshing. Thereafter, at a period of about one second, the smaller eccentrically-located relief pipe belches relief gas each time the liquid surface falls, thus producing regular combustion pulsation. The smaller relief pipe is normally submerged only a few inches, so small sloshing motions are sufficient. Further, being lightly damped (in older seal drum designs), the motion (and the resultant combustion pulses) may persist for a long time after the initial gas surge.

Seal drum sloshing can be eliminated by eliminating the seal drum, of course. In some systems it may be reasonable to use a gas seal instead of a liquid seal. Gas seals depend on vapor density differences and require continuous purging.

Perforated, anti-slosh baffles can be placed in the equilibrium liquid surface. These provide lots of viscous damping, and make any persistant sloshing virtually impossible. Most commercially-available seal drums now use this principle to combat seal drum sloshing.

Low flow instability. Another source of unsteady combustion in elevated flares is associated with low flow rates. Recent experiments[12] show that slow, upward flows of buoyant gas will induce a downward flow of air along the walls of the stack tip that can support combustion. Apparently, when flame dip occurs, a diffusion flame propagates into this region and is quenched at the wall. Then another downward flow of air is induced, and so forth, leading to periodic flame dip.

Tests on 12-inch and 18-inch stacks flaring hydrogen clearly show the pulsating nature of this instability, with two to one (or greater) flame length fluctuations occuring at a frequency of two to three Hertz.[12] Other tests show similar cycling of flame shape for smaller flames and other fuels at a frequency of 10 to 15 Hertz, with the frequency being independent of the type of gas.[13, 14]

It appears that the frequency should be related to the flame speed divided by the thickness of the annular region through which the flame propagates inside the pipe, and since the flame speed doesn't change much from fuel to fuel, it seems reasonable that no marked effect of the fuel on the frequency of pulsation is observed. Further, since we would expect the thickness of the annular region to increase with stack diameter, it seems reasonable to expect decreasing frequency with increasing stack diameter, as observed.

The experiments and analysis of Reference 12 suggest a flame dip limit for large stacks of the order of one to three feet per second. When low relief rates cause the stack tip velocity to fall below this limit, pulsating combustion caused by flame dip at a frequency of about once or twice per second is likely.

Acoustically related unsteady combustion. Two processes leading to unsteady combustion have been discussed so far—seal drum sloshing and flame dip. Neither process is fundamentally related to acoustics, though both can produce annoying noise (and light) fluctuations. However, the frequency of these disturbances can coincide with the "organ pipe" standing wave frequency in the flare stack, which for typical stacks is around two Hertz. Resonance presumably would augment the acoustical disturbance, but it does not appear that it would particularly affect the two driving mechanisms discussed thus far.

Under certain conditions, strong pulsations do arise from interaction of the combustion process with natural response frequencies of the enclosure. For this to work well, the flame generally needs to be near the point of maximum fluctuating pressure (antinode) of the standing wave in the enclosure that is involved. In an elevated flare, the flame is located almost exactly at the pressure node of the standing wave in the stack, which would seem to make effective interaction virtually impossible. However, when the flame is located at the base of the flare stack, as it is in the grade-level flare, combustion-driven pulsations are possible.

The grade-level flare would seem to be a promising means of reducing light and noise in flare systems, but some configurations have been subject to combustion-driven pulsations. Though complex and incompletely understood, a brief discussion of this interesting phenomenon, and a specific example, may be of interest.

The general features of "vibrations maintained by heat" were understood by the mid-1800's. Rayleigh[15] recognized the crucial phase relationship between the communication of heat and the vibration in the resonator. The classical demonstration consisted of a hydrogen flame burning inside an open tube. Rayleigh recognized that because of the pressure variations in the tube (caused by vibration), the flow of gas varies (because of back-pressure variation), and therefore that the heat release varies during the vibration. If the product of heat release and pressure, integrated over a cycle of the vibration, is positive, the vibration in general will be maintained.[16] In other words, if the fluctuating heat release is more in phase than out of phase with the vibration, conditions are right for feeding energy into the vibration, thus maintaining it.

As a practical example, consider the actual case of a typical grade-level flare that is 100-feet tall and 18 feet in diameter. It is open at the top, and at the bottom there are five vertical slots (about 2-feet wide by 10-feet high) into which protrude the gas supply pipes and through which passes the combustion air. In operation, this system occasionally exhibited strong vibrations at a frequency of about seven Hertz. Application of the Rayleigh criterion predicts that combustion-driven oscillations should not be sustained until the flue gas temperature reaches approximately 1,500° F, corresponding to a hydrocarbon flow rate about 75 percent of the maximum design load. Observed conditions during pulsation confirmed these predictions. Subsequently, new burners supplied by the flare manufacturer successfully eliminated the pulsation problem in tests of this grade-level flare. Under a variety of loads and operating conditions, including sustained operation at 27,000 lbs./hr. (about 108 percent of design capacity), no pulsation behavior could be produced. While we currently lack a theoretically satisfying explanation for the improved behavior, the new burners do seem to have eliminated (or sufficiently suppressed) the source of excitation for the pulsation. It is known that rather subtle changes at the supply/resonator interface (here, the burners) are sometimes significant. For an interesting discussion of operating experience with this flare, see Reference 17.

Besides burner modifications, pressure relief is another means that has been successfully used to suppress combustion-driven pulsations in grade-level flares. The stack can be modified part-way up to provide an opening to relieve and damp the standing pressure wave.

Grade-level flares currently are in wider use in Europe than in the United States. The modern grade-level flare is refractory-lined and equipped with many burners at the base. In some configurations, the stacks reach heights of 100 feet. Its capacity is limited, but the grade-level flare can be integrated (at substantial extra cost) into an elevated flare system to take care of the great majority of reliefs that are relatively small, and reserve the elevated flare for outright emergencies.

An elevated flare may be little more than a long piece of pipe with a flame burning on the end of it. The turbulent burning zone, in plain view, is a source of noise and light. Common sense suggests that enclosing the turbulent burning zone at grade, as in the grade-level flare, should reduce the radiated light and noise. However, there is not much directly comparable data to augment common sense. Furthermore, some configurations have been subject to severe combustion-driven pulsation, as previously described.

Available grade-level flare data is compared with elevated flare data in Fig. 6. To get even this limited comparison, some distance extrapolation was required. Nevertheless, the measurements suggest that, in the absence of combustion-driven pulsation, the grade-level flare is likely to be about 10 decibles quieter at the same load than its elevated counterpart.

The grade-level flare would seem to be a promising means of reducing light and noise in flare systems,[17] but some configurations have been subject to severe combustion-driven pulsations, and the cost is high.

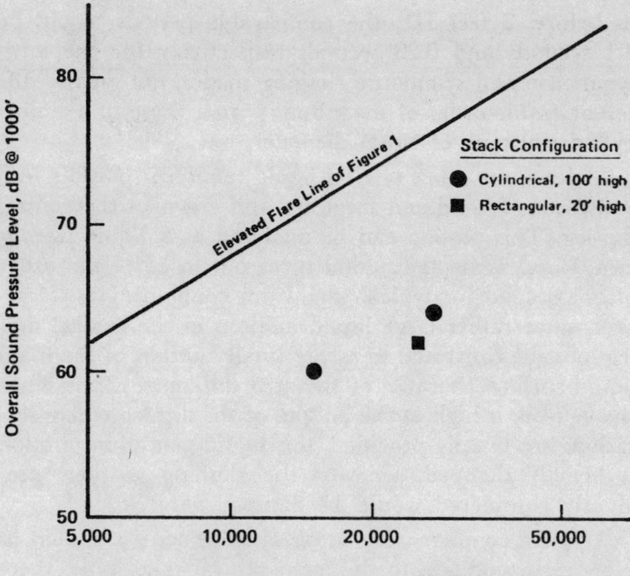

Fig. 6—Grade-level vs. elevated flare noise.

Originally presented as a paper to the ASME Petroleum Mechanical Engineering Conference, New Orleans, La., Sept. 18-21, 1972.

LITERATURE CITED

1 Bragg, S. L., "Combustion Noise", J. Inst. Fuel, Jan. 1963, p. 14, eqn. 4.
2 Giammar, R. D., and Putnam, A. A., "Combustion Roar from Experimental and Commercial Gas Burners", A Phase Report on AGA Basic Research Project BR-3-5, Battelle Memorial Institute, March 1970, pp. 24, 32.
3 Smith, T. J. B., and Kilham, J. K., "Noise Generation of Open Turbulent Flames", J. Acoust. Soc. Am., Vol. 35, 1963.
4 Smithson, R. N., and Foster, P. J., "Combustion Noise from a Meker Burner", Combustion and Flame, Vol. 9, Dec. 1965.
5 Giammar, R. D., and Putnam, A. A., "Combustion Roar of Turbulent Diffusion Flames", J. Eng. Power, Trans. ASME, Paper No. 69-WA/FA-3, 1969.
6 Seebold, J. G., "Combustion Noise and Its Control in Process Plant Furnaces", ASME Petroleum Mechanical Engineering Conference, Paper No. 71-Pet-6, Sept. 1971; see also Oil and Gas Journal, Jan 3, 1972, p. 48.
7 Seebold, J. G., and A. S. Hersh, "Refinery Flare Steam Injectors Redesigned for Noise Control", ASME Winter Annual Meeting, Paper No. 70-WA/Pet-4, Dec. 1970; see also Hydrocarbon Processing, Feb 1971, pp. 140.
8 Hersh, A. S., and J. G. Seebold, "Noise Control of a Flare Stack", 79th Meeting of the Acoustical Society of America, April 1970.
9 "*Compressible Fluid Flow*", A. H. Shapiro, Vol. I, The Ronald Press Company, New York, 1953, p. 203-205.
10 *Hydrodynamics*, H. Lamb, Dover Publications, New York, 1945, art. 191, pp. 284-288.
11 *Engineering Vibrations*, L. S. Jacobsen and R. S. Ayre, McGraw-Hill, New York, 1958, pp. 10.
12 Grumer, J., et al, "Hydrogen Flare Stack Diffusion Flames" Report of Investigations 7457, Bu. Mines, U.S. Dept. Int., Dec. 1970.
13 Barr, J., "Diffusion Flames", Proceedings of the 4th Internation Symposium on Combustion", Williams and Wilkins Co., Baltimore, Md., 1953, pp. 765-771.
14 Maklakov, A. I., "Oscillation of Diffusion Flames Appearing When There Is a Laminar Flow of the Ignitable Medium", Zh. Fiz. Khim, v. 30, 1956, p. 708.
15 Strutt, J. W. (Lord Rayleigh), "The Explanation of Certain Acoustical Phenomena", Roy. Inst. Proc., VIII, pp. 536-542.
16 *Combustion-Driven Oscillations in Industry*, A. A. Putnam, American Elsevier Publishing Company, New York, 1971, p. 3.
17 Powell, D. T., and R. Schwartz, "Operating Experience With a Low-level-type Flare," Paper No. 59-72, 37th midyear meeting of the American Petroleum Institute's Division of Refining, New York, May 11, 1972.

Flare Noise: Another Opinion

"Flare noise: Causes and Cures" is very interesting when compared with data obtained from other sources, as shown on the accompanying chart. If the line of Fig. 1 is replotted using steam flow rate as half the hydrocarbon rate, as suggested in the test, and the data from Fig. 2 is inserted using the same symbols there are some discrepancies which can be cancelled using the

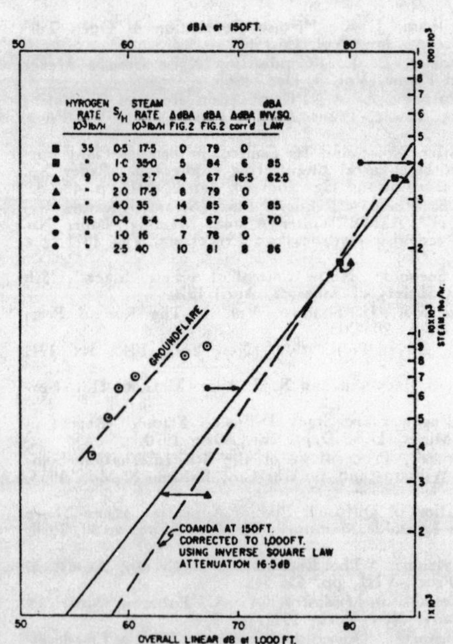

	HYROGEN RATE 10³ lb/H	S/H	STEAM RATE 10³ lb/H	ΔdBA FIG. 2	dBA FIG. 2	ΔdBA corr'd	dBA INV. SQ. LAW
■	35	0·5	17·5	0	79	0	
■	"	1·0	35·0	5	84	6	85
▲	9	0·3	2·7	−4	67	16·5	62·5
▲	"	2·0	17·5	8	79	0	
▲	"	4·0	35	14	85	6	85
●	16	0·4	6·4	−4	67	8	70
●	"	1·0	16	7	78	0	
●	"	2·5	40	10	81	8	86

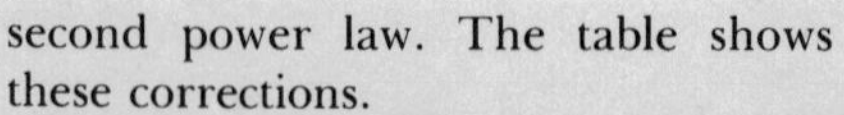

second power law. The table shows these corrections.

Along side the Seebold line is shown a "Coanda" line of test results obtained on flaretips using this method of steam injection and showing good agreement!

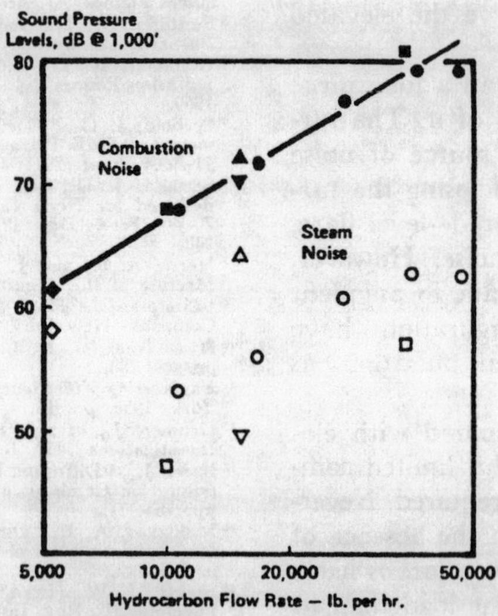

▲ Elevated Flare with Single Port Steam Injection Nozzles
◆ Elevated Flare with Single Port Steam Injection Nozzles
● Elevated Flare with Coanda Effect Nozzles
■ Elevated Flare with Multiport Nozzles
▼ Elevated Flare with Multiport Nozzles

Open Symbols Represent Overall Noise Levels Measured on the A-Weighting Network; Closed Symbols are C-Weighted Levels.

This rearranged data can now be examined in correct sound technology. The acoustical power of the emitting source in decibels is a logarithmic function of the energy of the source, i.e., $PWL = 10 \log_{10} W/W_o$. This power is transmitted by a sound wave and at the receiver is proportional to the square of the pressure variations, i.e., $S.P.L. = 20 \log_{10} P/P_O$. Also, the fraction of the total energy emitted follows the inverse square law. There is no need, therefore, to go into convoluted argument to prove these fundamentals. For example, Bragg's formula $W = 1055G$ (lb/sec) H (btu/lb) boils down to Briffa's (Institute of Fuel Symposium 7/1/72) expression of thermo-acoustic efficiency which is the fraction of heat energy appearing as sound. How the author can arrive at the second power of the heat release escapes me.

Thus, ignoring air/gas ratios, etc., the total energy of the source should be considered and the fraction received is proportional to this energy squared. This is a reasonable assumption since it relates all modes of energy including the degradation of steam energy to air plus steam energy to mixing energy to combustion energy.

Theory then fits the facts since the coincidence of two sets of unrelated data cannot be ignored.

Finally, on the graph, some data on groundflares has been inserted which seems to corroborate the statement that it is 10 decibels quieter. It would be interesting to apply the formula in Heitner's paper (December 1968 HP, Page 67) for enclosures.

A few postscripts will cover the other points of contention.

Page 210 steam injection does not help control C/H—this depends solely on the composition of the excess process gas. In practice it is

the motive fluid for primary air.

Page 211 (Fig. 4) Mollier chart. Generally the steam supplied to flares is superheated—the example shown is not really valid.

Page 212. There is a seal drum (patented) that avoids "sloshing" quite simply—it temporarily removes the water from the gas path.

General. To be of any use to the uninitiated, sound pressure levels should be expressed in dBA. Most noise specifications including Walsh-Healey require this for the very good reason that it is compatible to the receiver—the ear!

R. D. Wilkinson
Pinner, Middlesex,
England

Author's reply. Before commenting on the efficiency matter, let me just touch on your other points.

With respect to controlling the C/H ratio, the presence of more "water" enhances the water-gas reaction that plays a role in removing free carbon. (Indeed, water sprays are effective smoke suppressors.) Naturally, improved mixing and reduced temperature (to suppress cracking) are fundamental, too.

With regard to the quality of steam supplied for injection, maybe you have never seen wet steam supplied to a flare, but I have. In any case, we agree that it shouldn't be!

Regarding seal drum sloshing, we agree here, too. There are several ways available to combat this problem, not the least of which is to dispense altogether with the liquid seal and use a "molecular" (density difference) seal.

Generally, I agree with your concluding comment on the use of dBA levels. However, I feel the low-frequency combustion roar is not adequately taken into account by using A-weighted levels, and prefer overall (dBC). Perhaps it would have been better to give both (as in the accompanying figure), but I felt the figure then became too "busy."

Now let's talk about thermo-acoustic efficiency. Bragg's original paper (my Reference 1) expresses it basically as a product of fuel mass flow rate and mixing velocity squared divided by the calorific value, remaining parameters not varying much for most common fuels. For most common fuels, the calorific value doesn't change much either, on a mass basis. The thermal energy is proportional, of course, to the mass flow rate of fuel, so the thermal energy converted to acoustical energy, expressed as the product of thermal energy and efficiency of conversion, should be roughly proportional to fuel flow and mixing velocity, both squared. These expectations are pretty well borne out by Fig. 1 and 2. Briefly, that is how one arrives at the second power of heat release. It is really quite common —see, for example, the recent articles by Putman (Combustion and Flame, Vol. 18, 1972) and Swithenbank (AIChE Journal, Vol. 18, No. 3, May 1972).

The facts that the energy of an acoustic wave is proportional to the mean-square pressure variation, and that the decay follows the inverse square law, have nothing at all to do with the physics of energy conversion. A common example is the different powers of flow velocity with which jet noise varies depending on flow regime. In all cases, wave energy proportional to mean-square pressure variation and inverse square decay obtain in the far-field, but the dependence on flow velocity is different because the dominant physical mechanism of energy conversion is different in different flow regimes.

Reduce Noise from Pulsating Combustion in Elevated Flares

Process equipment should be designed for quiet but when existing units are noisy, some unique problems must be solved. Here's what one company did when faced with such a problem

James G. Seebold, Standard Oil Co. of California, San Francisco

ELEVATED FLARES are familiar sights in oil refineries and other industrial plants. They are used to safely dispose of excess gas that is generated from time to time by process upsets. In principle, these flares are nothing but a long pipe with a flame burning at the top, some 100 to 200 feet in the air. At the base, they are often provided with a water seal to prevent the entry of air and possible development of an explosive mixture in the main plant piping.

This article briefly discusses a very special flare noise problem—pulsating combustion due to seal drum sloshing. Other sources of pulsating combustion and the general problem of the noise of elevated flares, are not discussed. This article illustrates the frequently interdisciplinary nature of noise control. The problem is basically one of fluid mechanics, but its manifestation is the disturbing sound (and light) of pulsating combustion in an elevated flare.

No sound data are presented concerning this unusual flare problem. Detailed sound data are not really germane to the subjective reaction of people to the intermittent light and sound disturbance caused by pulsating combustion. Though the sound pulsations easily protrude more than 10 decibels above the steady combustion noise levels, it is the intermittent nature of the disturbance that really annoys people.

A schematic drawing typical of many of the older seal drums in use around the world today is shown in Fig. 1.

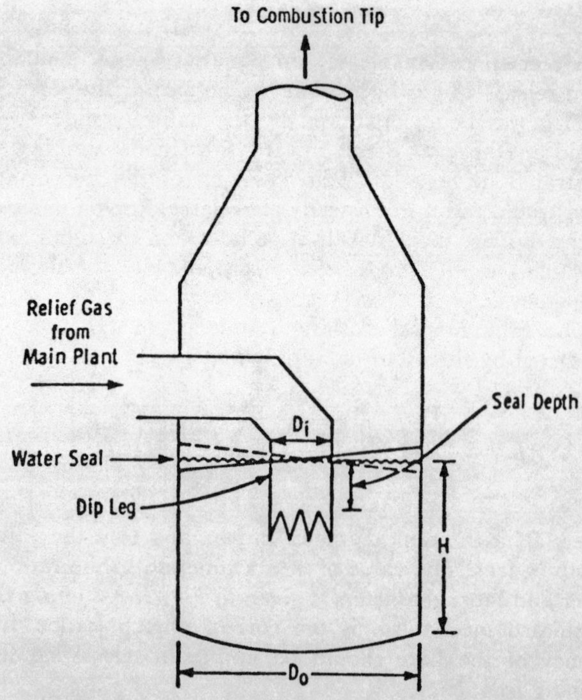

Fig. 1—Seal drum schematic showing first mode sloshing.

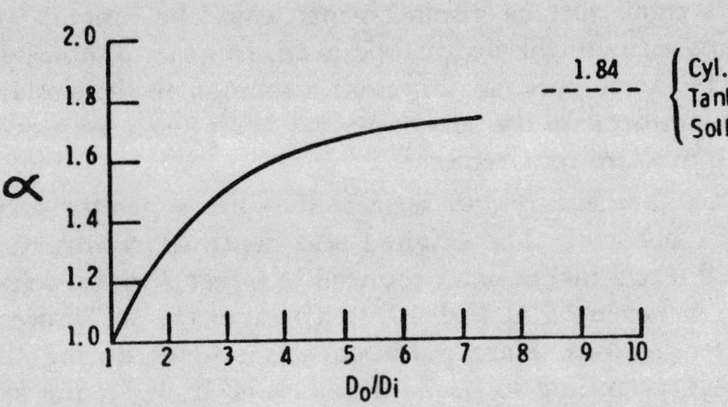

Fig. 2—Parameter α.

In normal operation, relief gas bubbles up all around the serrated dip leg in a consistent but random manner. The serrations are intended to promote this uniform gas flow.

Seal drum sloshing is a potential source of unsteady combustion in elevated flares. Sloshing of the water level in a seal drum can lead to periodic surges in the gas supplied to the combustion tip. The result is a periodic combustion "whump."

For typical seal drums, the period of the fundamental sloshing motion is on the order of one second. In-plant observers of the flare problem described here indicated that, when it occurred, the pulsation was very regular at "something like once per second." Consequently, an investigation was launched to relate the observed behavior with the sloshing characteristics of the seal drum and to develop a means of suppressing seal drum sloshing.

MECHANISM

It seemed reasonable to suppose that energy available from release of gas bubbles at the bottom of the serrated dip leg was being organized by the natural tendency of liquid in the seal drum toward first mode sloshing as illustrated in Fig. 1. Cyclic lowering of the hydrostatic head would result in unsteady gas release, first on one side of the dip leg, then (a half-cycle later) on the other side, producing a pulsating release of gas at twice the sloshing frequency.

The fundamental sloshing frequency in Hz (cps) is predicted by theory to be approximately

$$f = 1.277\sqrt{\left(\frac{\alpha}{D_o}\right)\tanh\left(\frac{2H\alpha}{D_o}\right)}$$

where D_o is the tank diameter in feet, and H is the water depth in feet. The value of α as a function of the ratio of outer and inner diameters is given in Fig. 2. Assuming the postulated mechanism is the correct one, pulsation frequency of the flare should be approximately twice this sloshing frequency.

Flame pulsations were timed and counted while the flare was operating. Periodic pulsations occurred at a frequency of 1.1 Hz.

The seal drum was 8 feet in diameter, with a dip leg 3 feet in diameter and a water depth of 5 feet. From the formula given above, the expected sloshing frequency is 0.526 Hz. Combustion pulsation is expected at twice the sloshing frequency, or 1.05 Hz (compared with the observed 1.1 Hz). This agreement with observation was encouraging.

ONE-QUARTER SCALE TESTS

Verification tests were conducted in a ¼-scale transparent seal drum. The transparent test vessel was 24 inches in diameter with an 8-inch central dip leg, similar in every respect to the prototype seal drum. At maximum sloshing response, water depth was 1½ feet and depth of the water seal above the top of the dip leg serrations was 9½ inches. The observed sloshing frequency of 1.2 Hz agreed well with the calculated 1.09 Hz.

Variation in sloshing response (directly-measured total amplitude of the sloshing wave at the wall) with seal depth and air rate is shown in Fig. 3. The peak in the response curves of Fig. 3 may be explained as follows.

Air is released at the bottom of the dip leg when the liquid surface is depressed. Sometime later the liquid surface on that side will be at its highest point. For greatest response, air bubbles released earlier should burst through the surface at this time, shoving more liquid up on the wall and adding potential energy when the slosh is stationary at its highest point. Thus, for maximum response, the bubble rise time should be equal to half the sloshing period. Rising bubbles literally lift or pump water up in phase with the wave motion, if they rise to the surface in half a cycle.

The observed variation in sloshing response suggests that if seal depth is made small enough (or large enough), sloshing can be effectively suppressed. This seems reasonable. At small seal depths, the disturbance is a random

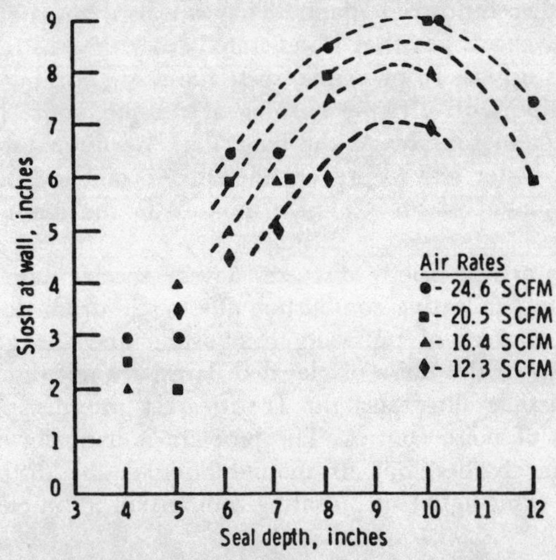

Fig. 3—Sloshing response vs. seal depth and air rate.

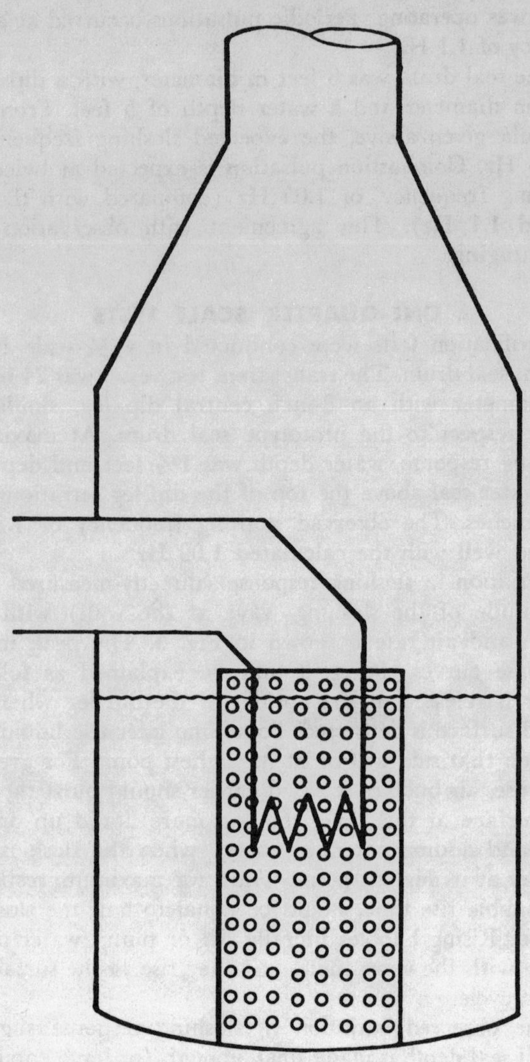

Fig. 4—Seal drum schematic showing perforated baffle.

frothing which has little chance to couple effectively to sloshing modes. At sufficiently large seal depths, the relatively small surface displacements would be insignificant compared with the hydrostatic head. In general, however, this behavior does not represent a solution to the sloshing problem because the seal depth is usually fixed by process back pressure requirements.

Past flare history does seem to show the suggested small depth behavior. The original seal depth at construction was 9 feet 2 inches, later reduced to 6 feet 7 inches sometime between 1959 and 1971, apparently for process-related reasons. Flare pulsation was evident during this period, according to in-plant observers. In 1971, the seal depth was reduced to 2 feet 2 inches, and again pulsation was evident. Subsequently, a test was run to determine what effect reducing the water seal depth would have on flare pulsation. Pulsation was evident down to a 4-inch seal, below 4 inches pulsation was not evident.

SOLUTION

The use of perforated baffles to provide viscous damping is a well known technique for suppressing sloshing motions. First, to prove this technique, a perforated 16-inch diameter concentric cylindrical baffle was installed in the test vessel as shown in Fig. 4. This diamond-pattern baffle with ½-inch holes separated approximately 1.4 inches (on diagonal centers) completely suppressed sloshing. The ½-inch holes in the test baffle were retained in the prototype baffle, but hole spacing was scaled up to 3 inches. Baffle diameter was scaled up to 5 feet 4 inches.

Prior to modification in March 1973, this flare had a long history of pulsation during moderate to heavy gas releases. Since the addition of the anti-slosh baffle and under the full range of gas flow conditions, combustion has proceeded smoothly with no evidence of pulsation.

ACKNOWLEDGMENT

Abstracted from a paper that appeared in *Noise Control Engineering*, September 1974.

Part VII.

Spill Control

Oil Spill Control: Prevention

Check this analysis to see what a complete program should include

Peter L. D'Alessandro, Jr., and **Calvin B. Cobb,**
The Pace Co., Houston

OIL SPILL CONTROL is comprised of two activities; prevention and counter-measure. Prevention is pre-planned, voluntary action accomplished under controlled conditions. Countermeasure is an involuntary reaction accomplished under emergency conditions over which one exerts very limited or no control.

The initial task that a company must perform is to assess the situation. One must first identify critical systems and then establish priorities for corrective action. Ideally this should be done by an engineering team—structural for storage tanks, civil for drainage, electrical for fail-safe systems, etc. If operational procedures have periodically been upgraded and maintenance schedules adhered to, this assessment will be relatively easy.

Once a comprehensive assessment has been completed, the problem then becomes one of training, testing and maintenance and inspection. These three elements when carefully developed adequately describe a spill prevention program.

TRAINING

The success of oil spill prevention depends upon people. This, however, is a most difficult factor to control. It can be approached only through adequate and thorough training organized to build confidence, insure competence and provide motivation. To meet these objectives, personnel training must be carried out on a continual basis, both formally and informally, and result in a personal commitment to spill prevention. This can evolve only through job familiarity, knowledge of company policy and procedures and adequate recognition of achievement. One way to accomplish this is to involve the operating personnel in the development of a spill control plan. If this phase is conducted with obvious management support, all personnel will understand the importance of spill prevention.

Meetings should be established on a monthly or bimonthly schedule. Training personnel should consider periodic integration of safety and spill training into one meeting. Films and slides should be used to supplement lectures and, most important of all, field demonstrations or practice sessions should be conducted frequently. These will be closely coordinated with the countermeasures portion of a spill control plan. If classroom sessions are well planned, they can transfer the technology associated with spill prevention and render any plan that much more effective.

TESTING AND MAINTENANCE

Existence of fail-safe devices tends to create an air of complacency. It is, therefore, prudent to periodically create situations in which these fail-safe systems must operate. This can be accomplished through pressure testing of selected pipes, valves and fittings, as well as through hydrostatic testing of storage tanks and pressure vessels.

Although testing will reveal spill potential, maintenance is necessary to prevent its occurrence. It is the key factor in elimination of equipment failures. Since maintenance should be scheduled to impose minimal operational disruptions, it will require some planning. This planning will manifest itself in the preparation of inspection forms. Not only will these forms list possible leak sources, but they will detail the required frequency as well as the optimum timing for each inspection. Preferably, these should be composed by operational personnel since they would be most familiar with spill sources and scheduling requirements. This should be done at the direction of one person, the plan coordinator. He should also insure that final check lists are developed, inspections are carried out on a periodic basis and all records of such are retained for future use.

INSPECTION

The unique aspect of each installation prevents a general solution to spill prevention. However, there are many

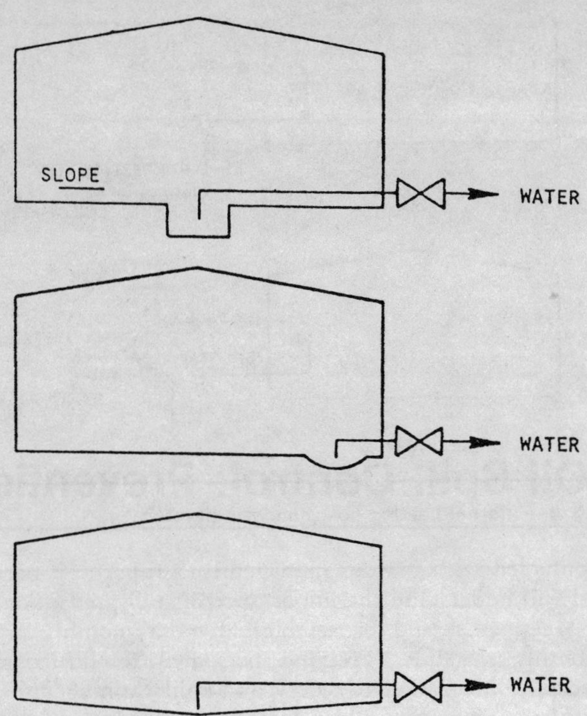

Fig. 1—Tank bottom drainage systems

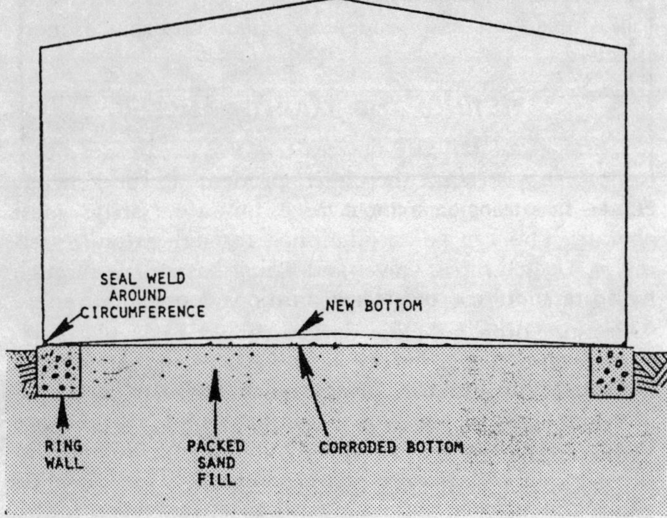

Fig. 2—Tank bottom replacement.

good engineering practices which do have general application.

Tank leaks

- Insure that the structural materials are compatible with the material being stored.

There are many references which list materials that are compatible with the potential corrosion conditions in storage tanks. Important variables are: liquid and vapor properties; temperature; liquid velocity at tank wall, etc. Engineering standards for materials of construction for tanks are published by numerous trade associations.

- Assess structural integrity by conformance to code construction. Tanks should conform to industry, engineering and/or company codes. (Example: API Standard 12 B, D, E, F applies to production tanks.)

- Determine practices regarding draining of water from tank bottom. Contained water promotes corrosion.

  Fig. 1 shows several commonly used methods for draining tank bottoms. It is also possible to develop automatic methods employing oil/water interface sensors such as density sensors, conductivity sensors and dielectric constant sensors.

- Repair of leaks, corrosion, etc. no matter how minor must be prompt. This is evidence of primary consideration of the environment rather than economics.

  Leaks may be repaired by patching while the tank is in service and numerous products are commercially available for patching.

- Buried carbon steel tanks should be coated, wrapped and lined. Depending on the nature of the soil, cathodic protection may be appropriate. Partially buried carbon steel tanks can set up galvanic corrosion and increase the rate of corrosion at the soil/air interface.

- Tanks should be examined periodically for evidence of external leakage (especially bottoms). This examination may be through visual inspection, hydrostatic testing and/or nondestructive shell thickness testing.

  Shell thickness may be measured by ultrasonic analysis. Inspection records should be kept on a frequency basis that is consistent with the historical failure rate of tanks in the same service.

- Corroded tanks should be lined and coated with epoxy. This treatment fills small pits and crevices and prevents inside corrosion.

  Normally, tanks are sandblasted to remove rust, dirt and scale which not only prevents product contamination but prepares the interior surface for epoxy coating. The coating needs to be selected for its compatibility with the material stored. X-ray analysis will locate pits and crevices.

- Deteriorated bottoms should be replaced with inverted cone type bottoms.

  Fig. 2 illustrates one technique for replacement of tank bottoms.

- Mobile storage tanks should be isolated from navigable waters by positioning and containment construction.

  One of the most common sources of leaks and spills are mobile storage tanks such as diesel fuel tanks used for construction machinery. It is usually a simple task to dig a small pit or construct temporary dikes around the tank.

- The condition of foundation and supports of tanks should be assessed regularly.

  In order to allow for adequate inspections and possible structural calculations, up-to-date drawings of the tank, foundation and structure must be maintained. Records of inspection should be kept for future reference.

- If a tank has internal heating coils, the condensate from these coils must be monitored for oil content.

  Condensate oil content can be monitored visually or automatically. Fig. 3 depicts the visual method using an inspection sump and the automatic method using a conductivity probe.
- Condensate from heating coils should be directed to oil/water separator or similar systems.
- Heating coils should be tested, maintained and replaced as needed.
- External heating systems are preferable to internal heating coils.

  Typical external systems use plate coils which are placed on the outside of the tank near the bottom. Plate coils are bolted together and equipped with a band that can apply pressure to the contact surface between tank and coil for improved heat transfer.

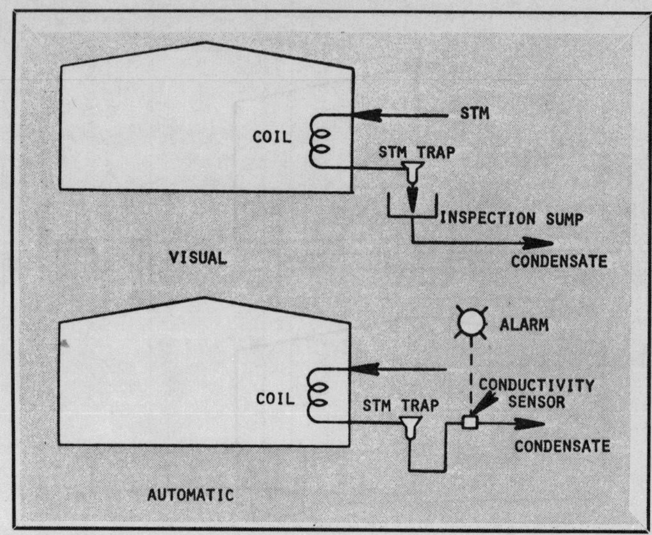

Fig. 3—Internal heating coil monitoring system.

Overfilling tanks

- Tanks should be carefully gauged before filling.
- High level alarms and shutoff devices to pumps should be in place.

  Fig. 4 shows a control system that will automatically stop a tank from overfilling. The signal generated by the level alarm can be used to close the inlet valve, stop the pump or both.
- Connecting overflow pipes to adjacent tanks should be in place.
- Automatic gauges and fail-safe devices must be tested periodically.
- A communication system between pump operation and tank gauging operation should be available.

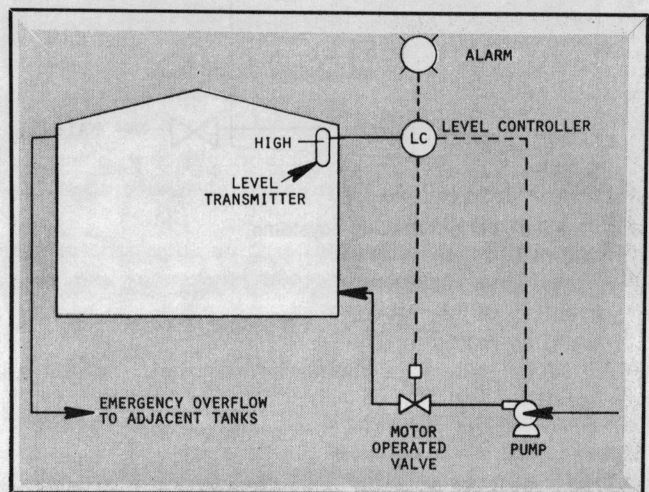

Fig. 4—Tank filling control system.

Tank rupture and boilover

- Insure structural integrity by code construction.
- Relief valves for excessive pressure and vacuum should be in place.

  Many types of relief devices are possible. One of the most common is a pressure release manhole in the tank top which provides a large opening that can quickly relieve any pressure buildup.
- Safety relief provisions should be tested periodically.
- Adequate fire protection facilities must be available.

Pipe, valve and fitting leaks (buried)

- Corrosion resistant pipe is preferable.
- Carbon steel pipe should be coated and wrapped (coal tar, asphalt, waxes, resins, fiberglass, asbestos, etc.).
- Cathodic protection system should be in place where surrounding soils contain organic or carbonaceous matter such as coke, cinders, coal, acid wastes or other conditions. A soil resistivity survey may be in order.

  There are companies which specialize in cathodic systems and they provide routine inspection services.
- Corrosion inhibitors should be used in piped products where internal corrosion is found and the inhibitor is compatible with the product.
- Marking lines should be obvious to prevent damage by third party excavators.
- Block valves should be located at strategic locations and periodically checked for operability.
- Insure that pipe meets specifications codes.
- Pressure drop fail-close devices should be in place. If the pressure in a line changes then alarms can be activated and shutdown procedures initiated.
- Check valves to insure one-way flow should be in place where required.
- Rate of flow indicators should be in use.
- Pipe corridors should be inspected visually.
- Pipe lines should be hydrostatically tested periodically.
- Acoustical or magnetic testing equipment should be used to check for leaks.
- Condition of pipe should be checked and recorded when construction activities expose buried lines.
- Inventory of emergency repair equipment and fittings should be maintained.
- All abandoned lines should be removed, plugged or capped.

Pipe, valve and fitting leaks (above ground)

- Frequent inspection should be made.
- Protection from vehicle collision should be used.
- Abrasion around pipe supports should be controlled.

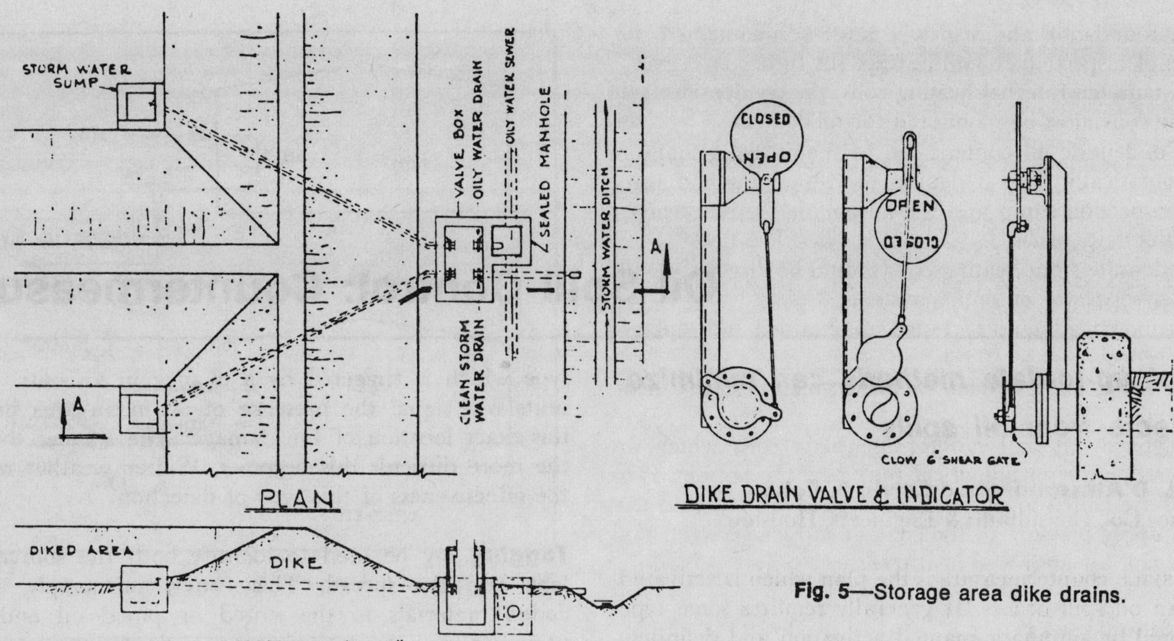

Fig. 5—Storage area dike drains.

- Periodic pressure testing of the system should be conducted.
- Pressure drop fail-close devices should be in place.
- Block valves should be installed at strategic locations.
- Rate of flow indicators should be in place.
- A preventive maintenance program should be in force.
- Inventory of emergency repair equipment and fittings should be maintained.
- All abandoned lines should be removed, plugged or capped.

Inadequate dike volume

- Dikes must be modified and reconstructed to achieve adequate volume.
- Adequate freeboard allowance for rainfall retention is imperative.
- Limit storage in tanks as a temporary measure if volume within diked area is inadequate.

Containment dike leaks

- Dikes should be stabilized with impervious coating (asphalt, clay, concrete, etc.).
- Dikes may be reconstructed with erosion resistant materials.
- A program of dike inspection and maintenance should be in force.
- Vegetation on earth dikes should be controlled.
- Through dike pipes no longer in use should be removed or plugged.
- Breaches in dikes for maintenance purposes should be minimized. Build ramps for vehicle access.

Containment area drainage valve leaks

- Flapper type valves should be removed and replaced with positive shutoff valves.
- Full operational range (positive open and closed) should be assured.
- Valves should be locked in closed position.
- All weather operation should be assured.
- No debris should be present in valve area.
- Contaminated water should go to oily water sewer (oil/water separator system).
- Visual indicator should be installed in drainage system.
- Easy access to drainage valves should be maintained.

Dike drainage

- Retained water should be checked before release (Fig. 5).
- Insure release valve is operational, and is locked in the closed position.
- Oily water should be directed to separator system.
- Records of drainage operations must be kept.

Water drawoff and tank cleaning

- Water drawoff from crude storage should go to oil/water separator or oily sewer system.
- Water drawoff must be accomplished under controlled conditions with fail-safe devices, direct supervision, visual inspection, etc.
- Tank bottoms (sludge) during cleanout maintenance should be disposed of promptly.
- Internal condition of tank should be checked during every cleanout maintenance.
- Temporary containment should be provided for bottom sludge.

Miscellaneous

- A closed drainage system should be installed at sample connections.
- A maintenance and housekeeping program for drainage ditches and sewer inlets should be followed.
- Flooding of separator facilities must be precluded by retention, designing separator for storm water flow and install connected spare pumping capacity. Such designs may be governed by NPDES regulations.
- Procedures for minimizing concentrated oil dumps to the separator (sample coolers, bleed valves, etc.) must be followed.
- Absorbents are preferable to flushing to sewer during maintenance of piping and equipment.
- Security should be observed through limited access, lighting, fencing, patrols, alarms, etc.

Oil Spill Control: Countermeasures

Use of up-to-date methods can minimize ill effects from oil spills

Peter L. D'Alessandro and **Calvin B. Cobb**
The Pace Co., Consultants & Engineers, Houston

OIL SPILL countermeasure is the plan which is activated when an oil spill occurs. It generally requires some type of manual or automatic-manual activation and definitely requires a commitment of manpower and resources. Contigency plans are procedures which are activated once oil reaches navigable waters. It is often difficult to separate the actions involved in countermeasures and contingency plans. Many of the provisions presented in the following apply to both.

There are four phases to countermeasure activity. They are sequential in implementation—detection, containment, recovery and disposal.

DETECTION

Spills must be detected quickly. Although catastrophic events usually receive immediate attention, this may not be true with smaller spills, whether continuous or intermittent. Frequently, they go unnoticed and unreported unless suitable detection methods are used. Some of these methods lie midway between prevention and detection.

Visual inspection. Periodic inspections are essential. A complete survey can identify potential danger areas for periodic or continuous surveillance. As an example, inspections along pipe lines may be conducted by fixed-wing aircraft or helicopter. The advantages are obvious in marshy and remote areas. Target locations should include heavily eroded stream banks where pipe line crossings occur, points where the pipeline has become exposed, and any area where construction or excavation work is in progress. Inspection frequency is a matter of judgment and depends on terrain and the condition of the system.

Generally, observation methods are marginally effective. Major damage often cannot be detected promptly nor can minor damage be located accurately. Visual detection depends on outflow, the soil structure, the position of the damage, and weather conditions: therefore, companion methods should be used.

Probes. Oil sensitive probes can be located throughout a drainage system of a potential spill. When a spill occurs, feedback to a central control panel will immediately identify the location. Two types of probes predominate: (1) a conductivity type which depends on an induced change in the dielectric constant, and (2) an ultrasonic type which is triggered by a change in viscosity. These units will signal the presence of oil in an area but not the exact location of the damage. The smaller the leak the more difficult this becomes. Winter weather reduces the effectiveness of this type of detection.

Tagging my be used to identify both the source of a leak and the spread. This consists of simply adding coded materials to the stored or piped oil and then periodically analyzing drainage samples for their presence. The materials used must satisfy at least the following criteria:

- Physically and chemically stable
- Readily identifiable
- No effect on commercial uses of oil
- Soluble or dispersible in oil, yet insoluble and nondispersible in water
- Inexpensive.

Examples of tagging substances include halogenated aromatics, nitrous oxide and radiochemicals. The latter are usually neutron-activated, that is, nonradioactive unless made so for an analysis. Tagging has not been widely used because of cost and complicating factors.

Routine analyses. As an example, corroded admiralty bundles are frequent sources of oil contamination. Therefore, the normally oil-free effluent water, particularly if it bypasses oil/water separators, should be analyzed periodically.

CONTAINMENT

The first priority in a control program is to limit the spread of an oil mass, since the greater the area covered, the more difficult and costly the cleanup program becomes. Containment is a process of isolation and/or diversion.

In the event that oil tanks are undiked, the drainage system should flow through a catchment basin which contains an oil trap. This should be designed to contain at least the amount of stored oil plus sufficient excess capacity to insure complete interception of all oil. The primary separation of the oil from the water should be accomplished as early in the system as possible so that the problem of handling large volumes of oil/water mix is minimized.

Should a spill take place outside the confines of a drainage system, it is imperative to construct either a temporary dike or diversion trench. The location would depend on expected direction and rate of flow. Information concerning these two factors, and in particular their

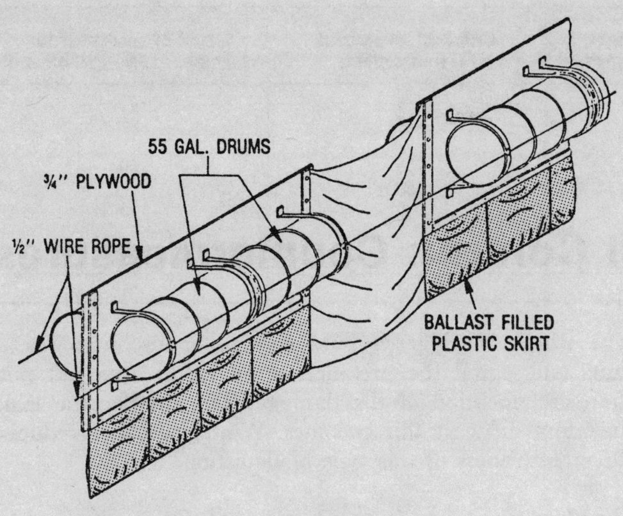

Fig. 6—"Navy" boom (curtain type)

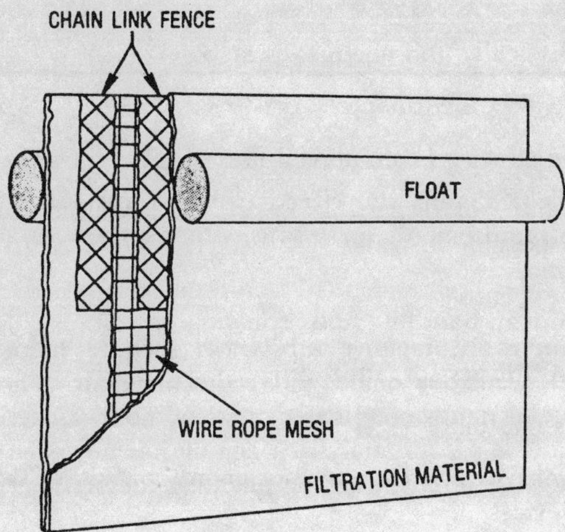

Fig. 7—Kain boom (fence type)

relationship to temperature, should be included in a reaction plan.

Materials and equipment necessary for the construction of diversion or holding structures should be on hand. Barriers may be manufactured or improvised from a wide variety of materials including wood, plastics and metal. In some cases, locally available materials such as haybales and sandbags will suffice. Aside from the requirement for mechnical strength, other considerations would include susceptibility to heat in the event of a fire and softening or cracking in the presence of some mineral oil components. Equipment that should be available would include standard excavation machinery and tools, and commercially marketed booms.

Booms. Two basic types of mechanical booms are available—curtain booms and fence booms.

Curtain booms include a surface float which acts as a barrier on the surface and which supports a subsurface curtain. The curtain is flexible and provides a barrier

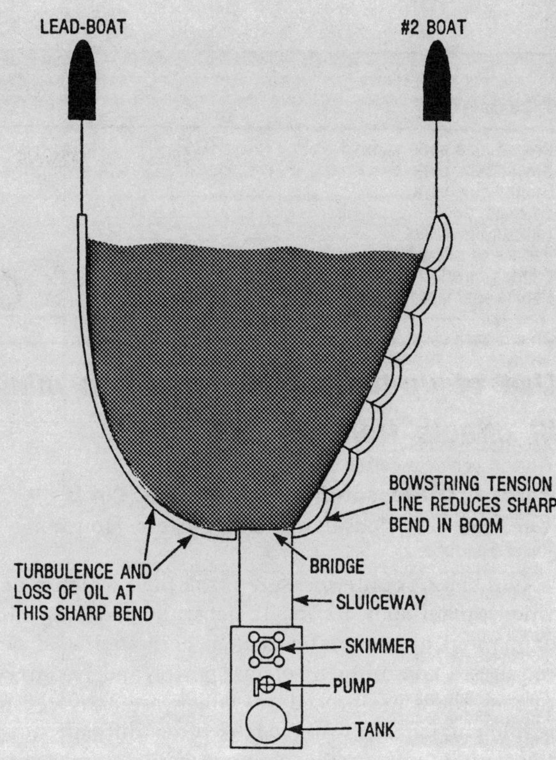

Fig. 8—Boom/skimmer configuration for oil spill cleanup

1 to 2 feet below the surface. Weights may be attached for stabilization. Fig. 6 is an example of this type.

Fence booms differ only in that they have a rigid curtain or panel both below and above the surface. Flotation supports the "fence." These types are normally employed in deeper and rougher waters. Fig. 7 is an example of this type.

Fig. 8 shows one method of using a boom in an oil spill. Generally a boom will perform the following two tasks:

Compact widely scattered puddles of oil to facilitate skimming operations.

Manipulate slicks away from sensitive areas or towards fixed removal installations.

With proper planning, the required booms should be readily available. In the event that such preparations have not been made, booms can be improvised through the use of materials such as hoses, bladders, tires, pipe and drums with attached planks.

Air curtains produced by an air supply to perforated hoses and pipes may also serve as a containment apparatus under specific conditions. A schematic of an air barrier is shown in Fig. 9.

Surface active agents. Surface tension modifiers inhibit the spread of oil in water. When relatively small quantities of these chemicals are placed on the surface next to the floating oil, the oil is repulsed and tends to agglomerate. Application is simple; only a small amount need be used, applied as a coarse spray on the water at the edge of the spill. The effect lasts only a matter of hours, so cleanup plans should be implemented as soon as possible. As with any chemical, approval for the use of

TABLE 1—Sorbents relative effectiveness

Type material	Pick-up ratio—weight oil pick up/weight absorbent	Unit cost. absorbent ($/T absorbent)	$ cost of absorbent for cleanup of 1,000-gal. oil spill
Ground pine bark, undried	0.9	6	27
Ground pine bark, dried	1.3	15	47
Ground pine bark	3	....	....
Sawdust, dried	1.2C	15	50
Industrial sawdust	....	56	....
Reclaimed paper fibers, dried, surface treated	1.7	30	75
Fibrous, sawdust and other	3	....	....
Porous peat moss	1.0	....	....
Ground corn cobs	5	30	21
Straw	3—5D	30	27
Chrome leather shavings	10	Indicated comparable to straw in cost.	....
Asbestos, treated	4	500	440
Fibrous, perlite and other	5	416	290
Perlite, treated	2.5	230	320
Talcs, treated	2	70—120	120—210
Vermiculite, dried	2	....	....
Fuller's earth	....	25	....
Polyester plastic shavings	3.5—5.5	100	80
Nylon-polypropylene rayon	6—15	....	....
Resin type	12	3,100	900
Polyurethane foam	70	20,000	1,000
Polyurethane foam	15	4,500	1,050
Polyurethane foam	70	....	....
Polyurethane foam	40	2,260	195
Polrurethane foam	80	1,200	55

C Another reference gives a ratio of 20 D Other reports have indicated ratios of 20 and above

a surface tension modifier must be obtained from appropriate governmental agencies.

RECOVERY

Numerous harvesting devices and various removal techniques exist for handling harbor and inland spills. Consider the following:

Sorbents. Sorbents are oil spill scavengers, cleanup agents which adsorb and/or absorb oil. Based on origin, sorbents may be divided into three classes:

Natural products include those derived from vegetative sources such as straw, seaweed and sawdust; mineral sources such as clays, vermiculite and asbestos; and animal sources such as wool wastes, feathers and textile wastes.

Modified natural products include expanded perlite, charcoal, silicone coated sawdust and surfactant treated asbestos.

Synthetic products include a vast array of rubber, foamed plastics and polymers.

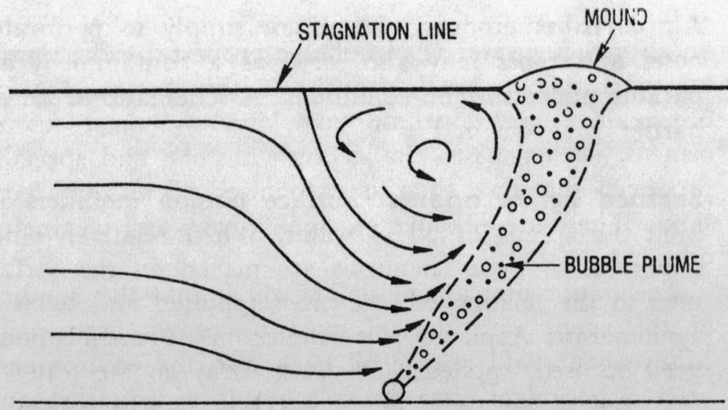

Fig. 9—Circulation pattern upstream of an air barrier in a current

Table 1 details the effectiveness of various materials. The requirements for a satisfactory sorbent include the following:

- Aids in handling and removing oil
- Minimizes spread of oil
- Is nontoxic
- Enhances performance of booms and other skimming devices

Skimming. Removal of oil on water may require skimming. Skimmers may be purchased commercially or built for a particular application. Additionally, they may be floating, fixed or mobile (mounted on boats, barges, trucks, etc.). The type of skimmer depends on its probable application. Of primary concern is its capacity in terms of total fluid handling volume, recovered oil volume, and pumping rate. These factors should be compatible with the expected utilization. Of secondary importance is the size, seaworthiness, speed, maneuverability, and other skimmer characteristics. Fig. 10 shows the classes of skimmers.

Vacuuming. Once the spill has been contained it is usually removed and disposed of with a vacuum truck. One or more of these should be permanently assigned to the installation. If this is not the case, outside contractors must be identified and made familiar with the site facilities.

Excavating. If an underground water supply is endangered by a land spill, a considerable portion of the oil can be removed by excavating the contaminated soil before the oil has reached great depths. The extent or depth to which this would be economically feasible would

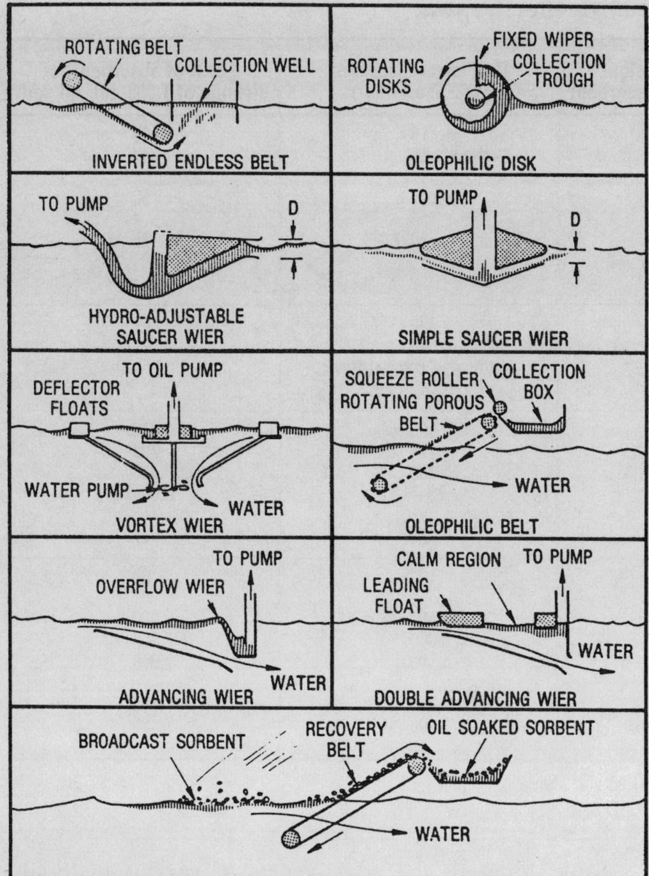

Fig. 10—Classes of skimmers

be a function of the type oil as well as the underlying soil structure.

Pumping. In the event that oil from a land spill has reached the water table, if its viscosity is not too high, large amounts can be recovered by pumping. A well is drilled, centered in the spill and screened at a depth no further than the oil/groundwater interface. In the pumping process, a cone of depression of the oil/water interface will be formed and will prevent oil from spreading further. At first the pump should extract oil exclusively and then later progressively more and more water. The amount of pumping is a function of recovered oil, spilled oil, and oil retained by the soil. Generally, one should stop when the oil-water ratio becomes less than 1%.

Treating. Spills eventually reach the plant drainage system. Therefore, the site treatment facilities play a significant role in the oil removal process. Various type separators are in use. Besides the classical API type, there are gravity plate separators and a host of multistage separators, some equipped with coalescence filters. In addition, there are still other devices which employ proprietary methods ranging from ultrasonic treatment to polyelectrolyte injection.

Separation is ideally followed by physical-chemical treatment. This will incorporate some sequence of coagulation, flocculation, sedimentation and possibly air flotation. The remaining petroleum fraction can be removed by biological treatment. The activated sludge process is commonly used, often in conjunction with an aerated lagoon and a trickling filter. Following a dewatering step, the sludge can either be incinerated or hauled off for landfilling operations by a local contractor.

DISPOSAL

Slop oil which has been recovered prior to reaching the drainage system or which has been separated in the initial step of the treatment system can be disposed of in several ways that would include the following:

Recycling recovered oil back into the plant process is the most common and the most economical. This is done by bleeding the slop oil into the feedstock over a period of time. Any impurities picked up during recovery of the spill are removed along with the usual basic sediment and water (BS&W). Any emulsions which have been formed can be broken using chemical agents and heat. As long as extensive "weathering" (evaporation of volatile components) has not significantly affected the fuel quality, this method can be used.

Reclaiming recovered oil for other uses is a less desirable alternative. It is economically feasible only when the oil is not amenable to blending with the feedstock. This is normally done by an outside contractor equipped with appropriate re-refining facilities. These might include steaming, filtering and additive rebalancing. Such contractors frequently also specialize in storage tank and sump cleanout operations.

Burning is another method of final disposal of oil, particularly nonreclaimable sludges. Large amounts can efficiently be disposed of in this manner with the help of combustion agents or by blending with lighter grades of fuel such as kerosine. The mixture is then atomized and burned. This course of action requires careful control to obtain complete combustion to avoid air pollution.

Dispersing. Dispersants are chemical agents which emulsify or solubilize oil in water. Their use is governed by Annex X of the National Contingency Plan. They should not be employed except when other methods are inadequate or infeasible.

Sinking. As oil weathers and becomes more dense, there is a natural tendency of the residual fraction to sink. This phenomenon depends, of course, on the type of oil involved in the spill. Oil can be made to sink by application of a nucleus of high density material having an affinity for the oil (oleophilic property) and not having an affinity for water (hydrophobic property). The resulting mass of material then settles to the bottom. Typical oil sinking agents include sand, fly ash, lime, stucco, cement, volcanic ash, chalk, crushed stone and specially produced materials such as carbonized-silicanized-waxed sands. These are effective on thick, heavy and weathered oil slicks.

The major problem in sinking oil is that the bonding of the agent with the oil must be nearly permanent. Many agents will release oil back into the environment after a period of time or as a result of agitation and turbulence. Microbial action on the oil-soaked particles also produces gaseous byproducts which give the particle a tendency to float.